Digital Signal Processing
in Modern Communication Systems
(Edition 2.0)

Andreas Schwarzinger
Lake Mary, Florida 32746
andreas_dsp@hotmail.com

Cover Image of the Planet Mars: provided by NASA

ISBN: 978-0-9888735-1-3

Authored and Published by Andreas O. Schwarzinger, Lake Mary, Florida 32746

This book is dedicated to the memory of my mother

Klara Maria Schwarzinger

I would also like to thank my wonderful wife and sons

Laura Beatriz Schwarzinger
Roger Andreas Schwarzinger
Andreas Martin Schwarzinger

Introduction

As author and publisher of this text, I would like to welcome you to the second edition of *Digital Signal Processing in Modern Communication Systems*, a book that is geared to both students and practicing engineers. This introduction will familiarize you with background material regarding the origin, intended audience, general structure, and purpose of this book. There are many textbooks on both digital signal processing and communication system engineering to choose from, and I hope that the following pages will help you differentiate the unique approach of this text from that of others.

This textbook originated in the need to expand my own practical knowledge of digital signal processing beyond basic principles to address various hardware and software design challenges in the area of wireless communication systems. A great number of textbooks teach the basic principles of digital signal processing; some do so eloquently and some do not. Unfortunately, even the better ones avoid advanced topics such as optimization and estimation, which are always left to advanced, hard-to-read textbooks written by specialists. Therefore, the first of two major goals of this text is to present both basic and advanced digital signal processing concepts in the same easy-to-read manner that can be found in high quality introductory DSP books. I have tried to avoid being overly general in the way the mathematics are presented and included as many steps as possible during the derivation of various formulas. Nothing is more frustrating than reading through an important derivation and getting stuck because the author assumed you can simply imagine the steps he just skipped. My second goal is to apply both the basic and advanced techniques of digital signal processing to the communication engineering discipline, which, over the last three decades, has benefited tremendously from this union. The concepts in this text are both interesting and important to us as DSP and communication engineers, but it is the way they are explained that will, I hope, differentiate this book from the many others out there.

Organization and General Structure

Digital Signal Processing in Modern Communication Systems is divided into four major parts:

Part one of this book (chapters 1 and 2) presents a review of basic math and introduces the reader to digital signal processing (DSP) concepts in a straight forward and understandable manner. We start by being reacquainted with complex numbers, matrix algebra, and statistics, before learning fundamentals including convolution, correlation, the Fourier transform, and digital filtering concepts.

Part two of this book (chapters 3 and 4) expands the horizon by covering more advanced concepts, including optimization, approximation, equalization, and adaptive signal processing.

Part three of this book (chapters 5 and 6) focuses entirely on the role of signal processing in wireless communication systems. Chapter 5 introduces us the transmission technologies used in wireless communication systems including modulation techniques (single-tone QAM modulation, MSK, FSK, GMSK, GFSK, OFDM and SC-FDMA) as well as error handling procedure such as CRC (cyclic redundancy checks), BCC (binary convolutional coding), LDPC (low density parity check coding), and Polar coding. In chapter 6, we cover the effects of distortion and noise on received signals.

Part four of this book (chapters 7 and 8) focus entirely on the design of OFDM and SC-FDMA based communication systems with emphasis on its implementation in 802.11a/g, 4G LTE, and LTE C-V2X. We provide an in-depth discussion not just on the physical layer aspects presented in the specifcations but also on the signal processing techniques used in the receiver that those same specifications conveniently avoid mentioning. Chapter eight finishes us off with multiple antenna techniques that have become a standard approach to enhancing the throughput and robustness of wireless OFDM communication systems.

The Role of MatLab

If you take the time to look over some of the examples in this book, you will notice a great deal of MatLab code embedded throughout all chapters. Although you can readily assimilate the material presented in this text without reference to MatLab, I highly recommend that you invest in a copy of this rather extraordinary software. Not using MatLab during a DSP course is akin to taking a class in auto repair but never entering the shop to open the hood of a car. Sure, it can be done, but the material you learn never quite comes alive the way it should to ensure lasting comprehension. The code segments not only will allow you to step through the different processing operations at a pace that is right for you but also serve as a starting point for your own experiments once you start to feel comfortable with the material.

Feedback and the Website

While my editors and I have made every effort to present the material in this book with the fewest possible errors, we are only human, which makes perfection beyond our grasp. If you come across any errors, whether conceptual, mathematical, grammatical, or orthographic in nature, please let me know at the email address listed below. Also, feel free to contact me in case you feel that certain sections were inadequately explained, you would like a soft copy of the MatLab examples, or you simply have general questions regarding the material. I may not be able to answer every question or comment, but I will make every effort to complement your reading experience of this textbook.

andreas_dsp@hotmail.com

Additionally, you may find the books MatLab source code as well as a large host of additional discussion at the following website that supports this book.

www.signal-processing.net

About the Author

Andreas O. Schwarzinger received a Bachelor of Science in E.E. in 1991 at Case Western Reserve University in Cleveland, Ohio, and a Master of Science in E.E. in 1994 at the University of Central Florida in Orlando, Florida. He started his 25 year professional career as an RF design engineer, migrated to analog IC design, and finally settled on digital signal processing, where he has worked on baseband processors, other DSP related ASIC modules as well as software defined radio projects. He currently works for Rohde and Schwarz USA.

Table of Content

PART 1

Basic Digital Signal Processing and the Underlying Mathematical Foundations.

1   A Review of Basic Concepts

To prepare ourselves for the journey through the subject of digital signal processing as it applies to communication systems, it is helpful to review some of the mathematical concepts that form the basis of our endeavor. These concepts are not unique to signal processing or communication engineering, but find utility in many areas of science and engineering. Specifically, we will take a look at complex numbers, linear algebra, statistics and waveform sampling.

Complex Numbers

Complex numbers are extraordinarily important to us for many reasons. First and foremost, the analysis of signals and systems makes extensive use of sinusoids. Sinusoids feature two independent parameters called magnitude and phase, which lend themselves wonderfully to mathematical manipulation via complex numbers. Your first exposure to sinusoids probably came in a course on *linear circuits* where you used complex valued *phasors* to parameterize and study their behavior. In communication systems, the ability to simultaneously transmit real, or *I*, information and imaginary, or *Q*, information requires that we process complex information in baseband and DSP processors. A mastery of complex numbers is therefore essential when working with any communication related signal processing system.

Linear Algebra

Although we don't require knowledge of linear algebra to understand the fundamental topics of signal processing and communication systems, some familiarity with it is required as we move into chapters 3 and beyond where we encounter more advanced subjects like approximation and equalizer design. Moreover, MatLab natively expresses data in matrix form, so a good grasp of linear algebra will be beneficial when it comes to using this programming language efficiently.

Statistics

In communication systems, the use of statistics allows us to understand engineering problems that involve very long number sequences, which appear random in nature. Statistics further provides us with ways of determining important metrics of both random and deterministic signals such as mean and variance. The more sophisticated concepts discussed in this text require the fusion of linear algebra and statistics, which leads us to methods of estimating received data symbols and characterizing the behavior of communication channels that distort them.

Waveform Sampling

Chapter 1 concludes by examining the limitations of signals that are expressed discretely rather than continuously. We examine the role of Nyquist's sampling theory, which sets the highest frequency that may be expressed and processed at a particular sampling rate. Signal conditioning for analog-to-digital and digital-to-analog conversion is discussed in later chapters.

1.1 Complex Numbers

1.1.1 The Development of Numbers

Complex numbers have confounded the greatest minds in the history of mathematics for the very same reason they confound us today. It is difficult to picture a complex amount of stuff. In our mind we can imagine 4 potatoes and 2½ oranges, but $2 + j$ lemons just doesn't make much sense. Even the idea of zero lemons becomes awkward. After all, numbers were invented to express the amount of something that we can see, touch, or otherwise appreciate.

Mathematicians brought complex numbers to life in their quest to find solutions to algebraic equations. To be more specific, they tried to find solutions to polynomial equations of the following form.

$$a_n x^n + a_{n-1} x^{n-1} + \ldots + a_1 x^1 + a_0 = 0$$

In the expression above, x represents the variable that we wish to find, while the quantities a_0 through a_n are constants. The degree, or order, of the polynomial is defined as the highest exponent, n, in the expression. For lower orders, polynomials reduce to simple algebraic expressions used for everyday calculations. Here are examples of first and second order polynomial equations.

$$x + 4 = 0$$

$$x^2 + 2x + 1 = 0$$

To get a better appreciation of numbers in general and complex numbers in particular, let's examine how their use evolved over time and how they helped find the solution for polynomial equations with which mathematicians seemed to be so enamored.

Natural or Whole Numbers

Earlier cultures required a number system that lent itself to counting items such as life stock, the years of one's life, or the number of gold coins tucked away under the mattress. If you knew that your daughter could be married off at the age of 15, and 12 summers have already passed since her birth, then the following algebraic equation could be used to find out how many more years she would need to remain at home.

$$x + 12 = 15$$

Whole numbers were restricted to positive integers such as 1, 2, 3, 4 … etc., and the use of the number zero was avoided since the idea of zero goats or zero oranges had no concrete meaning.

Integers

Integers expand the whole number domain by adding zero and appending negative numbers. Although negative numbers first appeared in Chinese text around 200 BC, the first comprehensive rule set regarding the mathematical treatment of integers wasn't composed until around AD 620 in India, when Brahmagupta explained how to handle situations of debt [1]. In Europe, some mathematicians treated the idea of negative numbers with suspicion until well into the 18th century. However, expanding our number set to integers allowed the solution to equations of the following form.

$$x + 12 = 2 \quad \text{and} \quad x^2 - 1 = 0$$

Real Numbers

Real numbers expanded the realm of integers to include rational numbers, which are fractions that can take the form of l/m (both l and m must be integers), as well as irrational numbers that do not fit this simple ratio format such as $\sqrt{2}$ and π. Real numbers allowed the solution of the following type of algebraic polynomial. The solution to x is $\pm\sqrt{2}$.

$$x^2 - 2 = 0$$

Complex Numbers

However, mathematicians quickly found out that even real numbers were ill-suited for finding the solution to some polynomials. Whereas it was assumed that a polynomial of degree n should have n roots or solutions, only one ($x = -2$) could be easily found for the following expression.

$$x^3 + 2x^2 + x + 2 = 0$$
$$or$$
$$(x^2 + 1)(x + 2) = 0$$

The introduction of complex numbers of the form $a + jb$, consisting of two real quantities, a and b, and the complex operator $j = \sqrt{-1}$, provided the other two roots as $\pm j$. Complex numbers supplied the final piece of the puzzle and led to the fundamental theorem of algebra, which now definitively states that a polynomial of degree n, where a_n must be non-zero, features n roots that obey the complex number format. In general, a complex number c (rectangular format), is said to be composed of a real part a and an imaginary part jb.

$$c = a + jb$$

1.1.2 Expanding Our Concept of Numbers as Mere Quantities

The fact that complex numbers have given closure to the task of finding all roots of polynomials may be reassuring to a whole host of long–dead mathematicians, but it still doesn't help us appreciate the meaning of negative two sacks of grain or $2 + j$ dollars. Luckily, the work of English mathematician John Wallis (1616–1703) paved the way for the use of the number line, which provides numbers with geometric meaning that had not previously been available to the layperson.

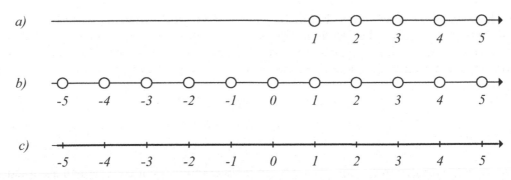

Figure 1-1: Whole Numbers, Integers, and Real Numbers

The number lines *a* through *c* above take us from the original *counting* or whole numbers through integers and finally to real numbers. While we still can't put a concrete meaning to negative numbers, as the integers and real number lines seem to allow, we have nevertheless seen and dealt with these lines often enough to feel comfortable handling the numbers that lie on them.

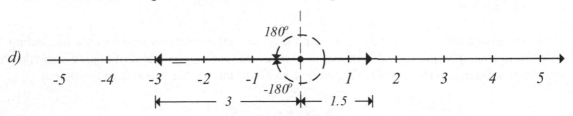

Figure 1-2: Thinking of Real Numbers as Vectors

Progression to Negative Numbers

To better understand positive and negative numbers, we think of them as vectors, which feature two separate parameters: a magnitude and a direction. The two vectors lying on number line *d* represent numbers 1.5 and -3. Their magnitudes, 1.5 and 3, are quantities that have a direct and palpable meaning that we as human beings can understand. Their direction conveys a quality about the object that is abstract in nature. The direction in the case of minus 2 sacks of grain could indicate their state of ownership. The positive sign means that you own it, while the negative sign indicates that you owe it. Just take a look at the negative sign of your bank account balance after you've spent more money than you should have. Clearly, in the realm of integers and real numbers there are only two directions, featuring opposite bearings. Since a direction in 2D space is equivalent to an angle, we assign 0 degrees to the positive sign and 180/-180 degrees to the negative sign. Let's reformulate the two numbers 1.5 and -3 in terms of magnitude and direction (angle).

$$+1.5 = 1.5\angle 0^{o} \qquad\qquad -3.0 = 3.0\angle 180^{o}$$

Progression to Complex Numbers

What if the direction spanned a continuous domain, as is suggested in number line *e*? The physical significance of such a number is hard to imagine, but so was the meaning of negative numbers hundreds of years ago. By switching from positive counting numbers to integers, we allowed numbers to take on an abstract quality that goes beyond what we can easily appreciate with our senses. By moving from integers to real numbers, we let go of the notion of discrete quantities and allowed for a continuous number domain. Finally, we let go of the discrete nature of the direction, which was limited to angles of 0 and ±180 degrees, and allowed it to vary continuously over the domain of 0 to 360 degrees or equivalently from 0 to -360 degrees.

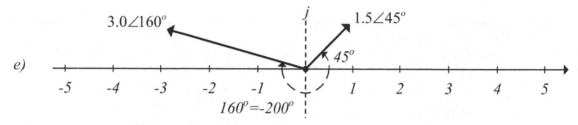

Figure 1-3: The Complex Plane

The journey of moving from whole to complex numbers was one of adding abstraction and removing restrictions to arrive at quantities featuring much greater scope than those with which we started. Increasing the scope of numbers to include abstract qualities was necessary to solve mathematical and engineering problems with reasonable effort.

In conclusion, numbers not only qualify the quantity of an object but also may be used to describe its state. Think of the position of a dart on a dart board. The position features a distance from the board's center that can be likened to a quantity. The position also contains a phase, and it is this additional quality that allows the position (or state) to be represented by a complex number. A dart position of $0.1 \angle 120^\circ$ makes sense, but $0.1 \angle 120^\circ$ or $2 + j$ lemons doesn't, since their description does not require the additional quality that a complex number

1.1.3   The Polar Format of Complex Numbers

From our discussion above, we may summarize that a complex number may be described in terms of magnitude and phase, *magnitude $\angle$ phase*, as the expression below for the number c indicates. This convention of describing complex numbers in terms of magnitude and phase is called the *polar format*.

$$c = Mag_c \angle \theta_c$$

The magnitude extends from zero to positive infinity, while the angle is restricted to a range of 0 to 2π radians. The polar format of complex numbers is exceptionally well suited to the purposes of multiplication and division. As an example, let's multiply +2 and -1 to arrive at -2, which we already know equals $2 \angle 180^\circ$.

$$+2 \cdot -1 = -2 = 2 \angle 180^\circ$$

Expressing the original numbers +2 and -1 in terms of magnitudes and angles points us to the rule of complex multiplication and division.

$$2\angle 0^{o} \cdot 1\angle 180^{o} = 2\cdot 1\angle(0^{o}+180^{o}) = 2\angle 180^{o}$$

As the expressions suggest, when multiplying two complex numbers expressed in the polar format, the lengths or magnitudes are multiplied in the traditional sense, while the angles are added. The rules for multiplication and division are summarized below.

Arithmetic Operation	Syntax
Multiplication	$Mag_1\angle\theta_1 \cdot Mag_2\angle\theta_2 = Mag_1 \cdot Mag_2\angle(\theta_1+\theta_2)$
Division	$\dfrac{Mag_1\angle\theta_1}{Mag_2\angle\theta_2} = \dfrac{Mag_1}{Mag_2}\angle(\theta_1-\theta_2)$

Example 1.1: *The Square Root of -1*

Now, as we all know from college, complex numbers are inextricably tied to the square root of negative 1. In light of our new polar format, what would the square roots of -1 be? Our task is to find a number that when multiplied by itself yields negative 1 or better $1\angle 180^{o}$.

$$\rightarrow 1\angle 90^{o} \cdot 1\angle 90^{o} \quad = 1\cdot 1\angle(90^{o}+90^{o}) \quad = 1\angle 180^{o} \quad = -1$$
$$\rightarrow 1\angle -90^{o} \cdot 1\angle -90^{o} = 1\cdot 1\angle(-90^{o}+-90^{o}) = 1\angle -180^{o} = -1$$

Clearly, the solution is two vectors with length 1 and angles equal to plus and minus 90 degrees.

Example 1.2: *Cube Roots of 1*

Let's now crank it up a notch by calculating the three cube roots of 1. The cube roots are three vectors with unity magnitude and angles of 0, -120, and 120 degrees respectively.

$$\rightarrow 1\angle 0^{o} \cdot 1\angle 0^{o} \cdot 1\angle 0^{o} \qquad = 1\cdot 1\cdot 1\angle(0^{o}+0^{o}+0^{o}) = 1\angle 0^{o} = 1$$
$$\rightarrow 1\angle -120^{o} \cdot 1\angle -120^{o} \cdot 1\angle -120^{o} = 1\cdot 1\cdot 1\angle(-360) \qquad = 1\angle 0^{o} = 1$$
$$\rightarrow 1\angle 120^{o} \cdot 1\angle 120^{o} \cdot 1\angle 120^{o} \qquad = 1\cdot 1\cdot 1\angle(360) \qquad = 1\angle 0^{o} = 1$$

Example 1.3: *The Square Root of a Complex Number*

Although it sounds strange, it is just as straight-forward to compute the two square roots of a complex number such as $4\angle 120^{o}$. The square roots are $2\angle 60^{o}$ and $2\angle -120^{o}$.

$$\to 2\angle 60^{o} \cdot 2\angle 60^{o} \qquad = 2 \cdot 2\angle(60^{o} + 60^{o}) \qquad = 4\angle 120^{o}$$

$$\to 2\angle -120^{o} \cdot 2\angle -120^{o} = 2 \cdot 2\angle(-120^{o} - 120^{o}) = 4\angle -240^{o} = 4\angle 120^{o}$$

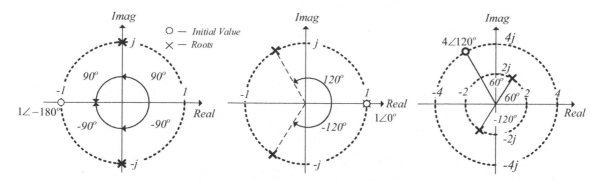

Figure 1-4: Graphical Solution to Problems 1.1, 1.2, and 1.3 (from Left to Right)

The most important lesson we take away from the polar format is that complex multiplication involves rotation, as the last couple of examples clearly show. However, while multiplication and division are a breeze, addition and subtraction of complex numbers requires the introduction of the rectangular format of complex numbers. In day to day operation, the rectangular format is preferred to its polar counterpart, because all four arithmetic operations (multiplication, division, subtraction, and addition) are natively supported, as we will see next.

1.1.4 The Rectangular Format of Complex Numbers

As mentioned above, the rectangular format of complex numbers is preferred in most engineering applications, because of the ease with which addition and subtraction operations may be evaluated. The figure below shows the *real* axis along which the traditional numbers that we know and love reside, as well as a perpendicular *imaginary* axis that provides the complex numbers with the angular quality we have seen earlier. We use the symbol j [2] to indicate the part of the complex number that lies on this new imaginary axis.

$$c = a + jb$$

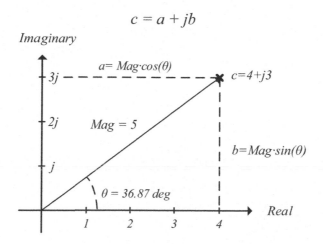

Figure 1-5: The Number c=4+j3 in the Complex Plane

The conversion between the rectangular and polar formats follows the rules of right triangles.

Polar to Rectangular Conversion	Rectangular to Polar Conversion
$a = Magnitude \cdot cos(\theta)$ $b = Magnitude \cdot sin(\theta)$	$Magnitude = \sqrt{a^2 + b^2}$ $\theta = atan2(b, a)$

How to Think About the Number j

It is helpful to think of a number like $1j$ as a real value 1 that has been rotated by 90 degrees onto the imaginary axis. Multiplying any number by j or $-j$ is equivalent to rotating the number by 90 degrees in the positive (counter-clockwise) and negative (clockwise) directions. Therefore, the quantity j^2 is equivalent to a rotation by 180 degrees in the polar format and a sign inversion in the rectangular format.

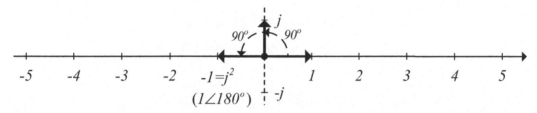

Figure 1-6: Multiplications by j Are Rotations by $\pi/2$ or 90 Degrees

The expressions below match some numbers in the rectangular format to their polar counterparts.

$$1 = 1\angle 0^o \qquad -1 = 1\angle 180^o \qquad j^2 = 1\angle 90^o \cdot 1\angle 90^o = 1\angle 180^o = -1$$

$$j = 1\angle 90^o \qquad -j = 1\angle -90^o \qquad -j \cdot j = 1\angle -90^o \cdot 1\angle 90^o = 1\angle 0^o = 1$$

Arithmetic Rules for Complex Numbers in Rectangular Format

Arithmetic Operation	Syntax
Addition and Subtraction	$(a_1 + jb_1) \pm (a_2 + jb_2) = a_1 \pm a_2 + j(b_1 \pm b_2)$
Multiplication	$(a_1 + jb_1) \cdot (a_2 + jb_2) = a_1 \cdot a_2 - b_1 \cdot b_2 + j(a_1 \cdot b_2 + a_2 \cdot b_1)$
Division	$\dfrac{a_1 + jb_1}{a_2 + jb_2} = \dfrac{a_1 \cdot a_2 + b_1 \cdot b_2}{a_2^2 + b_2^2} + j\dfrac{a_2 \cdot b_1 - a_1 \cdot b_2}{a_2^2 + b_2^2}$

Example 1.4: Multiplication and Division in the Rectangular Domain

In example 1.3, we found the two square roots of the complex number $4 \angle 120^o$ to be $2 \angle 60^o$ and $2 \angle -120^o$. In this example, we will verify this result by converting the roots into the rectangular format, squaring them, and restoring the result to its polar value of $4 \angle 120^o$.

Root 1 $(2\angle 60^o) \quad \rightarrow \quad a = 2 \cdot cos(60^o) = 1 \qquad b = 2 \cdot sin(60^o) = 1.732$

Root 2 ($2\angle -120^{o}$) → $\qquad a = 2 \cdot cos(-120^o) = -1 \qquad b = 2 \cdot sin(-120^o) = -1.732$

The two roots convert to the rectangular format as $r_1 = 1 + j1.732$ and $r_2 = -1 - j1.732$. The square of these roots must yield the rectangular equivalent of $4\angle 120^o$.

Square of Root 1 → $\quad c = (1 + j1.732) \cdot (1 + j1.732)$
$$= 1 + j^2 \cdot 1.732^2 + j(2 \cdot 1.732) = -2 + j3.464$$

Square of Root 2 → $\quad c = (-1 - j1.732) \cdot (-1 - j1.732)$
$$= 1 + j^2 \cdot (-1.732)^2 + j(2 \cdot 1.732) = -2 + j3.464$$

Converting back to the polar format yields the original value of $4\angle 120^o$.

$$Mag = \sqrt{(-2)^2 + 3.464^2} = 4$$
$$\theta = atan2(3.464, -2.0) = 120^o$$

1.1.5   Roots of Polynomials

The two expressions below show a quadratic (degree 2) and a cubic (degree 3) polynomial function. The roots or solutions to these expressions may be found by setting them to zero and solving for the dependent variable x, or by plotting them and observing where $f(x)$ crosses zero.

$$f(x) = a \cdot x^2 + b \cdot x + c \qquad \leftarrow quadratic$$
$$f(x) = a \cdot x^3 + b \cdot x^2 + c \cdot x + d \qquad \leftarrow cubic$$

In the examples below we factor two polynomials thus revealing the roots algebraically and then verify those results graphically.

Example 1.5: *The Quadratic Formula*

Plot and examine the following polynomial expressions.

$$f(x) = x^2 - 4x + 3 \qquad = (x-1) \cdot (x-3)$$
$$f(x) = x^3 - 6x^2 + 11x - 6 \qquad = (x-1) \cdot (x-2) \cdot (x-3)$$

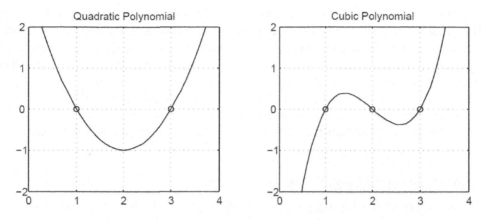

Figure 1-7: Roots of a Quadratic and Cubic Polynomial

The roots, which are clearly revealed when we factor the polynomials, naturally fall at the zero crossings of the curves. While the above quadratic is easy to factor using mental math, more sophisticated expressions require the use of the well-known quadratic formula.

$$az^2 + bz + c = 0 \quad \rightarrow \quad Roots = \frac{-b \pm \sqrt{b^2 - 4ac}}{2a}$$

All was well in the world of mathematics until someone tried to apply the equation above to an innocuous expression like $f(x) = s^2 - 4s + 8$. The initial curiosity was to fathom why the curve never crosses zero; moreover, applying the coefficients $a = 1$, $b = -4$, and $c = 8$ to our quadratic equation forces us to take the square root of a negative number.

$$Roots = \frac{+4 \pm \sqrt{-16}}{2} = \frac{+4 \pm 4\sqrt{-1}}{2} = 2 \pm j2$$

Given what we have learned so far, accepting the fact that roots can be complex is no longer an obstacle, but the fact that the curve does not cross zero remains confusing.

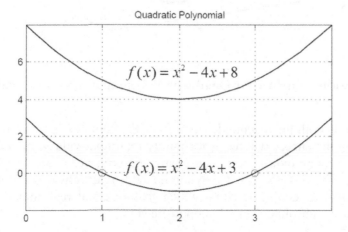

Figure 1-8: Quadratic Polynomials with Real and Complex Roots Side by Side

The explanation for this mystery is that the polynomial functions are actually surfaces rather than curves. By choosing complex values for the independent variable x, the dependent result $f(x)$ also becomes complex. Therefore, rather than choosing a real range of x, say 0 to 4 in steps of 0.1, we pick a grid like pattern that equally exercises the imaginary dimension. We may then plot a surface where we examine the real, imaginary, magnitude or phase components of $f(x)$. In the MatLab code that follows, we establish an *XY* grid where *X* represents *real(x)* while *Y* represents *imag(x)*. To test our belief that the magnitude of $f(x)$ goes to zero at the roots and verify our theory, we will plot the magnitude surfaces for the quadratic functions shown in figure 1-7.

Example 1.6: Picturing Polynomial Functions as Surfaces

Given a complex range for the independent variable x, we will show the magnitude surface for the two polynomial functions below.

$$f_1(x) = x^2 - 4x + 3 \; \leftarrow real\,roots \qquad f_2(x) = x^2 - 4x + 8 \; \leftarrow comlex\,roots$$

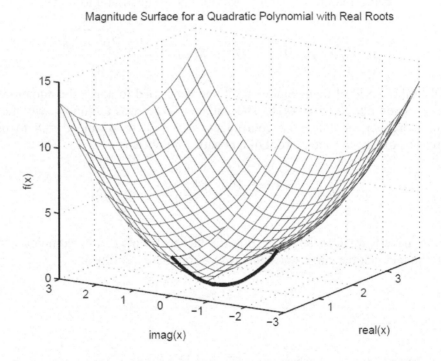

Magnitude Surface for a Quadratic Polynomial with Real Roots

Figure 1-9: Magnitude Surface for Quadratic Polynomial $f_1(x)$ with Real Roots at x = [1, 3]

The two figures show both the magnitude surfaces of the complex input valued polynomials, $f_1(x)$ and $f_2(x)$, as well as the curves that result for purely real input values. In both cases the roots, or positions where the magnitude is zero, can clearly be seen at [1, 3] for $f_1(x) = s^2$-4s+3 and at [2+*j*2, 2-*j*2] for $f_2(x) = s^2$-4s+8. For a polynomial with real valued coefficients (*a*, *b*, *c*...), all complex poles appear as conjugate pairs, which feature equal real components and imaginary components with equal absolute values but opposite sign.

```
%% Plotting the surfaces
x = 0:.25:4; y = -3:.25:3;
[X,Y] = meshgrid(x,y);

Mag1 = zeros(size(X));   %% Magnitude for polynomial with real roots
Mag2 = zeros(size(X));   %% Magnitude for polynomial with complex roots

for c1 = 1:size(X,1)
    for c2 = 1:size(X,2)
        Number      = X(c1,c2) + j*Y(c1,c2);
        Mag1(c1,c2) = abs(Number^2 - 4*Number + 3);
        Mag2(c1,c2) = abs(Number^2 - 4*Number + 8);
    end
end

figure(1);
h = mesh(X,Y,Mag1); hold on;
plot3(u, zeros(1,length(u)), Quadratic,'k', 'LineWidth' ,2);
xlabel('real(x)'); ylabel('imag(x)'); zlabel('f(x)');
title('Magnitude Surface for Quadratic with Real Roots'); hold on;
axis([0 4 -3 3 0.0 15]);
colormap(gray); caxis([0 40]);
set(gca, 'XTick', [1 2 3]);
set(h,'FaceAlpha',.5);
```

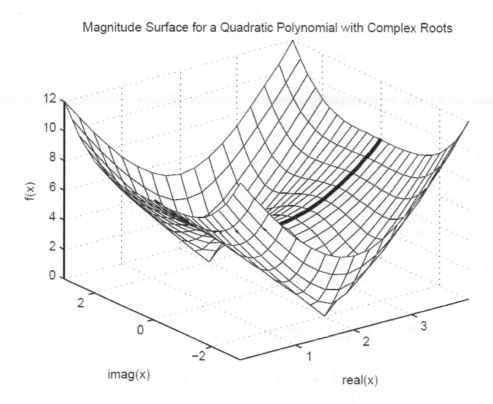

Figure 1-10: Magnitude Surface for Quadratic Polynomial $f_2(x)$ with Complex Roots at
x=[2+2j, 2-2j]

1.1.6   Complex Exponentials and Euler's Formulas

In the study of communication systems and digital signal processing, few equations are as mysterious and useful as Euler's formula, which establishes a relationship between trigonometric and complex exponential functions. His formula states the following.

$$e^{j\theta} = \cos(\theta) + j\sin(\theta)$$

and therefore

$$Mag \cdot e^{j\theta} = \text{Mag} \cdot (\cos(\theta) + j\sin(\theta))$$

This formula allows us to express a complex number in polar format, $Mag \angle \theta$, as a value, $Mag \cdot e^{j\theta}$, that can be easily manipulated in equations. It simplifies complex multiplication, since multiplying exponential functions of the form $e^{ja} \cdot e^{jb} = e^{j(a+b)}$ involves the mere addition of exponents.

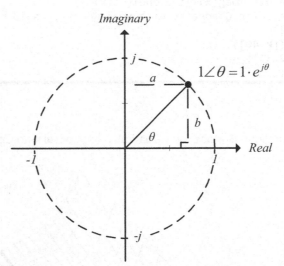

Figure 1-11: Reformulation of Complex Numbers

How an exponentially increasing function can be linked to trigonometric expressions (sine and cosine) that are oscillatory in nature is mysterious indeed. Although we won't recreate his approach used to arrive at the formula, we will show that it is true.

Leonhard Euler not only introduced the formula, for which he is now famous but also established the number e, which he defined as a series [3].

$$e = 1 + \frac{1}{1} + \frac{1}{1 \cdot 2} + \frac{1}{1 \cdot 2 \cdot 3} + \frac{1}{1 \cdot 2 \cdot 3 \cdot 4} \cdots = \sum_{n=0}^{\infty} \frac{1}{n!} = 2.7182818$$

The Taylor series expansions for the more general case of e^{θ}, as well as for $\sin(\theta)$ and $\cos(\theta)$ are shown next and can be looked up in any calculus text book [3].

$$e^{\theta} = 1 + \theta + \frac{\theta^2}{2!} + \frac{\theta^3}{3!} + \frac{\theta^4}{4!} + \frac{\theta^5}{5!} \cdots = \sum_{n=0}^{\infty} \frac{\theta^n}{n!}$$

$$\sin(\theta) = \theta - \frac{\theta^3}{3!} + \frac{\theta^5}{5!} - \frac{\theta^7}{7!} + \frac{\theta^9}{9!} \cdots = \sum_{n=0}^{\infty} \frac{(-1)^n}{(2n+1)!} \theta^{2n+1}$$

$$\cos(\theta) = 1 - \frac{\theta^2}{2!} + \frac{\theta^4}{4!} - \frac{\theta^6}{6!} + \frac{\theta^8}{8!} \cdots = \sum_{n=0}^{\infty} \frac{(-1)^n}{(2n)!} \theta^{2n}$$

For the term e^{θ}, we now replace the exponent θ with $j\theta$ and reformulate the expansion as follows.

$$e^{j\theta} = 1 + j\theta - \frac{\theta^2}{2!} - j\frac{\theta^3}{3!} + \frac{\theta^4}{4!} + j\frac{\theta^5}{5!} - \frac{\theta^6}{6!} - j\frac{\theta^7}{7!} + \frac{\theta^8}{8!} + j\frac{\theta^9}{9!} \cdots = \sum_{n=0}^{\infty} \frac{(j\theta)^n}{n!}$$

If we sum the expansions of $\cos(\theta)$ and $j\sin(\theta)$, they will indeed yield that of $e^{j\theta}$.

$$\cos\theta + j\sin\theta = 1 + j\theta - \frac{\theta^2}{2!} - j\frac{\theta^3}{3!} + \frac{\theta^4}{4!} + j\frac{\theta^5}{5!} - \frac{\theta^6}{6!} - j\frac{\theta^7}{7!} + \frac{\theta^8}{8!} + j\frac{\theta^9}{9!} \cdots = e^{j\theta}$$

Using these same series we introduce two more useful equations that express the sine and cosine functions in terms of the complex exponential. Prove them to yourself as an exercise.

$$\boxed{\begin{aligned} \cos(\theta) &= \frac{1}{2}(e^{j\theta} + e^{-j\theta}) \\ \sin(\theta) &= \frac{1}{2j}(e^{j\theta} - e^{-j\theta}) \end{aligned}}$$

You may similarly replace $e^{\pm j\theta}$ with $\cos(\theta) \pm j\sin(\theta)$ to prove that the expressions above are true. In the figure below, we illustrate how the individual complex exponentials sum to form the cosine and sine functions. Note that the term $1/j$ is equivalent to $-j$.

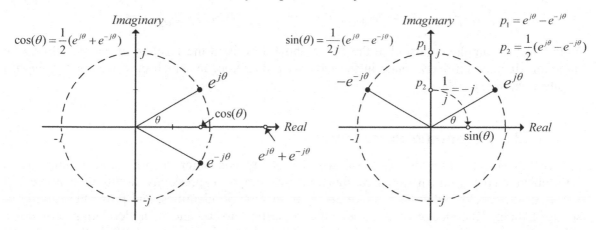

Figure 1-12: Graphical Recreation of cos(θ) and sin(θ) from Constituent Complex Exponentials

1.1.7   The Complex Sinusoid

As a natural extension to Euler's formula, let us introduce what is arguably the most important function of time in the field of digital signal processing: the complex sinusoid. A complex sinusoid is a complex function of unity magnitude and a phase parameter that changes at a constant rate. The complex sinusoid is defined as follows.

$$ComplexSinusoid(f,t) = e^{j2\pi ft} = cos(2\pi ft) + j\,sin(2\pi ft)$$

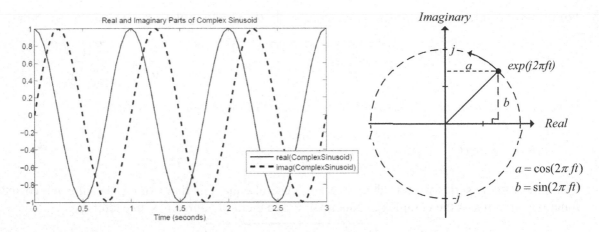

Figure 1-13: Time and Phasor Domain Representation of Complex Sinusoid

One of the important goals of digital signal processing is to express or parameterize a test signal, *Test*(*t*), as a linear combination of basis functions. In the field of communication engineering the most useful basis function is the complex sinusoid. We use techniques like the Fourier transform to determine how combinations of these basis functions can best approximate the original signal, *Test*(*t*). A typical approximation would try to find the magnitude and phase of complex sinusoids at different discrete frequencies.

$$Test(t) \approx a_0 e^{j(2\pi f_0 t + \theta_0)} + a_1 e^{j(2\pi 2 f_0 t + \theta_1)} + a_2 e^{j(2\pi 3 f_0 t + \theta_2)} + a_3 e^{j(2\pi 4 f_0 t + \theta_3)} \ldots$$

Note that the complex sinusoid in the figure above travels in the positive or counter clockwise direction. It may similarly rotate in the clockwise direction, in which case we show it with a negative exponent, $e^{-j2\pi ft}$.

1.1.8   Trigonometric Identities

As we advance through this text, we will frequently encounter situation that require us to reformulate challenging trigonometric functions into simpler expressions. In this section, we will introduce these reformulations, also known as trigonometric identities, and prove the validity of some of them. Knowledge of these identities is critical to us, and their derivation provides a splendid opportunity to work with Euler's formula and complex sinusoids. While Wikipedia and other websites present extensive lists of trigonometric identities, in this section, we discuss only those identities that directly concern the applications covered in this book.

Angle Sums and Differences

$$\cos(x \pm y) = \cos(x)\cos(y) \mp \sin(x)\sin(y)$$

$$\sin(x \pm y) = \sin(x)\cos(y) \pm \cos(x)\sin(y)$$

Prove the Angle Sums and Differences Identity

From Basic Calculus $\rightarrow$ $e^{jx} \cdot e^{jy} = e^{j(x+y)}$

$$e^{jx} \cdot e^{jy} = (\cos(x) + j\sin(x)) \cdot (\cos(y) + j\sin(y))$$

$$= \cos(x)\cos(y) - \sin(x)\sin(y) + j(\sin(x)\cos(y) + \cos(x)\sin(y))$$

$$e^{j(x+y)} = \cos(x + y) + j\sin(x + y)$$

The identity is proven by realizing that $real(e^{jx} \cdot e^{jy}) = \cos(x) \cdot \cos(y) - \sin(x)\sin(y) = \cos(x+y) = real(e^{j(x+y)})$. Comparing the imaginary portions of both exponentials provides proof of the second portion of the identity which expands $\sin(x \pm y)$.

Pythagorean Identity

$$\cos^2(x) + \sin^2(x) = 1$$

Double Angle Formulas

$$\sin(2x) = 2\sin(x)\cos(x)$$

$$\cos(2x) = \cos^2(x) - \sin^2(x)$$

Power Reduction Formulas

$$\sin^2(x) = \frac{1 - \cos(2x)}{2}$$

$$\cos^2(x) = \frac{1 + \cos(2x)}{2}$$

Product to Sum Rules

$$\cos(x)\cos(y) = \frac{\cos(x-y) + \cos(x+y)}{2}$$

$$\sin(x)\sin(y) = \frac{\cos(x-y) - \cos(x+y)}{2}$$

$$\sin(x)\cos(y) = \frac{\sin(x-y) + \sin(x+y)}{2}$$

Prove: $sin^2(x) = (1/2)\cdot(1 - cos(2x))$	Prove: $cos(x)\cdot cos(y)=(1/2)\cdot(cos(x-y) + cos(x+y))$
$$sin^2(x) = \frac{1}{2j}(e^{jx} - e^{-jx}) \cdot \frac{1}{2j}(e^{jx} - e^{-jx})$$	$$cos(x)cos(y) = \frac{1}{2}(e^{jx} + e^{-jx}) \cdot \frac{1}{2}(e^{jy} + e^{-jy})$$
$$= -\frac{1}{4}(e^{jx} - e^{-jx}) \cdot (e^{jx} - e^{-jx})$$	$$= \frac{1}{4}(e^{jx} + e^{-jx}) \cdot (e^{jy} + e^{-jy})$$
$$= -\frac{1}{4}(e^{j2x} - e^{j0} - e^{j0} + e^{-j2x})$$	$$= \frac{1}{4}(e^{j(x+y)} + e^{j(x-y)} + e^{j(y-x)} + e^{-j(x+y)})$$
$$= -\frac{1}{4}(-e^{j0} - e^{j0}) - \frac{1}{4}(e^{j2x} + e^{-j2x})$$	$$= \frac{1}{4}(e^{j(x-y)} - e^{-j(x-y)}) + \frac{1}{4}(e^{j(x+y)} + e^{-j(x+y)})$$
$$= \frac{1}{2} - \frac{1}{2}cos(2x)$$	$$= \frac{1}{2}cos(x - y) + \frac{1}{2}cos(x + y)$$

Trigonometric Identities Used in IQ Modulation

In wireless communication systems, information is transmitted over the air via an RF carrier signal, $RF(t)$. In the past, transmitters have modified either the amplitude, $M(t)$, or the angle, $\theta(t)$, of this RF carrier to convey this information.

$$RF(t) = M(t) \cdot cos(2\pi ft + \theta(t))$$

With the advent of powerful baseband processors and accurate digital-to-analog conversion modules, it has become practical to modulate both the magnitude, $M(t)$, and the angle, $\theta(t)$, of the RF carrier at the same time. The quantities $M(t)$ and $\theta(t)$ make up what RF engineers call the complex envelope of the carrier. Using trigonometric identities, we will show how to control this complex envelope – $M(t)$ and $\theta(t)$ – using two baseband signal, $I(t)$ and $Q(t)$, which modulate carrier signals, $cos(2\pi ft)$ and $sin(2\pi ft)$. The figure below claims that the $RF(t)$ signal, $M(t)$ $cos(2\pi ft+ \theta(t))$, may be equivalently synthesized as $I(t)\cdot cos(2\pi ft) + Q(t)\cdot cos(2\pi ft)$. On the next page we prove that the two formulations are in fact equivalent.

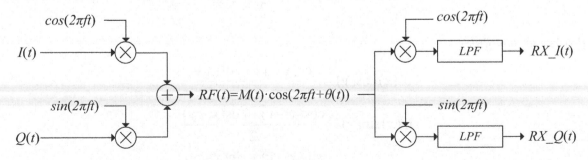

Figure 1-14: Simplified Zero-IF Transceiver Chain

$$\text{Prove:} M(t) \cdot \cos(2\pi ft + \theta(t)) = I(t) \cdot \cos(2\pi ft) + Q(t) \cdot \sin(2\pi ft)$$

given $\quad \cos(x+\theta) = \cos(x)\cos(\theta) - \sin(x)\sin(\theta) \leftarrow$ *angle sums identity*

then $\quad M \cdot \cos(x+\theta) = (M\cos(\theta)) \cdot \cos(x) - (M\sin(\theta)) \cdot \sin(x)$

now by setting $\quad I = M\cos(\theta) \quad$ *and* $\quad Q = -M\sin(\theta)$

then $\quad M \cdot \cos(x+\theta) = I \cdot \cos(x) + Q \cdot \sin(x)$

substituting $2\pi ft$ *for x yields*

$$M \cdot \cos(2\pi ft + \theta) = I \cdot \cos(2\pi ft) + Q \cdot \sin(2\pi ft)$$

In conclusion, forcing the magnitude and phase of the RF carrier's complex envelope, $M(t)$ and $\theta(t)$, is a simple matter of setting the baseband signals $I(t)$ and $Q(t)$ to $M(t) \cdot \cos(\theta(t))$ and $-M(t) \cdot \sin(\theta(t))$, respectively.

Trigonometric Identities used in IQ Demodulation

The previous figure illustrating the Zero-IF transceiver chain suggests that the $I(t)$ and $Q(t)$ transmit signals may be recovered by simply low pass filtering the products $RF(t) \cdot \cos(2\pi f_o t)$ and $RF(t) \cdot \sin(2\pi f_o t)$. The received quantities $RX\_I(t)$ and $RX\_Q(t)$ are merely scaled versions of $I(t)$ and $Q(t)$ as is proven below using the power reduction and product to sum identities.

$$\text{Prove: } RX\_I(t) = LPF[\ RF(t) \cdot \cos(2\pi f_o t)\ ] = a \cdot I(t) \leftarrow a \text{ is a constant}$$

$$
\begin{aligned}
RX\_I(t) &= LPF[(\ I(t) \cdot \cos(2\pi f_o t) + Q(t) \cdot \sin(2\pi f_o t)\ ) \cdot \cos(2\pi f_o t)] \\
&= LPF[\ I(t) \cdot \cos(2\pi f_o t) \cdot \cos(2\pi f_o t) + Q(t) \cdot \sin(2\pi f_o t)\ ) \cdot \cos(2\pi f_o t)] \\
&= LPF[\ I(t) \cdot \cos(2\pi f_o t) \cdot \cos(2\pi f_o t)] + LPF[Q(t) \cdot \sin(2\pi f_o t) \cdot \cos(2\pi f_o t)] \\
&= LPF[\ I(t) \cdot (1/2 + 1/2\cos(4\pi f_o t))] + LPF[Q(t) \cdot \sin(0) \cdot \sin(2\pi f_o t)] \\
&\qquad \textit{power reduction formula} \qquad\qquad \textit{product to sum rule} \\
&= 1/2 \cdot I(t)
\end{aligned}
$$

$$RX\_Q(t) = 1/2 \cdot Q(t)$$

1.2 Review of Linear Algebra (Part 1)

Linear algebra is a topic in mathematics that deals with calculations involving linear systems of equations. Although linear algebra is not required to understand the fundamentals of digital signal processing and communication engineering, a thorough command of its arithmetic and properties is paramount as we advance our skills beyond basic DSP concepts. The concept of optimization, which includes equalization, approximation, and optimal filter design, as well as the spatial multiplexing MIMO techique are built upon linear algebra.

While in-depth treatments of linear algebra are available in several text books [4], it is the goal of this section to review only those concepts that we will encounter in later chapters. Let's start by considering a simple linear system of equations describing two lines.

$$a_{11} \cdot x + a_{12} y = b_1$$
$$a_{21} \cdot x + a_{22} y = b_2$$

Although the equation of a line is usually shown in y-intercept form ($y = mx + b$), the formulation above, called the standard form, will be more convenient to work with. However, to get the first equation back to the more familiar y-intercept form, simply subtract $a_{11} \cdot x$ from both sides of the equation and divide by a_{12}.

$$y = -\frac{a_{11}}{a_{12}} x + \frac{b_1}{a_{12}}$$

For now, we will stay with the standard form and reformulate the systems of equations into an expressions using matrices.

$$\begin{bmatrix} a_{11} & a_{12} \\ a_{21} & a_{22} \end{bmatrix} \cdot \begin{bmatrix} x \\ y \end{bmatrix} = \begin{bmatrix} b_1 \\ b_2 \end{bmatrix}$$

One of the chief goals of linear algebra is to find the solution to such a system of equations, which, in the case above, means finding the x and y coordinate where the two lines cross (thus satisfying both equations simultaneously). We can rewrite the expression above by replacing the matrices with variables A, X, and B. Finding X obviously requires that both matrices A and B are known.

$$A \cdot X = B$$

The variable A represents a two row by two column (or 2x2) matrix while the other two variables are column vectors of dimension 2x1. Vectors are matrices that feature either one row or one column.

$$A = \begin{bmatrix} a_{11} & a_{12} \\ a_{21} & a_{22} \end{bmatrix} \qquad X = \begin{bmatrix} x \\ y \end{bmatrix} \qquad B = \begin{bmatrix} b_1 \\ b_2 \end{bmatrix}$$
$$\quad\; 2x2 \qquad\qquad\qquad 2x1 \qquad\qquad 2x1$$

From now on, we will represent matrices that feature two or more rows and two or more columns via capitalized variables. The variable notation for vectors may or may not appear capitalized, and you will have to examine the context to identify them.

Application of Linear Algebra in MIMO Systems

Before we dive into the more academic review of matrices and their properties, let's first explore how certain applications in the area of communication systems and DSP benefit from the use of linear algebra. The first application we will examine falls under the chategory known as MIMO, which represents a group of techniques that employ multiple antennas at both the transmitter and receiver to increase the robustness and throughput of a wireless communication link. The 4/5G cellular standards as well as WLAN implementations starting with the 802.11n specification use MIMO to increase the information throughput. In the figure below, we take a look at a MIMO scenario involving the transmission of independent symbols from two transmit antennas and their reception at two receive antennas. The input vector X holds the transmit symbols, whereas the output vector Y features the values detected at the receive antennas.

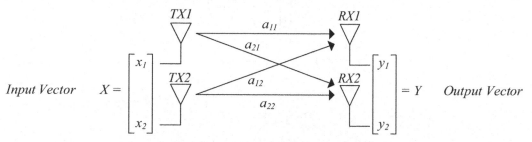

Figure 1-15: 2x2 MIMO Setup

As is evident form the figure above, the information arriving at receive antenna 1 and 2 will contain some amount of both input symbols. The amounts of symbols x_1 and x_2 arriving at the receive antennas is defined by the channel matrix A.

$$A = \begin{bmatrix} a_{11} & a_{12} \\ a_{21} & a_{22} \end{bmatrix}$$

The output vector, Y, is dependent on the input vector, X, via the expression $Y = A \cdot X$.

$$\begin{bmatrix} y_1 \\ y_2 \end{bmatrix} = \begin{bmatrix} a_{11} & a_{12} \\ a_{21} & a_{22} \end{bmatrix} \cdot \begin{bmatrix} x_1 \\ x_2 \end{bmatrix}$$

Now sit back for a moment and try to guess which coefficient values residing in our channel matrix, A, would be beneficial to our link and which ones wouldn't be. Imagine a scenario in which both a_{12} and a_{21} are equal to zero, and a second scenario in which $a_{11} = a_{21}$ and $a_{12} = a_{22}$. In those cases, the system of equations looks as follows.

$$\text{Scenario 1:} \quad \begin{bmatrix} y_1 \\ y_2 \end{bmatrix} = \begin{bmatrix} a_{11} & 0 \\ 0 & a_{22} \end{bmatrix} \cdot \begin{bmatrix} x_1 \\ x_2 \end{bmatrix} \quad \rightarrow \quad \begin{aligned} y_1 &= a_{11} \cdot x_1 \\ y_2 &= a_{22} \cdot x_2 \end{aligned}$$

$$\text{Scenario 2:} \quad \begin{bmatrix} y_1 \\ y_2 \end{bmatrix} = \begin{bmatrix} a_{11} & a_{12} \\ a_{11} & a_{12} \end{bmatrix} \cdot \begin{bmatrix} x_1 \\ x_2 \end{bmatrix} \quad \rightarrow \quad \begin{aligned} y_1 &= a_{11} \cdot x_1 + a_{12} \cdot x_2 \\ y_2 &= a_{11} \cdot x_2 + a_{12} \cdot x_2 \end{aligned}$$

In scenario one, the best possible case, the input symbols are neatly represented in the output symbols, whereas scenario two features the worst possible interference forcing the two output symbols to be equal to one another no matter what the values of the input symbols are.

Assuming that we know the values of the output symbols and the coefficients of the channel matrix, then solving for the input symbols is trivial in scenario one and impossible in scenario two. The channel matrix A of scenario one is said to be *well-behaved*, whereas the one in scenario two is called *singular* (or 'very badly behaved' in normal English).

Geometric Interpretation

The MIMO application and its matrix setup is identical in structure to the system of equations representing the two straight lines we mentioned earlier. Whereas the idea of a well behaved and singular matrix might not have been easy to grasp in the MIMO example, it will become blatantly obvious when examining two lines. Let's look at the two scenarios side by side for comparison.

$$\begin{bmatrix} a_{11} & a_{12} \\ a_{21} & a_{22} \end{bmatrix} \cdot \begin{bmatrix} x_1 \\ x_2 \end{bmatrix} = \begin{bmatrix} y_1 \\ y_2 \end{bmatrix} \quad \leftarrow \quad MIMO$$

$$A \quad \cdot \quad X \quad = \quad B$$

$$\begin{bmatrix} a_{11} & a_{12} \\ a_{21} & a_{22} \end{bmatrix} \cdot \begin{bmatrix} x \\ y \end{bmatrix} = \begin{bmatrix} b_1 \\ b_2 \end{bmatrix} \quad \leftarrow \quad Straight\ Lines$$

Note that the nomenclature of the vector X might be a little confusing for the straight line case, but let's simply assume that the vector X contains the variable for which we want to solve, and B features known quantities. To illustrate how different matrices affect the solution vector X, consider the following cases.

$$A = \begin{bmatrix} 1 & 1 \\ 0 & 1 \end{bmatrix} \quad B = \begin{bmatrix} 1 \\ 1 \end{bmatrix} \quad \leftarrow Case1$$

$$A = \begin{bmatrix} 1 & 1 \\ 2 & 1.8 \end{bmatrix} \quad B = \begin{bmatrix} 1 \\ 1 \end{bmatrix} \quad \leftarrow Case2$$

$$A = \begin{bmatrix} 1 & 1 \\ 2 & 2 \end{bmatrix} \quad B = \begin{bmatrix} 1 \\ 1 \end{bmatrix} \quad \leftarrow Case3$$

$$A = \begin{bmatrix} 1 & 1 \\ 2 & 2 \end{bmatrix} \quad B = \begin{bmatrix} 1 \\ 2 \end{bmatrix} \quad \leftarrow Case4$$

In a moment, we will show some simple MatLab code that converts the two lines for each combination of A and B into the y-intercept form. From there we plot the lines for x values of -5 to 5 in steps of 0.5 and examine the solution vector, which represents the coordinate at which the lines intersect.

The accompanying figure shows the two lines for each case where the first line can be identified by the circular markers. Both case 1 and 2 feature well defined solutions (points of intersection) at coordinates (0, 1) and (-4, 5) respectively. Case 3 has no solutions, whereas case 4 features an infinite number of them. Cases 3 and 4 are just like scenario 2 of our Mimo 2x2 example as no unique solution may be found, and the matrix A used in these cases is said to be *singular*. The matrix A is singular if the slopes of the two lines are equal. Below, we show the y-intercept format for both lines and their associated slopes $-a_{11}/a_{12}$ and $-a_{21}/a_{22}$. Take another look at the matrix A for case 3 and 4 and convince yourself that both slopes are equal to -1.

$$y_{Line1} = -\frac{a_{11}}{a_{12}}x + \frac{b_1}{a_{12}} \qquad y_{Line2} = -\frac{a_{21}}{a_{22}}x + \frac{b_2}{a_{22}}$$

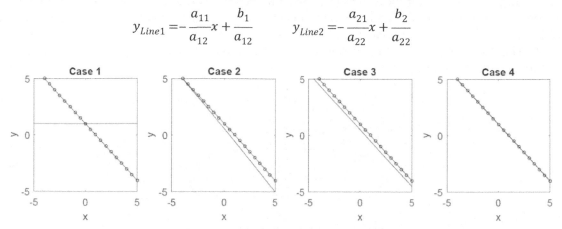

Figure 1-16: Behavior of Straight Lines for Cases 1 Through 4

Whereas both cases 1 and 2 have a unique solution, only case 1 is *well behaved*. The matrix A used for case 2 is said to be *poorly behaved* as the slopes of its two lines are almost the same. Assume for a moment that we had to estimate the values in matrix A before using them rather than knowing them exactly. This is in fact what happens in our MIMO example where the receiver guesses at the channel matrix by comparing certain reference symbols it receives under potentially noisy conditions to the correct values with which they were originally transmitted. Small errors in the coefficient estimates of A, and therefore the slope of either one of the lines in case 2, can cause large changes in the solution (point of intersection). The situation in case one will not feature this type of sensitivity. In our review of linear algebra, we will be interested not only in finding the solution vector X but also analyzing matrices to see how robust the results are when we work with them. In the MatLab code below, all four scenarios above are covered. Now make small changes to *A1* and *A2* to verify that the solution in vector X is far more volatile for case 2 than case 1.

```
A1 = [1 1; 0 1];      % Matrix A for case 1
A2 = [1 1; 2 1.8];    % Matrix A for case 2
A3 = [1 1; 2 2];      % Matrix A for cases 3 and 4
B1 = [1; 1];          % Vector B for cases 1 through 3
B2 = [1; 2];          % Vector B for case 4

x   = -5:.5:5;        % Evaluate lines for these x values

Y1  = ComputeLines(A1, B1, x);   % Compute Y coordinates of both lines for case 1
Y2  = ComputeLines(A2, B1, x);   % Compute Y coordinates of both lines for case 2
Y3  = ComputeLines(A3, B1, x);   % Compute Y coordinates of both lines for case 3
Y4  = ComputeLines(A3, B2, x);   % Compute Y coordinates of both lines for case 4
```

```
function Y = ComputeLines(A, B, x)
y1 = -( A(1,1)/A(1,2) ) * x + B(1,1)/A(1,2); % Change to y intercept form and calculate
y2 = -( A(2,1)/A(2,2) ) * x + B(2,1)/A(2,2); % y values for line 1 and 2 given x
Y = [y1; y2];
```

Application of Linear Algebra in Digital Filter Analysis

One of the most important tools in digital signal processing is the use of digital filters to alter the frequency content of signals and to model processes that occur in communication systems. Let's take a look at the most popular digital filter in use today: the finite impulse response, or FIR, filter. The filter structure features delay elements, implemented as registers in digital logic, multipliers and a summing element.

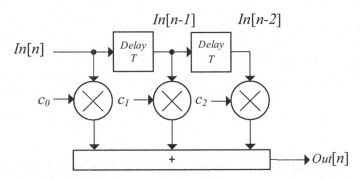

Figure 1-17: Structure of a Finite Impulse Response Filter

Whereas the coefficients of the filter, c_0 through c_2, remain constant, the input and output values, $In[n]$ and $Out[n]$, change as the discrete time variable, n, increments. Note that the quantities $In[n\text{-}1]$ and $In[n\text{-}2]$ represent values that appeared at the input T and $2T$ seconds ago, where T is the sampling period of the filter. But our goal here is not to dwelve too deeply into the inner workings of digital filters, but to show how its inputs, outputs and coefficients may be readily expressed in matrix form. For this purpose, let's write the linear equation relating the input and output at time n.

$$c_0 \cdot In[n] + c_1 \cdot In[n\text{-}1] + c_2 \cdot In[n\text{-}2] = Out[n]$$

Let us take a step further now and write a system of linear equations forming the output values as the discrete time index n increases from 0 to 3.

$$c_0 \cdot In[0] + c_1 \cdot In[\text{-}1] + c_2 \cdot In[\text{-}2] = Out[0]$$

$$c_0 \cdot In[1] + c_1 \cdot In[0] + c_2 \cdot In[\text{-}1] = Out[1]$$

$$c_0 \cdot In[2] + c_1 \cdot In[1] + c_2 \cdot In[0] = Out[2]$$

$$c_0 \cdot In[3] + c_1 \cdot In[2] + c_2 \cdot In[1] = Out[3]$$

Rewriting the system of linear equations in matrix form turns into a trivial task, which in later chapter will bring us untold advantages.

$$
\overset{\displaystyle A}{\begin{bmatrix} In[0] & In[-1] & In[-2] \\ In[1] & In[0] & In[-1] \\ In[2] & In[1] & In[0] \\ In[3] & In[2] & In[1] \end{bmatrix}} \cdot \overset{\displaystyle X}{\begin{bmatrix} c_0 \\ c_1 \\ c_2 \end{bmatrix}} = \overset{\displaystyle Y}{\begin{bmatrix} Out[0] \\ Out[1] \\ Out[2] \\ Out[3] \end{bmatrix}}
$$

1.2.1   Basic Matrices and Matrix Operations

1.2.1.1 Basic Matrix Arithmetic

In this subsection, we will review basic matrix arithmetic, which you have likely seen in high school level math classes.

Matrix Addition and Subtraction

Matrix addition and subtraction happen on an entry-by-entry basis. Only matrices of equal size may be used for these operations.

$$A \pm B = \begin{bmatrix} a_{11} & a_{12} & a_{13} \\ a_{21} & a_{22} & a_{23} \end{bmatrix} \pm \begin{bmatrix} b_{11} & b_{12} & b_{13} \\ b_{21} & b_{22} & b_{23} \end{bmatrix} = \begin{bmatrix} a_{11} \pm b_{11} & a_{12} \pm b_{12} & a_{13} \pm b_{13} \\ a_{21} \pm b_{21} & a_{22} \pm b_{22} & a_{23} \pm b_{23} \end{bmatrix}$$

Matrix Multiplication

For matrix multiplication to be possible, the number of columns of the first matrix must be the same as the number of rows of the second matrix. The resultant matrix will have the same number of rows as the first matrix and the same number of columns as the second matrix. Thus an $a \, x \, b$ sized matrix multiplied by a $b \, x \, c$ sized matrix will yield a product with dimensions $a \, x \, c$.

$$Size = 2x3 \qquad Size = 3x2$$

$$A \cdot B = \begin{bmatrix} a_{11} & a_{12} & a_{13} \\ a_{21} & a_{22} & a_{23} \end{bmatrix} \cdot \begin{bmatrix} b_{11} & b_{12} \\ b_{21} & b_{22} \\ b_{31} & b_{32} \end{bmatrix}$$

$$Size = 2x2$$

$$A \cdot B = \begin{bmatrix} a_{11} \cdot b_{11} + a_{12} \cdot b_{21} + a_{13} \cdot b_{31} & a_{11} \cdot b_{12} + a_{12} \cdot b_{22} + a_{13} \cdot b_{32} \\ a_{21} \cdot b_{11} + a_{22} \cdot b_{21} + a_{23} \cdot b_{31} & a_{21} \cdot b_{12} + a_{22} \cdot b_{22} + a_{23} \cdot b_{32} \end{bmatrix}$$

Note that the resultant matrix is composed of the sums of the entry-by-entry products of the rows of A and columns of B. Because of the restriction that the multiplication operation places on the matrix dimensions, matrix multiplication is *not* commutative (i.e. $A \cdot B \neq B \cdot A$). Whereas $A \cdot B$ is a 2x2 matrix, $B \cdot A$ would be a 3x3 matrix. Furthermore, assume that a matrices A and B feature dimensions of 2x3 and 3x4 respectively. The resulting product $A \cdot B$ is a 2x4 matrix, but $B \cdot A$ can't be computed since the number of columns of B do not equal the number of rows of A.

The Matrix Transpose

The matrix transpose interchanges the rows and columns of a matrix and conjugates each entry.

$$A = \begin{bmatrix} a_{11} & a_{12} & a_{13} \\ a_{21} & a_{22} & a_{23} \end{bmatrix} \rightarrow A^T = \begin{bmatrix} conj(a_{11}) & conj(a_{21}) \\ conj(a_{12}) & conj(a_{22}) \\ conj(a_{13}) & conj(a_{23}) \end{bmatrix} = \begin{bmatrix} a_{11}^* & a_{21}^* \\ a_{12}^* & a_{22}^* \\ a_{13}^* & a_{23}^* \end{bmatrix}$$

Many engineering texts omit the conjugation operation, indicated by the asterisk, when explaining the entry reordering process of the transpose. This is accurate only in situations where we are dealing exclusively with real variables, and since communication systems use complex (IQ) modulation, we formulate the transpose in its more general complex form. This general formulation is also called the Hermetian transpose and may be indicated by a superscript H (i.e. A^H). In this text, we use the superscript T and H notation interchangably.

The Dot Product and Vector Norm

Given two N element vectors $A = [a_1\ a_2\ a_3\ ...\ a_N]$ and $B = [b_1\ b_2\ b_3\ ...\ b_N]$, we define their dot product for both real and complex valued entries in the expression below.

$$Dot(A,B) = a_1 b_1 + a_2 b_2 + a_3 b_3 + ... + a_N b_N \quad real\ valued$$
$$= a_1 b_1^* + a_2 b_2^* + a_3 b_3^* + ... + a_N b_N^* \quad complex\ valued$$

The length of a vector or its norm is defined as follows.

$$Norm(A) = ||A|| = Dot(A,A) = sqrt(a_1 a_1^* + a_2 a_2^* + a_3 a_3^* + ... + a_N a_N^*)$$

In our section on principal component analysis, we will discuss vector projection that makes use of the geometric interpretation (in Euclidean space) of the dot product, which states the following.

$$Dot(A,B) = ||A|| \cdot ||B|| \cdot cos(\theta)$$

As we will show later, θ represent the ange between the vectors A and B.

Scalar Multiplication

The scalar multiplication of any matrix, A, involves the multiplication of each one of its entries by a scalar term.

$$if\ A = \begin{bmatrix} a_{11} & a_{12} & a_{13} \\ a_{21} & a_{22} & a_{23} \end{bmatrix} then \quad c \cdot A = \begin{bmatrix} c \cdot a_{11} & c \cdot a_{12} & c \cdot a_{13} \\ c \cdot a_{21} & c \cdot a_{22} & c \cdot a_{23} \end{bmatrix} = A \cdot c$$

1.2.1.2 The Identity Matrix

The identity matrix serves the same function as the scalar 1 in traditional algebra. If we multiply a matrix by its appropriately sized identity matrix, the result will be the original matrix. The column vectors of a $2x2$ identity matrix are shown in the left-most plot of the next figure.

$$I = \begin{bmatrix} 1 & 0 & \cdots & 0 \\ 0 & 1 & \cdots & 0 \\ \vdots & \vdots & \ddots & \vdots \\ 0 & 0 & \cdots & 1 \end{bmatrix}$$

An example of multiplication using the identity matrix can be seen below.

$$A \cdot I = \begin{bmatrix} a_{11} & a_{12} & a_{13} \\ a_{21} & a_{22} & a_{23} \end{bmatrix} \cdot \begin{bmatrix} 1 & 0 & 0 \\ 0 & 1 & 0 \\ 0 & 0 & 1 \end{bmatrix} = \begin{bmatrix} a_{11} & a_{12} & a_{13} \\ a_{21} & a_{22} & a_{23} \end{bmatrix} = A$$

1.2.1.3 Orthogonal Vectors and Matrices

Orthogonal matrices are composed of column vectors that are themselves orthogonal. Geometrically speaking, orthogonal vectors in two and three dimensional space, R^2 and R^3, feature directions that are *perpendicular* to one another. The same is true for higher dimensional space, but it is more difficult to visualize. A more generalized terms, two vectors, v_1 and v_2, of length N are orthogonal if the sum of their entry by entry products is equal to zero.

$$\sum_{n=0}^{N-1} v_1[n] \cdot v_2[n] = 0$$

The matrices below feature column vectors that are orthogonal.

$$\begin{array}{ccc} \overset{v_1 \quad v_2}{A_1 = \begin{bmatrix} 1 & 0 \\ 0 & 1 \end{bmatrix}} & \overset{v_1 \quad v_2}{A_2 = \begin{bmatrix} 1 & -1 \\ 1 & 1 \end{bmatrix}} & \overset{v_1 \quad v_2 \quad v_3}{A_3 = \begin{bmatrix} 1 & -1 & 0 \\ 1 & 1 & 0 \\ 0 & 0 & 1 \end{bmatrix}} \end{array}$$

The orthogonal column vectors of each matrix - let's call them v_1, v_2, and v_3 – are seen in the two and three dimensional coordinate systems below. A_1 is an *identity matrix*, which leaves all input vectors that it transforms unchanged. Matrix A_2 would cause an input vector to be rotated by 45 degrees and stretched by a factor equal to the square root of 2. Similarly, A_3 produces a 45 degree rotation around the z axis, stretches the x and y components of an input vector by the square root of 2 but leaves the z component unchanged. However, regardless of how each matrix affects its input vector, the column vectors in each matrix are perpendicular and therefore orthogonal.

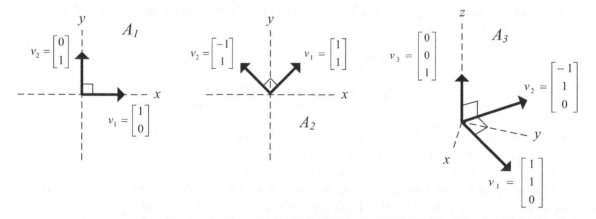

Figure 1-18: Geometric Interpretation of Orthogonal Matrices A₁, A₂, and A₃

Orthogonal matrices, whose column vectors feature unit-length are called *orthonormal*. Notice that of the three matrices shown above only A_1 is orthonormal. We will meet orthonormal matrices later on in this chapter and discover their interesting properties.

Example 1.7: *Proving Orthogonality*

In the following example we will prove that matrices A_1, A_2, and A_3 are orthogonal by showing that the dot products between all their column vectors are zero. Matrix A_3 features three columns and every possible inner product must be zero.

$$For\ A_1 \rightarrow \begin{bmatrix} 1 \\ 0 \end{bmatrix}^{\text{T}} \cdot \begin{bmatrix} 0 \\ 1 \end{bmatrix} = \begin{bmatrix} 1 & 0 \end{bmatrix} \cdot \begin{bmatrix} 0 \\ 1 \end{bmatrix} = 0 \cdot 1 + 1 \cdot 0 = 0 \qquad For\ A_2 \rightarrow \begin{bmatrix} -1 \\ 1 \end{bmatrix}^{\text{T}} \cdot \begin{bmatrix} 1 \\ 1 \end{bmatrix} = \begin{bmatrix} -1 & 1 \end{bmatrix} \cdot \begin{bmatrix} 1 \\ 1 \end{bmatrix} = -1 + 1 = 0$$

$$\underset{Col1 \quad Col2}{\qquad\qquad} \qquad\qquad\qquad \underset{Col1 \quad Col3}{\qquad} \qquad \underset{Col2 \quad Col3}{\qquad}$$

$$For\ A_3 \rightarrow \begin{bmatrix} 1 \\ 1 \\ 0 \end{bmatrix}^{\text{T}} \cdot \begin{bmatrix} -1 \\ 1 \\ 0 \end{bmatrix} = \begin{bmatrix} -1 & 1 & 0 \end{bmatrix} \cdot \begin{bmatrix} 1 \\ 1 \\ 0 \end{bmatrix} = -1 + 1 = 0 \quad similarly \quad \begin{bmatrix} 1 \\ 1 \\ 0 \end{bmatrix}^{\text{T}} \cdot \begin{bmatrix} 0 \\ 0 \\ 1 \end{bmatrix} = 0 \ and \ \begin{bmatrix} -1 \\ 1 \\ 0 \end{bmatrix}^{\text{T}} \cdot \begin{bmatrix} 0 \\ 0 \\ 1 \end{bmatrix} = 0$$

1.2.2   Matrices as Linear Transformations

Every matrix, A, may be thought of as a linear transformation that takes an input vector, *In*, with n entries (in vector space R^n) and maps it to an output vector, *Out*, with m entries (in vector space R^m).

$$Out = A \cdot In$$

Matrices allow arbitrary transformations to be conveniently represented and used during computation. We may even concatenate transformations by multiplying matrices. Let's review some simple types of transformations and examine their resulting matrix structure.

1.2.2.1 The Scaling and Rotational Transformation

The *scaling transformation* stretches or compresses the length of each entry in the vector. Scaling is usually done from R^n to R^n space where the input and output vectors feature the same number of entries. In this case, we construct the transformation matrix A by separately scaling the entries of the identity matrix.

$$In = \begin{bmatrix} x \\ y \end{bmatrix}$$

$$Out = A \cdot In = \begin{bmatrix} \alpha & 0 \\ 0 & \beta \end{bmatrix} \cdot \begin{bmatrix} x \\ y \end{bmatrix} = \begin{bmatrix} \alpha x \\ \beta y \end{bmatrix}$$

In the following example, we illustrate the scaling transformation by selecting a set of unit-length input vectors that form a circle around the origin. Once we transform these input vectors using the matrix A, the geometric properties of the transformation become especially clear. To understand the *rotational transformation*, recall from our discussion on complex numbers that we may rotate a point, $In = x + jy$, in the complex plane by multiplying it by $e^{j\theta} = \cos(\theta) + j \cdot \sin(\theta)$, where θ is our angle of rotation.

$$Out = e^{j\theta} \cdot In = (\cos\theta + j\sin\theta) \cdot (x + jy)$$

$$= (\cos\theta \cdot x - \sin\theta \cdot y) + j(\sin\theta \cdot x + \cos\theta \cdot y)$$

The rotational transformation is easily captured in matrix notation. Note that the rotation transformation matrix is orthonormal (its column vectors are of unit length and perpendicular).

$$Out = \begin{bmatrix} x_{out} \\ y_{out} \end{bmatrix} = \begin{bmatrix} cos(\theta) & -sin(\theta) \\ sin(\theta) & cos(\theta) \end{bmatrix} \cdot \begin{bmatrix} x \\ y \end{bmatrix}$$

Scaling and rotation are the two most important types of transformations that we will cover, and as it turns out, all other transformation matrices can be represented as a product of these two.

Example 1.8: The Scaling and Rotational Transformation in Action

In this example, we show how a set of unit-length input vectors are transformed via two scaling matrices A_1, A_2 and a rotational matrix A_3. We further select four unique input vectors at angles of $\pi/4$, $3\pi/4$, $5\pi/4$, and $7\pi/4$ as shown by the different markers in the figure below. Including these markers helps us illustrate the geometric process during the different transformations. Note how matrix $A2$ reduces the size of the input vector and reflects the input points around the x axis.

$$A_1 = \begin{bmatrix} 1 & 0 \\ 0 & 1.5 \end{bmatrix} \quad A_2 = \begin{bmatrix} 0.5 & 0 \\ 0 & -0.5 \end{bmatrix} \quad A_3 = \begin{bmatrix} cos(\frac{\pi}{8}) & -sin(\frac{\pi}{8}) \\ sin(\frac{\pi}{8}) & cos(\frac{\pi}{8}) \end{bmatrix}$$

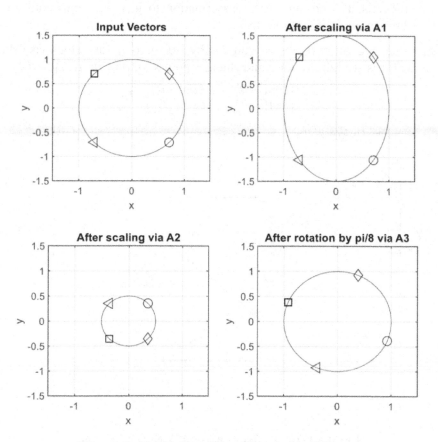

Figure 1-19: Scaling Transformation Examples

Notice the elliptical shape that is produced as the two dimensions are scaled differently by matrix A_1. A_2 on the other hand shrinks the input vectors and reflects them around the x-axis. The matrix A_3 produces a positive rotation by $\pi/8$ or 22.5 degrees.

```
%% Defining the Matrix A (Size 2x2)
A1  = [1.0 0; 0 1.5];
A2  = [0.5 0; 0 -0.5];
A3  = [cos(pi/8), -sin(pi/8); sin(pi/8), cos(pi/8)];
%% Defining Circular Unit length Input Vectors
Angles     = 0:0.02:2*pi;
Input      = [cos(Angles); sin(Angles)];
Markers_In = [cos([pi/4 3*pi/4 5*pi/4 7*pi/4]); sin([pi/4 3*pi/4 5*pi/4 7*pi/4])];
%% Transforming the Input and Markers into the Output
Output1    = A1*Input;  Markers_Out1 = A1*Markers_In;
Output2    = A2*Input;  Markers_Out2 = A2*Markers_In;
Output3    = A3*Input;  Markers_Out3 = A3*Markers_In;
```

1.2.2.2 Reflection as a Compound Transform

Reflection is a wonderful example of how we can express a sophisticated linear transformation as the product of simpler ones. Take a look at the graph at the top left of the figure below. Points In_1 and In_2 are reflected about a straight line that crosses the origin and features an orientation of θ from the positive x axis. It's not an easy transformation to just write out until we realize that reflection can be broken down into three steps.

Step 1: In step one, we rotate both In_1 and In_2 by $-\theta$, or θ in the clockwise direction. The transformation matrix is straight forward and changes points In_1 and In_2 to T_1 and T_2.

$$T = \begin{bmatrix} cos(-\theta) & -sin(-\theta) \\ sin(-\theta) & cos(-\theta) \end{bmatrix} \cdot In$$

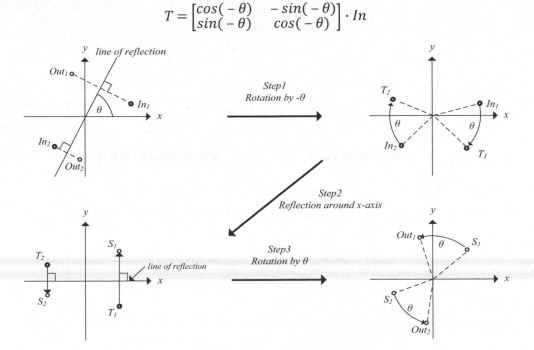

Figure 1-20: Reflection about a Straight Line

Step 2: In step two, we cause a reflection about the x-axis which is trivial to do using a scaling matrix. Here the intermediary points T_1 and T_2 are reflected onto S_1 and S_2.

$$S = \begin{bmatrix} 1 & 0 \\ 0 & -1 \end{bmatrix} \cdot T$$

Step 3: In step three, we undo the original rotation to arrive at the final points Out_1 and Out_2.

$$Out = \begin{bmatrix} cos(\theta) & -sin(\theta) \\ sin(\theta) & cos(\theta) \end{bmatrix} \cdot S$$

The reflection matrix, A, can thus be represented as the product of three separate transformations as shown below.

$$Out = \begin{bmatrix} \cos(\theta) & -\sin(\theta) \\ \sin(\theta) & \cos(\theta) \end{bmatrix} \cdot \begin{bmatrix} 1 & 0 \\ 0 & -1 \end{bmatrix} \cdot \begin{bmatrix} \cos(-\theta) & -\sin(-\theta) \\ \sin(-\theta) & \cos(-\theta) \end{bmatrix} \cdot In$$

$$= \begin{bmatrix} \cos(\theta) & \sin(\theta) \\ \sin(\theta) & -\cos(\theta) \end{bmatrix} \cdot \begin{bmatrix} \cos(-\theta) & -\sin(-\theta) \\ \sin(-\theta) & \cos(-\theta) \end{bmatrix} \cdot In$$

$$= \begin{bmatrix} \cos(\theta) \cdot \cos(-\theta) + \sin(\theta) \cdot \sin(-\theta) & -\cos(\theta) \cdot \sin(-\theta) + \sin(\theta) \cdot \cos(-\theta) \\ \sin(\theta) \cdot \cos(-\theta) - \cos(\theta) \cdot \sin(-\theta) & -\sin(\theta) \cdot \sin(-\theta) - \cos(\theta) \cdot \cos(-\theta) \end{bmatrix} \cdot In$$

Now, if you go back a few pages and review the angle sum and difference trigonometric identities, you will see that the unwieldly expression above reduces to the neat equation below.

$$Out = \begin{bmatrix} cos(2\theta) & sin(2\theta) \\ sin(2\theta) & -cos(2\theta) \end{bmatrix} \cdot In$$

Any linear transformation may be decomposed into a rotation, followed by a scaling operation and another rotation. In retrospect, finding the individual transformations for the reflection operation was not an impossible feat, but if you were simply given some arbitrary matrix A and asked to find the three transformations that make up A, then it may be time to duck for cover. We will soon cover a technique called *singular value decomposition*, or *SVD* for short, which computes these three subtransformations for any matrix. The *SVD* is one of the crown jewels of linear algebra, and is the key operation for the spatial multliplexing operation in MIMO.

Example 1.9: *Transformation as Stretching Along Different Axis*

In this example we will take another step forward in our understanding of matrix transformation by examining the following matrix A, which we show here in its transform equation.

$$A \quad \cdot \quad In \quad = \quad Out$$

$$\begin{bmatrix} 1 & | & 0 \\ 0.4 & | & 1.5 \end{bmatrix} \cdot \begin{bmatrix} x_{In} \\ y_{In} \end{bmatrix} = \begin{bmatrix} x_{Out} \\ y_{Out} \end{bmatrix}$$

$$\qquad V_1 \qquad V_2$$

The matrix A features two columns vectors V_1 and V_2. Note from the equivalent system of equations below, that vectors V_1 and V_2 appear to be stretched by x_{In} and y_{In} respectively. Rewriting the matrix expression on the right confirms this fact. It is as if V_1 and V_2 are establishing new coordinate axis that happen to have different lengths.

$$
\begin{matrix} V_1 & V_2 \end{matrix} \qquad\qquad \begin{matrix} V_1 & V_2 \end{matrix}
$$

$$
\begin{aligned}
1.0 \cdot x_{In} + 0.0 \cdot y_{In} &= x_{out} \\
0.4 \cdot x_{In} + 1.5 \cdot y_{In} &= y_{out}
\end{aligned}
\quad \text{or} \quad
\begin{bmatrix} 1 \\ 0.4 \end{bmatrix} \cdot x_{In} + \begin{bmatrix} 0 \\ 1.5 \end{bmatrix} \cdot y_{In} = \begin{bmatrix} x_{Out} \\ y_{Out} \end{bmatrix}
$$

In the next figure, we once again use a set of unit-length input vectors. The end points of the input vectors lie on the unit circle as attested to in the left most plot. The x coordinate of any input point, In, stretches vector V_1 while the y coordinate stretches V_2. Notice the ellypical output shape that always results when we set up the inputs in this way.

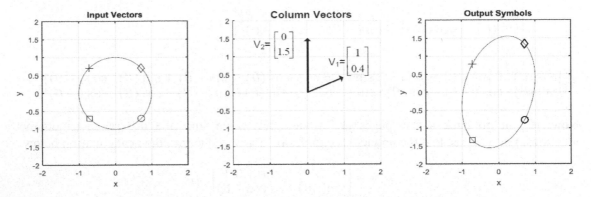

Figure 1-21: Transformation of an Arbitrary Matrix A

1.2.3   Introducing the Matrix Inverse

At the beginning of this section we introduced a simple system of equations describing two lines in its traditional and matrix form.

$$
a_{11} \cdot x + a_{12} y = b_1
$$
$$
a_{21} \cdot x + a_{22} y = b_2
$$
$$
\begin{bmatrix} a_{11} & a_{12} \\ a_{21} & a_{22} \end{bmatrix} \cdot \begin{bmatrix} x \\ y \end{bmatrix} = \begin{bmatrix} b_1 \\ b_2 \end{bmatrix}
$$
$$
A \quad \cdot \quad X = B
$$

Solving this system of equations (finding X) involves dividing by matrix A or multiplying by its reciprocal, A^{-1}, or inverse, $inv(A)$.

$$
A^{-1} \cdot A \cdot X = A^{-1} \cdot B
$$
$$
I \cdot X = A^{-1} \cdot B
$$
$$
X = A^{-1} \cdot B = inv(A) \cdot B
$$

If the inverse of *A* exists, then one unique solution vector *X* may be computed. If the inverse of *A* does not exist, then either an infinite number of solution vectors *X* exist or none at all. In terms of transformation, the inverse is very easy to understand. Assume a point *In* is transformed via matrix *A* to the point *Out*. Can we come up with a matrix, *D*, that will transform the point *Out* back to the point *In*? If such a matrix *D* exists, then *D* is the inverse of *A* and their product will result in the identity matrix *I*. Matrix invertibility may be determined by calculating the determinant or by proceeding with the inverse calculation directly. If the matrix is not invertible, the determinant will be zero and the inverse calculation will suffer an internal divide-by-zero error. We will meet the determinant calculation next and the Gaussian elimination, which we use to compute the inverse, in a later section. Let's summarize the basic rules regarding the inverse of a square matrix *A*.

1 → If $AD = I$ is true, then *D* is said to be the inverse of the square matrix *A* and $AD = DA$. The inverse of *A* exists only if the determinant of *A* is not equal to zero.

2 → If $AX = 0$, and an inverse of the square matrix *A* exists, then each element of the solution vector *X* must be equal to zero.

3 → If $AX = 0$, and an inverse of the square matrix *A* does not exist (*A* is said to be singular and its determinant = 0), then an infinite number of solutions for vector *X* exist or none at all.

4 → The inverse may be computed only for square matrices.

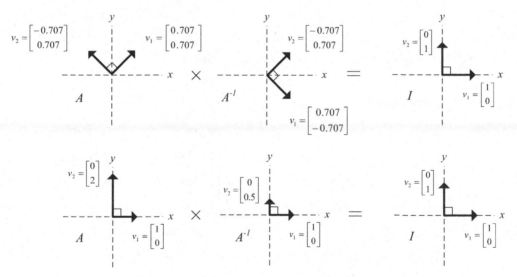

Figure 1-22: Transformation Matrix A and its Inverse, A⁻¹, in 2D Geometric Representation

1.2.4 The Determinant

The determinant is a metric that indicates whether a square matrix is invertible or not. If the determinant of matrix *A* is *not zero*, then a unique solution vector, *X*, exists for the set of linear equations $A \cdot X = B$ that are being modeled and the matrix *A* has an inverse. If the determinant of matrix *A* is *zero*, then there may be no or an infinite number of solutions to the set of linear equations and the matrix cannot be inverted. The determinant calculations for a 2x2, 3x3, and 4x4 matrix are shown next, while those for larger square matrices may be reviewed here [4].

$$A_1 = \begin{bmatrix} a_1 & b_1 \\ a_2 & b_2 \end{bmatrix} \qquad A_2 = \begin{bmatrix} a_1 & b_1 & c_1 \\ a_2 & b_2 & c_2 \\ a_3 & b_3 & c_3 \end{bmatrix} \qquad A_3 = \begin{bmatrix} a_1 & b_1 & c_1 & d_1 \\ a_2 & b_2 & c_2 & d_2 \\ a_3 & b_3 & c_3 & d_3 \\ a_4 & b_4 & c_4 & d_4 \end{bmatrix}$$

Determinant for a 2x2 Matrix

$$\det(A_1) = |A_1| = \begin{bmatrix} a_1 & & b_1 \\ & \ddots & \\ a_2 & & b_2 \end{bmatrix} - \begin{bmatrix} a_1 & & b_1 \\ & \iddots & \\ a_2 & & b_2 \end{bmatrix} = a_1 b_2 - a_2 b_1$$

Determinant for a 3x3 Matrix

$$\det(A_2) = |A_2| = \begin{bmatrix} a_1 & b_1 & c_1 \\ a_2 & b_2 & c_2 \\ a_3 & b_3 & c_3 \end{bmatrix} \begin{matrix} a_1 & b_1 \\ a_2 & b_2 \\ a_3 & b_3 \end{matrix} - \begin{bmatrix} a_1 & b_1 & c_1 \\ a_2 & b_2 & c_2 \\ a_3 & b_3 & c_3 \end{bmatrix} \begin{matrix} a_1 & b_1 \\ a_2 & b_2 \\ a_3 & b_3 \end{matrix}$$

$$= (a_1 b_2 c_3 + b_1 c_2 a_3 + c_1 a_2 b_3) - (a_3 b_2 c_1 + b_3 c_2 a_1 + c_3 a_2 b_1)$$

Determinant for a 4x4 Matrix

$$\det(A_3) = |A_3| = \begin{vmatrix} a_1 & & & \\ & b_2 & c_2 & d_2 \\ & b_3 & c_3 & d_3 \\ & b_4 & c_4 & d_4 \end{vmatrix} - \begin{vmatrix} & b_1 & & \\ a_2 & & c_2 & d_2 \\ a_3 & & c_3 & d_3 \\ a_4 & & c_4 & d_4 \end{vmatrix} + \begin{vmatrix} & & c_1 & \\ a_2 & b_2 & & d_2 \\ a_3 & b_3 & & d_3 \\ a_4 & b_4 & & d_4 \end{vmatrix} - \begin{vmatrix} & & & d_1 \\ a_2 & b_2 & c_2 & \\ a_3 & b_3 & c_3 & \\ a_4 & b_4 & c_4 & \end{vmatrix}$$

$$= a_1 \cdot \begin{vmatrix} b_2 & c_2 & d_2 \\ b_3 & c_3 & d_3 \\ b_4 & c_4 & d_4 \end{vmatrix} - b_1 \cdot \begin{vmatrix} a_2 & c_2 & d_2 \\ a_3 & c_3 & d_3 \\ a_4 & c_4 & d_4 \end{vmatrix} + c_1 \cdot \begin{vmatrix} a_2 & b_2 & d_2 \\ a_3 & b_3 & d_3 \\ a_4 & b_4 & d_4 \end{vmatrix} - d_1 \cdot \begin{vmatrix} a_2 & b_2 & c_2 \\ a_3 & b_3 & c_3 \\ a_4 & b_4 & c_4 \end{vmatrix}$$

$$= a_1 \cdot Minor(a_1) - b_1 \cdot Minor(b_1) + c_1 \cdot Minor(c_1) - d_1 \cdot Minor(d_1)$$

Computing the determinant for the 2x2 and 3x3 matrices is relatively straight forward, but calculating it for the 4x4 matrix requires some explanation. The determinant of A_3 is computed by multiplying the entries of the first row by the determinant of 3x3 submatrices. The determinants of these submatrices are called minors. For example, the minor belonging to the entry a_1 (of row 1 and column 1) is the determinant of a 3x3 matrix containing all entries of A_3 not belonging to row 1 and column 1. This methos of computing the determinant can easily be expanded to arbitrarily sized square matrices and becomes very useful to us in our section on eigenvalues and eigenvectors.

1.2.5   Select Matrices and Their Properties

In this section, we introduce several interesting matrices and some of their associated properties. Not just are these matrixes and properties useful for topics in later chapters of this text, they are a convenient way to strengthen some of the matrix algebra concepts that we have already learned.

1.2.5.1 Symmetric Matrices

If a matrix S is equal to its transpose S^T, then S is said to be symmetric. The following matrix is symmetric as $S = S^T$, and its symmetry becomes evident as the entries at the top right are a mirrow image of those at the bottom left. They appear reflected about the main diagonal. Note that column one and row one have identical entries, as do column 2 / row 2 and column 3 / row 3.

$$S = \begin{bmatrix} 1 & 2 & 3 \\ 2 & 4 & 5 \\ 3 & 5 & 8 \end{bmatrix} = \begin{bmatrix} 1 & 2 & 3 \\ 2 & 4 & 5 \\ 3 & 5 & 8 \end{bmatrix} = S^T$$

➔ *Property1: Symmetry of $A^T \cdot A$*

In some of our later exercises we come across the product of the transpose of a matrix A and the matrix A itself. This product will be symmetric for any matrix A, and we illustrate this fact via a simple example.

Example 1.10: *Find $A^T \cdot A$ for the Following Matrix*

$$A = \begin{bmatrix} 1 & 2 & 3 \\ 4 & 5 & 6 \end{bmatrix}$$

$$A^T \cdot A = \begin{bmatrix} 1 & 4 \\ 2 & 5 \\ 3 & 6 \end{bmatrix} \cdot \begin{bmatrix} 1 & 2 & 3 \\ 4 & 5 & 6 \end{bmatrix} = \begin{bmatrix} 17 & 22 & 27 \\ 22 & 24 & 36 \\ 27 & 36 & 45 \end{bmatrix}$$

The rows and columns of $A^T A$ feature the same entries and the product is in fact symmetric.

➔ *Property2: The Transpose of a Product of Matrices*

The transpose of the product of two matrices is equal to the product of the transposes of these two matrices in reverse order. This fact is formulated below and illustrated in the following example.

$$(A \cdot B)^T = B^T \cdot A^T$$

Example 1.11: *The Transpose of the Product of Matrices*

Compute $(A \cdot B)^T$ and $B^T \cdot A^T$ given the two matrices below.

$$A = \begin{bmatrix} 1 & 2 & 3 \\ 4 & 5 & 6 \end{bmatrix} \quad B = \begin{bmatrix} 1 & 0 & 1 \\ 0 & 1 & 2 \\ 1 & 1 & 1 \end{bmatrix}$$

Let's substitute the two matrices into the expression we want to prove and see if it holds true.

$$A \cdot B = \begin{bmatrix} 1 & 2 & 3 \\ 4 & 5 & 6 \end{bmatrix} \cdot \begin{bmatrix} 1 & 0 & 1 \\ 0 & 1 & 2 \\ 1 & 1 & 1 \end{bmatrix} = \begin{bmatrix} 4 & 5 & 8 \\ 10 & 11 & 20 \end{bmatrix} \qquad (A \cdot B)^T = \begin{bmatrix} 4 & 10 \\ 5 & 11 \\ 8 & 20 \end{bmatrix}$$

$$B^T \cdot A^T = \begin{bmatrix} 1 & 0 & 1 \\ 0 & 1 & 1 \\ 1 & 2 & 1 \end{bmatrix} \cdot \begin{bmatrix} 1 & 4 \\ 2 & 5 \\ 3 & 6 \end{bmatrix} = \begin{bmatrix} 4 & 10 \\ 5 & 11 \\ 8 & 20 \end{bmatrix}$$

Therefore, the rule holds and we may now trivially expand it to larger matrix products as follows. Here we look at the product of three and four matrices.

$$(A \cdot B \cdot C)^T = (A \cdot (B \cdot C))^T = (B \cdot C)^T \cdot A^T = C^T \cdot B^T \cdot A^T$$

$$(A \cdot B \cdot C \cdot D)^T = ((A \cdot B) \cdot (C \cdot D))^T = (C \cdot D)^T \cdot (A \cdot B)^T = D^T \cdot C^T \cdot B^T \cdot A^T$$

> The rule regarding the transpose of the product of any number of matrices becomes apparent rather quickly. The transpose of a product of several matrices is equal to the product of the transpose of each matrix arranged in opposite order.

→ *Property3: The Inverse of a Product of two Matrices*

Property three claims that the inverse of the product of matrices A and B is the product of the inverses of A and B in reverse order. To state the property formally, we claim the following.

$$(A \cdot B)^{-1} = B^{-1} \cdot A^{-1}$$

To prove this, we remind ourselves that the product of a matrix and its inverse is equal to the identity matrix, I. Thus, $(A \cdot B) \cdot (A \cdot B)^{-1} = I$ and if our property is indeed true, then the following substitution must be valid.

$$(A \cdot B) \cdot (B^{-1} \cdot A^{-1}) = I$$

We can prove this to be correct by simply rearranging the parenthesis as follows.

$$A \cdot (B \cdot B^{-1}) \cdot A^{-1} = I$$

$$A \cdot (I) \cdot A^{-1} = I$$

$$A \cdot A^{-1} = I$$

$$I = I$$

→ *Property4: The Inverse of the Transpose of a Matrix*

In this property, we claim that the inverse of the transpose of a matrix A is equal to the transpose of the inverse of A. This tunguetwister can be expressed as follows.

$$(A^T)^{-1} = (A^{-1})^T$$

Just looking at this expression, the steps leading to its proof may not necessarily jump at us. We begin by reminding ourselves that the product of any matrix and its inverse is the identity matrix, I. Therefore, we know that $(A^T)^{-1} \cdot (A^T) = I$.

$$(A^T)^{-1} \cdot A^T = I$$

Because our property claims that $(A^T)^{-1}$ is equal to $(A^{-1})^T$, we should be able to substitute $(A^{-1})^T$ for $(A^{-1})^T$ in the last expression and still produce I.

$$(A^{-1})^T \cdot A^T = I$$

From property two we remember that $(A \cdot B)^T = B^T \cdot A^T$ which allows us to rewrite the expression.

$$(A \cdot (A^{-1}))^T = I$$

$$I^T = I$$

$$I = I$$

1.2.5.2 Orthonormal Matrices

Orthogonal matrices with unit length column vectors are also known as orthonormal matrices. As covered earlier, orthogonal matrices feature column vectors that are themselves orthogonal. Geometrically speaking, orthogonal vectors are perpendicular to one another, while algebraically, their inner product is zero. Matrices A_1, A_2, and A_3 are orthonormal. Geometrically speaking, an orthonormal matrix causes a rotation and a possible reflection about the X, Y or Z axis.

$$A_1 = \begin{bmatrix} 1 & 0 \\ 0 & 1 \end{bmatrix} \qquad A_2 = \begin{bmatrix} 0.707 & -0.707 \\ 0.707 & 0.707 \end{bmatrix} \qquad A_3 = \begin{bmatrix} 0.707 & -0.707 & 0 \\ 0.707 & 0.707 & 0 \\ 0 & 0 & 1 \end{bmatrix}$$

→ *Property 5: The Inverse of an Orthonormal Matrix is its Transpose*

An interesting property of orthonormal matrices is that their inverse and transpose are always equal. For this to be true, the product of any orthonormal matrices, A, and their transpose, A^T, must be equal to the identity matrix.

$$A \cdot A^T = A \cdot A^{-1} = I$$

Let's examine matrix A_2 to find out if the property holds true.

$$A_2 \cdot A_2^T = \begin{bmatrix} 0.707 & -0.707 \\ 0.707 & 0.707 \end{bmatrix} \cdot \begin{bmatrix} 0.707 & 0.707 \\ -0.707 & 0.707 \end{bmatrix} = \begin{bmatrix} 0.5 + 0.5 & 0.5 - 0.5 \\ 0.5 - 0.5 & 0.5 + 0.5 \end{bmatrix} = \begin{bmatrix} 1 & 0 \\ 0 & 1 \end{bmatrix}$$

1.2.5.3 A Summary of Matrix Arithmatic and Properties

At this point of our review regarding linear algebra, it is appropriate to present a summary of basic matrix arithmetic and properties. For the sake of completeness, we will present matrix operations and properties that we have already seen as well as some that we have yet to cover. Return to this page as you work through some of the more advanced material in later chapters of this text.

Table 1-1: Basic Arithmetic Rules for Matrices

	MATRIX OPERATIONS	COMMENTS
1	$A \cdot B \neq B \cdot A$	Matrix multiplication is not commutative
2	$A + B = B + A$	Matrix addition is commutative
3	$A \cdot (B \cdot C) = (A \cdot B) \cdot C$	Matrix multiplication is associative
4	$A \cdot (B + C) = A \cdot B + A \cdot C$	Distributive Rule
5	$c \cdot (A + B) = c \cdot A + c \cdot B$	Multiplication by an arbitrary scalar, c
6	$(A + B)^T = A^T + B^T$	Transpose of the sum of two matrices

Table 1-2: Basic Matrix Properties

	MATRIX PROPERTIES	COMMENTS
1	$A^T \cdot A$ and $A \cdot A^T$	These two products are symmetric. They have the same eigenvalues, which are both positive and real.
2	$(A \cdot B)^T = B^T \cdot A^T$	Extends easily to larger number of submatices
3	$(A \cdot b)^T = conj(b) \cdot A^T$	Extension to the product of a matrix A and a scalar b
4	$(A \cdot B)^{-1} = B^{-1} \cdot A^{-1}$	The inverse of the product of two matrices
5	$(A^T)^{-1} = (A^{-1})^T$	The inverse of the transpose of a matrix
6	$A^T = A^{-1}$	This is **only** true for orthonormal matrices

Table 1-3: List of Matrix Decompositions

	DECOMPOSITIONS	COMMENTS
1	$A = V \cdot D \cdot V^{-1}$	Eigenvalue Decomposition (D is diagonal)
2	$A = O \cdot S$	Polar Decomposition (O is orthogonal and S is Symmetric)
3	$A = U \cdot S \cdot V^T$	Singular Value Decomposition (U and V are orthonormal and S is diagonal)

1.2.6 Gaussian Elimination

A system of linear equations may be solved via an algorithm called Gauss elimination [5], shown in the computational flow below. We start by declaring a system of three equations $AX=B$.

$$a_{11}x_1 + a_{12}x_2 + a_{13}x_3 = b_1$$

$$a_{21}x_1 + a_{22}x_2 + a_{23}x_3 = b_2$$

$$a_{31}x_1 + a_{32}x_2 + a_{33}x_3 = b_3$$

$$
\begin{bmatrix} a_{11} & a_{12} & a_{13} \\ a_{21} & a_{22} & a_{23} \\ a_{31} & a_{32} & a_{33} \end{bmatrix} \cdot \begin{bmatrix} x_1 \\ x_2 \\ x_3 \end{bmatrix} = \begin{bmatrix} b_1 \\ b_2 \\ b_3 \end{bmatrix}
$$
$$ A \qquad \cdot \quad X \ = \ B $$

The principle used for the elimination processes is based on two simple facts. First, we may scale any one of the linear equations, meaning we multiply both sides by a constant scalar, without changing the solution vector, X, of the system. Furthermore, we may add an equation, or its scaled version, to any of the others without changing the solution vector X.

We therefore proceed with the following steps called *forward elimination*.

1. Scale equation 1 by $-a_{21}/a_{11}$ and add it to equation 2. (Equation 2 is now updated)

2. Scale equation 1 by $-a_{31}/a_{11}$ and add it to equation 3. (Equation 3 is now updated)

3. Scale equation 2 by $-a_{32}'/a_{22}'$ and add it to equation 3. (Equation 3 is updated again)

The systems on the left and right show the state of our setup after steps two and three respectively. Note that the hyphens behind the coefficients indicate the number of times their values have been altered by the addition / subtraction operations.

$$a_{11}x_1 + a_{12}x_2 + a_{13}x_3 = b_1$$
$$0 + a_{22}'x_2 + a_{23}'x_3 = b_2'$$
$$0 + a_{32}'x_3 + a_{33}'x_3 = b_3'$$

$\rightarrow$

$$a_{11}x_1 + a_{12}x_2 + a_{13}x_3 = b_1$$
$$0 + a_{22}'x_2 + a_{23}'x_3 = b_2'$$
$$0 + 0 + a_{33}''x_3 = b_3''$$

$$
\begin{bmatrix} a_{11} & a_{12} & a_{13} \\ 0 & a_{22}' & a_{23}' \\ 0 & 0 & a_{33}'' \end{bmatrix} \cdot \begin{bmatrix} x_1 \\ x_2 \\ x_3 \end{bmatrix} = \begin{bmatrix} b_1 \\ b_2' \\ b_3'' \end{bmatrix}
$$

At the end of the forward elimination step, we reach what is called the *row echelon form* of matrix A. In this form, the leading non-zero element in each row will have zeros placed in the column entries beneath it. The row echelon form also requires that the first non-zero element in any particular row must be positioned to the right compared to the non-zero element of the row above it. The next step is to solve for x_3 and back substitute that value into the second equations. To complete the Gaussian elimination procedure, we find x_2 and back-substitute both x_2 and x_3 into equation one to solve for x_1.

Gauss-Jordan Elimination

The Gauss-Jordan version of Gaussian elimination uses *backward elimination* rather than substitution to finish solving the linear system of equations. Let begin by scaling the first, second, and third equations by $1/a_{11}$, $1/a_{22}'$, and $1/a_{33}''$ respectively, to produce coefficient values equal to 1 along the diagonal of the matrix.

$$\begin{bmatrix} 1 & a_{12}' & a_{13}' \\ 0 & 1 & a_{23}'' \\ 0 & 0 & 1 \end{bmatrix} \cdot \begin{bmatrix} x_1 \\ x_2 \\ x_3 \end{bmatrix} = \begin{bmatrix} b_1' \\ b_2'' \\ b_3''' \end{bmatrix}$$

We begin the backward elimination process by appropriately scaling equation three and subtracting it from equations two and one to eliminate a_{23}'' and a_{13}'. We finally scale equation two and subtract it from equation one to eliminate a_{12}'.

$$\begin{bmatrix} 1 & a_{12}' & 0 \\ 0 & 1 & 0 \\ 0 & 0 & 1 \end{bmatrix} \cdot \begin{bmatrix} x_1 \\ x_2 \\ x_3 \end{bmatrix} = \begin{bmatrix} b_1'' \\ b_2''' \\ b_3''' \end{bmatrix} \quad \rightarrow \quad \begin{bmatrix} 1 & 0 & 0 \\ 0 & 1 & 0 \\ 0 & 0 & 1 \end{bmatrix} \cdot \begin{bmatrix} x_1 \\ x_2 \\ x_3 \end{bmatrix} = \begin{bmatrix} b_1''' \\ b_2''' \\ b_3''' \end{bmatrix}$$

$$I \quad \cdot \quad X \quad = \quad B'''$$
$$X \quad = \quad B'''$$

Pivoting

It is possible that during the forward elimination process one of the diagonal elements (a_{11}, a_{22}', a_{33}''...) sums to zero. In this case, the next scaling operator is infinity and a pivoting operation must be executed to remedy the problem. If, for example, a_{22}' computed as zero, we would simply exchange rows two and three and continue with the elimination process. If repeated pivoting does not solve the issue, then the determinant of the matrix is zero and no unique solution exists for the vector X.

Examples 1.12: *Gauss-Jordan Elimination with Pivoting*

Use Gauss-Jordan elimination and pivoting to solve the system of equations marked 'original'.

$$\begin{bmatrix} 1 & 1 & 4 \\ 1 & 1 & 5 \\ 2 & 6 & 4 \end{bmatrix} \cdot \begin{bmatrix} x_1 \\ x_2 \\ x_3 \end{bmatrix} = \begin{bmatrix} 1 \\ 2 \\ 10 \end{bmatrix} \quad \leftarrow \text{ \textit{Original}}$$

$$\begin{bmatrix} 1 & 1 & 4 \\ 0 & 0 & 1 \\ 0 & 4 & -4 \end{bmatrix} \cdot \begin{bmatrix} x_1 \\ x_2 \\ x_3 \end{bmatrix} = \begin{bmatrix} 1 \\ 1 \\ 8 \end{bmatrix} \quad \leftarrow \text{ \textit{After Step}1}$$

$$\begin{bmatrix} 1 & 1 & 4 \\ 0 & 4 & -4 \\ 0 & 0 & 1 \end{bmatrix} \cdot \begin{bmatrix} x_1 \\ x_2 \\ x_3 \end{bmatrix} = \begin{bmatrix} 1 \\ 8 \\ 1 \end{bmatrix} \qquad \leftarrow \quad \textit{After Pivoting Rows 2 and 3}$$

In the forward elimination step we multiply the top equation by 1 and subtracted it from equation two. Unfortunately, the first two entries, a_{21}' and a_{22}', in that row both yield values of zero and our computerized elimination algorithm cannot proceed. We therefore pivot rows two and three and continue the elimination operation to reach the row echelon form.

$$\begin{bmatrix} 1 & 1 & 4 \\ 0 & 1 & -1 \\ 0 & 0 & 1 \end{bmatrix} \cdot \begin{bmatrix} x_1 \\ x_2 \\ x_3 \end{bmatrix} = \begin{bmatrix} 1 \\ 2 \\ 1 \end{bmatrix} \qquad \leftarrow \quad \textit{row echelon form}$$

$$\begin{bmatrix} 1 & 1 & 0 \\ 0 & 1 & 0 \\ 0 & 0 & 1 \end{bmatrix} \cdot \begin{bmatrix} x_1 \\ x_2 \\ x_3 \end{bmatrix} = \begin{bmatrix} -3 \\ 3 \\ 1 \end{bmatrix} \qquad \leftarrow \quad \textit{Step1 of Backward Elimination}$$

$$\begin{bmatrix} 1 & 0 & 0 \\ 0 & 1 & 0 \\ 0 & 0 & 1 \end{bmatrix} \cdot \begin{bmatrix} x_1 \\ x_2 \\ x_3 \end{bmatrix} = \begin{bmatrix} -6 \\ 3 \\ 1 \end{bmatrix} \qquad \leftarrow \quad \textit{Final Result}$$

We now normalized row two and continue with the backward elimination step of the Gauss-Jordan algorithm. In step 1 we simply add row three to row two and subtract 4 times row three from row one. To reach the final result, we subtract row two from row one.

1.2.7 Computing the Matrix Inverse

If a matrix X exists such that $AX = XA = I$, then X is said to be the inverse of A. The traditional inverse matrix solution is only defined for square matrices. To find the inverse of A, we replace the vector B of our previous example with the identity matrix, I, and the solution vector X with the solution matrix X. Each column in the matrix X, is the solution to the similarly positioned column in I.

$$A \cdot A^{-1} = A \cdot X = I$$

$$\begin{bmatrix} a_{11} & a_{12} & \cdots & a_{1N} \\ a_{21} & a_{22} & \cdots & a_{2N} \\ \vdots & \vdots & \ddots & \vdots \\ a_{N1} & a_{N2} & \cdots & a_{NN} \end{bmatrix} \cdot \begin{bmatrix} x_{11} & x_{12} & x_{13} & \cdots & x_{1N} \\ x_{21} & | & x_{22} & | & x_{23} & \cdots & x_{2N} \\ \vdots & | & \vdots & | & \vdots & \ddots & \vdots \\ x_{N1} & x_{N2} & x_{N3} & \cdots & x_{NN} \end{bmatrix} = \begin{bmatrix} 1 & 0 & 0 & \cdots & 0 \\ 0 & | & 1 & | & 0 & \cdots & 0 \\ \vdots & | & \vdots & | & \vdots & \ddots & \vdots \\ 0 & 0 & 0 & \cdots & 1 \end{bmatrix}$$

Examples 1.13: *Inverse for Two by Two Matrices*

Find the inverse of a generic two by two matrix A.

$$\begin{bmatrix} a_{11} & a_{12} \\ a_{21} & a_{22} \end{bmatrix} \cdot \begin{bmatrix} x_{11} & | & x_{12} \\ x_{21} & | & x_{22} \end{bmatrix} = \begin{bmatrix} 1 & | & 0 \\ 0 & | & 1 \end{bmatrix}$$

The solution for a two by two matrix is simple and does not require us to go through the Gaussian elimination process.

$$A^{-1} = \frac{1}{a_{11}a_{22} - a_{12}a_{21}} \begin{bmatrix} a_{22} & -a_{12} \\ -a_{21} & a_{11} \end{bmatrix}$$

Examples 1.14: *Finding the Inverse via Gauss-Jordan Elimination*

In this example we take the square $3x3$ matrix shown in the last section and pivot rows two and three before starting the elimination process. We use the Gauss-Jordan elimination to solve for the matrix X, which is equal to the A^{-1}.

$$A \cdot X = I$$

$$\begin{bmatrix} a_{11} & a_{12} & a_{13} \\ a_{21} & a_{22} & a_{23} \\ a_{31} & a_{32} & a_{33} \end{bmatrix} \cdot \begin{bmatrix} x_{11} & | & x_{12} & | & x_{13} \\ x_{21} & | & x_{22} & | & x_{23} \\ x_{31} & | & x_{32} & | & x_{33} \end{bmatrix} = \begin{bmatrix} 1 & | & 0 & | & 0 \\ 0 & | & 1 & | & 0 \\ 0 & | & 0 & | & 1 \end{bmatrix}$$

$$\begin{bmatrix} 1 & 1 & 4 \\ 2 & 6 & 4 \\ 1 & 1 & 5 \end{bmatrix} \cdot \begin{bmatrix} x_{11} & | & x_{12} & | & x_{13} \\ x_{21} & | & x_{22} & | & x_{23} \\ x_{31} & | & x_{32} & | & x_{33} \end{bmatrix} = \begin{bmatrix} 1 & | & 0 & | & 0 \\ 0 & | & 1 & | & 0 \\ 0 & | & 0 & | & 1 \end{bmatrix} \leftarrow Start$$

In step one, we execute the following mathematical operations:

→ *Row*2 = *Row*2 - 2·*Row*1

→ *Row*3 = *Row*3 - 1·*Row*1

$$\begin{bmatrix} 1 & 1 & 4 \\ 0 & 4 & -4 \\ 0 & 0 & 1 \end{bmatrix} \cdot \begin{bmatrix} x_{11} & | & x_{12} & | & x_{13} \\ x_{21} & | & x_{22} & | & x_{23} \\ x_{31} & | & x_{32} & | & x_{33} \end{bmatrix} = \begin{bmatrix} 1 & | & 0 & | & 0 \\ -2 & | & 1 & | & 0 \\ -1 & | & 0 & | & 1 \end{bmatrix} \leftarrow Step1$$

Before normalizing row 2 via a division by 4 (step 3), we execute the following mathematical opertions in step2.

→ *Row*2 = *Row*2 + 4·*Row*3

→ *Row*1 = *Row*1 - 4·*Row*1

$$\begin{bmatrix} 1 & 1 & 0 \\ 0 & 4 & 0 \\ 0 & 0 & 1 \end{bmatrix} \cdot \begin{bmatrix} x_{11} & x_{12} & x_{13} \\ x_{21} & x_{22} & x_{23} \\ x_{31} & x_{32} & x_{33} \end{bmatrix} = \begin{bmatrix} 5 & 0 & -4 \\ -6 & 1 & 4 \\ -1 & 0 & 1 \end{bmatrix} \leftarrow Step2$$

→ *Row2 = Row2 /4*

$$\begin{bmatrix} 1 & 1 & 0 \\ 0 & 1 & 0 \\ 0 & 0 & 1 \end{bmatrix} \cdot \begin{bmatrix} x_{11} & x_{12} & x_{13} \\ x_{21} & x_{22} & x_{23} \\ x_{31} & x_{32} & x_{33} \end{bmatrix} = \begin{bmatrix} 5 & 0 & -4 \\ -1.5 & 0.25 & 1 \\ -1 & 0 & 1 \end{bmatrix} \leftarrow Step3$$

The final step sees us subtracting row 2 from row one to get to the equation for the matrix X, which is the inverse of matrix A.

→ *Row1 = Row1 – Row2*

$$\begin{bmatrix} 1 & 0 & 0 \\ 0 & 1 & 0 \\ 0 & 0 & 1 \end{bmatrix} \cdot \begin{bmatrix} x_{11} & x_{12} & x_{13} \\ x_{21} & x_{22} & x_{23} \\ x_{31} & x_{32} & x_{33} \end{bmatrix} = \begin{bmatrix} 6.5 & -0.25 & -5 \\ -1.5 & 1 & 1 \\ -1 & 0 & 1 \end{bmatrix} \leftarrow Step4$$

In the first example of the Gauss-Jordan elimination we only had one column vector positioned on the right hand side of the equation. In the inverse operation, we simply have a square identity matrix on the right hand side and follow the same operations as before on each on of its columns.

$$inv(A) = \begin{bmatrix} 6.5 & -0.25 & -5 \\ -1.5 & 1 & 1 \\ -1 & 0 & 1 \end{bmatrix}$$

When writing MatLab or C++ code for the Gaussian-Jordan elimination process we need to continually check that none of the diagonal terms reduce to zero. For example, when eliminating the term a_{21}, we check to see whether a_{22} was also eliminated. If it was eliminated as well, then we need to pivot and continue the forward elimination process. If the problem can't be resolved by repeated pivoting attempts, then the code should return a message stating that the matrix in not invertible.

Example 1.15: Simple Curve Fitting

This example demonstrates a typical application in which matrix inversion is used. In this application, called curve fitting, the goal is to calculate the coefficients, a_0 through a_3, of a cubic polynomial that passes through four XY coordinates on a two-dimensional plot. The equation for a cubic polynomial is given below, along with the coordinate values.

$$y = c_0 + c_1 x + c_2 x^2 + c_3 x^3$$

$$P1 = [1, 2], P2 = [2, 2], P3 = [3, 3], P4 = [4, 1]$$

Since the cubic polynomial must satisfy all *XY* coordinates given, we formulate a system of equations that must be solved to arrive at the proper coefficients c_0, c_1, c_2 and c_3.

$$c_0 + c_1 x_1 + c_2 x_1^2 + c_3 x_1^3 = y_1$$

$$c_0 + c_1 x_2 + c_2 x_2^2 + c_3 x_2^3 = y_2$$

$$c_0 + c_1 x_3 + c_2 x_3^2 + c_3 x_3^3 = y_3$$

$$c_0 + c_1 x_4 + c_2 x_4^2 + c_3 x_4^3 = y_4$$

The system of equations is then transformed into matrix notation and solved by taking the inverse of *A*. Try not to get confused by the notation, but the solution vector *C* will contain the coefficients.

$$
\begin{bmatrix}
1 & x_1 & x_1^2 & x_1^3 \\
1 & x_2 & x_2^2 & x_2^3 \\
1 & x_3 & x_3^2 & x_3^3 \\
1 & x_4 & x_4^2 & x_4^3
\end{bmatrix}
\cdot
\begin{bmatrix}
c_0 \\
c_1 \\
c_2 \\
c_3
\end{bmatrix}
=
\begin{bmatrix}
y_1 \\
y_2 \\
y_3 \\
y_4
\end{bmatrix}
$$

$$A \qquad \cdot \quad C \ = \ Y$$

Finally,

$$inv(A) \cdot A \cdot C = inv(A) \cdot Y$$

$$C = inv(A) \cdot Y$$

The MatLab code below computes the solution vector *C* to be [7.0 -8.333 4.5 -0.666] and plots the resulting cubic polynomial over an *X* coordinate range of 0 through 4.5.

```
%% Defining the Problem
P1 = [1,2]; P2 = [2,2]; P3 = [3,3]; P4 = [4,1];
X  = [P1(1,1); P2(1,1); P3(1,1); P4(1,1)];  %% the x coordinates
Y  = [P1(1,2); P2(1,2); P3(1,2); P4(1,2)];  %% the y coordinates


%% Solving the Problem
A  = [ones(4,1) X X.^2 X.^3];
C  = inv(A)*Y


%% Plotting the Cubic Polynomial
x  = (0:.01:4.5)';
y  = [ones(length(x),1) x x.^2 x.^3]*C;

plot(x, y, 'k'); grid on; hold on; title('Cubic Polynomial');
plot(X, Y, 'ko'); xlabel('X'); ylabel('Y');
```

The resulting polynomial equation and its graphical representation can then be summarized as follows.

$$y = 7.0 - 8.333x + 4.5x^2 - 0.666x^3$$

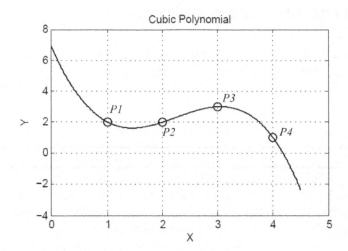

Figure 1-23: Illustration of the Resulting Cubic Polynomial

1.3 A Review of Linear Algebra (Part 2)

The second part of our review of linear algebra is dedicated to covering advanced topics in linear algebra such as the eigenvalue and singular value decompositions as well as algorithms that build on these topics. The following table introduces the topics and explains how we will use them in the rest of this text.

Table 1-4: Topics in

TOPIC	COMMENTS
The Pseudo Inverse (Curve Fitting) (FIR Filter design) (Tracking of Timing Drift LTE) (Frequency Offset computation LTE)	The Pseudo Inverse is used in situations where we need to solve an overdetermined linear system of equations featuring a non-square matrix that must apparently be inverted. Assume that in our last example, we are required to find a cubic polynomial that passes through five points rather than four. Mathematically, you can't find an exact polynomial that does this, but you can find one that passes by the five points as closely as possible. This concept is one of the cornerstones of our discussion in chapter three on optimization. It is incredibly useful.
The Gradient	The gradient operation is used during our derivation of the least squares error solution of an overdetermined system of equations. We mention the gradient here on its own as it will make the derivation in chapter three easier to understand.
The Eigenvalue Decomposition (Used in the Singular Value Decomposition) (Used in Principle Component Analysis) (Used everywhere in engineering)	The eigenvalue decomposition finds the eigenvalues and eigenvectors of a square matrix. This technique is incredibly important in all sorts of engineering applications, but we will use it in this text to compute certain operations on matrices that would otherwise be impossible. We also use the eigenvector decomposition to arrive at the singular value decomposition and principle component analysis.
The Singular Value Decomposition	The singular valued decomposition, or SVD, is one of the true pearls of linear algebra. All major search engines use the SVD as a learning algorithm that finds proper matches for your search queries. The use of this technique in engineering is ubiquitous. In this text we use the SVD to understand multiple antenna communication, or MIMO.
The Principle Component Analysis	The principle component analysis is another incredibly important algorithm used in machine learning and pattern recognition. In this text, we use the principle component analysis to compute the signal to noise ratio of received wireless communication signals and to estimate the channel distortion these signals experience during transmission.

1.3.1   Pseudo Inverse for Overdetermined Systems of Linear Equations

In our last example, we found a polynomial curve that passed through four arbitrary points of our choosing. Solving the problem involved finding the inverse of the square matrix A. You may ask yourself whether there are situations in which the matrix A would not be square and just what the meaning of such a configuration would be. This situation occurs when the system of linear equations is overdetermined. Such a configuration is actually rather common in the field of DSP and numerical methods. Chapter 3 on optimization will go into great detail regarding the treatment and solution of over-determined systems of linear equations. A simple example will give us a wonderful insight into an application that requires the solution of a linear system of equations that is overdetermined.

Example 1.16: *Linear Regression*

Linear regression is a technique in which we attempt to fit a line to a set of observed test values. Let's assume that we are taking pictures of an object moving across the sky with a digital camera. We know that the path of the object is a straight line across the sky but try as we want, the vibration of the ground, perhaps due to nearby traffic, is causing small amounts of shaking when the camera takes its pictures. Therefore, the position that the camera determined with each exposure includes a random error, which we can't eliminate. We use linear regression to guess at the true parameters of the line given potentially noisy observations of the objects position in the sky.

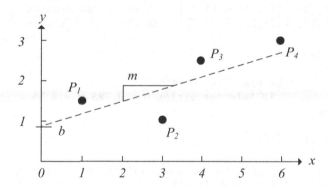

Figure 1-24: Observed Positions as XY Coordinates in the Picture Frame

Our goal is to determine the y-intercept, b, and the slope, m, to be able to completely parameterize the equation of straight line.

$$y = b + m \cdot x$$

Given the noisy corrupted coordinates $P_1 = [1, 1.5]$, $P_2 = [3, 1]$, $P_3 = [4, 2.5]$, and $P_4 = [6, 3]$, we may write the following equations and matrix expression.

$$b + m \cdot x_1 = y_1$$

$$b + m \cdot x_2 = y_2$$

$$b + m \cdot x_3 = y_3$$

$$b + m \cdot x_4 = y_4$$

$$\begin{bmatrix} 1 & x_1 \\ 1 & x_2 \\ 1 & x_3 \\ 1 & x_4 \end{bmatrix} \cdot \begin{bmatrix} b \\ m \end{bmatrix} = \begin{bmatrix} y_1 \\ y_2 \\ y_3 \\ y_4 \end{bmatrix} \rightarrow \begin{bmatrix} 1 & 1 \\ 1 & 3 \\ 1 & 4 \\ 1 & 6 \end{bmatrix} \cdot \begin{bmatrix} b \\ m \end{bmatrix} = \begin{bmatrix} 1.5 \\ 1 \\ 2.5 \\ 3 \end{bmatrix}$$

$$A \quad \cdot \quad C \ = \ Y \qquad\qquad A \quad \cdot \quad C \ = \ Y$$

The system is over–determined since it features more constraints, namely four equations, than degrees of freedom, namely two variables. There is no solution for X since A^{-1} is only defined for square matrices and clearly we can't draw a straight line through more than two coordinates.

However, we can find a solution vector C, which satisfies the system of equations with the least amount of error. In English, we can determine the y-intercept, b, and slope, m, of a line that passes by the four coordinates as closely as possible. The equation below provides the pseudo inverse of A, PIA, which we will derive in section 3.1.

$$PIA = inv(A^T \cdot A) \cdot A^T$$

$$C = PIA \cdot Y = \begin{bmatrix} b \\ m \end{bmatrix} = \begin{bmatrix} 0.788 \\ 0.35 \end{bmatrix}$$

```
%% Defining the Problem
P1 = [1,1.5]; P2 = [3,1]; P3 = [4,2.5]; P4 = [6,3];
X  = [P1(1,1); P2(1,1); P3(1,1); P4(1,1)];   %% the x coordinates
Y  = [P1(1,2); P2(1,2); P3(1,2); P4(1,2)];   %% the y coordinates


%% Solving the Problem
A  = [ones(4,1) X];
C  = inv(A'*A)*A'*Y;        %% PIA = inv(A'*A)*A'
                            %% Solution yields b = 0.788 m = 0.35
%% Plotting the Line
x  = (0:.01:6)';
y  = [ones(length(x),1) x]*C;


figure(2); plot(x, y, 'k'); grid on; hold on; title('Best Fit Line');
plot(X, Y, 'ko'); xlabel('X'); ylabel('Y');
```

1.3.2   The Gradient

The gradient of a scalar function $\Re$ with respect to the vector W is denoted by $\nabla_W(R)$, where ∇ is the differential operator. The idea is straight forward. We simply take the partial derivative of $\Re$ with respect to each member variable of the vector W. So let's examine how the gradient is officially defined for some arbitrary function $\Re$.

$$\nabla_W(R) = \frac{dR}{dW} = \begin{bmatrix} \dfrac{dR}{dw_0} \\ \dfrac{dR}{dw_1} \\ \vdots \\ \dfrac{dR}{dw_{N-1}} \end{bmatrix}$$

The scalar function $\Re$ is quite often the product of several matrices, one of them being the vector W. The two examples that follow are borrowed from the derivation of the least squares optimization algorithm. Least squares optimization is the pivotal concept at the center of adaptive and optimal signal processing, which we discuss soon.

Example 1.17: Gradient Example 1

In this example $\Re$ is the product of a transposed column vector P and the column vector w.

$$\Re = P^T \cdot W = \begin{bmatrix} 1 & 3 & 2 \end{bmatrix} \cdot \begin{bmatrix} w_0 \\ w_1 \\ w_2 \end{bmatrix} = 1w_0 + 3w_1 + 2w_2$$

The scalar $\Re$ is the sum of quantities that depend on the variables in the vector W. Therefore, according to the definition, we formulate the gradient as follows.

$$\nabla_W(\Re) = \frac{\partial(P^T \cdot W)}{\partial W} = \begin{bmatrix} \dfrac{\partial(1w_0 + 3w_1 + 2w_2)}{\partial w_0} \\ \dfrac{\partial(1w_0 + 3w_1 + 2w_2)}{\partial w_1} \\ \dfrac{\partial(1w_0 + 3w_1 + 2w_2)}{\partial w_2} \end{bmatrix} = \begin{bmatrix} 1 \\ 3 \\ 2 \end{bmatrix} = P$$

In this case the partial differentiation operations above are quite simple, so let's turn our attention to the more sophisticated example.

Example 1.18: Gradient Example 2

The matrix R, in the example below, is an autocorrelation matrix which, as mentioned earlier, appears in the least squares optimization derivation of chapter 3.

$$\Re = W^T \cdot R \cdot W = \begin{bmatrix} w_0 & w_1 & w_2 \end{bmatrix} \begin{bmatrix} r_{00} & r_{01} & r_{02} \\ r_{10} & r_{11} & r_{12} \\ r_{20} & r_{21} & r_{22} \end{bmatrix} \begin{bmatrix} w_0 \\ w_1 \\ w_2 \end{bmatrix}$$

$$\Re = W^T \cdot R \cdot W = \begin{bmatrix} w_0 & w_1 & w_2 \end{bmatrix} \begin{bmatrix} w_0 r_{00} + w_1 r_{01} + w_2 r_{02} \\ w_0 r_{10} + w_1 r_{11} + w_2 r_{12} \\ w_0 r_{20} + w_1 r_{21} + w_2 r_{22} \end{bmatrix}$$

$$= w_0{}^2 r_0 r_0 + w_0 w_1 r_{01} + w_0 w_2 r_{02} + w_1 w_0 r_{10} + w_1{}^2 r_{11} + w_1 w_2 r_{12} + w_2 w_0 r_{20} + w_2 w_1 r_{21} + w_2{}^2 r_{22}$$

As we apply the gradient, the expressions start looking a little more orderly.

$$\nabla_w (\Re) = \begin{bmatrix} \dfrac{\partial(W^T \cdot R \cdot W)}{\partial w_0} \\ \dfrac{\partial(W^T \cdot R \cdot W)}{\partial w_1} \\ \dfrac{\partial(W^T \cdot R \cdot W)}{\partial w_2} \end{bmatrix} = \begin{bmatrix} 2w_0 r_{00} + w_1 r_{01} + w_2 r_{02} + w_1 r_{10} + w_2 r_{20} \\ w_0 r_{01} + w_0 r_{10} + 2w_1 r_{11} + w_2 r_{12} + w_2 r_{21} \\ w_0 r_{02} + w_0 r_{20} + w_1 r_{12} + w_1 r_{21} + 2w_2 r_{22} \end{bmatrix}$$

To simplify things somewhat, let's assume that the vector x, from which the correlation matrix R was generated, is real. Therefore, $r_{ab} = r_{ba}$ and we can simplify to the following expression.

$$R = \begin{bmatrix} r_{00} & r_{01} & r_{02} \\ r_{10} & r_{11} & r_{12} \\ r_{20} & r_{21} & r_{22} \end{bmatrix} = \begin{bmatrix} r_{00} & r_{01} & r_{02} \\ r_{01} & r_{11} & r_{12} \\ r_{02} & r_{12} & r_{22} \end{bmatrix}$$

$$\nabla_w (\Re) = \begin{bmatrix} 2w_0 r_{00} + 2w_1 r_{01} + 2w_2 r_{02} \\ 2w_0 r_{01} + 2w_1 r_{11} + 2w_2 r_{12} \\ 2w_0 r_{02} + 2w_1 r_{12} + 2w_2 r_{22} \end{bmatrix} = 2 \cdot \begin{bmatrix} r_{00} & r_{01} & r_{02} \\ r_{01} & r_{11} & r_{12} \\ r_{02} & r_{12} & r_{22} \end{bmatrix} \begin{bmatrix} w_0 \\ w_1 \\ w_2 \end{bmatrix} = 2 \cdot R \cdot W$$

1.3.3   Eigenvectors, Eigenvalues and Matrix Decompositions

For some of the linear transformations that we have covered in the previous section, it may be possible to just look at their matrices and make an educated guess as to what they do. However, in most practical applications, the transformation matrix will not reveal its secrets so easily, and we need a more rigorous analysis approach that yields parameters indicating the transformation's underlying mechanisms. One such analysis is the calculation of the eigenvalues and eigenvectors of a transformation matrix. These eigen-parameters are important in a variety of engineering disciplines including mechanical and chemical engineering, but they will also play a part in our later discussion on MIMO (multiple-input multiple-output) communication systems. After having found the eigenvalues and vectors, it is a simple matter to complete a mathematic process called the eigenvalue decomposition. The eigenvalue decomposition manages to break the original transformation matrix apart to produce three separate matrices, whose product yields the original transformation matrix. Beside the fact that we can immediately read off the eigenvalues and eigenvectors from the three matrices, the decomposition itself enables us to undertake certain mathematical operations that would have been difficult to do on the original transformation matrix. There are in fact several additional decompositions including the polar and singular value decompositions, and each allows us unique insight into the underlying mechanisms of the associated transformation.

What are Eigenvalues and Eigenvectors?

In mathematics, an *eigenvector* of a transformation is a non-zero vector that remains unchanged in direction by that transformation. The factor, by which the transformation scales the magnitude of the eigenvector, is called the *eigenvalue*. Eigenvectors and eigenvalues are *only* defined for transformation matrices that are square.

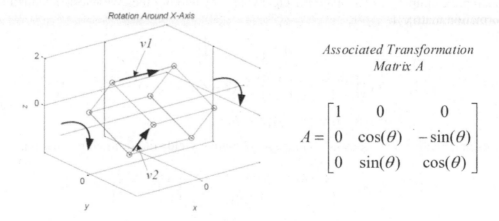

$$A = \begin{bmatrix} 1 & 0 & 0 \\ 0 & \cos(\theta) & -\sin(\theta) \\ 0 & \sin(\theta) & \cos(\theta) \end{bmatrix}$$

Figure 1-25: Cube Rotating around the X-Axis

The previous figure shows an object being rotated around the *x*-axis. While the start and stop positions of vector v_1 change, its direction remains constant since it lies parallel to the axis of rotation. Additionally, a simple rotation such as this will not change the magnitude of v_1, resulting in an associated eigenvalue of 1.0. The direction of vector v_2, however, changes with the angle of rotation and thus fails our eigenvector condition. Of the two, only v_1 is an eigenvector.

As the reader can readily imagine, there are actually an infinite number of possible vectors that we could draw both in the plus and minus x-axis sense whose direction would not be affected by the rotation. These vectors would simply differ in length. The collection of all these vectors is called the eigenspace, and their transformation scales each one of them by the eigenvalue, $\lambda_1=1$. The idea that an eigenvector, v, is a vector remaining unchanged in direction after a transformation via the matrix A and scales in magnitude by the eigenvalue, λ, is expressed as follows.

$$Av = \lambda v$$

To find a solution for v, we will rewrite the equation where the identity matrix I and A have the same dimension.

$$Av = \lambda v$$
$$(A - \lambda I)v = 0$$

In section 1.2.2.3, under the subheading called '*The Matrix Inversion*', we look back at rules 2 and 3 to realize that if $(A - \lambda I)$ is invertible, then v must be the zero vector and as mentioned earlier, [0 0 .. 0] is not a valid eigenvector. We must therefore choose the eigenvalues (λ) such that $(A - \lambda I)$ cannot be inverted which means that its determinant must be zero.

$$det(A - \lambda I) = 0$$

The eigenvalues of A are simply the roots of that characteristic polynomial, $p(\lambda) = det(A - \lambda I)$. Now as we already know, if the determinant is zero, there will be either no or an infinite number of solutions. This supports our understanding that there may be an infinite number of eigenvectors in a particular eigenspace associated with one eigenvalue.

Example 1.19: Eigenvalue Example 1

In this simple example, we calculate the eigenvalues by solving the characteristic equation of the transformation matrix A.

$$A = \begin{bmatrix} 2 & -4 \\ -1 & -1 \end{bmatrix}$$

$$p(\lambda) = det\left(\begin{bmatrix} 2-\lambda & -4 \\ -1 & -1-\lambda \end{bmatrix}\right) = (2-\lambda)(-1-\lambda) - (-4)(-1)$$

$$= \lambda^2 - \lambda - 6 = (\lambda - 3)(\lambda + 2)$$

The roots (eigenvalues) are thus $\lambda_1 = 3$ and $\lambda_2 = -2$, which yields the following situation.

$$(A - \lambda_1 I) \cdot v_1 = \begin{bmatrix} 2-\lambda_1 & -4 \\ -1 & -1-\lambda_1 \end{bmatrix} \cdot v_1 = \begin{bmatrix} -1 & -4 \\ -1 & -4 \end{bmatrix} \cdot v_1 = 0$$

$$(A - \lambda_2 I) \cdot v_2 = \begin{bmatrix} 2-\lambda_2 & -4 \\ -1 & -1-\lambda_2 \end{bmatrix} \cdot v_2 = \begin{bmatrix} 4 & -4 \\ -1 & 1 \end{bmatrix} \cdot v_2 = 0$$

From here it is readily apparent that v_1 is any scalar multiple of $[-4 \ 1]^T$ whereas v_2 is any scalar multiple of $[1 \ 1]^T$. Again, we could normalize the eigenvector to unit length, in which case they become $[-4/sqrt(17) \ 1/sqrt(17)]^T$ and $[0.707 \ 0.707]^T$ respectively. Note that the eigenvalues and eigenvectors may be complex.

$$v_1 = \begin{bmatrix} \dfrac{-4}{\sqrt{17}} \\ \dfrac{1}{\sqrt{17}} \end{bmatrix} \quad v_2 = \begin{bmatrix} \dfrac{1}{\sqrt{2}} \\ \dfrac{1}{\sqrt{2}} \end{bmatrix}$$

Example 1.20: Eigenvalue Example 2

Let us next look at an example with complex eigenvalues. Let

$$A = \begin{bmatrix} -4 & -17 \\ 2 & 2 \end{bmatrix}$$

and proceeding as before we find the following characteristic equation.

$$p(\lambda) = \det\left(\begin{bmatrix} -4-\lambda & -17 \\ 2 & 2-\lambda \end{bmatrix}\right) = (-4-\lambda)(2-\lambda) + 34$$

$$= \lambda^2 + 2\lambda + 26 = (\lambda + (1 + 5j))(\lambda + (1 - 5j))$$

The final eigenvalues are -1 + 5*i* and -1 - 5*i*.

For two by two matrices, the eigenvalues and eigenvectors of both real and complex matrices can be readily found using the quadratic formula for the eigenvalues and the above common-sense approach to find the eigenvectors. For larger matrices, dedicated numerical methods libraries should be used. See the documentation of the MatLab function '*eig*' for more information.

Example 1.21: Eigenvalues and Eigenvectors of the Reflection Transformation

In the figure below, we illustrate several points being reflected around a line crossing the origin.

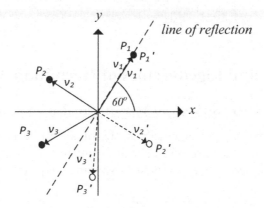

Figure 1-26: Reflection around a Straight Line and its Transformation Matrix, A

The illustration shows three coordinates, P_n, and associated vectors, v_n, which when reflected about the straight line form a new set of points and vectors named P_n' and v_n'. For v_n to be an eigenvector it may differ from its transformed counterpart, v_n', only by a scalar. Of the three points, v_1 remains entirely unchanged and v_2 when multiplied by the scalar -1 yields v_2'. Therefore, v_1 and v_2 are the eigenvectors of the transformation and the scalars $\lambda_1 = 1$ and $\lambda_2 = -1$ represent their corresponding eigenvalues.

Note that v_1 and v_2 are orthogonal. Furthermore, v_1 and v_2 may be of any length and still be eigenvectors. We therefore chose the vectors to be unit length, which results in a pair of orthonormal eigenvectors shown below.

$$v_1 = \begin{bmatrix} cos(60^o) \\ cos(60^o) \end{bmatrix} \quad v_2 = \begin{bmatrix} cos(150^o) \\ cos(150^o) \end{bmatrix}$$

Hermitian Matrices

In preparation for our discussion on matrix decompositions, we will cover two important properties regarding symmetric and hermitian matrices. We already know that a matrix is symmetric if it remains unchanged after the transpose operation (i.e. $A = A^T$). A matrix is hermitian if some of its entries are complex and the matrix remains unchanged after the hermitian transpose (i.e. $A = A^H$). The following matrix is hermitian.

$$A = \begin{bmatrix} 2 & 1-j \\ 1+j & -1 \end{bmatrix}$$

$$A^H = conj\left(\begin{bmatrix} 2 & 1+j \\ 1-j & -1 \end{bmatrix}\right) = \begin{bmatrix} conj(2) & conj(1+j) \\ conj(1-j) & conj(-1) \end{bmatrix} = \begin{bmatrix} 2 & 1-j \\ 1+j & -1 \end{bmatrix} = A$$

In this text, we have always stipulated that the transpose operation indicated by the superscript T notation includes the conjugation of all entries. Therefore, the superscript T and H notations indicate identical operations. Some texts make a distinction between the two operators, but we do not. Note that for a matrix to be hermitian, its diagonal elements must be real, and those entries reflected about the main diagonal, may be complex conjugate pairs. The two properties proven below, illustrate interesting features, which significantly simplify our work when we cover matrix decomposition in the next section.

1.3.3.1 Eigenvalues and Eigenvectors of Hermitian Matrices

The Eigenvalues of a Hermitian (Symmetric) Matrices are Real (May be Positive or Negative)

In this property, we claim that the eigenvalues, λ_n, of a hermitian (symmetric) matrix A will always be real. If the eigenvalues are real then their imaginary parts are equal to zero and $\lambda_n = conj(\lambda_n)$. Therefore, the following two expressions should be equivalent.

$$A \cdot v_n = \lambda_n \cdot v_n \tag{1}$$

$$A \cdot v_n = \lambda_n^* \cdot v_n \tag{2}$$

To prove this property, we will manipulate equation 2 until it becomes equation 1. Note that in this derivation, we will use the superscript H notation to indicate the Hermitian transpose, and the superscript * to indicate the complex conjugate of a scalar. We start by multiplying the transpose of the eigenvector v_n on both sides of equation 2.

$$v_n^T \cdot A \cdot v_n = v_n^T \cdot \lambda_n^* \cdot v_n$$

Because the matrix A is hermetian, $A=A^H$, we proceed as follows.

$$v_n{}^T{\cdot}A^T{\cdot}v_n = v_n{}^T{\cdot}\lambda_n{}^*{\cdot}v_n$$

Note that both $v_n{}^H{\cdot}A^H$ and $v_n{}^H{\cdot}\lambda_n{}^*$ are row vectors and must be equal to one another.

$$v_n{}^T{\cdot}A^T = v_n{}^T{\cdot}\lambda_n{}^*$$

If we now take the hermetian transpose of both sides, equation 1 starts to take shape.

$$\left(v_n{}^T{\cdot}A^T\right)^T = \left(v_n{}^T{\cdot}\lambda_n{}^*\right)^T$$

$$A{\cdot}v_n = \lambda_n{\cdot}v_n$$

If the steps in the proof above are a bit challenging, then review the matrix properties in section 1.2.3.

The Eigenvectors of a Hermitian (Symmetric) Matrices are Orthogonal

In this property, we claim that the eigenvectors of a symmetric matrix are mutually orthogonal. To start our discussion, we recall that for a symmetric matrix A, $A=A^T$, and the transpose of a product of two matrices, A and B, follows the rule $(A{\cdot}B)^T = B^T{\cdot}A^T$. Let's now pick the case of a 2x2 matrix A, featuring two unique eigenvalues, λ_1 and λ_2, as well as two eigenvectors, v_1 and v_2. Remembering that by default the v_1 and v_2 are column vectors. We now write the following well known expressions.

$$A{\cdot}v_1 = \lambda_1{\cdot}v_1 \tag{1}$$

$$A{\cdot}v_2 = \lambda_2{\cdot}v_2 \tag{2}$$

Recall that the vectors v_1 and v_2 are orthogonal if their inner product, $v_1{}^T{\cdot}v_2$ or $v_2{}^T{\cdot}v_1$, is a scalar equal to zero. In order to make an inner product appear in one of the expressions above, we multiply $v_1{}^T$ on both sides of equation 2.

$$v_1{}^T{\cdot}A{\cdot}v_2 = v_1{}^T{\cdot}\lambda_2{\cdot}v_2$$

And given that λ_2 is a scalar, we may move the eigenvalue λ_2 to the left of $v_1{}^T$.

$$v_1{}^T{\cdot}A{\cdot}v_2 = \lambda_2{\cdot}v_1{}^T{\cdot}v_2$$

$$\left(v_1{}^T{\cdot}A\right){\cdot}v_2 = \lambda_2{\cdot}v_1{}^T{\cdot}v_2$$

Because A is symmetric, $A = A^T$, then from our rule of the transpose of the product of two matrices we proceed to write the following.

$$\left(v_1{}^T{\cdot}A^T\right){\cdot}v_2 = \lambda_2{\cdot}v_1{}^T{\cdot}v_2$$

$$\left(A{\cdot}v_1\right)^T{\cdot}v_2 = \lambda_2{\cdot}v_1{}^T{\cdot}v_2$$

From equation 1, we know that $A{\cdot}v_1 = \lambda_1{\cdot}v_1$, which we readily substitute into the above expressions to yield.

$$\left(\lambda_1{\cdot}v_1\right)^T{\cdot}v_2 = \lambda_2{\cdot}v_1{}^T{\cdot}v_2$$

$$\lambda_1 \cdot v_1^T \cdot v_2 = \lambda_2 \cdot v_1^T \cdot v_2$$

Unless $\lambda_1 = \lambda_2$, the only way the above expression can be true is if the inner product of v_1 and v_2, $v_1^T \cdot v_2$, is equal to 0. Given that two vectors whose inner product is zero are orthogonal, we have proven our property. Remember that we may always scale our eigenvectors to have unit length in which case v_1 and v_2 are orthonormal.

Example 1.20: *Eigenvalues of the Hermitian (Symmetric) Matrices $A^T A$ and AA^T are Non-negative*

In this example we will illustrate that the matrices $A^T A$ and AA^T are square and symmetric for any matrix A and feature identical non-zero eigenvalues. We already know that the eigenvalues of a symmetric matrix are real, but here they are also non-negative as the example below illustrates.

$$A = \begin{bmatrix} 1 & -j & 1 \\ j & 2j & 1-j \end{bmatrix}$$

$$AA^T = \begin{bmatrix} 1 & -j & 1 \\ j & 2j & 1-j \end{bmatrix} \cdot \begin{bmatrix} 1 & -j \\ j & -2j \\ 1 & 1+j \end{bmatrix} = \begin{bmatrix} 3 & -1 \\ -1 & 7 \end{bmatrix}$$

$$A^T A = \begin{bmatrix} 1 & -j \\ j & -2j \\ 1 & 1+j \end{bmatrix} \cdot \begin{bmatrix} 1 & -j & 1 \\ j & 2j & 1-j \end{bmatrix} = \begin{bmatrix} 2 & 2-j & -j \\ 2+j & 5 & -2-j \\ j & -2+j & 3 \end{bmatrix}$$

They characteristic polynomial produced by $det(AA^T - \lambda I)$ is a straight forward affair and with the help of the quadratic formula yields the following results.

$$\det(AA^T - \lambda I) = det\begin{bmatrix} 3-\lambda & -1 \\ -1 & 7-\lambda \end{bmatrix} = (3-\lambda) \cdot (7-\lambda) - (-1 \cdot -1)$$

$$= \lambda^2 - 10\lambda + 20$$

$$= (\lambda - 7.2361) \cdot (\lambda - 2.7369)$$

While not quite as straight forward, the non-zero eigenvalues of 7.2361 and 2.7369 also appear in the following characteristic equation.

$$\det(A^T A - \lambda I) = det(\begin{bmatrix} 2-\lambda & 2-j & -j \\ 2+j & 5-\lambda & -2-j \\ j & -2+j & 3-\lambda \end{bmatrix}) = \lambda(\lambda - 7.2361) \cdot (\lambda - 2.7369)$$

1.3.3.2 Eigenvectors in Pictures

In previous examples, we have examined how different matrices transform input vectors and often attempted to guess at the eigenvectors from the shape of the set of output vectors. In the following figure, we show a set of input vectors which lie on a unit circle and sets of output vectors that illustrate the input vectors' transformation via the matrices H1, H2 and H3. In each case, you may use the MatLab function [V, D] = eig(Matrix) to find both eigenvalues and eigenvectors of the three matrices.

$$H1 = \begin{bmatrix} 1 & -2 \\ -0.5 & -0.5 \end{bmatrix} \quad H2 = \begin{bmatrix} -0.2 & 1 \\ 1 & -1.8 \end{bmatrix} \quad H3 = H1 \cdot H1^T = \begin{bmatrix} 5 & 0.5 \\ 0.5 & 0.5 \end{bmatrix}$$

Eigenvalues: 1.5   -1.0   -2.28   0.28   2.24   0.66

Eigenvectors: $\begin{bmatrix} 0.97 \\ -0.24 \end{bmatrix}, \begin{bmatrix} 0.707 \\ 0.707 \end{bmatrix}$ $\begin{bmatrix} -0.433 \\ 0.901 \end{bmatrix}, \begin{bmatrix} 0.901 \\ -0.433 \end{bmatrix}$ $\begin{bmatrix} 0.994 \\ 0.109 \end{bmatrix}, \begin{bmatrix} -0.109 \\ 0.994 \end{bmatrix}$

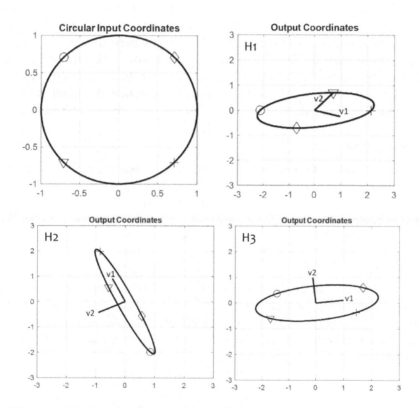

Figure 1-27: Transformation Result for Matrices H1, H2, and H3

Notice first, that matrices H2 and H3 are symmetric and therefore, their eigenvectors are orthogonal as expected from our discussion on the last few pages. Notice further that the eigenvalues of H2 may be both negative and positive, whereas those of H3 must be non-negative given that it was formed as H1·H1^T. Lastly, notice that the eigenvectors of the two symmetric matrices indicate the direction of **expansion** and **contraction** of the transformation. This effect, which does not hold for H1, will come in quite handy during out later discussion on **principle component analysis**. Only symmetric matrices have this convenient property.

1.3.4 Matrix Decomposition

Any matrix, whether real or complex valued square or rectangular, may be decomposed into a certain number of simpler matrices that provide us with a better understanding of the associated transformation.

The Eigenvalue Decomposition

The first decomposition that we will examine is the eigenvalue decomposition, at which we can arrive using a trivial extension to the discussion of the last section. The eigenvalue decomposition shows that a square matrix may be broken apart into the product of a matrix V, containing the eigenvectors of A in its columns, a scaling (diagonal) matrix Λ, containing the eigenvalues of A, and the inverse of V. The general expression for the eigenvalue decomposition is shown below.

$$A = V \cdot \Lambda \cdot inv(V)$$

The eigenvalue decomposition provides us with three benefits. As mentioned above, it breaks apart a sophisticated matrix into three simpler ones that provide a unique insight into the underlying transformation. Additionally, the only way to perform certain complicated matrix operations is to first compute the eigenvalue decomposition of A, perform the operation on Λ, and finally recompose the matrix. This sounds odd at first but becomes very clear as we proceed through this section. The third benefit is its invaluable assistance in computing the singular value decomposition, which by most accounts is one of the crown jewels of linear algebra.

The Polar Decomposition

The polar decomposition breaks apart a real or complex valued matrix A to produce a symmetric matrix S and orthonormal matrix O. By itself, the polar decomposition does not see excessive use, but its derivation lays the ground work for the more illustrious singular value decomposition.

$$A = O \cdot S$$

The Singular Value Decomposition

The singular value decomposition illustrates a rather unexpected and very useful result in linear algebra. It shows that any matrix A, whether real, complex, square, or rectangular, may be decomposed into an orthonormal matrix U, a diagonal matrix Σ with real valued non-negative entries, and an additional orthonormal matrix V. The singular value decomposition, or SVD, is expressed as follows.

$$A = U \cdot \Sigma \cdot V^T$$

Recall that an orthonormal matrix causes a rotation, with a possible reflection, whereas the diagonal matrix represents a simple scaling transformation as we have seen in section 1.2.2.1. In a nutshell, the SVD claims that any matrix A is at its core composed of a rotation/reflection, a real positive valued scaling transformation, followed by an additional rotation/reflection. The SVD finds use in many engineering applications, but in this text, we will use it to understand spatial multiplexing in MIMO antenna systems, which we discuss in the last chapter.

1.3.4.1 The Eigenvalue Decomposition

To demonstrate the eigenvalue decomposition, we continue our earlier example, which calculated the eigenvalues and eigenvectors of the transformation matrix, A, shown below. We find the eigenvalues by solving for λ in the polynomial $p(\lambda) = det(A - \lambda I)$.

$$A = \begin{bmatrix} 2 & -4 \\ -1 & -1 \end{bmatrix}$$

$$p(\lambda) = det\begin{bmatrix} 2-\lambda & -4 \\ -1 & -1-\lambda \end{bmatrix} = (2-\lambda)(-1-\lambda)-(-4)(-1) = \lambda^2 - \lambda - 6 = (\lambda-3)(\lambda+2)$$

In example 1.20, we found the eigenvectors to be $v_1 = [-4\ 1]^T$ for $\lambda_1 = 3$ and $v_2 = [1\ 1]^T$ for $\lambda_2 = -2$. Recall the original requirement of $Av_n = \lambda_n v_n$ for each eigenpair.

$$A \quad \cdot \quad v_n \quad = \lambda_n \cdot\ v_n$$

$$\begin{bmatrix} 2 & -4 \\ -1 & -1 \end{bmatrix} \cdot \begin{bmatrix} -4 \\ 1 \end{bmatrix} = 3 \cdot \begin{bmatrix} -4 \\ 1 \end{bmatrix}$$

$$\begin{bmatrix} 2 & -4 \\ -1 & -1 \end{bmatrix} \cdot \begin{bmatrix} 1 \\ 1 \end{bmatrix} = -2 \cdot \begin{bmatrix} 1 \\ 1 \end{bmatrix}$$

We can easily combine these two systems of equations into one using notation V to represent the matrix of eigenvectors and Λ for the eigenvalues.

$$A \qquad v_1\ |\ v_2 \qquad 3\cdot v_1 \quad |\quad -2\cdot v_2$$

$$\begin{bmatrix} 2 & -4 \\ -1 & -1 \end{bmatrix} \cdot \begin{bmatrix} -4 & 1 \\ 1 & 1 \end{bmatrix} = \begin{bmatrix} 3 \cdot \begin{bmatrix} -4 \\ 1 \end{bmatrix} & -2 \cdot \begin{bmatrix} 1 \\ 1 \end{bmatrix} \end{bmatrix}$$

We now reformulate the expression.

$$A \quad \cdot \quad V \quad = \quad V \quad \cdot \quad \Lambda$$

$$\begin{bmatrix} 2 & -4 \\ -1 & -1 \end{bmatrix} \cdot \begin{bmatrix} -4 & 1 \\ 1 & 1 \end{bmatrix} = \begin{bmatrix} -4 & 1 \\ 1 & 1 \end{bmatrix} \cdot \begin{bmatrix} \lambda_1 = 3 & 0 \\ 0 & \lambda_2 = -2 \end{bmatrix}$$

At this point, we isolate the transformation matrix A on the left-hand side of the equation by multiplying both sides by the inverse of V. By convention, we arrange Λ such that the largest eigenvalue is located at the top left and the smallest at the bottom right.

$$A \quad = \quad V \quad \cdot \quad \Lambda \quad \cdot \quad inv(V)$$

$$\begin{bmatrix} 2 & -4 \\ -1 & -1 \end{bmatrix} = \begin{bmatrix} -4 & 1 \\ 1 & 1 \end{bmatrix} \cdot \begin{bmatrix} 3 & 0 \\ 0 & -2 \end{bmatrix} \cdot \begin{bmatrix} -0.2 & 0.2 \\ 0.2 & 0.8 \end{bmatrix}$$

$$A = \begin{bmatrix} v_1 & v_2 \end{bmatrix} \cdot \begin{bmatrix} \lambda_1 & 0 \\ 0 & \lambda_2 \end{bmatrix} \cdot inv\left(\begin{bmatrix} v_1 & v_2 \end{bmatrix}\right)$$

The general formula for the eigenvalue decomposition is shown above. We should note, that given this decomposition, it is possible for us to synthesize a transformation matrix A that features eigenvectors and eigenvalues of our choice. Just place the eigenvectors in V, the eigenvalues in Λ, find the inverse of V and compute their product to produce the desired transformation matrix A. Remember that there are an infinite number of eigenvector in the eigenspace and they differ in length. We are thus at liberty to choose eigenvectors of unit length (as we did in example 1.20) which make V orthonormal. From section 1.2.3.2 we showed that for orthonormal matrices $inv(V)$ = V^T, and for that case, the eigen-decomposition reduces to the following.

$$A = V \cdot \Lambda \cdot V^T$$

As a matter of fact, the MatLab function $[V, D]$ = eig(A) will always yield a matrix V that is orthonormal. The matrix D in this case represents Λ.

The Power of the Eigenvalue Decomposition

There are mathematical challenges which require that we compute an equation that takes a matrix as an argument. Can you image computing $A^{\frac{1}{2}}$, or e^A? Nothing that we have learned up to now allows us to even guess at how to compute the quantities above. To get a handle on these challenges, we will use the eigenvalue decomposition to compute the quantity A^2.

$$A \cdot A = (V \cdot \Lambda \cdot V^{-1}) \cdot (V \cdot \Lambda \cdot V^{-1})$$

Note that the quantities V^{-1} and V appear next to one another and their product reduces to the identity matrix, I.

$$A^2 = V \cdot \Lambda \cdot I \cdot \Lambda \cdot V^{-1}$$

$$A^2 = V \cdot \Lambda \cdot \Lambda \cdot V^{-1}$$

$$A^2 = V \cdot \Lambda^2 \cdot V^{-1}$$

This rather interesting manipulation proves that we can square a matrix A by firsts computing its eigenvalue decomposition, squaring the diagonal entries in Λ and multiplying out $V \cdot \Lambda^2 \cdot V^{-1}$. Without a doubt, it would have been easier to just compute A^2 directly. But this example is pointing us into an interesting direction. Could we compute the square root of A, or equivalently $A^{\frac{1}{2}}$, in a similar manner?

Example 1.21: *Prove that $A^{\frac{1}{2}} = V \cdot \Lambda^{\frac{1}{2}} \cdot V^{-1}$*

To prove the validity of this statement we simply square both sides, which will produce the matrix A on the left hand side.

$$A^{\frac{1}{2}} \cdot A^{\frac{1}{2}} = V \cdot \Lambda^{\frac{1}{2}} \cdot V^{-1} \cdot V \cdot \Lambda^{\frac{1}{2}} \cdot V^{-1}$$

$$A^{(\frac{1}{2}+\frac{1}{2})} = V \cdot \Lambda^{\frac{1}{2}} \cdot I \cdot \Lambda^{\frac{1}{2}} \cdot V^{-1}$$

$$A^{(1)} = V \cdot \Lambda^{(\frac{1}{2}+\frac{1}{2})} \cdot V^{-1}$$

$$A = V \cdot \Lambda \cdot V^{-1}$$

Clearly, the last expression is the eigenvalue decomposition itself and we know it to be true. We can go a step further now and prove that the inverse of A is computed similarly.

Example 1.22: *Prove that $A^{-1} = V \cdot \Lambda^{-1} \cdot V^{-1}$*

Let's compute one last example, that of the inverse of A, before we move on to the more complicated expressions. To complete this proof, we realize that for the hypothesis above to be true, the following statement must hold.

$$A^{-1}A = (V \cdot \Lambda^{-1} \cdot V^{-1}) \cdot (V \cdot \Lambda \cdot V^{-1}) = I$$

We again rearrange the parenthesis to collapse $(V \cdot \Lambda^{-1} \cdot V^{-1}) \cdot (V \cdot \Lambda \cdot V^{-1})$ into a simpler form, which hopefully will be the identity matrix I.

$$V \cdot \Lambda^{-1} \cdot (V^{-1} \cdot V) \cdot \Lambda \cdot V^{-1} = I$$

$$V \cdot \Lambda^{-1} \cdot I \cdot \Lambda \cdot V^{-1} = I$$

$$V \cdot (\Lambda^{-1} \cdot \Lambda) \cdot V^{-1} = I$$

$$V \cdot (I) \cdot V^{-1} = I$$

$$V \cdot V^{-1} = I$$

By definition of the inverse of a matrix, $V \cdot V^{-1}$ is indeed equal to the identity matrix and the proof is complete. Later on, we will show how the use of the eigenvalue decomposition to compute the square root of A will help us with other decomposition problems. But for now, let's add here that other functions that take a matrix A as an argument are defined using the eigenvalue decomposition. For example, matrix A may appear as an exponent of a constant as shown here.

$$e^A = V \cdot e^\Lambda \cdot V^{-1}$$

In one of our earlier examples we had computed the eigenvalue decomposition of a matrix A and produced the following result.

$$
\begin{array}{ccccccc}
A & = & V & \cdot & \Lambda & \cdot & V^{-1}
\end{array}
$$

$$
\begin{bmatrix} 2 & -4 \\ -1 & -1 \end{bmatrix} = \begin{bmatrix} -4 & 1 \\ 1 & 1 \end{bmatrix} \cdot \begin{bmatrix} 3 & 0 \\ 0 & -2 \end{bmatrix} \begin{bmatrix} -0.2 & 0.2 \\ 0.2 & 0.8 \end{bmatrix}
$$

The exponential function of A can be shown as follows for the case above.

$$
\begin{array}{ccccccc}
e^A & = & V & \cdot & e^\Lambda & \cdot & V^{-1}
\end{array}
$$

$$
\exp\left(\begin{bmatrix} 2 & -4 \\ -1 & -1 \end{bmatrix} \right) = \begin{bmatrix} -4 & 1 \\ 1 & 1 \end{bmatrix} \cdot \begin{bmatrix} e^3 & 0 \\ 0 & e^{-2} \end{bmatrix} \cdot \begin{bmatrix} -0.2 & 0.2 \\ 0.2 & 0.8 \end{bmatrix}
$$

Example 1.23: *Using the Eigenvalue Decomposition to Illustrate Reflection*

In section 1.2.2.2, we concluded that we may arrive at the transformation matrix A, representing reflection about a straight line passing through the origin, by the cascading of a rotational matrix, a scaling matrix and another rotation matrix. To jog our memory, these three submatrices, $M1$ through $M3$, are shown below.

$$
A = \begin{bmatrix} \cos(2\theta) & \sin(2\theta) \\ \sin(2\theta) & -\cos(2\theta) \end{bmatrix} = \overset{M1}{\begin{bmatrix} \cos(\theta) & -\sin(\theta) \\ \sin(\theta) & \cos(\theta) \end{bmatrix}} \cdot \overset{M2}{\begin{bmatrix} 1 & 0 \\ 0 & -1 \end{bmatrix}} \cdot \overset{M3}{\begin{bmatrix} \cos(-\theta) & -\sin(-\theta) \\ \sin(-\theta) & \cos(-\theta) \end{bmatrix}}
$$

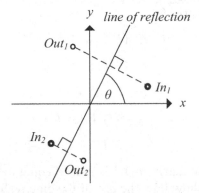

Figure 1-28: Reflection Transformation about a Straight Line Passing Through the Origin

What we had arrived at using common sense in section 1.2.2.2, we now reproduce using the eigenvalue decomposition. The eigenvalue decomposition of A yields three matrices as shown in the generalized formula below.

$$ A = V \cdot \Lambda \cdot V^{-1} $$

Thus, our goal is to show that $V = M1$, $\Lambda = M2$ and $inv(V) = M3$. As we had done before, we first find the eigenvalues by solving the characteristic polynomial produced by $p(\lambda) = det(A - \lambda I)$.

$$
\begin{aligned}
p(\lambda) &= \begin{bmatrix} \cos(2\theta) - \lambda & \sin(2\theta) \\ \sin(2\theta) & -\cos(2\theta) - \lambda \end{bmatrix} \\
&= (\cos(2\theta) - \lambda) \cdot (-\cos(2\theta) - \lambda) - \sin^2(2\theta) \\
&= -\cos^2(2\theta) - \lambda^2 - \sin^2(2\theta) \\
&= -1 - \lambda^2
\end{aligned}
$$

To find the eigenvalues, we set $p(\lambda) = -1 - \lambda^2 = 0$, which produces two solutions of value ± 1. In the case that the eigenvalues are real, which happens rather frequently, we will assign the largest most positive eigenvalue to variable λ_1 and the most negative eigenvalue to the last variable λ_N. In our case, with $N = 2$ eigenvalues, Λ can thus be written as follows and is indeed equal to $M2$ as we had hoped.

$$\Lambda = \begin{bmatrix} \lambda_1 & 0 \\ 0 & \lambda_2 \end{bmatrix} = \begin{bmatrix} 1 & 0 \\ 0 & -1 \end{bmatrix}$$

Finding the eigenvectors is equally straight forward. However, in order not to overgeneralize the problem, let's compute the eigenvectors for an angle of 60°. Remember, to be called an eigenvector, v must satisfy the following equation.

$$Av = \lambda v$$

$$(A - I\lambda)v = 0$$

Recall that the matrix $A - I\lambda$ is singular and both rows are linearly dependent. We thus only need to evaluate one of the two rows (row one in this case) to find a relationship between x and y.

$$\begin{bmatrix} \cos(2\theta) - \lambda_1 & \sin(2\theta) \\ \sin(2\theta) & -\cos(2\theta) - \lambda_1 \end{bmatrix} v_1 = \begin{bmatrix} \cos(2 \cdot 60^o) - 1 & \sin(2 \cdot 60^o) \\ \sin(2 \cdot 60^o) & -\cos(2 \cdot 60^o) - 1 \end{bmatrix} \begin{bmatrix} x_1 \\ y_1 \end{bmatrix} = 0$$

$$(\cos(2 \cdot 60^o) - 1) \cdot x_1 + \sin(2 \cdot 60^o) \cdot y_1 = 0$$

$$x_1 = -\frac{\sin(2 \cdot 60^o)}{(\cos(2 \cdot 60^o) - 1)} \cdot y_1$$

$$x_1 = 0.577 \cdot y_1$$

There are an infinite number of eigenvectors, all of them pointing in the same direction but different in length. We are at liberty to choose an eigenvector that we find convenient, and we pick one that features unit length. Let's proceed to find an expression governing eigenvector v_2.

$$\begin{bmatrix} \cos(2\theta) - \lambda_2 & \sin(2\theta) \\ \sin(2\theta) & -\cos(2\theta) - \lambda_2 \end{bmatrix} v_2 = \begin{bmatrix} \cos(2 \cdot 60^o) + 1 & \sin(2 \cdot 60^o) \\ \sin(2 \cdot 60^o) & -\cos(2 \cdot 60^o) + 1 \end{bmatrix} \begin{bmatrix} x_2 \\ y_2 \end{bmatrix} = 0$$

$$(\cos(2 \cdot 60^o) + 1) \cdot x_2 + \sin(2 \cdot 60^o) \cdot y_2 = 0$$

$$x_2 = -\frac{\sin(2 \cdot 60^o)}{(\cos(2 \cdot 60^o) + 1)} \cdot y_2$$

$$x_2 = -1.7 \cdot y_2$$

To find the eigenvectors, we simply choose $y_1 = 1$ and $y_2 = 1$, producing $x_1 = 0.577$ and $x_2 = -1.7$ before normalizing the eigenvectors to length to unity.

$$v_1 = \begin{bmatrix} x_1 \\ y_1 \end{bmatrix} = \begin{bmatrix} 0.577 \\ 1 \end{bmatrix} \cdot \frac{1}{sqrt(1 + (0.577)^2)} = \begin{bmatrix} 0.5 \\ 0.866 \end{bmatrix}$$

$$v_2 = \begin{bmatrix} x_2 \\ y_2 \end{bmatrix} = \begin{bmatrix} -1.7 \\ 1 \end{bmatrix} \cdot \frac{1}{sqrt(1 + (-1.7)^2)} = \begin{bmatrix} -0.866 \\ 0.5 \end{bmatrix}$$

We now insert the eigenvectors and eigenvalues into the general decomposition expression (repeated below for completeness), at which we arrived earlier. Note that the eigenvectors are orthonormal, a fact that is by no means guaranteed by the decomposition. However, because A happens to be symmetric, they are orthonormal making the inverse and transpose of the eigenvector matrix identical.

$$A = \begin{array}{cc} v_1 & v_2 \\ \begin{bmatrix} x_1 & x_2 \\ y_1 & y_2 \end{bmatrix} \end{array} \cdot \begin{array}{c} \Lambda \\ \begin{bmatrix} \lambda_1 & 0 \\ 0 & \lambda_2 \end{bmatrix} \end{array} \cdot inv \left(\begin{array}{cc} v_1 & v_2 \\ \begin{bmatrix} x_1 & x_2 \\ y_1 & y_2 \end{bmatrix} \end{array} \right)$$

Inserting the eigenvalues and eigenvectors into the generalized eigenvalue decomposition expression, produces the equation below. Note that because the original matrix A is symmetric, the eigenvectors are guaranteed to be orthogonal.

$$A = \begin{array}{ccc} V & \Lambda & inv(V) \\ \begin{bmatrix} 0.5 & -0.866 \\ 0.866 & 0.5 \end{bmatrix} & \begin{bmatrix} 1 & 0 \\ 0 & -1 \end{bmatrix} & \begin{bmatrix} 0.5 & 0.866 \\ -0.866 & 0.5 \end{bmatrix} \end{array} = \begin{bmatrix} -0.5 & 0.866 \\ 0.866 & 0.5 \end{bmatrix}$$

If we now go back to the original submatrices $M1$ and $M3$ and evaluate them for an angle $\theta = 60^o$, we will find that they are indeed equal to V and $inv(V)$.

$$M1 = \begin{bmatrix} \cos(60^o) & -\sin(60^o) \\ \sin(60^o) & \cos(60^o) \end{bmatrix} = \begin{bmatrix} 0.5 & -0.866 \\ 0.866 & 0.5 \end{bmatrix} = V$$

$$M3 = \begin{bmatrix} \cos(-60^o) & -\sin(-60^o) \\ \sin(-60^o) & \cos(-60^o) \end{bmatrix} = \begin{bmatrix} 0.5 & 0.866 \\ -0.866 & 0.5 \end{bmatrix} = inv(V)$$

The following simple MatLab code breaks the matrix A into the constituent matrices $M1$, $M2$ and $M3$, which correspond to V, Λ, and $inv(V)$.

```
%% The Eigenvalue Decomposition in MatLab
A = [-0.5 0.866; ...
    0.855   0.5];

[V, D] = eig(A);      % In our example M1 = V, M2 = D, and M3 = inv(V)
```

1.3.4.2 The Polar Decomposition

The eigenvalue decomposition broke apart a matrix A into three separate submatrices with unique properties. The second matrix Λ was diagonal, and the first and third, V and V^1, were inverses of one another. The matrix of eigenvectors, V, was not necessarily orthogonal or symmetric. We now introduce a different method of breaking apart a matrix A into two submatrices O and S, where O is orthonormal and S is symmetric.

$$A = O \cdot S$$

To find the matrix S, we multiply both sides by their transposes.

$$A^T A = (O \cdot S)^T \cdot O \cdot S = S^T \cdot O^T \cdot O \cdot S$$

Remember that the transpose of an orthonormal matrices is identical to its inverse causing the product $O^T \cdot O$ to collapse to the identity matrix I. Furthermore, by definition, the transpose of a symmetric matrix S is S itself. We may thus simply the expression above as follows.

$$A^T A = S \cdot I \cdot S = S^2$$

Therefore, to find S we take the square root of $A^T A$ using the eigenvalue decomposition as we demonstrated earlier.

$$S = \sqrt{A^T A}$$

Remember that $A^T A$ has non-negative eigenvalues allowing us to easily take the square root of the matrix Λ, which features those eigenvalues in its diagonal. By convention, we will choose only the positive results of the square root operation of each eigenvalue of $A^T A$. Using the negative results isn't wrong, it simply produces a different matrix O, which we find as follows. Multiply S^{-1} on both sides of the equation $A = O \cdot S$ to find O.

$$O \cdot S \cdot S^{-1} = A \cdot S^{-1}$$
$$O \cdot I = A \cdot S^{-1}$$
$$O = A \cdot S^{-1}$$

We can prove that the matrix O is in fact orthonormal by multiplying both sides of the equation above by O^T. If O is orthonormal, then $O^T \cdot O = O^{-1} \cdot O = I$, and $(A \cdot S^{-1})^T \cdot A \cdot S^{-1}$ must also equal I.

$$O^T \cdot O = (A \cdot S^{-1})^T \cdot A \cdot S^{-1}$$
$$O^T \cdot O = (S^{-1})^T \cdot A^T \cdot A \cdot S^{-1}$$

Note that the inverse of a symmetric matrix is itself symmetric and $(S^{-1})^T = (S^{-1})$. Furthermore, remember that $A^T A = S^2$ which we showed previously.

$$O^T \cdot O = (S^{-1}) \cdot A^T \cdot A \cdot S^{-1}$$
$$O^T \cdot O = S^{-1} \cdot S^2 \cdot S^{-1}$$
$$O^T \cdot O = S^{-1} \cdot S \cdot S \cdot S^{-1}$$
$$O^T \cdot O = I$$

1.3.4.3 The Singular Value Decomposition (SVD)

The second technique, called singular value decomposition, break the transformation into the product of two orthonormal matrices and one diagonal matrix, whose entries are both real and non-negative. Before we start, let's recall a few useful concepts from section 1.3.1.

→ The eigenvectors of a Hermitian/Symmetric matrix may be scaled to be orthonormal.
→ $A^T A$ is a symmetric matrix and its eigenvalues are real and non-negative.
→ The product of two orthonormal matrices is itself orthonormal.
→ The transpose of an orthonormal matrix is equal to the inverse of that matrix.

The singular value decomposition is written as follows. It claims that any matrix A, whether square, diagonal, real or complex valued, may be broken apart into the product of an orthonormal matrix U, a diagonal matrix Σ featuring non-negative entries and an orthonormal matrix V^T. As V^T is orthogonal, it is in fact the same as V^{-1}.

$$A = U \cdot \Sigma \cdot V^T$$

To get us started with an explanation of the singular value decomposition, we recall the polar decomposition of the last section which states the following.

$$A = O \cdot S$$

Remember that O was an orthonormal matrix causing either rotation (with a possible reflection), and S was a symmetric matrix equal to the square root of $A^T A$. From the eigenvalue decomposition we know that $A^T A$ and the square root of $A^T A$ break apart as follows.

$$A = V \Lambda V^{-1} = V \Lambda V^T$$
$$S = \sqrt{A^T A} = V \sqrt{\Lambda} V^{-1} = V \sqrt{\Lambda} V^T$$

Because $A^T A$ is symmetric, its eigenvectors in V are orthonormal and $V^{-1} = V^T$. Further, because Λ is a diagonal matrix with real, non-negative numbers, it's easy to find its square root. Just take the square root of each one of the diagonal entries. By convention, we only choose the positive root of the entries in Λ. Note further, that the square root of Λ is usually referred to as Σ. We thus rewrite the expression above as follows.

$$S = \sqrt{A^T A} = V \Sigma V^T$$

We now substitute the equivalent expression for S into the polar decomposition. Remember that the matrix $O = A \cdot S^{-1}$.

$$A = O \cdot S$$
$$A = O \cdot V \cdot \Sigma \cdot V^T$$

We now rename the matrix $O \cdot V$ as U, which yield the singular value decomposition.

$$\boxed{A = U \cdot \Sigma \cdot V^T}$$

Let's summarize what we have found out regarding the singular value decomposition. We know that because O and V are both orthonormal, the product U is orthonormal as well. Remember that orthonormal matrices are rotations with a possible reflection about one or both of the axis thrown in. The matrix Σ is diagonal and contains the singular values of the matrix A. Because Σ is the square root of V, the singular values are in fact the square roots of the eigenvalues of $A^{T}A$. Finally, because V^{T} contains the eigenvectors of the symmetric matrix $A^{T}A$, it too is orthonormal.

The singular value decomposition thus breaks any matrix apart into a rotation (and possible reflection), a real valued, positive scaling operation followed by another rotation (and possible reflection).

In the expressions below, we illustrate both the diagonal matrices Λ *and* Σ. Given that the eigenvalues, λ_1 through λ_N, of $A^{T}A$ are non-negative, the singular values, σ_1 through σ_N, which are the square roots of the eigenvalues can also be chosen to be non-negative. Remember, there are two square roots of a positive number. One is positive the other negative and we simply chose the positive root. Choosing negative roots isn't wrong per say, it simply results in a different matrix V. By convention, we chose positive singular values and arrange them such that σ_1 is the largest and σ_N is the smallest value.

$$\Lambda = \begin{bmatrix} \lambda_1 & 0 & 0 & 0 \\ 0 & \lambda_2 & 0 & 0 \\ \vdots & \vdots & \ddots & \vdots \\ 0 & 0 & 0 & \lambda_N \end{bmatrix} \qquad \Sigma = \begin{bmatrix} \sigma_1 & 0 & 0 & 0 \\ 0 & \sigma_2 & 0 & 0 \\ \vdots & \vdots & \ddots & \vdots \\ 0 & 0 & 0 & \sigma_N \end{bmatrix}$$

To summarize, computing the three matrices of the eigenvalue decomposition thus involves the following steps.

1. Compute the eigenvalue decomposition of $A^{T}A$ and scale the eigenvectors in V to unit length. V is thus orthonormal. Computing V and Λ is the hard part.

2. Take the square roots of the eigenvalues in Λ to produce the diagonal matrix Σ featuring the singular values.

3. $S = V \cdot \Sigma \cdot V^{T}$ has to be inverted now which is straight forward: $S^{-1} = V \cdot \Sigma^{-1} \cdot V^{T}$, where taking the inverse of Σ simply reduces to computing the reciprocal of each singular value ($1/\sigma_1$ through $1/\sigma_N$).

4. Compute O, which is the matrix product $A \cdot S^{-1}$ or alternatively $A \cdot V \cdot \Sigma^{-1} \cdot V^{T}$.

5. Compute U as $O \cdot V$, and we are done.

Example 1.24: Using the Singular Value Decomposition

The singular value decomposition and its purpose in mathematics is not easily understood. In this example we show you the geometric meaning of the decomposition. We will pick a set of input vectors that lie on the unit circle and feature incrementing angles from 0 to 2π. In particular, we pick four symbols positioned at angles $\pi/4$, $3\pi/4$, $5\pi/4$, and $7\pi/4$, to illustrate the transformation. Let's get to know our transformation matrix A which features the following entries.

$$A = \begin{bmatrix} 0.5 & 0.4 \\ 0.2 & 1.4 \end{bmatrix}$$

Take a look at the following figure which shows the set of input values as well as the output values.

$$A \quad \cdot \quad In \quad = \quad Out$$

$$\begin{bmatrix} 0.5 & 0.4 \\ 0.2 & 1.4 \end{bmatrix} \cdot \begin{bmatrix} x_{in} \\ y_{in} \end{bmatrix} = \begin{bmatrix} x_{out} \\ y_{out} \end{bmatrix}$$

Notice that the set of input values, *In*, has been transformed into an ellipse. As a matter of fact, any transformation matrix *A* will squeeze the input values on the unit circle into an ellipse.

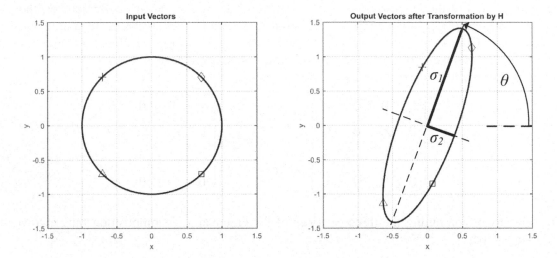

Figure 1-29: Input and Output Values of the Transformation via Matrix *A*

Note, there are three apparent transformations in the graph of the output vector.

$$H \qquad \cdot \; In = Out$$
$$U \cdot \Sigma \cdot V^T \quad \cdot \; In = Out$$

→ Notice that the markers initially positioned at angles $\pi/4$, $3\pi/4$, $5\pi/4$, and $7\pi/4$ are not symmetrically placed on the ellipse as was the case for the input vectors. Matrix V^T is responsible for this rotation.

→ The circle has been stretched into an ellipse. The ellipse, which still lies on perpendicular basis vectors, is stretch according to the singular values found in matrix Σ. The singular values, σ_n, are in fact the lengths of the output vectors along these axes.

→ Finally, the ellipse appears to have been rotated counterclockwise by θ. Matrix U is responsible for this rotation.

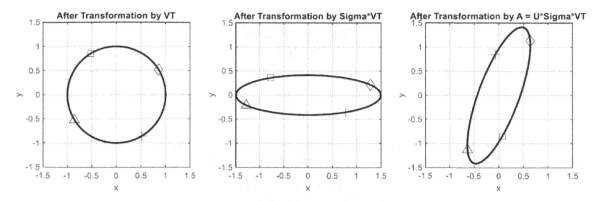

Figure 1-30: Transformations via V^T, $\Sigma \cdot V^T$, and $U \cdot \Sigma \cdot V^T$

The figure above shows the impact on the input vectors after the multiplication by V^T, by $\Sigma \cdot V^T$ and by $U \cdot \Sigma \cdot V^T$. Let's actually calculate the singular value decomposition of A according to the steps enumerated on the last page.

→ Compute the eigenvalue decomposition of $A^T A$ (steps 1 and 2 from two pages ago).

$$A^T A = \begin{bmatrix} 0.29 & 0.48 \\ 0.48 & 2.12 \end{bmatrix}$$

We find the eigenvalues by taking the determinant of $H^T \cdot H$ and setting it equal to zero.

$$det(A - \lambda I) = 0$$

$$det \begin{bmatrix} 0.29 - \lambda & 0.48 \\ 0.48 & 2.12 - \lambda \end{bmatrix} = (0.29 - \lambda) \cdot (2.12 - \lambda) - 0.48 \cdot 0.48$$

$$= \lambda^2 - 2.41\lambda + 0.3844$$

$$= (\lambda - 2.2383)(\lambda - 0.1717)$$

The eigenvalues are $\lambda_1 = 2.2383$ and $\lambda_2 = 0.1717$, whereas the singular values are their positive square roots which equal $\sigma_1 = 1.4961$ and $\sigma_2 = 0.4144$.

$$\Lambda = \begin{bmatrix} \lambda_1 & 0 \\ 0 & \lambda_2 \end{bmatrix} = \begin{bmatrix} 2.2383 & 0 \\ 0 & 0.1717 \end{bmatrix}$$

$$\Sigma = \begin{bmatrix} \sigma_1 & 0 \\ 0 & \sigma_2 \end{bmatrix} = \begin{bmatrix} 1.4961 & 0 \\ 0 & 0.4144 \end{bmatrix}$$

We find the eigenvectors v_1 and v_2 as follows.

$$\left(H^T H - \lambda_1 I\right) \cdot v_1 = \begin{bmatrix} 0.29 - \lambda_1 & 0.48 \\ 0.48 & 2.12 - \lambda_1 \end{bmatrix} \cdot v_1 = \begin{bmatrix} -1.948 & 0.48 \\ 0.48 & -0.118 \end{bmatrix} \cdot \begin{bmatrix} x_1 \\ y_1 \end{bmatrix} = 0$$

$$\left(H^T H - \lambda_1 I\right) \cdot v_1 = \begin{bmatrix} 0.29 - \lambda_2 & 0.48 \\ 0.48 & 2.12 - \lambda_2 \end{bmatrix} \cdot v_2 = \begin{bmatrix} 0.119 & 0.48 \\ 0.48 & 1.949 \end{bmatrix} \cdot \begin{bmatrix} x_2 \\ y_2 \end{bmatrix} = 0$$

As the two matrices are singular, we need only get a relationship between entries of v_1 and v_2.

$$1.9483 \cdot x_1 = 0.48 \cdot y_1$$
$$x_1 = 0.2464 \cdot y_1$$
and
$$0.119 \cdot x_2 = -0.48 \cdot y_2$$
$$x_2 = -4.0336 \cdot y_2$$

We now pick values for x_1 and y_1 that make the vector have unit length. We do the same for x_2 and y_2. Note that there is some flexibilities with the signs of the coefficients.

$$v_1 = \begin{bmatrix} x_1 \\ y_1 \end{bmatrix} = \begin{bmatrix} 0.2392 \\ 0.971 \end{bmatrix}$$

$$v_2 = \begin{bmatrix} x_2 \\ y_2 \end{bmatrix} = \begin{bmatrix} 0.971 \\ -0.2392 \end{bmatrix}$$

$$V = \begin{bmatrix} 0.2392 & 0.971 \\ 0.971 & -0.2392 \end{bmatrix} = V^T$$

We have now found two of the matrices, Σ and v^T, needed for the singular value decomposition.

→ Compute the quantity $S^{-1} = V \cdot \Sigma^{-1} \cdot V^T$ (step 3)

$$S^{-1} = \begin{bmatrix} 0.2392 & 0.971 \\ 0.971 & -0.2392 \end{bmatrix} \cdot \begin{bmatrix} 1/1.4961 & 0 \\ 0 & 1/0.4144 \end{bmatrix} \cdot \begin{bmatrix} 0.2392 & 0.971 \\ 0.971 & -0.2392 \end{bmatrix}$$
$$= \begin{bmatrix} 2.3133 & -0.4053 \\ -0.4053 & 0.7682 \end{bmatrix}$$

→ Compute O, which is the matrix product $H \cdot S^{-1}$ (step 4)

$$O = H \cdot S^{-1} = \begin{bmatrix} 0.5 & 0.4 \\ 0.2 & 1.4 \end{bmatrix} \cdot \begin{bmatrix} 2.3133 & -0.4053 \\ -0.4053 & 0.7682 \end{bmatrix} = \begin{bmatrix} 0.9945 & -0.1047 \\ -0.1047 & 0.9945 \end{bmatrix}$$

→ Compute U as $O \cdot V$ (step 5)

$$U = O \cdot V = \begin{bmatrix} 0.9945 & -0.1047 \\ -0.1047 & 0.9945 \end{bmatrix} \cdot \begin{bmatrix} 0.2392 & 0.971 \\ 0.971 & -0.2392 \end{bmatrix}$$

$$= \begin{bmatrix} 0.3395 & 0.9406 \\ 0.9406 & -0.3395 \end{bmatrix}$$

The steps we have shown here entail a reasonable effort. However, as we move to larger matrices, Eigenvalue and Singular Value decompositions are computes using specialized numerical methods libraries. Use the MatLab code below to check the answers above.

```
%% Verifying the Example
A = [0.5 0.4; 0.2 1.4];
[U, S, V] = svd(A);
```

1.3.5   Principal Component Analysis

Principal component analysis, or PCA, is a statistical procedure that uses orthogonal transformation to convert a set of observations of possibly correlated variables into a set of linearly uncorrelated variables called principal components. The transformation provides successive principle components, with the first one defining the largest possible data extension whereas the last defines the smallest data extension. Let's go over principal component analysis one step at a time to make sense of our definition.

What are some of the Application of Principal Component Analysis?

In this text, we will use principle component analysis to remove noise from a set of observations. As you continue your journey through this book, you will find out that information transmitted over a wired or wireless link suffers distortion which the receiver attempts to estimate. This distortion caused by the channel changes with frequency and its estimate is called the frequency response of the channel. Once the receiver has estimated this frequency response it attempts to undo it in a process called equalization. If the distance between the transmitter and receiver is large, the transmit signal will be weak and naturally occurring noise at the receiver will make the transmit signal hard to detect and the frequency response more difficult to estimate. We will use principle component analysis to basically filter away some of the noise that the receiver has to deal with when making its estimate of the frequency response. As we will see in the following example, PCA is also used in data compression.

Example 1.25: *Observing Current and Voltage across a Resistor*

Imagine an electrical engineering student repeatedly making current measurement through a 1 Ohm resistor as he randomly changes the voltage across it. From Ohms law, we are all familiar with the fact that voltage, V, and current, I, are linearly related through the resistance, R, via $V/I = R$. Let's assume now that the voltage and current meter used by the student has seen better days and provides reading that are both above and below the correct values. The recorded readings are shown in the figure below.

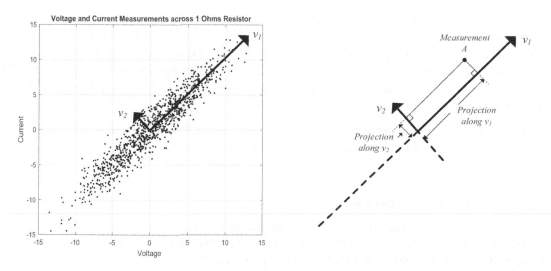

Figure 1-31: Voltage and Current Readings across a 1 Ohm Resistor

Our expectation was to see a straight line with a slope equal to one and an intercept point of 0 volts. Clearly, this didn't entirely happen, and whereas the data naturally extends along the vector v1, which we expected, the data points' extension along the vector v2 is clearly due to the noise or inaccuracy in the measurement.

Correlation

Each observation that the student took consists of a voltage/current coordinate, which we can express in terms of vectors.

$$\begin{bmatrix} v_n \\ i_n \end{bmatrix} for\ n = 0...N-1$$

Let's now express the voltage and current in terms of lengths along their basis vectors $\begin{bmatrix} 1 \\ 0 \end{bmatrix}$ and $\begin{bmatrix} 0 \\ 1 \end{bmatrix}$.

$$\begin{bmatrix} v_n \\ i_n \end{bmatrix} = v_n \begin{bmatrix} 1 \\ 0 \end{bmatrix} + i_n \begin{bmatrix} 0 \\ 1 \end{bmatrix}$$

Given Ohms law, the voltage and current values, v_n and i_n, are directly related through the resistance, and these values are said to be highly correlated. As illustrated in the right portion of the last figure, one of these measurements, call it point *A*, can also be broken up in terms of lengths, or projections, along v_1 and v_2 and the associated unit vectors of v_1 and v_2.

$$\begin{bmatrix} v_n \\ i_n \end{bmatrix} = ProjA\_v1_n \cdot v1_{Unit} + ProjA\_v2_n \cdot v2_{Unit}$$

Take a close look and realize that the projected lengths $ProjA\_v1_n$ and $ProjA\_v2_n$ are completely unrelated or uncorrelated. The projected length $ProjA\_v1_n$ tells us nothing about $ProjA\_v2_n$.
Thus, at this point we need to ask ourselves two important questions.

→ Given that the data points' variation along vector, $v2$, is due to noise or inaccuracies, should we possibly find a way to get rid of it? This train of thought leads us to the idea of noise reduction.
→ Let's assume next that the noise or inaccuracies in the measurements were very small, and the extension along v_2 is less than in the figure above. You may then ask yourself whether you can again eliminate the extensions along v_2, given that it doesn't tell a potential user anything interesting and thus reduce the amount of information that you have to store on your hard drive to archive the data set. We simply retain the vector v_1 and the set of projected lengths $ProjA\_v1_n$. This train of thought leads us to data compression.

Whatever our motivation, our goal is to find the two vectors that characterize the extension of the data and compute the projections of each observation against them.

Finding the Principal Components

To find the principal components, the unit vector of v_1 and v_2, return to section 1.3.1.2 which illustrated the eigenvectors of different matrices. Notice that symmetric matrices stretch their data along different directions and these directions are characterized by the eigenvectors of that matrix. It is our goal to generate a symmetric, or Hermitian symmetric, matrix given the data set at hand. Fortunately, the covariance matrix of the data set provides exactly what we needed. If you are unfamiliar with concepts of variance and covariance, please jump ahead to section 1.4.2 and 1.5.3 to get a quick preview. However, we can easily define the variance and covariance calculations of a sample set as follows. Given a sequence of N numbers, $x[n]$, where $n = 0, 1, \ldots N\text{-}1$, the variance is computed using the following expression. Remember that the asterisk indicates conjugation and $\bar{x}$ indicates mean of the variable x.

$$\sigma_x^2 = \frac{1}{N} \sum_{n=0}^{N-1} (x[n] - \bar{x}) \cdot (x[n] - \bar{x})^*$$

The covariance between two sequences, $x[n]$ and $y[n]$, is similarly defined as follows.

$$\sigma_{xy} = \frac{1}{N} \sum_{n=0}^{N-1} (x[n] - \bar{x}) \cdot (y[n] - \bar{y})^*$$

The covariance matrix is a hermitian symmetric matrix and obeys the following expression. The eigenvectors of this matrix provide us with the principle components, $ProjA\_v1_n$ and $ProjA\_v2_n$, that we are looking for. Note that the sequences $x[n]$ and $y[n]$ in our example above represent the voltage and current values of each observation, n.

$$C_{xy} = \begin{bmatrix} \sigma_x^2 & \sigma_{xy} \\ \sigma_{yx} & \sigma_y^2 \end{bmatrix}$$

Understanding Projection

To find the projections of the measurement along principle component vectors, we recall the geometric interpretation of the dot product of two vectors A and B, which states the following.

$$Dot(A,B) = a_1 v_1^* + a_2 b_2^* + a_3 b_3^* + \ldots + a_N b_N^* = \|A\| \cdot \|V\| \cdot cos(\theta)$$

Observe the figure below that illustrates the geometric interpretation of the dot product using the vectors A (one of our measurements), B (one of the two principle component vectors), and ProjA_B·B$_{unit}$ (A projected onto B). Simple algebra proves that the length ProjA_B is equal to the $\|A\| \cdot cos(\theta)$, which leads us to the equation for the projection of A onto B. Note that the unit vector of B = B/$\|B\|$, where $\|B\|$ represents the length of norm of B.

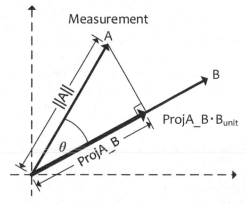

Figure 1-32: Projection of Vector A onto Vector B

The projection of the measurement, A, onto the principle component vector, B, has the following form, which is valid for both real and complex vectors.

$$ProjA\_B \cdot B_{Unit} = ||A|| \cdot cos(\theta) \cdot \frac{B}{||B||}$$

$$= ||A|| \cdot cos(\theta) \cdot \frac{||B||}{||B||} \cdot \frac{B}{||B||}$$

$$= \frac{||A|| \cdot ||B|| \cdot cos(\theta)}{||B|| \cdot ||B||} \cdot B$$

$$= \frac{Dot(A,B)}{Dot(B,B)} \cdot B$$

Example 1.25 (Continued): A MatLab Example

We continue our example by generating the measurement data set that we showed a few pages ago. We accomplish this by transforming random data via a matrix H into simulated measurement data.

```
%% Generate the observed data
H = [2.5, 3.5; 3.5, 2.5];

% The random input data set
Input = [randn(1, 1000); randn(1, 1000)];

% The output data set
Output = H*Input;

% The eigenvectors (principle components are in the columns of V_Eig1, whereas the
stretch or
% extension along these principle components are in the diagonal of D_Eig1.
[V_eig1, D_eig1] = eig(H);

% The eigenvectors (principle components are in the columns of V_Eig1, whereas the
stretch or
% extension along these principle components are in the diagonal of D_Eig1.
[V_eig1, D_eig1] = eig(H);
```

$$H = \begin{bmatrix} 2.5 & 3.5 \\ 3.5 & 2.5 \end{bmatrix}$$

As the matrix is symmetric, the eigenvectors and eigenvalues are the principle components and the data stretch along those components. Unfortunately, we can't get to this matrix directly, so we compute the covariance matrix, which has the same eigenvectors and eigenvalues that are the square of the eigenvalues of H. The figure below illustrates the simulated measurement set as uncorrelated random data transformed by H as well as the reconstituted data set which has be generated by removing the extensions along the principle component vector v_2.

```
%% Compute the covariance matrix
x  = Output(1,:);  % The voltage values
y  = Output(2,:);  % The current values
C  = [Covariance(x,x) Covariance(x,y); Covariance(y,x) Covariance(y,y)];

% ------------------------------------------------------------------------------------
```

```
% Compute the eigenvectors, which are the same as those of matrix H, and the accompanying
% eigenvalues which are the square of those of matrix H.
[V_eig2, D_eig2] = eig(C);
v1 = V_eig2(:,2);   % The eigenvectors are in the columns of V_eig2
v2 = V_eig2(:,1);   % v1 is the vector pointing in the larger extension

% -----------------------------------------------------------------------------
%  Find the projected vectors P1 = ProjA_v1·v1_unit and P2 = ProjA_v2·v2_unit of each
%  measurement along the principle component vectors
P1 = []; P2 = [];
for i = 1:length(Output)
   A  = Output(:,i);                      % The current measurement (voltage, current)
   P1 = [P1, (dot(A,v1)/dot(v1,v1))*v1];  % We plot P1 below
   P2 = [P2, (dot(A,v2)/dot(v2,v2))*v2];  % P2 will be absent thus reducing inaccuracy
end                                       % but only along the P2 direction.
% -----------------------------------------------------------------------------
function Output = Covariance(X, Y)
   X  = X(:);     Y  = Y(:);              % Ensure that X and Y are column vectors
   mx = mean(X); my = mean(Y);
   Output = (1/length(X))*((X - mx).' * conj(Y-my));
end
```

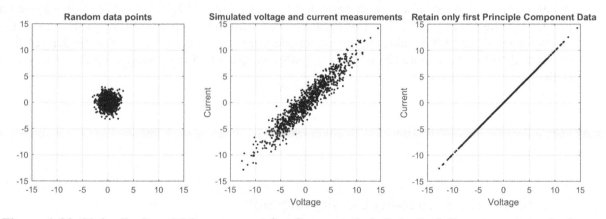

Figure 1-33: Noise Reduced Measurement Set Generated via Principal Component Analysis

1.4   Statistics and Random Variables

This section will provide an overview of random variables and the statistical metrics used to quantify their behavior. In simple terms, a random variable maps the outcome of an experiment of chance into numbers. Flipping a coin to get heads or tails can be mapped to a variable r which assumes the numbers 1 for heads and -1 for tails. In a game of dice, one of six unique faces will appear after each throw. As soon as we assign the numbers one through six to these unique faces, we may use a random variable to describe the experiment of chance. These random variables may be discrete in nature, as was the case for the coin flipping experiment where only two states are possible or continuous as would be the case for a measurement of the outside air temperature at 7:00 o'clock every morning. A random variable t, which represents early morning temperatures, will deviate around an average value in a manner that can be described probabilistically. How probable is it that the measurements end up between 10°C and 13°C, what would be their average and how far from that mean (average) do the values deviate? Finding mathematical descriptions that statistically parameterize random variables is the topic of this section.

A random variable is described by its probability density function, or PDF, which provides information regarding the probability of every possible outcome of our experiment of chance. There are several very important probability density functions that appear in communication systems, including the Gaussian, Uniform, and Raleigh PDF.

A complete description of a random variable as provided by a PDF may not be necessary in our day to day engineering tasks. Most of the time, statistical metrics providing less information like mean, variance, and RMS value are more than adequate. These metrics can either be derived from the PDF directly or computed by observing a large sample set of outcomes in which case we refer to them as sample statistics.

1.4.1   The Probability Density Function

The PDF or probability density function, *Prob(x)*, of a random variable x is used to calculate the likelihood of a random variable assuming a value between two limits x_1 and x_2. To calculate that probability, the PDF must be integrated between these two positions.

$$Probability(x > x_1 \ \& \ x < x_2) = \int_{x_1}^{x_2} Prob(x)dx$$

A game of darts can illustrate the common types of probability density functions. Imagine a vertical line running through the center of the dart board, and as we throw our darts we record the displacement, x, to the right of that vertical line. When the dart hits to the left of the vertical center line, the displacement we record is negative. The displacement, x, in this case is a random variable, which assumes values that are continuous in nature.

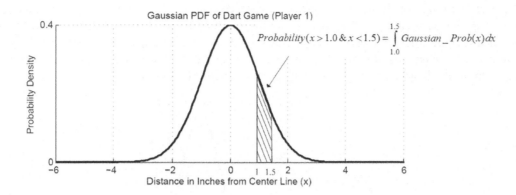

Figure 1-34: Gaussian PDF for Dart Player 1

The probability density function for this random variable is called a _Gaussian PDF_ [6] also known as the famous bell curve. It makes intuitive sense that the probability density should be higher in the center than further out. The equation for the Gaussian PDF must be integrated between two displacements to find the probability of the dart landing in between them.The equation for the Gaussian PDF is shown below.

$$Prob(x) = \frac{1}{\sqrt{2\pi\sigma_x^2}} exp(-\frac{(x - m_x)^2}{2\sigma_x^2})$$

There are several very important facts that we should mention about the Gaussian PDF.

1. The integral of the equation above has no closed form solution. To find the probability of the dart landing between two displacements, we use numerical methods.

2. The Gaussian PDF is completely defined by two variables, the mean m and the standard deviation σ. The standard deviation is a measure of uncertainty, or in this case indicates your skill at throwing darts. If you are new to the game of darts, the amount of uncertainty, σ, in your throws will be greater than that of a professional.

3. The Gaussian PDF is the most important and common probability density function in nature. There are random variables that are themselves made up of the summation of other random variables. The amount of money that a casino makes at the end of the day is really the summation of the amount earned at each table. The Dow Jones industrial average is another great example of a random number being made up of the fluctuations of the value of many different stocks. The central limit theorem states that a random variable that is the summation of other random variables will tend to feature a Gaussian PDF. This is the case even if the constituent random variables were not Gaussian in nature. Since most random events are actually a combination of many different random occurrences, it is no wonder that the Gaussian PDF is so common.

4. Finally, by definition, the total probability computed by integrating between negative and positive infinity must be equal to 1. If this is not the case, then the PDF must be properly rescaled until this condition is met.

$$TotalProbability = \int_{-\infty}^{\infty} Prob(x)dx = 1$$

In the earlier figure for dart player 1, the standard deviation was 1.0 and the mean equaled zero. Below is a second scenario, which shows a much tighter curve. We may assume that player 2 is better at this game than player 1, but we can see that on the average he tends to shoot slightly to the right. In this case, the standard deviation and mean were 0.5 and 1.0 respectively. (Perhaps he was playing outside and a steady wind was pushing the dart slightly to the right.)

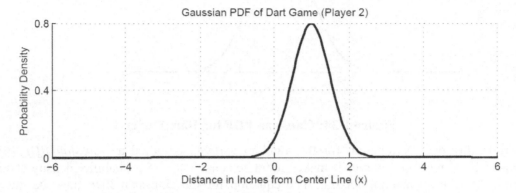

Figure 1-35: Gaussian PDF with Non-Zero Mean

The _Uniform PDF_ is another common probability density function that is often seen in engineering problems. If we picture the dart board as a polar plane, we can record the absolute distance from the center and the angle, θ, of the location where each dart hits the board. Disregarding any condition that could bias the trajectory of the dart, common sense dictates that the dart will not favor any particular angle. The random variable θ will therefore feature a flat or uniform PDF as seen in the figure below. In order for the total probability to be equal to 1.0, the level of the PDF must be constant at $1/2\pi$. The equation of the uniform PDF can be seen here for a lower limit L_1 (-π, in this case) and an upper limit L_2 (π, in this case).

$$Prob(\theta) = \begin{vmatrix} 1/(L_2 - L_1) & \text{for} & L_1 \leq x \leq L_2 \\ 0 & & \text{otherwise} \end{vmatrix}$$

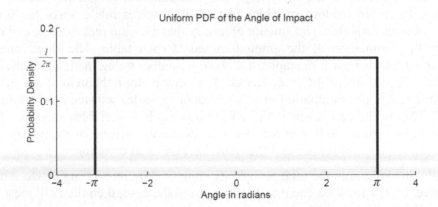

Figure 1-36: Uniform Probability Density Functions for the Random Variable θ

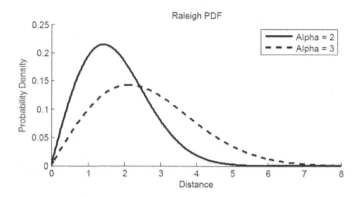

Figure 1-37: Rayleigh PDF

To get to our last PDF, we will make the assumption that the horizontal and vertical displacements of the dart's locations are random variables with Gaussian distributions. If we now consider the distance from the center as the square root of the sum of the squares of the horizontal and vertical displacements, we arrive at the _Rayleigh PDF_. Obviously, this distance can't be negative, and the PDF's mean and standard deviation cannot be chosen independently. The mean $E[x]$ and the standard deviation, σ_x, are a function of the factor α.

$$Prob(x) = \frac{x}{\alpha^2} \exp(-\frac{x^2}{\alpha^2})$$

$$E[x] = \bar{x} = \alpha\sqrt{\frac{\pi}{2}} \qquad \sigma_x = \sqrt{(2-\frac{\pi}{2})\alpha^2}$$

Rayleigh distributions are frequently used to model the effects of multipath channels in wireless communication systems as we will see in chapter 5.

Distribution Functions

There are two additional PDF related functions, which merit closer attention since they are frequently used by engineers to visualize the behavior of a random variable. These functions are called CDF, or cumulative distribution function, and CCDF, or complementary cumulative distribution function.

$$CDF_x(x) = \int_{-\infty}^{x} Prob_x(a)da$$

$$CCDF_x(x) = \int_{x}^{\infty} Prob_x(a)da$$

The CDF and CCDF integrate the PDF from the left and right side respectively and thereby give us an alternative view regarding the random variable's distribution.

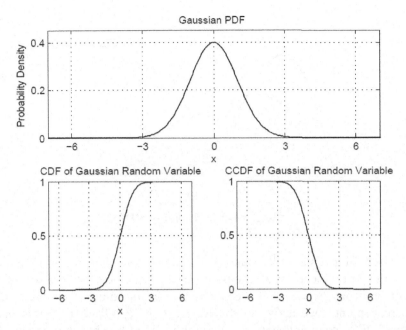

Figure 1-38: PDF, CDF, and CCDF of a Gaussian Random Variable

1.4.2 Statistical Measurements

There are two ways of finding a random variable's statistical descriptors such as the mean and the standard deviation. We may find them directly from the probability density function, or we compute them via a large set of measured outcomes. Measurements done on large sets of outcomes are called sample statistics. The sample statistics are valid not just for a set of outcomes of some random variable but for any discrete number sequence whether random or deterministic in nature. In day to day engineering work, it is far more likely that we will evaluate the statistical metrics of large number sequences than derive them from the PDF of the random variables that generated the sequences. In most cases, the PDF simply isn't available. Turn on any modern oscilloscope and you will be able to navigate to a menu that shows you the sample mean, sample variance and *rms* value of the waveform that the voltage probe is currently sampling. In this section we will cover both methods of computing these statistics. The variable N indicates the number of outcomes considered in the calculation of each metrics.

Sample Statistics

Mean → Given a large set of measured outcomes, $x[n]$, the sample mean or average value of the random variable that is described by its set of outcomes may be calculated by adding up the values of all outcomes and dividing that sum by the number of recorded outcomes, N.

$$Mean_x = \bar{x} = \frac{1}{N} \sum_{n=0}^{N-1} x[n]$$

Variance → Given a large set of measured outcomes, $x[n]$, the sample variance, or AC power, of that set or sequence may be calculated via the following equation.

$$Variance_x = ACPower_x = \sigma_x^2 = \frac{1}{N} \sum_{n=0}^{N-1} (x[n] - \bar{x}) \cdot (x[n] - \bar{x})^*$$

Standard Deviation → The standard deviation, which refers to the spread of the random variable around its mean, is defined as the square root of the variance. If the mean of a random variable or sequence of numbers is zero, then the standard deviation of that random variable or number sequence is equal to the RMS, or root mean square, value.

$$STD_x = \sigma_x = \sqrt{Variance_x}$$

Mean Square → Given a large set of measured outcomes, $x[n]$, the mean square, or total average power, of that set or sequence may be calculated as follows.

$$MeanSquare_x = TotalAveragePower_x = \frac{1}{N} \sum_{N=0}^{N-1} x[n] \cdot x[n]^*$$

Instantaneous Power → The instantaneous normalized power of any one sample $x[n]$ of a sequence or number set is defined as the product of $x[n]$ and its complex conjugate.

$$InstantaneousPower = x[n] \cdot x[n]^*$$

DC Power → The normalized DC power of a sequence of deterministic numbers or random outcomes is defined as the product of the mean and that mean's complex conjugate.

$$DCPower_x = \bar{x} \cdot \bar{x}^*$$

Root Mean Square → The root mean square, or RMS value, of a sequence of numbers is a statistical metric of its magnitude. The RMS value is not the average magnitude (mean root square) of the sequence. Mathematically, it is more convenient to work with the mean of the squared values of a sequence as we will see in the chapter on approximation. The RMS value is the square root of the total average power of a sequence.

$$RMS_x = \sqrt{\frac{1}{N} \sum_{n=0}^{N-1} x[n] \cdot x[n]^*}$$

Notice that we have assigned the concept of power to some of the statistical metrics discussed above. This nomenclature is more familiar to electrical engineers than it is to pure statisticians. The choice of wording makes most sense when thinking of a voltage waveform captured by a modern oscilloscope. The oscilloscope doesn't know the impedance across which it digitizes voltage, so for simplicity the resistance is normalized to 1 ohm and we subsequently speak of normalized powers. Keep in mind that these normalized powers are not always in watts and may in fact lack units altogether. They are simply terms that lend intuitive meanings to statistical metrics, which can then be better appreciated by engineers.

Peak to Average Power Ratio → The idea of power becomes even more concrete when considering our final metric, the peak to average power ratio. The PAPR represents the relationship between the maximum instantaneous power and the total average power of a number sequence or sampled waveform. RF engineers developing transmitter solutions require knowledge of this metric to properly choose the correct power amplifier technology. The larger the peak to average power ratio the more demanding the requirements are on the RF power amplifier.

$$PAPR_x = \frac{x_{max}[n] \cdot x_{max}[n]^*}{\frac{1}{N} \sum_{n=0}^{N-1} x[n] \cdot x[n]^*}$$

Statistics Derived from the PDF

If we know the PDF of a random variable, rather than a large set of outcomes, then the mean may be calculated differently. Let's say we know the weight, $x[n]$, of every person in North America to the nearest pound. Let's assume further that outcome, x_0, corresponds to 0 pounds, x_1 to 1 pound all the way until we reach $x_{500} = 500$ pounds, which we claim to be the largest weight. Out of the total number of people, N, in North America, N_0 weigh 0 pounds, N_1 weigh one pound and

N_{500} weigh 500 pounds. The mean, or expected value, $E(x)$, may then be equivalently defined as follows.

$$E(x) = \frac{N_0}{N}x_0 + \frac{N_1}{N}x_1 + \cdots + \frac{N_{500}}{N}x_{500}$$

$$= Prob_x(x_0)\cdot x_0 + Prob_x(x_1)\cdot x_1 + \cdots + Prob_x(x_{500})\cdot x_{500}$$

$$= \sum_{i=0}^{500} Prob_x(x_i)\cdot x_i$$

The equations below use the continuous PDF, $Prob_x(x)$, as well as the discretized version, $Prob_x(x_i)$, to calculate the mean (i indexes our 501 possible outcomes).

$$E(x) = \sum_i x_i \cdot Prob_x(x_i) \quad \leftarrow Discrete\ Case$$

$$E(x) = \int_{-\infty}^{\infty} x \cdot Prob_x(x)\cdot dx \quad \leftarrow Continuous\ Case$$

In statistics, we define the ath moment – and a^{th} central moment – of a random variable x via the following expression.

$$E(x^a) = \int_{-\infty}^{\infty} x^a \cdot Prob_x(x)\cdot dx \quad\quad \leftarrow a^{th}\ Moment$$

$$E[(x - E(x))^a] = \int_{-\infty}^{\infty} (x - E(x))^a \cdot Prob_x(x)\cdot dx \quad \leftarrow a^{th}\ Central\ Moment$$

The mean, or expected value of the random variable x, is in fact its first moment, while the second moment and second central moment refer to the total average power and variance respectively.

Example 1.26: *Computing Sample Statistics*

In this example we will compare the statistics of two separate number sequences, each featuring a Gaussian distribution. Sequence one is a complex number stream in which both the real and imaginary parts are generated to have a variance equal to 1. Sequence two is purely real and features a variance that is equal to 2 and a mean of 0.6. Additionally, in this example we will learn how to generate a discrete PDF for visualization purposes.

```
%% Defining our Waveform
x = randn(1, 1000000);      %% Return Gaussian distributed random numbers
y = randn(1, 1000000);      %% with variance = 1
Sequence1 = x + j*y;        %% Zero-mean, random complex number sequence
Sequence2 = x*sqrt(2) + .6; %% Random, real sequence with variance = 2
                            %% and mean = 0.6
%% Sequence 1
N1         = length(Sequence1);
Mean1      = (1/N1)*sum(Sequence1);
TotalPower1 = (1/N1)*sum(Sequence1.*conj(Sequence1));
```

```
RMS1          = sqrt(TotalPower1);
Var1          = (1/N1)*sum((Sequence1-Mean1).*conj(Sequence1-Mean1));
STD1          = sqrt(Var1);


%% Sequence 2
N2            = length(Sequence2);
Mean2         = (1/N2)*sum(Sequence2);
TotalPower2   = (1/N2)*sum(Sequence2.*conj(Sequence2));
RMS2          = sqrt(TotalPower2);
Var2          = (1/N2 )*sum((Sequence2-Mean2).*conj(Sequence2-Mean2));
STD2          = sqrt(Var2);

%% Plotting the PDF / Histograms
W = 0.1;                        %% Width of histogram bin
x = -4:W:4;
n1 = hist(real(Sequence1),x); %% Histogram --> Gaussian Distribution
n2 = hist(abs(Sequence1) ,x); %% Histogram --> Raleigh  Distribution
```

Clearly, sequences one and two are row vectors which, in the code above, we repeatedly multiply on an entry by entry basis. MatLab is natively built to work with matrices and therefore includes convenient notation for entry by entry arithmetic. Just add a period in front of the multiplication, division, addition and subtraction operators and 'voila.' Obviously, MatLab will complain bitterly if the dimensions of the matrices are not identical.

Table 1-5.1: Calculated and Theoretical Statistics

Statistic	Sequence S1	Sequence S2	Theoretical S1 / S2
Mean	0.0005 +j0.0004	0.5993	0.0 / 0.6
Total Power	2.0017	2.3629	2.0 / 2.36
RMS	1.4148	1.5372	1.41421 / 1.5362
Var	2.0017	2.0038	2.0 /2.0
STD	1.4148	1.4155	1.41421 / 1.41421

The table above shows the analysis results, and, as expected, the means of sequences one and two are close to zero and 0.6 respectively. The first interesting conclusion that we can make is that the total power of sequence one is the sum of the real and imaginary power.

$$TotalPowerS1 = \frac{1}{N}\sum_{n=0}^{N-1}(x[n] + j \cdot y[n]) \cdot (x[n] - j \cdot y[n])$$

$$= \frac{1}{N}\sum_{n=0}^{N-1}(x[n]^2 + j \cdot y[n]x[n] - j \cdot x[n]y[n] - j^2 \cdot y[n]^2)$$

$$= \frac{1}{N}\sum_{n=0}^{N-1}x[n]^2 + \frac{1}{N}\sum_{n=0}^{N-1}y[n]^2$$

Another curiosity is the fact that the total power of sequence two seems to be the sum of its variance, which we had set to 2.0, and its DC power of $0.6^2 = 0.36$. This too can easily be explained if we break $x[n] = x_{AC}[n] + 0.6$ into its AC and mean component and then calculate the total power.

$$TotalPowerS1 = \frac{1}{N}\sum_{n=0}^{N-1}(x_{AC}[n] + 0.6)\cdot conj(x[n]_{AC} + 0.6)$$

$$= \frac{1}{N}\sum_{n=0}^{N-1}x_{AC}[n]\cdot x_{AC}[n]^* + 0.6\cdot x_{AC}[n] + 0.6\cdot x_{AC}[n] + 0.6\cdot 0.6$$

$$= \sigma_x^2 + \frac{1}{N}\sum_{n=0}^{N-1}1.2\cdot x_{AC}[n] + \frac{1}{N}\sum_{n=0}^{N-1}0.36$$

$$= \sigma_x^2 + 0.36$$

The result is obvious once we realize that the mean of $0.6 \cdot x_{AC}[n]$ is equal to zero. The remaining results are quite straight forward, so let's have a look at how we would plot a probability density function of each sequence. MatLab gets us part of the way there as well, by providing the histogram function seen in the code. We start off by defining a series of bins, which feature a center location as well as a width. The center locations are defined as $x = -4\!:\!W\!:\!4$ (-4.0, -3.9, -3.8 ... etc.), where W is the step size and bin width. In the statement n1 = hist(real(Sequence1),x), we now count how many entries in sequence one fall within the first bin [-4.05 to -3.95], the second bin [-3.95 to -3.85], and so on. The results are then returned in the vector $n1$, which we will use to plot the PDF. The probability that any one sample will land in the first bin is equal to $n1(1,1)/N$, where N is the number of samples in sequence one. This probability is related to the PDF via the following equation.

$$\frac{n1(1,1)}{N} = \int_{-4.05}^{-3.95}PDF(x)dx$$

For the purposes of graphing the *PDF*, we may crudely assume that the *PDF*(x) is constant within any one bin thus simplifying our integration.

$$\frac{n1(1,1)}{N} = PDF(x_1)\Big|_{-4.05}^{-3.95} = PDF(x_1)\cdot 0.1$$

$$= PDF(x_1)\cdot W$$

$$\boxed{PDF(x_i) = \frac{n1(1,i)}{N}\cdot\frac{1}{W}}$$

The index i spans the number of bins that we had chosen to compute our histogram. We finish off this example by plotting the *PDF* of the real part and the magnitude of sequence 1. The first PDF shows a nice zero-mean bell curve that is indicative of the Gaussian distribution, while the second curve is the above–mentioned Rayleigh distribution.

```
PDF1 = (n1/length(Sequence1))*(1/W);    %% PDF of real(Sequence1)
PDF2 = (n2/length(Sequence1))*(1/W);    %% PDF of abs(Sequence1)
```

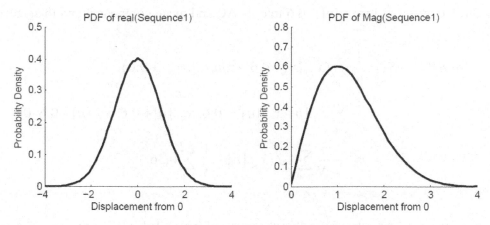

Figure 1-39: Gaussian and Rayleigh Distribution

1.4.3 Common Variation between Random Variables

In the last section we took a look at statistical metrics used to describe random variables or discrete number sequences. In this section we will explore how to quantify the dependence of one random variable on the outcome of another using the concepts of covariance and correlation.

Covariance

The covariance is a measure of the common variation from their mean of two random variables x and y. This variation from the mean is expressed as $x-E[x]$ and $y-E[y]$. If these variations from their mean tend to have the same sign, then the covariance equation below will average to a positive value, since the product of these variations will usually be positive. If a positive variation of x goes hand in hand with a negative variation of y, then the covariance function below will average to a negative number. A zero will indicate that there is no common variation from their mean and the two random variables are not correlated.

$$\boxed{c_{xy} = E[(x - E[x]) \cdot (y - E[y])]}$$

$$c_{xy} = E[xy - x \cdot E[y] - E[x] \cdot y + E[x]E[y]]$$
$$= E[xy] - E[x \cdot E[y]] - E[E[x] \cdot y] + E[E[x]E[y]]$$
$$= E[xy] - E[x] \cdot E[y] - E[x] \cdot E[y] + E[x]E[y]$$
$$= E[xy] - E[x]E[y]$$

The covariance may be normalized to produce the correlation coefficient p, which extends from − 1 to 1. A correlation coefficient of 1.0 indicates that the outcomes of random variables x and y are the same except for a potential difference in scaling. A correlation coefficient of -1.0 indicates the same except that X and Y always have opposite signs.

$$p = \frac{c_{xy}}{\sqrt{\text{var}(x) \cdot \text{var}(y)}}$$

> If the covariance and thus the correlation coefficient are equal to 0.0, then the two variables are said to be *uncorrelated*.

Correlation

The correlation is a measure of the common variation of two random variables x and y. The correlation between two random variables x and y is defined as the expected value of their product.

$$r_{xy} = E[x \cdot y]$$

The covariance considered only common variations from the respective mean of each random variable, whereas the correlation measures the common variation from zero.

> If the correlation equation for x and y equals 0.0, then the two random variables are said to be *orthogonal*.

Statistical Independence

Two random variables are said to be statistically independent if from the outcome of one variable we cannot draw any conclusion as to the outcome of the other. If two people that are standing back to back are both flipping coins to get heads or tails, then we may safely assume that the outcome of one flip in no way allows us an advantage in guessing at the outcome of the other.

Statistical Independence versus Orthogonality versus Uncorrelatedness

As you may have guessed, statistical independence is the most restrictive of the three metrics we discuss here. To unclutter some of the confusion regarding the three metrics of dependence, we make the following claims.

→ Two random variables or number sequences that are statistically independent are also orthogonal and uncorrelated.

→ Two random variables or number sequences that are orthogonal are also uncorrelated but may not necessarily be statistically independent.

→ Two random variables or number sequences that are uncorrelated are not necessarily orthogonal or statistically independent.

The equation for covariance, $c_{xy} = E[x \cdot y] - E[x] \cdot E[y]$, reduces to that of the correlation, $r_{xy} = E[x \cdot y]$, if either the mean of x or the mean of y or both are zero. Therefore, if two random variables x and y are uncorrelated and at least one of them has a mean equal to zero, then they are also orthogonal.

To understand how two random variables or number sequences may be orthogonal but not statistically independent, we consider the following example. Let's look at the two number sequences $x = \cos(\theta)$ and $y = \sin(\theta)$, where variable θ may monotonically increase from 0 to infinity or may be completely random. From our earlier discussion on trigonometric identities, we compute our correlation equation as follows.

$$E[\cos(\theta) \cdot \sin(\theta)] = E[\frac{\sin(\theta - \theta) + \sin(\theta + \theta)}{2}] = E[\frac{\sin(2\theta)}{2}] = 0$$

Clearly, the two number sequences are orthogonal, but they are not statistically independent. The fact that an outcome $x_i = \cos(\theta_i)$ is equal to 0.707 provides us with clear information as to what y_i can be. The angles that produced $x_i = 0.707$ are ±45 degrees and will yield values for $y_i = \sin(\pm 45)$ that are equal to ±0.707. There is a clear relationship between x and y that renders them statistically dependent. The expressions below summarize the requirements for the uncorrelatedness and orthogonality of two random variables x and y.

$$c_{xy} = E[xy] - E[x]E[y] = 0 \qquad \leftarrow uncorrelated$$

$$r_{xy} = E[xy] = 0 \qquad\qquad\quad \leftarrow orthogonal$$

For their complex counterparts the equations change only slightly.

$$c_{xy} = E[xy^*] - E[x]E[y^*] = 0 \qquad \leftarrow uncorrelated$$

$$r_{xy} = E[xy^*] = 0 \qquad\qquad\quad \leftarrow orthogonal$$

The concept of orthogonality may be difficult to appreciate when dealing with abstract quantities like random variable, but when we discuss it in the context of deterministic number sequences or waveform as is done in the examples on the following pages, it will become much more intuitive. Let's have a look at a few examples.

Example 1.27: Statistical Dependence

Comparing two random variables obviously involves some awkward nomenclature, so let us illustrate what is meant by them. Imagine two gentlemen – call them Eric and Michael – playing darts at the local pub while we sit on the side line recording how far to the left or right of center their darts hit the board. If Eric's dart hits the target slightly to the left of center, is there any reason to believe that Michael's will follow suit and tend to the left as well? Could his dart tend to prefer the other side of the board? In all likelihood, the outcome of Eric's throw in no way affects that of Michael. The two outcomes are said to be *statistically independent*. Not only are they statistically independent but they are also *uncorrelated* and *orthogonal*. If we were to record these errors all night and plot them, then their probability density function would look Gaussian with zero mean.

Now assume that there is a slight breeze in the pub blowing past the dart board from right to left. The *PDF* of each man's error will still be Gaussian, but both means will be on the left side. One man's throw will give you a hint as to the direction of the wind, a fact that will help you guess the likely error of the other's throw. The errors are no longer statistically independent. However, the errors (x for Eric and y for Michael) are still uncorrelated since the deviations from their means are not dependent. Mathematically, we state that the covariance $c_{xy} = E[(x - E[x]) \cdot (y - E[y])]$ is equal to zero. If we assume that the mean errors in both men's throws are not equal to zero, then clearly $r_{xy} = E(x \cdot y)$ will also not equal zero and the two errors are not orthogonal. Remember, two random variables that are uncorrelated will only be orthogonal if at least one is zero-mean. *Uncorrelatedness* places the least amount of restriction on the independence of two random variables, while *statistical independence* is the most demanding condition.

Example 1.28: Summing Statistically Independent Variables

Given two zero-mean, statistically independent complex random variables 'x' and 'y', we can prove that the variance of their sum, $z = x + y$, is equal to the sum of their individual variances.

$$\begin{aligned}
\sigma_z^{\,2} &= E((x+y)(x+y)^*) \\
&= E(xx^*) + E(xy^*) + E(yx^*) + E(yy^*) \\
&= E(xx^*) + E(yy^*) \\
&= \sigma_x^{\,2} + \sigma_y^{\,2}
\end{aligned}$$

1.4.4   Orthogonal Waveforms

The idea of orthogonal random variables may have been a bit abstract, but orthogonal waveforms are much more intuitive and it turns out they are extremely important in the realm of communication engineering. Understanding and using these signals have led to a significant increase in transmission throughput in modern communication links. Let's assume that we add two or more information carrying waveforms – via superposition – into a composite waveform for transmission. If these waveforms are orthogonal, then the information embedded in each may be independently detected at the receiver. The following two examples illustrate how the use of *orthogonal* signal sets accomplishes this feat.

$Composite(t) = Info1(t) \cdot OrthWaveform1(t) + Info2(t) \cdot OrthWaveform1(t) + \ldots$

Example 1.29: IQ Signaling

As we all remember from our introductory courses in communication systems, early radio systems relied on FM/PM and AM modulation techniques, which embedded information in either the phase or the amplitude of the carrier. With the advent of digital communication components such as dedicated baseband processors and high performance digital-to-analog converters, it became possible to accurately embed information in both the phase and the amplitude – or I and Q components – of an RF carrier. The only way that this is possible is if an RF carrier is in fact composed of two orthogonal signals, which can each carry an independent data stream. The signals in question, whose orthogonality is proven below, are $cos(2\pi f_0 t)$ and $sin(2\pi f_0 t)$.

$$E[\cos(2\pi f_0 t) \cdot \sin(2\pi f_0 t)] = E[\frac{\sin(0) + \sin(4\pi f_0 t)}{2}]$$

$$= E[\frac{\sin(4\pi f_0 t)}{2}]$$

$$= 0$$

In the all too familiar figure below we illustrate how the RF carrier signal is assembled from the two orthogonal waveforms, $cos(2\pi f_0 t)$ and $sin(2\pi f_0 t)$, and the associated IQ data streams riding on them. These modulating data streams are the familiar $I(t)$ and $Q(t)$ signals.

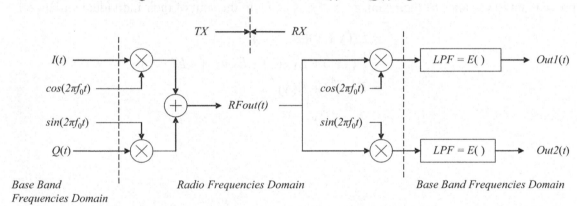

Figure 1-40: A Basic RF Communication Link from the RF Systems Viewpoint

The modern formulation above shows the RF carrier as the superposition of differently scaled cosine and sine waves rather than one sinusoid featuring a particular amplitude and phase.

$$RFout(t) = I(t) \cdot \cos(2\pi f_o t) + Q(t) \cdot \sin(2\pi f_o t)$$

Now, the fact that we can embed two separate data streams in one RF signal is useful to us only if we are able to recover the $I(t)$ and $Q(t)$ terms independently. Ironically enough, the receiver structure computes the very equations that we use to verify the orthogonality of two signals.

$$Out1(t) = LPF[RFout(t) \cdot \cos(2\pi f_o t)]$$

$$Out2(t) = LPF[RFout(t) \cdot \sin(2\pi f_o t)]$$

It is the low pass filter itself that provides the averaging or expectation process, which we show here for *Out2(t)*. (See Section 1.1.8 which shows the derivation of *Out1(t)*.)

$$Out2(t) = LPF[(\ I(t) \cdot \cos(2\pi f_o t) + Q(t) \cdot \sin(2\pi f_o t) \ ) \cdot \sin(2\pi f_o t)]$$
$$= LPF[\ I(t) \cdot \cos(2\pi f_o t) \cdot \sin(2\pi f_o t) + Q(t) \cdot \sin(2\pi f_o t) \ ) \cdot \sin(2\pi f_o t)]$$
$$= LPF[\ I(t) \cdot \cos(2\pi f_o t) \cdot \sin(2\pi f_o t)] + LPF[Q(t) \cdot \sin(2\pi f_o t) \cdot \sin(2\pi f_o t)]$$

Given the fact that $cos(2\pi f_0 t)$ and $sin(2\pi f_0 t)$ are orthogonal, the expression $LPF[I(t) \cdot sin(2\pi f_0 t) \cdot cos(2\pi f_0 t)]$ collapses to zero.

$$Out2(t) = LPF[\ Q(t) \cdot (1/2 - 1/2 \cos(4\pi f_o t))] + 0$$
$$= 1/2 \ Q(t)$$

In a similar fashion, we can trivially show that *Out1(t)* equals ½*I(t)*. In conclusion, IQ communication systems take advantage of the fact that an RF signal is composed of two separate and orthogonal carriers, which can both be used to independently transmit baseband information.

Example 1.30: OFDM

OFDM, or orthogonal frequency division multiplexing, is without a doubt one of the recent stars in the communication systems arena. An OFDM signaling scheme simultaneously transmits separate baseband information streams on tones that differ in frequency. Clearly, these tones interfere with one another, and the resulting composite signal looks quite messy. However, since the different sinusoids used in the scheme are orthogonal, we can conceive of a receiver that will detect the information on any one of the tones without seeing that of the others. Let's therefore consider two differently scaled cosine signals, one at 1MHz and the other at 2MHz. We add these signals and then try some manner of signal processing step that separately recovers the information of each one.

$$Orthogonal\_Tone1(t) = \cos(2\pi \cdot 1e6 \cdot t)$$
$$Orthogonal\_Tone2(t) = \cos(2\pi \cdot 2e6 \cdot t)$$

$$Info1(t) = 3$$
$$Info2(t) = -2$$

$$Composite(t) = Info1(t) \cdot Orthogonal\_Tone1(t) + Info2(t) \cdot Orthogonal\_Tone2(t)$$
$$= 3 \cdot \cos(2\pi \cdot 1e6 \cdot t) - 2 \cdot \cos(2\pi \cdot 2e6 \cdot t)$$

Note that the smallest time period at which both tones complete a full cycle is 1 microsecond. To recover the information on tone 1 from the composite, we in fact multiply the composite waveform by that tone and integrate or average over any integer multiple of 1 microsecond. The same holds true for tone 2.

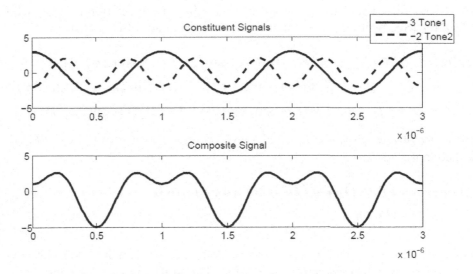

Figure 1-41: Superposition of Tones 1 and 2

$$RX1 = E[\ Composite(t) \cdot Tone1(t)\ ]$$
$$RX2 = E[\ Composite(t) \cdot Tone2(t)\ ]$$

$$RX1 = \frac{2}{1e-6} \int_{0}^{1e-6} Composite(t) \cdot Tone1(t)dt$$

$$= \frac{2}{1e-6} \int_{0}^{1e-6} [3 \cdot \cos(2\pi \cdot 1e6 \cdot t) - 2 \cdot \cos(2\pi \cdot 2e6 \cdot t)] \cdot \cos(2\pi \cdot 1e6 \cdot t)dt$$

$$= \frac{2}{1e-6} \int_{0}^{1e-6} 3 \cdot \cos^2(2\pi \cdot 1e6 \cdot t) - 2 \cdot \cos(2\pi \cdot 2e6 \cdot t) \cdot \cos(2\pi \cdot 1e6 \cdot t)dt$$

$$= \frac{2}{1e-6} \int_{0}^{1e-6} 3(\frac{1}{2} + \frac{1}{2}\cos(2\pi \cdot 2e6 \cdot t)) - 2 \cdot (\cos(2\pi \cdot 1e6 \cdot t) + \cos(2\pi \cdot 3e6 \cdot t))dt$$

$$= \frac{2}{1e-6} \int_{0}^{1e-6} \frac{3}{2} dt = 3$$

In the equations above, the waveforms $2 \cdot cos(2\pi2e6 \cdot t)$ and $cos(2\pi1e6 \cdot t)$ are orthogonal and integrate to zero, leaving behind only the composite information that held data from tone 1. The same procedure is repeated below to recover the scaling factor that was associated with tone 2. Clearly, the information that we are transmitting and recovering here is the *Info1* and *Info2* values.

$$RX2 = \frac{2}{1e-6} \int_{0}^{1e-6} Composite(t) \cdot Tone2(t)dt = -2$$

We could easily choose a larger set of tones including $cos(2\pi3e6 \cdot t)$, $cos(2\pi4e6 \cdot t)$ and so on; the trick is to make sure that the tones share a common time during which they have all completed an integer multiple of their period. In the case above, the 2–MHz tone completed two cycles in 1 microsecond while the 1 MHz tone completed just one. A real OFDM system like 802.11a/g features 52 complex tones to transmit and receive information. We will discuss OFDM systems in much greater detail in chapter six.

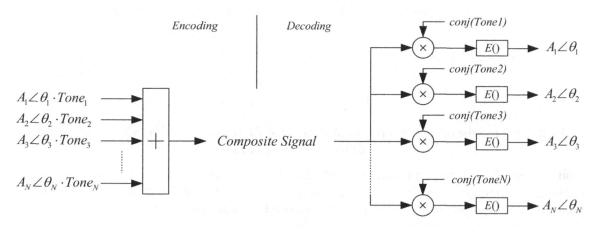

Figure 1-42: Generalized OFDM Encoding Decoding Structure

1.5   The Random Process and Statistical Signal Processing

Let us consider a symbol generator, or symbol source, that produces random sequences of 1s and negative 1s as a function of discrete time, n. Instead of a random symbol generator we could also imagine the digital part of a communication transmitter producing symbols that are eventually pulse shaped and transmitted by an analog transceiver. Either way, the casual observer sees a stream of what appear to be random symbols at the output of the generator.

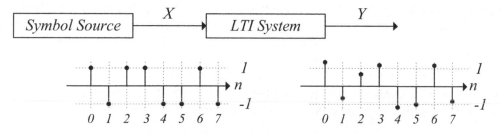

Figure 1-43: Random Symbol Source with LTI System

Assume, then, that we want to examine a certain number of consecutive symbols, which we will group into a vector called X. Imagine we can press a little button, and some device captures eight consecutive symbols for us anytime we want. There are only 256 possible outcomes that can be generated by the experiment. The random process maps these outcomes into discrete number sequences that are grouped into what is called an ensemble.

$$Ensemble(X) = \begin{bmatrix} 1 & 1 & 1 & 1 & 1 & 1 & 1 & 1 & \leftarrow Sequence_0 \\ 1 & 1 & 1 & 1 & 1 & 1 & 1 & -1 & \leftarrow Sequence_1 \\ 1 & 1 & 1 & 1 & 1 & 1 & -1 & 1 & \leftarrow Sequence_2 \\ \vdots & \vdots & \vdots & \vdots & \vdots & \vdots & \vdots & \vdots & \vdots \\ -1 & -1 & -1 & -1 & -1 & -1 & -1 & -1 & \leftarrow Sequence_{255} \end{bmatrix}$$

Figure 1-44: Ensemble for a Random Process of Eight Random Variables with Symbol Values Restricted to ±1

You can quickly see that the ensemble can take on huge dimensions as we increase the number of captured symbols or allow those random symbols to take values other than just ±1. Rather than picturing a random process as a mapping operation into or out of an ensemble, we may alternately think of it as a vector containing random variables.

$$X = \begin{bmatrix} x_0 & x_1 & x_2 & x_3 & x_4 & x_5 & x_6 & x_7 \end{bmatrix}$$

In this text, we think of a ***random (stochastic) process*** as a vector consisting of a certain number of random variables. In a communication setting, the random variables represent waveform samples that are defined or captured at even time steps. The group of possible number sequences that can be produced by the random process is called an ensemble.

These random symbols now pass through a linear time-invariant system, like a simple digital filter, or another processing block that can be modeled as such, and reappear somewhat distorted on the other side. We then group what we believe to be the corresponding eight symbols at the output into a new vector called Y.

$$Y = \begin{bmatrix} y_0 & y_1 & y_2 & y_3 & y_4 & y_5 & y_6 & y_7 \end{bmatrix}$$

We define a random process as a vector of random variables for a simple reason. We are interested in finding out whether any one of the variables in vector X influences another variable in X or some in the output process Y. If neighboring samples tend to be similar in value, we may assume that the waveform being sampled is slowly moving with low frequency content. Similarly, if x_1 and y_4 are always equal, then we can draw the conclusion that the LTI system features a delay equal to three samples. The point is that we are gaining insight into the nature of signals that appear random and LTI systems that are unknown via statistical observation. In the next section we will examine how to compute and quantify the interdependence of the various random variables within a single process and between two processes.

1.5.1 Dependencies between Variables within a Random Process

In the field of statistical signal processing, one of the basic questions asked is how the different random variables in a stochastic process are related to one another. To quantify this dependency, we introduce the autocorrelation and autocovariance functions. Auto- and crosscorrelation definitions may vary between text books; the definitions used in this text agree with [7].

The Autocorrelation Function

The *autocorrelation function* computes correlation of any two random variables that reside in the same random process. The idea is identical to the correlation r_{xy} that we discussed in the section on random variables. We call it a function owing to the presence of the independent indices, a and b, which select the random variables in the process X.

$$r_x(a,b) = E(x_a \cdot x_b^*)$$

Again, a and b are indices – from 0 to 7 – of the two random variables in our random process X.

How, you may ask, does the equation $E(x_a \cdot x_b^*)$ yield a value of dependence or similarity? If the outcomes of x_a and x_b are similar, then the product $x_a \cdot x_b^*$ will likely be located close to the positive real axis featuring a larger real than imaginary component. The expected value operation will produce a stable positive, real result indicating the similarity. If the outcomes tend to be opposites, then their product will more likely be located on the negative real axis and the expected value operation will produce a stable negative result. If there is no similarity at all, the autocorrelation will fizzle to zero during the expectation operation.

The Autocovariance Function

The *autocovariance function* computes the covariance between any two random variables that reside in the same random process.

$$c_x(a,b) = E([x_a - E(x_a)] \cdot [x_b - E(x_b)]^*)$$

The autocovariance function determines the common variation of two random variables away from their mean, whereas the autocorrelation function determines the common variation of two random variables away from zero. Therefore, if the random variables of our process are zero-mean, then the two functions are equivalent.

Clearly, these functions can be applied to either random process X or Y. We may also be interested in the common variation between a random variable in process X and one in Y and in that case we rename our statistical function as follows.

The Crosscorrelation Function

The *crosscorrelation function* computes the co-mean square between a variable in process X and one in process Y.

$$r_{xy}(a,b) = E(x_a \cdot y_b^*)$$

The Crosscovariance Function

The *crosscovariance function* computes the covariance between a variable in process X and one in process Y.

$$c_{xy}(a,b) = E([x_a - E(x_a)] \cdot [y_b - E(y_b)]^*)$$

The crosscorrelation and crosscovariance concepts are very important in the area of wireless communication systems. In later chapters, we will try to estimate the effects of multipath RF channels, which can be modeled as an LTI system. The statistical relationships between the received information and that transmitted lead us to an understanding of the unknown multipath channel. We will see the concepts discussed in this section again in chapter four when we develop optimal equalizer.

1.5.2   Stationary Random Processes

Let us once again consider our symbol generator and the accompanying random process X, which is composed of a certain number of random variables. Each one of these random variables has a mean, variance and potentially some effect on the other random variables in the process X. If these statistics don't change over time – meaning they are the same now, in five minutes and next month – then the process is said to be stationary. Furthermore, if the linear time–invariant or LTI system we showed earlier also remains unchanged, then the random process Y is also considered to be stationary.

> Assuming that a process is in some way **stationary** is a very reasonable hypothesis in many situations, and it leads to a simpler formulation of the different functions we have listed earlier. If, however, the characteristics of the LTI system were to change with time, then the process Y would no longer be stationary; dealing with such situations requires adaptive signal processing concepts, which we discuss in chapter four.

The autocorrelation and autocovariance functions for a stationary random process will no longer depend on the exact time index a and b inside its vector of random variables but on the time difference between them. The variable x_0 influences x_1 in the same way as variable x_{100} influences x_{101}. Thus, for any time offset of c the following expression holds true.

$$r_x(a,b) = r_x(a+c,b+c)$$

Better yet, we express the functions in terms of a difference in time index which we will call τ, where $\tau = a - b$. The location or index, a, of the random variable inside our vector is arbitrary.

$$r_x(\tau) = r_x(a, a+\tau) \quad \leftarrow \textit{Autocorrelation Function (stationary RP)}$$
$$c_x(\tau) = c_x(a, a+\tau) \quad \leftarrow \textit{Autocovariance Function (stationary RP)}$$

If the characteristics of the linear system are indeed time–invariant, then the processes X and Y are said to be joint stationary and we may define the following expressions.

$$r_{xy}(\tau) = r_{xy}(a, a + \tau) \qquad \leftarrow Crosscorrelation\ Function\ (stationary\ RP)$$

$$c_{xy}(\tau) = c_{xy}(a, a + \tau) \qquad \leftarrow Crosscovariance\ Function\ (stationary\ RP)$$

Example 1.31: Estimation of LTI System Characteristics

In this example, the generator produces symbols that are statistically independent of one another and are grouped into a stationary random process, X. The LTI system, which is represented by the simple mathematical construct shown below, gives rise to the random process Y.

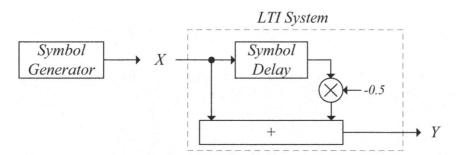

Figure 1-45: A Simple Linear Time–Invariant System with Associated Input and Output Random Processes X and Y

The system clearly indicates that the value of a random variable y_n will consist of the value of random variable x_n and half that of $-x_{n-1}$, which arrived one symbol period earlier. The MatLab function below creates 1,000 possible sequences of source symbols that are restricted to taking values of ± 1. To make the MatLab code more compact, the random processes X and Y are assumed to consist of only five random variables. The autocorrelation sequence we calculate here compares the random variable x_2 to x_0 through x_4. Unsurprisingly, x_2 is correlated only with itself.

$$r_x(a,b) = \begin{bmatrix} r_x(2,0) & r_x(2,1) & r_x(2,2) & r_x(2,3) & r_x(2,4) \end{bmatrix} \approx \begin{bmatrix} 0 & 0 & 1 & 0 & 0 \end{bmatrix}$$

```
Matrix = rand(1000,6);   %% A Matrix of 1000 rows and 6 columns
                         %% featuring random values between 0 and 1
Ensemble = 2*round(Matrix) - 1; %% limiting the values to +/- 1


Ensemble_X = Ensemble(:, 2:6);   %% Pick just 5 of the six columns


% Computing the autocorrelation function by comparing random
% variables X2 to X0, X1, X2, X3 and X4
r_x20 = mean(Ensemble_X(:, 2 + 1) .* conj( Ensemble_X(:, 0 + 1)) );
r_x21 = mean(Ensemble_X(:, 2 + 1) .* conj( Ensemble_X(:, 1 + 1)) );
r_x22 = mean(Ensemble_X(:, 2 + 1) .* conj( Ensemble_X(:, 2 + 1)) );
r_x23 = mean(Ensemble_X(:, 2 + 1) .* conj( Ensemble_X(:, 3 + 1)) );
r_x24 = mean(Ensemble_X(:, 2 + 1) .* conj( Ensemble_X(:, 4 + 1)) );


[r_x20 r_x21 r_x22 r_x23 r_x24] % r_x(a,b) - the autocorrelation function
```

```
% We now apply the effect of the LTI system and define the Y ensemble
Ensemble_Y = Ensemble(:, 2:6) - 0.5*Ensemble(:, 1:5);


% Let's now compute the crosscorrelation function by comparing
% random variables X2 to Y0, Y1, Y2, Y3 and Y4
r_xy20 = mean(Ensemble_X(:, 2 + 1).* conj( Ensemble_Y(:, 0 + 1)) );
r_xy21 = mean(Ensemble_X(:, 2 + 1).* conj( Ensemble_Y(:, 1 + 1)) );
r_xy22 = mean(Ensemble_X(:, 2 + 1).* conj( Ensemble_Y(:, 2 + 1)) );
r_xy23 = mean(Ensemble_X(:, 2 + 1).* conj( Ensemble_Y(:, 3 + 1)) );
r_xy24 = mean(Ensemble_X(:, 2 + 1).* conj( Ensemble_Y(:, 4 + 1)) );


[r_xy20 r_xy21 r_xy22 r_xy23 r_xy24] % r_xy(a,b) - the crosscorrelation function
```

Computing the crosscorrelation sequence unveils the characteristics of the LTI system's transfer function.

$$r_{xy}(a,b) = \begin{bmatrix} r_{xy}(2,0) & r_{xy}(2,1) & r_{xy}(2,2) & r_{xy}(2,3) & r_{xy}(2,4) \end{bmatrix} \approx \begin{bmatrix} 0 & 0 & 1 & -0.5 & 0 \end{bmatrix}$$

Given that our process is stationary, we reformulate our autocorrelation and autocovariance results in terms of τ, where $\tau = a - b = [-2\ -1\ 0\ 1\ 2]$.

$$r_x(\tau) \approx \begin{bmatrix} 0 & 0 & 1 & 0 & 0 \end{bmatrix}$$

$$r_{xy}(\tau) \approx \begin{bmatrix} 0 & 0 & 1 & -0.5 & 0 \end{bmatrix}$$

As an experiment you can change the MatLab code by augmenting the symbols with imaginary terms, such that the possible symbol combinations yield $\pm 1 \pm j$. You can easily do this by creating a second random ensembles matrix, then multiplying it by j and adding it to the Ensembles_X matrix. You will find that the auto/crosscorrelation sequences will be identical to the real number case.

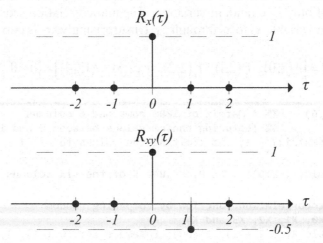

Figure 1-46: Auto/Crosscorrelation Functions $R_x(\tau)$ and $R_{xy}(\tau)$

1.5.3   Correlation and Covariance expressed as Matrices

The various correlation and covariance metrics may also be expressed in matrix form. We see this extensively when dealing with estimation, equalizers, and MIMO systems, as will be discussed in chapters 4 and 7. The autocorrelation matrix is defined as the expected value of the outer product of the random process vector X.

$$R_X = E[X \cdot X^*] = E\left[\begin{bmatrix} x_0 \\ x_1 \\ x_2 \\ x_3 \\ x_4 \end{bmatrix} \begin{bmatrix} x_0^* & x_1^* & x_2^* & x_3^* & x_4^* \end{bmatrix}\right]$$

$$= E\begin{bmatrix} x_0 x_0^* & x_0 x_1^* & x_0 x_2^* & x_0 x_3^* & x_0 x_4^* \\ x_1 x_0^* & x_1 x_1^* & x_1 x_2^* & x_1 x_3^* & x_1 x_4^* \\ x_2 x_0^* & x_2 x_1^* & x_2 x_2^* & x_2 x_3^* & x_2 x_4^* \\ x_3 x_0^* & x_3 x_1^* & x_3 x_2^* & x_3 x_3^* & x_3 x_4^* \\ x_4 x_0^* & x_4 x_1^* & x_4 x_2^* & x_4 x_3^* & x_4 x_4^* \end{bmatrix}$$

Once the expectation is taken of every position in the matrix, the elements take the form of $r_x(a,b) = E(x_a \cdot x_b^*)$. In the real number case, $r_x(a,b) = r_x(b,a)$, we reformulate the correlation matrix as follows.

$$R_X = \begin{bmatrix} r_x(0,0) & r_x(0,1) & r_x(0,2) & r_x(0,3) & r_x(0,4) \\ r_x(1,0) & r_x(1,1) & r_x(1,2) & r_x(1,3) & r_x(1,4) \\ r_x(2,0) & r_x(2,1) & r_x(2,2) & r_x(2,3) & r_x(2,4) \\ r_x(3,0) & r_x(3,1) & r_x(0,2) & r_x(3,3) & r_x(3,4) \\ r_x(4,0) & r_x(4,1) & r_x(4,2) & r_x(4,3) & r_x(4,4) \end{bmatrix} = \begin{bmatrix} r_x(0,0) & r_x(1,0) & r_x(2,0) & r_x(3,0) & r_x(4,0) \\ r_x(1,0) & r_x(1,1) & r_x(2,1) & r_x(3,1) & r_x(4,1) \\ r_x(2,0) & r_x(2,1) & r_x(2,2) & r_x(3,2) & r_x(4,2) \\ r_x(3,0) & r_x(3,1) & r_x(0,2) & r_x(3,3) & r_x(4,3) \\ r_x(4,0) & r_x(4,1) & r_x(4,2) & r_x(4,3) & r_x(4,4) \end{bmatrix}$$

If on top of that, the process X is stationary, the autocorrelation matrix is rewritten as seen below where the indices a and b are replaced by $\tau = a - b$.

$$R_X = E[X \cdot X^*] = \begin{bmatrix} r_x(0) & r_x(1) & r_x(2) & r_x(3) & r_x(4) \\ r_x(1) & r_x(0) & r_x(1) & r_x(2) & r_x(3) \\ r_x(2) & r_x(1) & r_x(0) & r_x(1) & r_x(2) \\ r_x(3) & r_x(2) & r_x(1) & r_x(0) & r_x(1) \\ r_x(4) & r_x(3) & r_x(2) & r_x(1) & r_x(0) \end{bmatrix} \quad \leftarrow real\ numbers$$

In the complex valued case, we take a closer look at the relationship of diagonal (bottom–left to top–right) terms in the matrix. Note that the terms $x_0 x_1^*$ and $x_1 x_0^*$ are not equal. As a matter of fact, for two arbitrary complex variables, x_0 and x_1, the following expression holds.

$$x_0 \cdot x_1^* = (x_1 \cdot x_0^*)^*$$

Therefore, if $E(x_a \cdot x_b{}^*) = r_x(a,b)$, then $E(x_b \cdot x_a{}^*) = E((x_a \cdot x_b{}^*)^*) = r_x{}^*(a,b)$ and the following reformulation is valid.

$$R_X = E[X \cdot X^*] = \begin{bmatrix} E(x_0 x_0^*) & E(x_0 x_1^*) & E(x_0 x_2^*) & E(x_0 x_3^*) & E(x_0 x_4^*) \\ E(x_1 x_0^*) & E(x_1 x_1^*) & E(x_1 x_2^*) & E(x_1 x_3^*) & E(x_1 x_4^*) \\ E(x_2 x_0^*) & E(x_2 x_1^*) & E(x_2 x_2^*) & E(x_2 x_3^*) & E(x_2 x_4^*) \\ E(x_3 x_0^*) & E(x_3 x_1^*) & E(x_3 x_2^*) & E(x_3 x_3^*) & E(x_3 x_4^*) \\ E(x_4 x_0^*) & E(x_4 x_1^*) & E(x_4 x_2^*) & E(x_4 x_3^*) & E(x_4 x_4^*) \end{bmatrix}$$

$$= \begin{bmatrix} r_x(0,0) & r_x(0,1) & r_x(0,2) & r_x(0,3) & r_x(0,4) \\ r_x(1,0) & r_x(1,1) & r_x(1,2) & r_x(1,3) & r_x(1,4) \\ r_x(2,0) & r_x(2,1) & r_x(2,2) & r_x(2,3) & r_x(2,4) \\ r_x(3,0) & r_x(3,1) & r_x(3,2) & r_x(3,3) & r_x(3,4) \\ r_x(4,0) & r_x(4,1) & r_x(4,2) & r_x(4,3) & r_x(4,4) \end{bmatrix} = \begin{bmatrix} r_x(0,0) & r_x^*(1,0) & r_x^*(2,0) & r_x^*(3,0) & r_x^*(4,0) \\ r_x(1,0) & r_x(1,1) & r_x^*(2,1) & r_x^*(3,1) & r_x^*(4,1) \\ r_x(2,0) & r_x(2,1) & r_x(2,2) & r_x^*(3,2) & r_x^*(4,2) \\ r_x(3,0) & r_x(3,1) & r_x(3,2) & r_x(3,3) & r_x^*(4,3) \\ r_x(4,0) & r_x(4,1) & r_x(4,2) & r_x(4,3) & r_x(4,4) \end{bmatrix}$$

If we assume that the random process X is stationary, then the individual correlations are dependent only on the difference in time index τ, where $\tau = a - b$. The autocorrelation matrix now simplifies as follows.

$$R_X = E[X \cdot X^*] = \begin{bmatrix} r_x(0) & r_x^*(1) & r_x^*(2) & r_x^*(3) & r_x^*(4) \\ r_x(1) & r_x(0) & r_x^*(1) & r_x^*(2) & r_x^*(3) \\ r_x(2) & r_x(1) & r_x(0) & r_x^*(1) & r_x^*(2) \\ r_x(3) & r_x(2) & r_x(1) & r_x(0) & r_x^*(1) \\ r_x(4) & r_x(3) & r_x(2) & r_x(1) & r_x(0) \end{bmatrix} \quad \leftarrow complex\ numbers$$

The autocovariance, crosscorrelation and crosscovariance matrices feature similar structures and are shown below in abbreviated form.

$$C_X = E[(X - E[X]) \cdot (X - E[X])^*] = \begin{bmatrix} c_x(0) & c_x^*(1) & \cdots \\ c_x(1) & \ddots & \ddots \\ \vdots & \ddots & \ddots \end{bmatrix} \quad \leftarrow Autocovariance\ Matrix$$

Note that the autocorrelation and autocovariance matrices feature constant descending diagonals. The values along diagonals that start at the top left and end at the bottom right are the same. This type of structure is called a *Toeplitz* matrix. Beyond this feature, we also note that certain symmetry exists in the ascending diagonals. If the entries in the top right half are complex conjugates of those in the bottom left half, then the structure is called a *Hermetian Toeplitz* matrix. If the entries in the two halves are actually the same, then the structure is called a *symmetric Toeplitz* matrix.

The autocorrelation and autocovariance matrices for real numbers are clearly *symmetric Toeplitz* in nature, while those for the complex valued case are *Hermetian Toeplitz*. This, however, is not the case for the crosscorrelation and crosscovariance matrices, as can be seen below.

$$R_{XY} = E[X \cdot Y^*] = \begin{bmatrix} E(x_0 y_0^*) & E(x_0 y_1^*) & E(x_0 y_2^*) & E(x_0 y_3^*) & E(x_0 y_4^*) \\ E(x_1 y_0^*) & E(x_1 y_1^*) & E(x_1 y_2^*) & E(x_1 y_3^*) & E(x_1 y_4^*) \\ E(x_2 y_0^*) & E(x_2 y_1^*) & E(x_2 y_2^*) & E(x_2 y_3^*) & E(x_2 y_4^*) \\ E(x_3 y_0^*) & E(x_3 y_1^*) & E(x_3 y_2^*) & E(x_3 y_3^*) & E(x_3 y_4^*) \\ E(x_4 y_0^*) & E(x_4 y_1^*) & E(x_4 y_2^*) & E(x_4 y_3^*) & E(x_4 y_4^*) \end{bmatrix}$$

If we once again assume that both processes are stationary, then we know that the diagonals from the top left to the bottom right will be constant and the matrix has *Toeplitz* structure. However, we can't apply any type of symmetry to the quantities alongside the other diagonals from the bottom left to the top right. For example, the values of $E(x_0 y_1^*)$ and $E(x_1 y_0^*)$ don't have a convenient relationship as was the case for the autocorrelation and autocovariance matrices. Therefore, the crosscovariance and crosscorrelation matrices are simply *Toeplitz* in nature.

$$R_{XY} = E(X \cdot Y^*) = \begin{bmatrix} r_{xy}(0,0) & r_{xy}(0,1) & \cdots \\ r_{xy}(1,0) & \ddots & \ddots \\ \vdots & \ddots & \ddots \end{bmatrix} \quad \leftarrow Crosscorrelation\ Matrix$$

$$C_{XY} = E[(X - E[X]) \cdot (Y - E[Y])^*] = \begin{bmatrix} c_{xy}(0,0) & c_{xy}(0,1) & \cdots \\ c_{xy}(1,0) & \ddots & \ddots \\ \vdots & \ddots & \ddots \end{bmatrix} \quad \leftarrow Crosscovariance\ Matrix$$

Example 1.32: *Correlation Matrices*

Continuing our last example, let's now compute the autocorrelation and crosscorrelation matrices of our five–element random processes X and Y, which we assume to be stationary. The MatLab code is simply extended to compute the remaining correlations, with the following result.

$$R_X = E[X \cdot X^*] = \begin{bmatrix} 1 & 0 & 0 & 0 & 0 \\ 0 & 1 & 0 & 0 & 0 \\ 0 & 0 & 1 & 0 & 0 \\ 0 & 0 & 0 & 1 & 0 \\ 0 & 0 & 0 & 0 & 1 \end{bmatrix}$$

$$R_{XY} = E[X \cdot Y^*] = \begin{bmatrix} 1 & -.5 & 0 & 0 & 0 \\ 0 & 1 & -.5 & 0 & 0 \\ 0 & 0 & 1 & -.5 & 0 \\ 0 & 0 & 0 & 1 & -.5 \\ 0 & 0 & 0 & 0 & 1 \end{bmatrix} \qquad R_Y = E[Y \cdot Y^*] = \begin{bmatrix} 1 & -.5 & 0 & 0 & 0 \\ -.5 & 1 & -.5 & 0 & 0 \\ 0 & -.5 & 1 & -.5 & 0 \\ 0 & 0 & -.5 & 1 & -.5 \\ 0 & 0 & 0 & -.5 & 1 \end{bmatrix}$$

Continuing the MatLab code from example 1.31, we write the correlation matrices as follows.

```
%% Computing autocorrelation and crosscorrelation matrices

RX  = zeros(5,5);
RXY = zeros(5,5);
for a = 1:5
    for b = 1:5
        RX(a,b)  = mean(Ensemble_X(:,a) .* conj(Ensemble_X(:,b)));
        RXY(a,b) = mean(Ensemble_X(:,a) .* conj(Ensemble_Y(:,b)));
    end
end
```

1.6   Miscellaneous Concepts

1.6.1   The Fixed Point Number Format and Quantization Noise

Since digital signal processing hardware cannot express numbers with infinite precision, computations are done on fixed point quantities featuring limited resolution. There are three parameters that define a fixed point number. First, the number may be either signed or unsigned (positive only). The second and third parameters are the total number of bits and fractional number of bits assigned to the number. The notation below is commonly used to define the fixed point format of a number or variable.

$$U[Total\ Bits,\ Fractional\ Bits] \rightarrow Unsigned$$
$$S[Total\ Bits,\ Fractional\ Bits] \rightarrow Signed$$

Therefore, $U[12,8]$ indicates an unsigned quantity with 12 total and 8 fractional bits. Clearly, the number of bits assigned to the integer portion of the value is the difference between the number of total and fractional bits. For the $U[12,8]$ format, only 4 integer bits are available, while the $S[12,8]$ format features 3 integer bits and the MSB is reserved to indicate the sign.

$$Integer\ Bits = Total\ Bits - Fractional\ Bits$$

Given the number of fractional bits, we may define the smallest possible step size of the number as follows.

$$StepSize = \Delta = \frac{1}{2^{FractionalBits}}$$

A $U[12,8]$ format number can assume a value as small as $StepSize$ and as large as $StepSize \cdot (2^{TotalBits}-1)$. The range of possible values for a $U[12,8]$ and $S[12,8]$ formatted number is as follows.

$$U[12,8] \rightarrow 0, \frac{1}{256}, \frac{2}{256} \dots \frac{4095}{256}$$

$$S[12,8] \rightarrow -\frac{2048}{256}, -\frac{2047}{256} \dots \frac{2047}{256}$$

Digital hardware and most hardware definition languages (VHDL and Verilog) operate only in terms of integers, which are expressed in binary, decimal, or hexadecimal form. A fixed point number may be expressed in native integer format by dividing the fixed point number by its step size.

$$MachineInteger = \frac{FixedPNumber}{StepSize} = FixedPNumber \cdot 2^{Fractional\ Bits}$$

The number $1/256$ with fixed point format $U[12,9]$ would translate into $(1/256) \cdot 2^9 = 2$. Digital hardware engineers who write Verilog often want to see numbers in hexadecimal format. Luckily, MatLab features a function that translates a fixed point number directly into the integer, hexadecimal, or binary formats for our convenience. These functions are called *num2int*, *num2hex*, and *num2bin* respectively. The MatLab fixed point toolbox allows us to define quantization objects that are used in combination with these fixed point functions. As an example, we define and exercise some of these functions below.

```
%% Using the Fixed Point Toolbox
U1208 = quantizer('ufixed','nearest','saturate', [12 08]);
S1208 = quantizer( 'fixed',  'ceil', 'saturate',  [12 08]);


Q_PI     = quantize(U1208, pi);       %% Quantizing the value of PI
Q_PI_Dec = num2int(U1208, Q_PI);      %% Produces Value as Integer
Q_PI_Hex = num2hex(U1208, Q_PI);      %% Produces Value as Hexadecimal String


%% Quantizing and Rounding without the Fixed Point toolbox
PI_RoundX08 = (round(pi*256)/256); %% Rounding to the nearest 1/256
PI_FloorX08 = (floor(-pi*256)/256);%% Rounding down to lower 1/256
PI_CeilX08  = (ceil(-pi*256)/256); %% Rounding up to higher 1/256
```

We may define how a quantizer rounds a floating point quantity like π to a fixed point number by specifying different rounding modes. Keywords such as '*nearest*', '*ceil*', and '*floor*' mimic how rounding algorithms operate in hardware. So what happens if you don't have the fixed point toolbox? Don't despair! In the code below, we create our own quantization function that can be used instead of those provided in the fixed point library.

```
function [Output Overflow] = QuantizeIt(Input, TotalBits, ... FractionalBits, Flag)

Overflow = 0;
IntegerBits = TotalBits - FractionalBits;  %% Number of integer bits
FM = 2^FractionalBits;                      %% Fractional Multiplier
InputSign = Input./abs(Input+1e-40);        %% We need to figure out the
                                            %% sign here, since we will be taking the
                                            %% absolute value soon and the sign is lost


Input          = abs(Input);
IntegerInput   = floor(Input);
FractionalInput = InputSign.*(Input - IntegerInput);

switch(lower(Flag))
    case 'nearest'
        QuantizedFractionalOutput = (1/FM)*round(FractionalInput*FM);
    case 'ceil'
        QuantizedFractionalOutput = (1/FM)*ceil(FractionalInput*FM);
    case 'floor'
        QuantizedFractionalOutput = (1/FM)*floor(FractionalInput*FM);
end

QuantizedFractionalOutput = InputSign.*QuantizedFractionalOutput;
Output = InputSign*(IntegerInput + QuantizedFractionalOutput);


%%  Saturate the result
MaxPositiveValue = 2^(IntegerBits-1) - 1/FM;
MaxNegativeValue = -2^(IntegerBits-1);

if(Output > MaxPositiveValue)
    Overflow = 1;
    Output = MaxPositiveValue;
end

if(Output < MaxNegativeValue)
    Overflow = 1;
    Output = MaxNegativeValue;
end
```

Clearly, rounding floating point quantities to fixed point numbers causes a loss of information which we model as an additive random noise value. In the example below we define the quantization noise power associated with rounding from floating point to fixed point formats.

Example 1.33: *Quantization Noise Power*

In the figure below we round two floating point numbers, A and B, to the nearest valid fixed point value, which will always be a multiple of the step size, Δ. Here, A rounds to 2Δ, while B is cast to 3Δ. The error or noise term, e, is generally distributed uniformly between $-\Delta/2$ and $\Delta/2$.

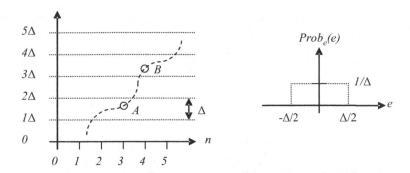

Figure 1-47: Quantization Noise

We compute the quantization noise power by using the formula for the second moment of the error variable. As we saw before the second moment yields the average power of a random variable.

$$E(e)^2 = \int_{-\infty}^{\infty} e^2 \cdot Prob_e(e) \cdot de = \int_{-\frac{\Delta}{2}}^{\frac{\Delta}{2}} e^2 \cdot \frac{1}{\Delta} \cdot de$$

$$= \frac{e^3}{3\Delta}\bigg|_{-\frac{\Delta}{2}}^{+\frac{\Delta}{2}} = \frac{1}{3\Delta}(\frac{\Delta^3}{8} + \frac{\Delta^3}{8}) = \frac{\Delta^2}{12}$$

1.6.2   Sampling of Continuous Signals

Although in the digital signal processing domain we exclusively deal with discrete sequences of numbers, those numbers often have their origin in a continuous counterpart. Whether we want to sample a true analog signal at the input of an analog-to-digital converter (ADC) or simply retain a valid sinusoidal signal in digital memory somewhere, we must ask ourselves how closely in time the waveform samples should be spaced. We investigate this question by sampling the most fundamental waveform that we will encounter in the text: the complex sinusoid. The waveform below, C_1, oscillates at 0.25 Hz and is being sampled four times per cycle, which corresponds to a sample rate of 1Hz. We now want to know whether any other complex sinusoid sampled at 1Hz could produce the same discrete samples.

$$C_1(t) = e^{j2\pi 0.25t}$$

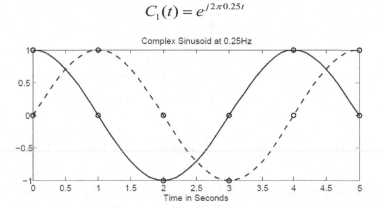

Figure 1-48: Basic 0.25Hz Complex Sinusoid Sampled Once per Second (Solid Line – Real Component / Dashed Line – Imaginary Component)

As it turns out, there is an infinite number of complex sinusoids whose samples are identical to those shown above. The frequencies of these alternate sinusoids are offset from the original 0.25Hz in the positive and negative frequency directions by integer multiples of 1Hz.

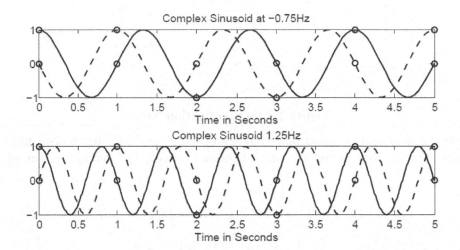

Figure 1-49: Complex Sinusoids at -0.75 and 1.25Hz Sampled Once per Second

The figure below shows four alternate complex sinusoids at -1.75, -0.75, 1.25 and 2.25Hz, which all feature the same discrete samples as the source waveform. As it turns out, for a sample rate of 1Hz, only complex sinusoids whose frequencies reside within the valid range of -0.5 and 0.5Hz may be uniquely expressed. Attempting to process signals beyond this range results in what we call aliasing.

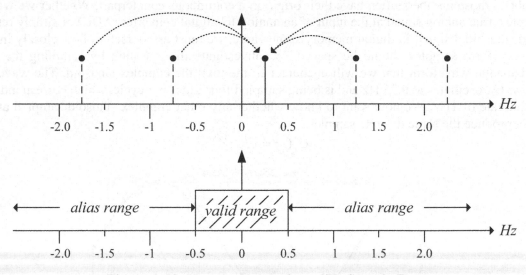

Figure 1-50: Nyquist Range for a 1Hz Sample Rate

The figure above shows a range that extends from -0.5Hz to 0.5Hz. A negative frequency should be interpreted here as complex sinusoids with positive frequencies moving in the clockwise or negative direction around the complex plane. It is only for convenience that we use the term *negative frequencies*.

> The Nyquist-Shannon sampling theory states that if a waveform has frequency content up to but not including B Hz, then that waveform is completely described by samples taken at time intervals of $1/(2B)$. Therefore, to correctly capture an analog waveform, the sampling rate must be larger than two times the highest frequency component in that waveform.

$$F_{Sampling} > 2B$$

The figure below shows the point where a communication signal moves from the analog to the digital domain. The anti-aliasing filter will reject all frequencies larger than *SampleRate*/2 to avoid aliasing high frequency content into our valid frequency range.

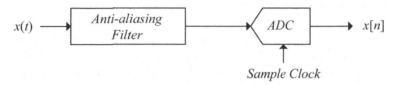

Figure 1-51: Anti-Aliasing Filter

References

[1] Boyer, C. B. (1991) *A History of Mathematics*, Second Edition, John Wiley & Sons, Hoboken, NJ, 218-221

[2] Nahin, P. J. (2007) *An Imaginary Tale – The Story Of $\sqrt{-1}$*, Princeton University Press, Princeton, NJ

[3] Stein, S. K. (1987) *Calculus and Analytic Geometry*, McGraw-Hill, New York, NY, 244, 531

[4] Anton , H. (2010) *Elementary Linear Algebra*, John Wiley & Sons, Hoboken, NJ

[5] Grcar , J. F., (2011b), "Mathematicians of Gaussian elimination", *Notices of the American Mathematical Society*, 58 (6): 782–792, http://www.ams.org/notices/201106/rtx110600782p.pdf

[6] Lathi, B. P. (1989) *Modern Digital and Analog Communication Systems*, Saunders College Publishing, Philadelphia, PA, 385 – 392

[7] Hayes, M. H. (1996) *Statistical Digital Signal Processing and Modeling*, John Wiley & Sons, Hoboken, NJ, 74 – 88

2 DSP Fundamentals

Now that we have greased our intellectual gears with the background material of chapter 1, we are ready to master the fundamentals of digital signal processing. Specifically, in this chapter, we will take a detailed look at convolution, correlation, the Fourier transform, and the design of finite impulse response – or FIR – filters.

The Fourier transform is part of a larger family of algorithms that analyze waveforms or number sequences. This family also includes techniques such as the correlation operation, the Laplace transform and Z transform. These operations don't alter the waveforms but simply express them in terms of different parameters. These new parameters reveal characteristics that weren't visible when the waveforms were looked at as sequences of time-indexed values. We loosely group these techniques under the heading of *signal analysis*.

The other heavyweight topic in this and subsequent chapters is FIR filter design. The digital filter is arguably the most glamorous and prominent representative of the DSP discipline, and the basic introduction given in this chapter will get us ready for prime-time signal processing applications. The associated convolution algorithm computes the output of a digital filter given an input sequence and a set of filter coefficients, which we produce in the design phase. Since digital filters and the associated convolution operation alter the input signal, we loosely group these topics under the heading of *signal manipulation*.

The figure below charts the course of study we will take through the next couple of chapters.

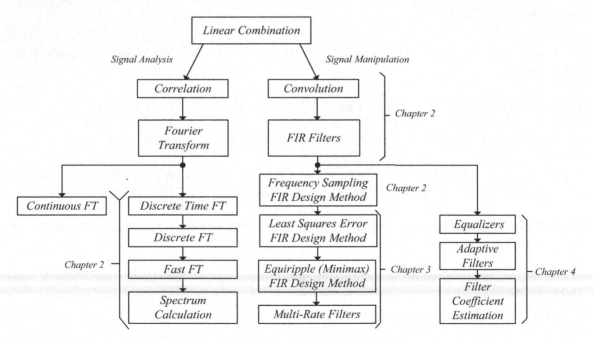

Figure 2-1: Signal Analysis and Manipulation at a Glance

The Linear Combination

The linear combination is at the heart of virtually all signal processing algorithms that we discuss in this chapter. The linear combination of two vectors, A and B, of length N is defined as follows.

$$A = \begin{bmatrix} a_0 & a_1 & \dots & a_{N-1} \end{bmatrix} \qquad B = \begin{bmatrix} b_0 & b_1 & \dots & b_{N-1} \end{bmatrix}$$

$$Linear\ Combination(A,B) = \sum_{n=0}^{N-1} a_n \cdot b_n = a_0 \cdot b_0 + a_1 \cdot b_1 + \dots + a_{N-1} \cdot b_{N-1}$$

For different signal processing techniques, the vectors A and B may represent varying entities, but the calculation involving point-by-point multiplications and a final summation step remains the same. We will see the linear combination in many sections, including the ones discussing correlation, convolution and the Fourier transform.

Signal Analysis

In signal analysis, we compute the similarity between a test waveform and certain analysis or basis functions, which have unique meaning to the current engineering applications. For example, a test waveform could be the voltage output of a circuit element, while the analysis function could be a sinusoid. We know that certain circuit elements and mechanical structure tent to vibrate, and it is of great utility to us as engineers to understand the different manners of how they vibrate. By computing the similarity between an observed or measured test waveform and sinusoids of different frequencies, new information about the circuit's or mechanical structure's performance is revealed to us.

In signal analysis, we go a step further and attempt to express a waveform under test as a summation of many simpler analysis or basis functions, which are scaled by parameters. Parameters in this case are scalar values that, when associated with their analysis function and summed up, will approximate the original waveform under test. The idea is quite similar to that of decomposing a color into red, green, and blue.

$$TestColor = A_0 \cdot red + A_1 \cdot green + A_2 \cdot blue$$

The test color was expressed using three analysis or basis colors, each one parameterized to express their amounts of red, green, and blue using the scalars A_0, A_1, and A_2. As you know, the color yellow is composed of an even amount of red and green, thus causing A_0 and A_1 to be set to 1.0 while A_2 vanishes to zero. Signal analysis differs from the example above in that the test and analysis quantities are functions of time.

$$\begin{aligned} TestWaveform(t) \approx{} & A_0 \cdot AnalysisFunction_0(t) + A_1 \cdot AnalysisFunction_1(t) + \dots \\ & + A_N \cdot AnalysisFunction_N(t) \end{aligned}$$

The type of analysis function changes from application to application. In the area of *communication engineering*, for example, we are interested in finding the frequency content, and subsequently the bandwidth, of the signal with which we work. The family of analysis functions that best suits this purpose is the complex sinusoid.

$$AnalysisFunction(f,t) = e^{j2\pi ft} = \cos(2\pi ft) + j\sin(2\pi ft)$$

In the area of *control engineering*, these analysis functions are inadequate, since they cannot represent a signal that is decaying to a steady state or increasing in an unstable fashion. We thus add the term $e^{\sigma t}$ to our previous analysis function, which now exponentially increases with time for a positive value of σ and decreases for a negative value. The term σ is commonly called the damping factor.

$$AnalysisFunction(\sigma, f, t) = e^{\sigma t}e^{j2\pi ft} = e^{(\sigma + j2\pi f)t} = e^{\sigma t}(\cos(2\pi ft) + j\sin(2\pi ft))$$

The analysis functions used in communication systems are actually a subset ($\sigma = 0$) of the ones used for the control engineering applications. The figure below shows examples of the different types of analysis functions that are used in the areas of communication systems, control systems, and curve fitting.

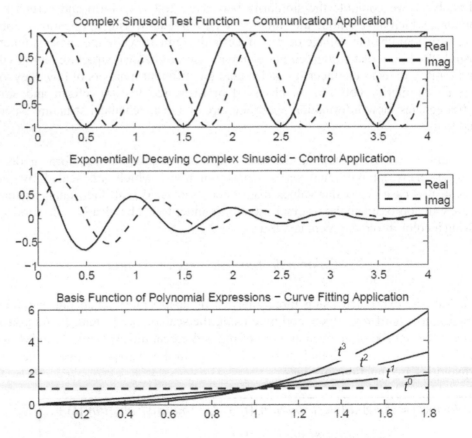

Figure 2-2: Two Different Types of Analysis Functions

The task of *curve fitting* using polynomials requires us to decompose a waveform under test into a superposition of analysis or basis functions of the following form.

$$AnalysisFunction(p,t) = t^p$$

The superposition approximates the original test waveform as follows.

$$TestWaveform(t) \approx A_0 \cdot 1 + A_1 \cdot t + A_2 \cdot t^2 + ... + A_N \cdot t^N$$

Using polynomial approximation in the manner described above shines in applications where waveform data is available only at certain discrete time instances. By approximating what the original continuous waveform looked like, we can easily compute waveform values at time instances for which no data was available. This technique, called interpolation, finds great utility in communication systems.

Now that we understand that signal analysis consists primarily of decomposing waveforms into simpler analysis functions, what type of algorithms would we use to find the coefficients A_0 through A_N associated with these analysis functions? The algorithm that primarily concerns us in this chapter is called *correlation*, which determines the presence of an analysis function in a test waveform by computing their similarity. In the realm of communication systems, we use the *Fourier transform*, which is directly based on the concept of correlation. A modified version of correlation operation is used as the *Laplace* and *Z transforms* to analyze the exponentially decaying and increasing nature of both the output waveforms of a control system and the structure of a control system itself. As mentioned earlier, the algorithm used to decompose curves or other sets of data into more basic analysis functions like polynomials is loosely called *curve fitting*.

Signal Manipulation

Signal manipulation is a very broad term that we use here to refer to the study of linear systems and their ability to modify signals. Whereas in later chapters we will see very sophisticated algorithms that manipulate signals in communication systems, in this chapter, we will limit ourselves to understanding how basic digital filters are specified, analyzed, and designed. The mathematical operation that computes the output of a linear system, such as a digital filter, given the input signal and system coefficients is called *convolution*.

The basic linear system that we will work with in this book is the finite impulse response, or *FIR*, filter. The most fundamental technique of FIR filter design is called the frequency sampling method, which we modify in later chapters to introduce optimal FIR filters. Note that we will not explore *IIR* (infinite impulse response) filters until a future revision of this text. Although they are more cumbersome to design and lack the flexibility of the FIR filter, they do require fewer hardware resources. However, with the advent of deep submicron processes, this advantage is nowadays much less pronounced.

2.1   Convolution

Convolution is the mathematical operation that computes the time domain output waveform, $y[n]$, of a LTI (linear time-invariant) system given an arbitrary time domain input signal, $x[n]$, and the impulse response, $h[n]$, of the system. LTI systems in the realm of discrete math are also called LSI (linear sample-invariant) systems.

2.1.1   The LTI Model

The model that we will use to express LTI systems needs to include three central features that make it suitable for our purposes.

1. _Memory_

The model used to approximate linear time-invariant systems must be able to represent processes that are differential / discrete in nature. Systems described by differential or discrete equations are unique in that they remember and use past inputs to produce current outputs. In electrical engineering, we may think of a lumped element circuit featuring capacitors and inductors whose voltages and currents depend on circuit conditions in the past. The model that we use must therefore have memory.

2. _Time Invariance_

To keep our model simple enough, we assume that the coefficients describing the differential / discrete equations do not change with time. The assumption that the capacitance, inductance, and resistance values of a circuit remain constant during operation is entirely reasonable. A system whose characteristics remain constant over time is called time-invariant. Systems whose characteristics change over time will be addressed in the section on adaptive signal processing.

3. _Linearity_

Our system model must be linear in order to keep the mathematical rigor involved in analyzing such systems at a level that is reasonable. Although virtually all systems become non-linear when their inputs are excessively large, many systems act linearly over their useful range of operation.

The structure we use to model a discrete LTI system is the transversal filter, shown next. It features time delays marked D (or z^{-1}), which are arranged serially and delay their input values by one sample period. The terminals of these delays, also called taps, are multiplied by constants h_0 through h_{N-1} and summed to yield the output $y[n]$. It is these delay elements that provide the model with the required memory. Further, the model is time-invariant if factors h_0 through h_{N-1} remain constant.

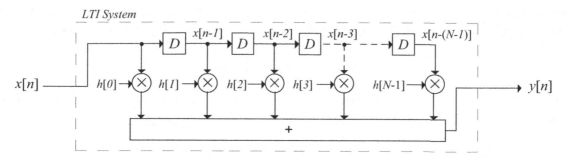

Figure 2-3: Linear Time-invariant System Model (The Transversal Filter)

At any one time instance n, the value $x[n]$ will reside at the input of our model, whereas $y[n]$ will reside at its output. Those samples that have appeared at the input in the past ($x[n-1]$, $x[n-2]$ ...) have already traveled down the delay line and are multiplied by the model's coefficient vector $h = [\,h[0],\ h[1]\ ...\ h[N-1]\,]$. As is evident in the figure above, $y[n]$ is calculated as follows.

$$y[n] = h[0]{\cdot}x[n] + h[1]{\cdot}x[n\text{-}1] + \cdots + h[N\text{-}1]{\cdot}x[n\text{-}(N\text{-}1)]$$

The simple expression above points us to the first formulation of the convolution operation for an LTI system with N model coefficients.

$$y[n] = x[n] \otimes h[n] = \sum_{k=-\infty}^{\infty} h[k] \cdot x[n-k] = \sum_{k=0}^{N-1} h[k] \cdot x[n-k]$$

The Unit Impulse Response

The unit impulse, $\delta[n]$, is defined as a discrete time sequence of samples that features zero values everywhere except at $n = 0$, where its value is unity.

$$\delta[n] = \begin{vmatrix} 1 & if\ n = 0 \\ 0 & otherwise \end{vmatrix}$$

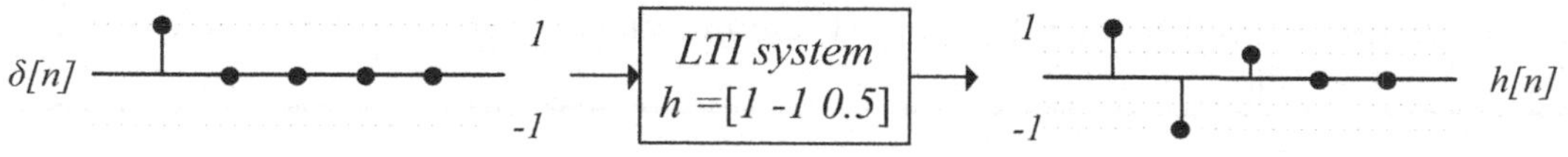

Figure 2-4: Unit Impulse and the Unit Impulse Response of a LTI/LSI System

Looking back at the structure of our linear time-invariant system model in Figure 2-3, we see that the coefficients $h[0]$ through h[N-1] are revealed at the output terminal if we stimulate the input terminal with a unit impulse $\delta[n]$. Simply imagine a value of 1 traveling along the delay path of our model. The resulting output, which is the set of internal model coefficients, $h[n]$, is appropriately called the *unit impulse response* of the system. In the equation below, we show the convolution of a time-shifted unit impulse $\delta[n-k]$ and our model coefficient vector $h[n]$.

$$y[n] = \delta[n-k] \otimes h[n] = h[n-k]$$

The impulse response is the coefficient vector $h[n]$, while the response of a time-shifted impulse $\delta[n-k]$ is a time-shifted version of the coefficient vector $h[n-k]$. No surprises here.

Convolution in Matrix Form

In later chapters, we will work with the convolution operation in matrix form, so let us reformulate our equation for N model coefficients accordingly. As done in the MatLab code, X and H are both defined as column vectors.

$$y[n] = X^T[n] \cdot H$$

$$y[n] = [x[n] \quad x[n-1] \quad \cdots \quad x[n-(N-1)]] \cdot \begin{bmatrix} h[0] \\ h[1] \\ \vdots \\ h[N-1] \end{bmatrix}$$

Expanding our formulation for M outputs at time indices $n = 0, 1 \ldots M\text{-}1$ changes the expression by expanding the number of columns of X from one to M. The matrix expression below computes all M outputs from $n = 0, 1 \ldots M\text{-}1$.

$$Y = X^T \cdot H$$

$$\begin{bmatrix} y[n=0] \\ y[n=1] \\ \vdots \\ y[n=M-1] \end{bmatrix} = \begin{bmatrix} x[0-0] & x[0-1] & x[0-2] & \cdots & x[0-(N-1)] \\ x[1-0] & x[1-1] & x[1-2] & \cdots & x[1-(N-1)] \\ \vdots & \vdots & \vdots & \ddots & \vdots \\ x[M-1-0] & x[M-1-1] & x[M-1-2] & \cdots & x[M-1-(N-1)] \end{bmatrix} \cdot \begin{bmatrix} h[k=0] \\ h[k=1] \\ h[k=2] \\ \vdots \\ h[k=N-1] \end{bmatrix}$$

$$Mx1 \qquad\qquad\qquad\qquad MxN \qquad\qquad\qquad\qquad Nx1$$

> The model of the linear time-invariant system with its delays, multipliers, and adders is also the structure we use to implement FIR filters. Therefore, the convolution expression computes the output of any FIR filter we can describe.

Example 2.1*: Convolution Example 1*

The code below executes the convolution operation for an input sequence $x[n]$ = [2 2 3 4 2 2 4] and a model coefficient vector h = [3 3 2]. The output produced by the code is $y[n]$ = [6 12 19 25 24 20 22].

```
%% Calculating convolution formulation 1
x   = [2 2 3 4 2 2 4];    %% Our input sequence x[n]
h   = [3 3 2];            %% Coefficients of our system h[n]
y   = zeros(1,length(x)); %% Output y[n]

DelayLine = zeros(1,3);
for n = 1:length(x)
    DelayLine(1,2:3) = DelayLine(1,1:2); %% Increment the delay line
    DelayLine(1,1)   = x(1,n);
    y(1,n)           = DelayLine*h.';
end
display(['Output y[n]: [ ' num2str(y) ' ]']);

%% Calculating convolution formulation 1 in matrix form
X   = [2; 2; 3; 4; 2; 2; 4];   %% Input sequence as column vector
h   = [3 3 2];                 %% Coefficients of our system h[n]
M = length(X);                 %% M = 7  -> Length of input/output sequence
N = length(h);                 %% N = 3  -> Length of the impulse response

H = toeplitz([h zeros(1, M - N)], [h(1,1) zeros(1, M-1)] );
Y = H*X;
display(Y);
```

To continue our example, let us set up the above matrix expression for our previous input sequence $x[n] = [2\ 2\ 3\ 4\ 2\ 2\ 4]$ and model coefficients $h = [3\ 3\ 2]$.

$$Y = X^T \cdot H$$

$$
\begin{bmatrix} y[0] \\ y[1] \\ y[2] \\ y[3] \\ y[4] \\ y[5] \\ y[6] \end{bmatrix} =
\begin{bmatrix} 2 & 0 & 0 \\ 2 & 2 & 0 \\ 3 & 2 & 2 \\ 4 & 3 & 2 \\ 2 & 4 & 3 \\ 2 & 2 & 4 \\ 4 & 2 & 2 \end{bmatrix} \cdot
\begin{bmatrix} 3 \\ 3 \\ 2 \end{bmatrix}
\quad \rightarrow \quad
\begin{bmatrix} y[0] \\ y[1] \\ y[2] \\ y[3] \\ y[4] \\ y[5] \\ y[6] \end{bmatrix} =
\begin{bmatrix} 6 \\ 12 \\ 19 \\ 25 \\ 24 \\ 20 \\ 22 \end{bmatrix}
$$

$$Mx1 \qquad MxN \qquad Nx1$$

2.1.2   The Traditional Formulation for Convolution

Assuming we know the impulse response of a system, it is our goal to express the input sequence $x[n]$ as a series of scaled unit impulses to facilitate the calculation of the final output $y[n]$. The quantity $x[k]$ in the expression below is a sequence of scaling factors, while $x[n]$ is a sequence of scaled unit impulses.

$$x[n] = \sum_{k=-\infty}^{\infty} x[k] \cdot \delta[n-k]$$

Linearity

Because our system is linear, we may apply the superposition principle, which states that an output $y[n]$ produced by a sum of several inputs $x_1[n]$, $x_2[n]$, ... $x_{M-1}[n]$ is the same as the sum of outputs $y_1[n]$, $y_2[n]$, ... $y_{M-1}[n]$ that were produced by separate inputs $x_1[n]$, $x_2[n]$, ... $x_{M-1}[n]$.

$$(a_1 \cdot x_1[n] + a_2 \cdot x_2[n] + \cdots) \otimes h[n] = a_1 \cdot x_1[n] \otimes h[n] + a_2 \cdot x_2[n] \otimes h[n] + \cdots$$

If the response of a single unit impulse $\delta[n]$ is $h[n]$, then the response of a sequence of scaled unit impulses will be a sequence of superimposed and scaled copies of $h[n]$.

$$y[n] = x[n] \otimes h[n] = (\sum_{k=-\infty}^{\infty} x[k] \cdot \delta[n-k]) \otimes h[n]$$

$$= \sum_{k=-\infty}^{\infty} x[k] \cdot (\delta[n-k]) \otimes h[n]) \leftarrow \textit{Due to linearity}$$

$$\boxed{y[n] = x[n] \otimes h[n] = \sum_{k=-\infty}^{\infty} x[k] \cdot h[n-k]}$$

The figure below clearly shows how $x[n]$ is actually made up of separately scaled and time shifted unit impulses, and they in turn cause separately scaled and time-shifted unit impulse responses. In this case, the unit impulse response is $h[n] = [\ 1\ \text{-}1\ \ 0.5]$, the input is $x[n] = [1\ \ 2\ \ \text{-}1\ \ 1]$ and the output of the convolution is $y[n] = [0\ \ 1\ \ 1\ \text{-}2.5\ \ 3\ \ \text{-}1.5\ \ 0.5]$.

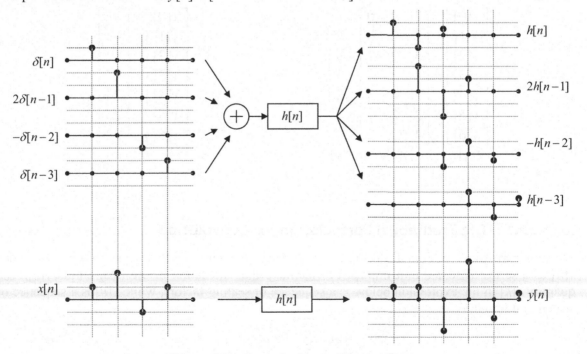

Figure 2-5: Convolution in Picture Form

As we did with the earlier formulation of the correlation operation, here too, we will use matrix notation to express the equation above. The impulse response, $h[n]$, is of length N, and the input sequence $x[n]$ is of length M.

$$Y = H \cdot X$$

$$
\begin{bmatrix} y[n=0] \\ y[n=1] \\ \vdots \\ y[n=M-1] \end{bmatrix} =
\begin{bmatrix}
h[0-0] & h[0-1] & h[0-2] & \cdots & h[0-(M-1)] \\
h[1-0] & h[1-1] & h[1-2] & \cdots & h[1-(M-1)] \\
\vdots & \vdots & \vdots & \ddots & \vdots \\
h[M-1-0] & h[M-1-1] & h[M-1-2] & \cdots & h[M-1-(M-1)]
\end{bmatrix} \cdot
\begin{bmatrix} x[k=0] \\ x[k=1] \\ x[k=2] \\ \vdots \\ x[k=M-1] \end{bmatrix}
$$

$$Mx1 \qquad\qquad\qquad\qquad MxM \qquad\qquad\qquad\qquad Mx1$$

Example 2.2: Convolution Example 2

We use the same impulse response and input sequence of the earlier convolution example and set up the matrix equation $Y = H \cdot X$ to solve for the output vector.

```
x   = [2 2 3 4 2 2 4];      %% Our input sequence x[n]
h   = [3 3 2];              %% Coefficients of our system h[n]
y   = zeros(1,length(x));   %% Output y[n]

M = length(x);              %% M = 7  -> Length of input/output sequence
N = length(h);              %% N = 3  -> Length of the impulse response

H = toeplitz([h zeros(1, M - N)], [h(1,1) zeros(1, M-1)] )
y = H*x.';
```

$$Y = H \cdot X$$

$$
\begin{bmatrix} y[0] \\ y[1] \\ y[2] \\ y[3] \\ y[4] \\ y[5] \\ y[6] \end{bmatrix} =
\begin{bmatrix}
3 & 0 & 0 & 0 & 0 & 0 & 0 \\
3 & 3 & 0 & 0 & 0 & 0 & 0 \\
2 & 3 & 3 & 0 & 0 & 0 & 0 \\
0 & 2 & 3 & 3 & 0 & 0 & 0 \\
0 & 0 & 2 & 3 & 3 & 0 & 0 \\
0 & 0 & 0 & 2 & 3 & 3 & 0 \\
0 & 0 & 0 & 0 & 2 & 3 & 3
\end{bmatrix} \cdot
\begin{bmatrix} 2 \\ 2 \\ 3 \\ 4 \\ 2 \\ 2 \\ 4 \end{bmatrix}
\rightarrow
\begin{bmatrix} y[0] \\ y[1] \\ y[2] \\ y[3] \\ y[4] \\ y[5] \\ y[6] \end{bmatrix} =
\begin{bmatrix} 6 \\ 12 \\ 19 \\ 25 \\ 24 \\ 20 \\ 22 \end{bmatrix}
$$

Take a good look at the expression above and you should readily see that the actual multiplication and additions being performed are identical to those of the first convolution equation and matrix expression. Although the impulse response $h = [\ h[0]\ h[1]\ h[2]\ ]$ features only three indices, we should think of the response as infinite in length, with response values featuring indices less than zero and larger than two being equal to zero. As in the first example, the column vector y yields the following result: $y[n]^T = [6\ 12\ 19\ 25\ 24\ 20\ 22]$.

2.2 The Correlation Algorithm

> The *correlation* operation computes the similarity between two number sequences. It does this by computing the mean of the point-by-point multiplications between the entries of sequence one and the conjugated entries of sequence two.

Given two discrete number sequences, $x[n]$ and $y[n]$, or continuous waveforms, $x(t)$ and $y(t)$, the correlation operation is defined as follows.

$$R_{xy} = \frac{1}{N} \sum_{n=0}^{N-1} x[n] \cdot y[n]^*$$

$$R_{xy} = \frac{1}{T} \int_{t\_start}^{t\_start+T} x(t) \cdot y(t)^* dt$$

When we look at these expressions, two questions should immediately come to mind. How do these expressions detect similarity, and what is that asterisk indicating complex conjugation doing next to $y[n]$ and $y(t)$?

First of all, we should recall that the product of two positive and the product of two negative numbers is always positive. For numbers featuring differing signs, the product will be negative. If $x[n]$ and $y[n]$ are not related, then the point-by-point products of their entries are positive just as often as they are negative and finally average to zero. As the similarity between $x[n]$ and $y[n]$ increases, the chances that both sequences at any one time instant have the same sign is better than 50%. In that case, the equations above will average to larger positive values indicating increased similarity.

The second peculiarity about the equations above is that they use the complex conjugate of the sequence $y[n]$. Remember that any complex number C that is multiplied by its complex conjugate C^* will yield the square of the absolute value of that complex number. $C \cdot C^* = |C|^2$ is always a positive real number. Now, if the sample $x[n]$ is similar to the sample $y[n]$, then $y[n]^*$ will rotate $x[n]$ back towards the positive real axis. Even if the samples are not quite the same, the product will likely have a larger positive real than imaginary part. The complex product, $x[n] \cdot y[n]^*$, whose real part is positive and significantly larger than its imaginary part thus indicates that $x[n]$ and $y[n]$ are similar.

Example 2.3: Correlation Example 1

This example illustrates very nicely why the second sequence must be conjugated in the correlation operation.

$$x[n] = \begin{bmatrix} 1 & i & -1 & -i \end{bmatrix}$$
$$y[n] = \begin{bmatrix} 1 & i & -1 & -i \end{bmatrix}$$

The average of the point-by-point multiplications with and without the conjugation is then computed as follows.

$$R_{xy} = \frac{1}{4}\sum_{n=0}^{3} x[n] \cdot y[n] = \frac{1}{4}\left(1 + i^2 + 1 + i^2\right) = 0$$

$$R_{xy} = \frac{1}{4}\sum_{n=0}^{3} x[n] \cdot y[n]^* = \frac{1}{4}\left(1 + i \cdot (-i) + 1 + (-i) \cdot i\right) = \frac{4}{4} = 1$$

Given that the complex sequences are identical, it should be quite clear that the expression without the conjugation of $y[k]$ is ill-suited for our purpose of determining similarity.

2.2.1   Using Correlation to Detect Sinusoids in Composite Waveforms

One of the major applications of the correlation operation is the detection of a analysis sequence $y[n]$ in a composite waveform $x[n]$. The composite waveform $x[n]$ could be a noise-corrupted copy of $y[n]$, a superposition of many different waveforms, or some other arbitrary number sequence that we wish to analyze. The most popular analysis waveform, $y[n]$, that we would want to find in a communication systems setting is the sinusoid. There are other specialized sequences that find utility in general direct sequence spread spectrum, DSSS, and code division multiple access, CDMA, systems. In the following sections, we will explore each of these applications.

Example 2.4: *Finding a Cosine Sequence in a Composite Waveform*

Find the presence of analysis waveform $y_1[n] = \cos(2\pi \cdot 0.1 \cdot n)$ in the composite signal $x[n] = 2 \cdot \cos(2\pi \cdot 0.1 \cdot n) + 1 \cdot \cos(2\pi \cdot 0.2 \cdot n) + 4 \cdot \cos(2\pi \cdot 0.3 \cdot n)$ of length $N = 60$.

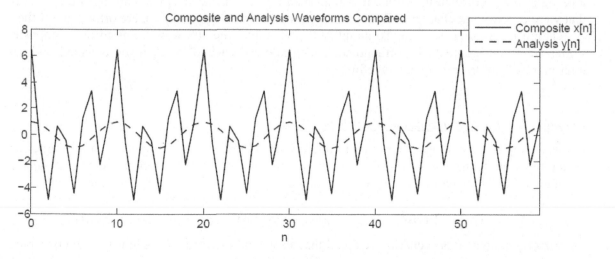

Figure 2-6: Composite and Analysis Waveforms

```
n = 0:59;
y = cos(2*pi*0.1*n);                         %% Analysis waveform
x = 2*cos(2*pi*0.1*n) + 1*cos(2*pi*0.2*n) ... %% Composite waveform
  + 4*cos(2*pi*0.3*n);
Corr = mean(x.*conj(y));                     %% Correlation result
```

The correlation result of 1.0 indicates the presence of the analysis waveform $y_1[n] = \cos(2\pi \cdot 0.1 \cdot n)$. Let us now rerun the test for the analysis waveforms $y_2[n] = \cos(2\pi \cdot 0.2 \cdot n)$, $y_3[n] = \cos(2\pi \cdot 0.3 \cdot n)$, $y_4[n] = \sin(2\pi \cdot 0.1 \cdot n)$ and $y_5[n] = \cos(2\pi \cdot 0.3 \cdot n)$.

$$R1_{xy} = \frac{1}{60} \sum_{n=0}^{59} x[n] \cdot \cos(2\pi \cdot 0.1 \cdot n) = 1.0$$

$$R2_{xy} = \frac{1}{60} \sum_{n=0}^{59} x[n] \cdot \cos(2\pi \cdot 0.2 \cdot n) = 0.5$$

$$R3_{xy} = \frac{1}{60} \sum_{n=0}^{59} x[n] \cdot \cos(2\pi \cdot 0.3 \cdot n) = 2$$

$$R4_{xy} = \frac{1}{60} \sum_{n=0}^{59} x[n] \cdot \sin(2\pi \cdot 0.1 \cdot n) = 0$$

$$R5_{xy} = \frac{1}{60} \sum_{n=0}^{59} x[n] \cdot \cos(2\pi \cdot 0.4 \cdot n) = 0$$

The correlation operation accurately predicts the presence of analysis waveforms one through three by producing a result that is equal to half the magnitude of each embedded sub sequence. The magnitude is only 0.5 of what we expected since $\cos^2(\theta)$ naturally decomposes into $\frac{1}{2} + \frac{1}{2}\cos(2\theta)$, and the second term vanishes in the averaging operation. The correlation produces zero for the last two test waveforms, since they are not present in the compose signal $x[n]$. When looking at these correlation results, it is important to recall the concepts regarding orthogonality that were reviewed in chapter 1. Remember, two waveforms, $x[n]$ and $y[n]$, are orthogonal if their correlation r_{xy} is equal to zero. To simplify this example, the sub-waveforms in our composite signal were configured to be orthogonal, which allowed each of them to be detected without interference from the other sub-waveforms.

Example 2.5: Finding a Sinusoid of Arbitrary Phase

In this example, we will use correlation to find a sinusoid that is neither an exact sine nor an exact cosine function. Unlike in the earlier example, we now want to determine both the magnitude and the angle of the sinusoid of frequency 0.1Hz that is hidden in the composite waveform $x[n]$.

$$x[n] = 2 \cdot cos(2\pi \cdot 0.1 \cdot n + \pi/4) + cos(2\pi \cdot 0.2 \cdot n) + 4 \cdot cos(2\pi \cdot 0.3 \cdot n)$$

To properly analyze this scenario, we recall that a sinusoid of arbitrary angle is the sum of a pure cosine and sine wave, whose respective amplitudes determine the size and angle of the sinusoid. We claim that

---

$$A \cdot cos(\theta) + B \cdot sin(\theta) = |M| \cdot cos(\theta + \varphi)$$

where

$$M = A + jB \quad \text{and} \quad \varphi = -angle(M)$$

Proof: In the derivation below, we will assume that $M = A + jB$.

$$A\cos(\theta) + B\sin(\theta) = \frac{A}{2}(e^{j\theta} + e^{-j\theta}) + \frac{B}{2j}(e^{j\theta} - e^{-j\theta})$$

$$= \frac{1}{2}[e^{j\theta}(A + \frac{B}{j}) + e^{-j\theta}(A - \frac{B}{j})] = \frac{1}{2}[e^{j\theta}(A - jB) + e^{-j\theta}(A + jB)]$$

$$= \frac{1}{2}[e^{j\theta} |M| e^{-j\angle M} + e^{-j\theta} |M| e^{j\angle M}] = \frac{|M|}{2}[e^{j(\theta - \angle M)} + e^{-j(\theta - \angle M)}]$$

$$= |M| \cos(\theta - \angle M)$$

Finding the parameters of the embedded sinusoid at 0.1Hz therefore involves correlating the composite waveform with the cosine and sine test waveforms. The code below computes the correlation for both test waveforms, scales the results by two, and computes the magnitude and negative of the phase as 2 and $\pi/4$ radians.

$$R1_{xy} = \frac{2}{60} \sum_{n=0}^{59} x[n] \cdot \cos(2\pi \cdot 0.1 \cdot n) = 1.414$$

$$R2_{xy} = \frac{2}{60} \sum_{n=0}^{59} x[n] \cdot \sin(2\pi \cdot 0.1 \cdot n) = -1.414$$

$$M = 1.414 - j\,1.414 \qquad |M| = 2 \qquad \varphi = -angle(M) = \pi/4$$

The signal component $2 \cdot cos(2\pi \cdot 0.1 \cdot n + \pi/4)$ was correctly identified by the correlation operation.

```
n  = 0:59;
y1 = cos(2*pi*0.1*n);                    %% Analysis waveform 1
y2 = sin(2*pi*0.1*n);                    %% Analysis waveform 2
x  = 2*cos(2*pi*0.1*n+pi/4) + ...        %% Composite waveform
     1*cos(2*pi*0.2*n) + 4*cos(2*pi*0.3*n);

R1 = mean(x.*conj(y1));                  %% Correlation result 1
R2 = mean(x.*conj(y2));                  %% Correlation result 2

M = 2*R1 + j*2*R2;
[abs(M) -angle(M)]
```

Example 2.6: Finding Complex Sinusoids

Let us take a closer look at the last embedded sinusoid of our composite waveform $x[n]$. From Euler's formula, we know that the term, $4 \cdot cos(2\pi \cdot 0.3 \cdot n)$, will break apart into two complex sinusoids.

$$x[n] = 2 \cdot \cos(2\pi \cdot 0.1 \cdot n + \pi/4) + \cos(2\pi \cdot 0.2 \cdot n) + 4 \cdot \cos(2\pi \cdot 0.3 \cdot n)$$

$$4 \cdot \cos(2\pi \cdot 0.3 \cdot n) = 2 \cdot e^{j2\pi \cdot 0.3 \cdot n} + 2 \cdot e^{-j2\pi \cdot 0.3 \cdot n}$$

It stands to reason that we should be able to find any one of those complex sinusoids, so let us pick $y[n] = exp(2\pi \cdot 0.3 \cdot n)$, and find out whether the correlation operation works here as well.

$$R1_{xy} = \frac{1}{60} \sum_{n=0}^{59} x[n] \cdot conj(e^{j2\pi \cdot 0.3 \cdot n}) = \frac{1}{60} \sum_{n=0}^{59} x[n] \cdot e^{-j2\pi \cdot 0.3 \cdot n} = 2.0$$

The correlation result of 2.0 does indeed match the magnitude of our embedded complex sinusoid. Without knowing it, we just calculated the *discrete time Fourier transform*, an algorithm that searches for complex sinusoids inside test signals. We will examine the Fourier transform in much greater detail shortly.

2.2.2 Using Correlation to Detect Specialized Codes in Received Waveforms

Certain communication systems require us to detect a specialized sequence in a received sample stream, *RXSignal*[n]. This specialized sequence may indicate the beginning of an information packet or indeed represent an information symbol, as in direct sequence spread spectrum communication. The 802.11b WLAN standard [1] uses such a specialized sequence called Barker code, B[n]. In this section, we show how to use correlation to find a Barker code in a noise-corrupted number sequence that is being sampled in real time.

$$B[n] = [1 \ 1 \ 1 \ -1 \ -1 \ -1 \ 1 \ -1 \ -1 \ 1 \ -1]$$

To set up this scenario, we will generate a long sequence of noise values, embed the code B[n] with a length of 11 symbols at some random location within the noise stream, and repeatedly use correlation to find B[n]. The code below inserts B[n] at sample 50 of the *RXSignal*[n] sequence.

```
SNR         = 10;                          %% dB
B           = [1 1 1 -1 -1 -1 1 -1 -1 1 -1]; %% Barker code of length = 11
PowerOfB    = (1/length(B))*mean(B*B');      %% Mean square of B -->
                                             %% Average power
SNRLinear   = 10^(SNR/10);
NoisePower  = PowerOfB/SNRLinear;
AWGNoise    = sqrt(NoisePower)*randn(1,200); %% Creating additive
                                             %% white Gaussian noise
RXSignal    = AWGNoise + [zeros(1,50) B zeros(1,139)];
```

In order for a receiver to detect the sequence B[n] in real time, we slide the incoming data stream *RXSignal*[n] into a shift register with multiplier taps that are equal to the complex conjugate of the sequence B[n]. This structure, which is shown in the figure below, is commonly referred to as a *sliding correlator*. Starting with *RXSignal*[0], we shift the incoming sequence into the shift register on a sample per sample basis and compute the correlation each time.

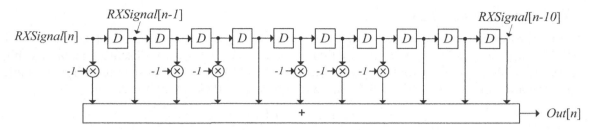

Figure 2-7: Sliding Correlator with Coefficients *C* = *B*[10 - *n*]*

This simple structure forms the corner stone of most synchronization algorithms used in modern base band processors. Note that the tap values *C* = [-1 1 -1 -1 1 -1 -1 -1 1 1 1] are actually flipped right to left compared to those of the sequence *B*[*n*]. When the sequence *B*[*n*], which is embedded in *RXSignal*[*n*], arrives at the shift register, the sample *B*[0] naturally enters the shift register first, followed by *B*[1], *B*[2], and so on. Once the entire sequence has been slid into the shift register, *B*[0] is located at the last tap, and for the entire sequence to be properly lined up with the sequence of tap values, *C* must be the mirror image of *B*[*n*].

```
ShiftRegister = zeros(1,11);
C             = fliplr(B);    %% Shift register tap values = B[10-n]

for i= 1:200
    ShiftRegister(1,2:11) = ShiftRegister(1,1:10); %% Shifting
    ShiftRegister(1,1)    = RXSignal(1,i);         %% operation
    Output(1,i)           = ShiftRegister*C';
end
```

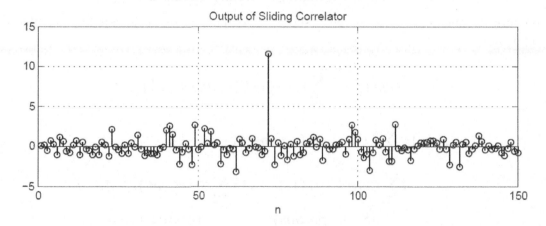

Figure 2-8: The Large Peak of the Sliding Correlator Output Indicates the Presence of B[n]

The large output sample at *n*=61 indicates that the embedded sequence *B*[*n*] has completely entered the shift register and has been detected by the sliding correlator. At that moment, the correlator output should equal 11 plus some small amount of noise. Notice that the specified signal to noise ratio of 10dB is really rather poor, but the correlator still manages to produce a surprisingly high peak at *n* = 61, compared to its output when only noise is present in its shift register. This phenomenon, called processing gain, is discussed next.

The Correlator Processing Gain

Let us for a moment consider a BPSK system transmitting symbols of value ±1 at a rate of 11 MSPS (mega symbols per second). In this scenario, each symbol represents one unique bit, therefore making the data rate and symbol rate equivalent. At the input of the receiver, we sample a combination of the intended symbol, $s[n]$, and superimposed Gaussian noise, $v[n]$.

$$Input[n] = s[n] + v[n]$$

Assuming that both the symbol stream and noise contributions are zero-mean, the signal to noise ratio reduces to the following simple expression.

$$SNR_{Input} = \frac{E(s \cdot s^*)}{E(v \cdot v^*)} = \frac{\sigma_s^2}{\sigma_v^2}$$

Now, rather than assigning one data bit to one symbol, we instead assign one data bit to our 11-symbol Barker code. While 11 million symbols are still transmitted per second, the data rate has fallen from 11Mbps to 1Mbps. Given that a lower data rate is not desirable, there needs to be a significant advantage to doing this. Every 11 symbols, the Barker sequence is properly aligned in the receive correlator, and it is at that instant, that the output features a signal to noise ratio that far exceeds that of the 11Mbps case. This increase in signal to noise ratio, called *processing gain*, enables the reception of much weaker signals.

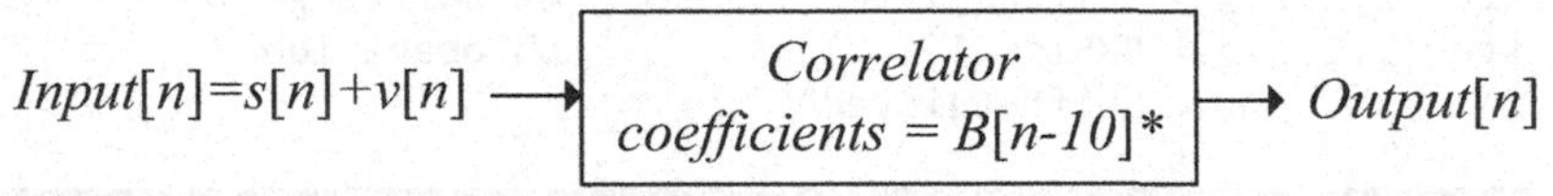

$$Input[n]=s[n]+v[n] \longrightarrow \boxed{\begin{array}{c} Correlator \\ coefficients = B[n\text{-}10]* \end{array}} \longrightarrow Output[n]$$

Figure 2-9: Sliding Correlator Scenario

To properly compute the signal to noise ratio at the output, we first calculate the convolution of the input and the correlator coefficients.

$$Output[n] = \sum_{k=0}^{N-1} CorrCoeff[k] \cdot Input[n\text{-}k]$$

$$Output[n] = \sum_{k=0}^{N-1} CorrCoeff[k] \cdot (s[n-k] + v[n-k])$$

$$= \sum_{k=0}^{N-1} CorrCoeff[k] \cdot s[n-k] + \sum_{k=0}^{N-1} CorrCoeff[k] \cdot v[n-k]$$

(Signal portion)   *(Noise portion)*

Each input symbol, $s[n]$, of value ±1 has an RMS value and standard deviation, σ_s, of 1.0. Every 11th symbol, the correlator output will produce a value equal to $11 \cdot \sigma_s$, and a variance of $(11 \cdot \sigma_s)^2$. The useful output signal power is therefore defined as $N^2 \cdot \sigma_s^2$, where N is the correlator length of 11.

To understand how the noise power increases during the correlation, we show how just two of the 11 noise quantities sum in the correlator.

$$z = v_1 + v_2$$

$$E(z^2) = E((v_1 + v_2)^2)$$
$$= E(v_1^2 + 2v_1 v_2 + v_2^2) = E(v_1^2) + E(2v_1 v_2) + E(v_2^2)$$
$$= \sigma_{v1}^2 + 0 + \sigma_{v2}^2$$
$$= 2\sigma_v^2$$

As we sum all N noise quantities the noise power linearly increases as follows.

$$z = v_1 + v_2 + \dots + v_N$$

$$E(z^2) = N\sigma_v^2$$

To define our processing gain we compare the input and output signal to noise ratios.

$$SNR_{Out} = 10 \cdot \log_{10}(\frac{N^2 \sigma_s^2}{N\sigma_v^2})$$

$$= 10 \cdot \log_{10}(N) + 10 \cdot \log_{10}(\frac{\sigma_s^2}{\sigma_v^2})$$

$$= 10 \cdot \log_{10}(N) + SNR_{In}$$

$$ProcessingGain = 10 \cdot \log_{10}(N) \ dB$$

Communication systems, like 802.11b wireless [1] LAN, exploit correlators when the received signal to noise ratio becomes poor. At low signal to noise ratios, the 802.11b modem will drop from its normal 11 or 22 MBPS speeds to use direct sequence spread spectrum, or DSSS, communication which uses the processing gain provided by its correlators to boost the SNR. The DSSS based throughput drops to 1 or 2 MBPS, but at least communication can continue.

2.2.3   Codes and Sequences Used in Communication Systems

Correlators are heavily used in direct sequence spread spectrum, or DSSS, communication systems, which have found great utility in military applications and later in communication standards such as CDMA, WCDMA and WLAN 802.11b.

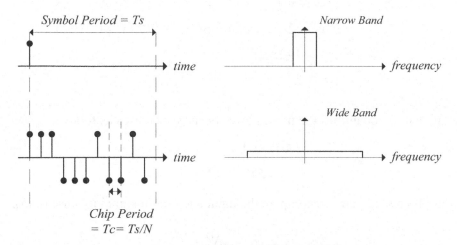

Figure 2-10: Direct Sequence Spread Spectrum Signal

The basic premise is to replace a data symbol, which extends over a period T_s, with a specific code sequence of N other symbols. These other symbols, also called chips, feature a chip period of T_s/N, which naturally leads to a spreading effect on the signal bandwidth, as seen in the figure above. The process of replacing a data symbol by such a sequence is called spreading, while the recovery of the original data symbol by the correlator is called despreading. What appears to be a waste of bandwidth to the casual observer was advantageous in the battlefield. The despreading process mitigated the effect of enemy jamming signals that had previously wiped out the narrow band version of the communication link.

Eventually, the technology made it into commercial cell phone systems that needed to share the same frequency band with older narrow band cell phone infrastructures. These older narrow band cell phone signals looked just like jammers to the spread DSSS waveforms. The commercial implementation, called code division multiple access, or CDMA, also allowed several users to communicate over the same spread bandwidth via orthogonal spreading codes.

The spread waveforms have a further advantage over their narrow band counterparts regarding multipath distortion. In mobile communication systems, transmitted RF signals tend to take different paths by bouncing off different objects to reach the receiver antenna. This multipath effect introduces linear distortion into the older narrow band signals and can only be counteracted by the use of sophisticated equalization schemes. Whereas such equalizers can at best attempt to mitigate multipath distortion, a device called a Rake receiver (used specifically for spread signals) can actually take advantage of the multiple incoming spread signals by cleverly combining them. Because of these and other advantages, wide band communication systems like CDMA and WCMDA all but took over the cell phone industry. Today, the new 4G cell phone architecture called LTE uses a different scheme called OFDM, or orthogonal frequency division multiplexing.

We will meet this new scheme, which will likely displace some of the CDMA infrastructures in the near future, later on in this text.

Example 2.7: *Walsh Codes in CDMA*

Code division multiple access, or CDMA, refers to a technique in which different users are assigned unique spreading codes that they use to communicate over the same frequency channel simultaneously. A popular commercial code family is called Walsh codes [2].

$$Walsh8 = \begin{bmatrix} W(0,8) \\ W(1,8) \\ W(2,8) \\ W(3,8) \\ W(4,8) \\ W(5,8) \\ W(6,8) \\ W(7,8) \end{bmatrix} = \begin{bmatrix} 1 & 1 & 1 & 1 & 1 & 1 & 1 & 1 \\ 1 & -1 & 1 & -1 & 1 & -1 & 1 & -1 \\ 1 & 1 & -1 & -1 & 1 & 1 & -1 & -1 \\ 1 & -1 & -1 & 1 & 1 & -1 & -1 & 1 \\ 1 & 1 & 1 & 1 & -1 & -1 & -1 & -1 \\ 1 & -1 & 1 & -1 & -1 & 1 & -1 & 1 \\ 1 & 1 & -1 & -1 & -1 & -1 & 1 & 1 \\ 1 & -1 & -1 & 1 & -1 & 1 & 1 & -1 \end{bmatrix}$$

Let us assume that a cellular base station needs to communicate with three users simultaneously and assigns these users codes $W(0,8)$, $W(1,8)$, and $W(2,8)$.

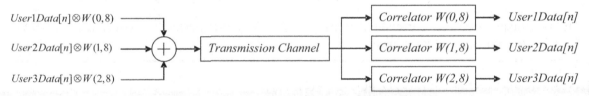

Figure 2-11: CDMA Channel Example

In the code below we will compose $User1Data[n] = [1\ -1\ \ 1]$, $User2Data[n] = [-1\ -1\ \ 1]$, and $User3Data[n] = [-1\ 1\ -1]$ via codes $W(0,8)$, $W(1,8)$, and $W(2,8)$ respectively. The three spread waveforms are combined into a composite signal and sent over the transmission channel.

```
W = [1  1  1  1  1  1  1  1;    %% Defining the Walsh code family
     1 -1  1 -1  1 -1  1 -1;
     1  1 -1 -1  1  1 -1 -1;
     1 -1 -1  1  1 -1 -1  1;
     1  1  1  1 -1 -1 -1 -1;
     1 -1  1 -1 -1  1 -1  1;
     1  1 -1 -1 -1 -1  1  1;
     1 -1 -1  1 -1  1  1 -1];
User1DataW0 = [ 1*W(1,:) -1*W(1,:)  1*W(1,:)];
User2DataW1 = [-1*W(2,:) -1*W(2,:)  1*W(2,:)];
User3DataW2 = [-1*W(3,:)  1*W(3,:) -1*W(3,:)];
Composite = User1DataW0 + User2DataW1 + User3DataW2;
```

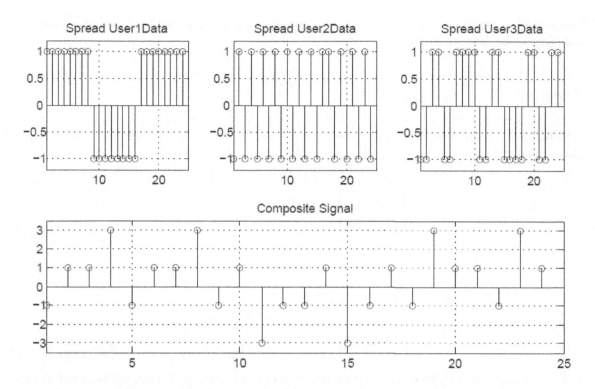

Figure 2-12: Individually Spread and Composite Chip Streams

The two dispreading (or correlation) steps in the code below perfectly recover the *User1Data*, *User2Data*, and *User3Data* sequences.

```
%% Despreading Step 1
T1 = Composite .* conj([W(1,:) W(1,:)  W(1,:)]);
T2 = Composite .* conj([W(2,:) W(2,:)  W(2,:)]);
T3 = Composite .* conj([W(3,:) W(3,:)  W(3,:)]);


%% Despreading Step 2
User1Out = (1/8)*[sum(T1(1,1:8)) sum(T1(1,9:16)) sum(T1(1,17:24))]
User2Out = (1/8)*[sum(T2(1,1:8)) sum(T2(1,9:16)) sum(T2(1,17:24))]
User3Out = (1/8)*[sum(T3(1,1:8)) sum(T3(1,9:16)) sum(T3(1,17:24))]
```

In chapter 1, we had established that two orthogonal random variables, X and Y, feature a common variation, $E(X \cdot Y)$, equal to zero. In the deterministic case here, each sequence in our Walsh code family is orthogonal to all remaining sequences, a fact we express mathematically as follows.

$$R(codeA, codeB) = \sum_{n=0}^{7} codeA[n] \cdot codeB[n]^* = 0$$

It is thanks to this feature of orthogonality that user B, whose receive correlator is programmed to the coefficients of *codeB*, will not experience interference due to users communicating via any of the remaining codes in the orthogonal code family.

Auto and Crosscorrelation Functions for Deterministic Sequences

Given two deterministic sequences $x[n]$ and $y[n]$, the autocorrelation and crosscorrelation functions may be defined as follows.

$$r_x[n] = \frac{1}{N} \sum_{k=-\infty}^{\infty} x[k] \cdot x[k+n]^* \quad \leftarrow Autocorrelation\ Function$$

$$r_{xy}[n] = \frac{1}{N} \sum_{k=-\infty}^{\infty} x[k] \cdot y[k+n]^* \quad \leftarrow Crosscorrelation\ Function$$

As n increases, the two respective sequences appear to slide past one another with the correlation being computed for every new n. When the sequences overlap perfectly, $n = 0$, the equations reduce to the normal correlation expressions we have seen previously. The limits on these equations should be taken with a pinch of salt since the sequences exist only over a finite length N. Once the start and stop indices of $x[n]$ and $y[n]$ are known, the limits can be changed from their values of positive and negative infinity. Sliding sequences past one another is precisely what happens in our sliding correlator. Where we need to detect header or training sequences at the beginning of data streams (using a correlator), we require special sequences, $x[n]$, that feature excellent autocorrelation properties. Good autocorrelation property is achieved if the autocorrelation function produces a large output only at $n = 0$ and nowhere else. For the purpose of comparison, let us take a look at the two 11-symbol sequences below.

$$B[n] = \begin{bmatrix} 1 & 1 & 1 & -1 & -1 & -1 & 1 & -1 & -1 & 1 & -1 \end{bmatrix}$$

$$C[n] = \begin{bmatrix} 1 & 1 & -1 & -1 & 1 & -1 & 1 & 1 & -1 & 1 & -1 \end{bmatrix}$$

As we can see next, the autocorrelation function of the Barker Code, $B[n]$, is extremely well behaved yielding a large result when it perfectly overlays with itself. The auto correlation function for $C[n]$ yields multiple peaks and, had we used it to indicate the start of a data stream, we would be unable to tell just when $C[n]$ has arrived in the sliding correlator that we use for timing synchronization in modems. Clearly, not all sequences lend themselves well to this task.

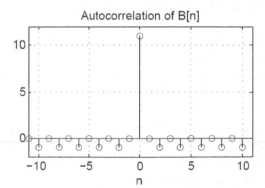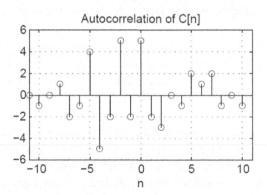

Figure 2-13: Autocorrelation Results of $B[n]$ and $C[n]$

2.3 The Fourier Transform

The Fourier transform [3], or *FT*, is one of the most powerful numerical techniques used in the field of digital signal processing and analysis. In the realm of communication engineering, its purpose is to find the frequency representation of a time domain test signal $x(t)$, which may be either real or complex. The Fourier transform accomplishes this task by detecting the presence of complex sinusoids, $e^{j2\pi ft}$, inside this time domain signal. The detection happens via the correlation process that was the focus of the last section.

$$X(f) = \int_{-\infty}^{\infty} x(t) \cdot (e^{j2\pi ft})^* dt$$

The *FT* output, $X(f)$, indicates the magnitude and orientation (phase) of the detected complex sinusoid. The Fourier transform calculation may be repeated at many different frequencies, f, to produce the spectrum of the signal under test $x(t)$. In the following sections, we will examine several variants of the Fourier transform.

CTFT → The continuous time Fourier transform is primarily an academic tool used to familiarize ourselves with the behavior and properties of the transform. Given that we generally find the transform result via closed form integration, it does not allow us to easily compute the spectrum of just any arbitrary test signal $x(t)$.

DTFT → The discrete time Fourier transform is the discretized counterpart to the CTFT. It requires a sampled input sequence, $x[n]$, which invariably changes the computational problem from one of integration to one of summation. Whereas the input sequence, $x[n]$, must be sampled at discrete time instances, the frequency f at which $X(f)$ is evaluated is continuous between the Nyquist frequency bounds. In plain English: Feel free to evaluate the DTFT at any frequency between $\pm F_{sampling}/2$. The DTFT is easily computed for any arbitrary test sequences, $x[n]$, given the fact that we use summation rather than the more cumbersome integration of the CTFT.

DFT → The discrete Fourier transform is a set of DTFT results, $X[m]$, that were evaluated at certain evenly spaced frequencies. For an input time sequence of length N, the DFT contains N DTFT results. The set of frequency output values, $X[m]$, may be used to recreate the original time based sequence, $x[n]$, via the inverse discrete Fourier transform, or IDFT. Whereas the DFT is used to calculate the spectrum of an arbitrary input sequence, it is the specialized frequency spacing between analysis tones and the relationship to the IDFT that make it unique.

FFT → The fast Fourier transform is an algorithm that efficiently calculates the DFT. Because of the computational structure of the FFT, the input length N must be a number equal to 2^n where n is any positive integer. The FFT and its inverse are immensely important in modem design and constitute the core of the OFDM transceiver.

The Fourier transform is also pivotal in the design of digital filters. Taking the DTFT of a digital filter's impulse response yields its frequency response. More importantly still, almost all FIR filter design methods use the inverse discrete Fourier transform in some way. Therefore, beyond being a signal analysis tool, the Fourier transform is a digital filter design tool.

2.3.1 The Continuous Time Fourier Transform, or CTFT

The continuous time Fourier transform multiplies an input/test signal, $x(t)$, by a complex analysis tone, $exp(-j2\pi ft)$, to form a waveform that is first integrated and then normalized. As mentioned before, the Fourier transform is in fact the correlation between $x(t)$ and $exp(j2\pi ft)$ at some particular test frequency f. The value T in the equation below indicates the length over which $x(t)$ is non-zero.

$$X(f) = \frac{1}{T} \int_{-\infty}^{\infty} x(t) \cdot (e^{j2\pi ft})^* dt$$

$$= \frac{1}{T} \int_{-\infty}^{\infty} x(t) \cdot e^{-j2\pi ft} dt$$

To make the equation more readable we may express the complex analysis tone in terms of the cosine and sine functions.

$$X(f) = \frac{1}{T} \int_{-\infty}^{\infty} x(t) \cdot e^{-j2\pi ft} dt$$

$$= \frac{1}{T} \int_{-\infty}^{\infty} x(t) \cdot \cos(2\pi ft) dt - j\frac{1}{T} \int_{-\infty}^{\infty} x(t) \cdot \sin(2\pi ft) dt$$

Even though practical input signals, $x(t)$, are always time-limited, the equation for the Fourier transform integrates the point-by-point multiplications from minus to plus infinity. This notation means to remind us that even if the integration limits change later, owing to the limited time range of $x(t)$, the analysis tones $exp(j2\pi ft)$ are defined for t equal to minus to plus infinity. Since they are defined as such, $exp(j2\pi ft)$ will be orthogonal to any other complex analysis waveform, $exp(j2\pi(f+f_o)t)$ no matter how small f_o turns out to be. Remember that two waveforms are orthogonal if their cross-correlation is equal to zero.

$$r_{xy} = E(e^{j2\pi(f+f_o)t} \cdot (e^{j2\pi ft})^*)$$

$$= E(e^{j2\pi(f+f_o)t} \cdot e^{-j2\pi ft}) = E(e^{j2\pi f_o t})$$

$$=_{lim\ T\to\infty} \frac{1}{T} \int_{Tstart}^{Tstart+T} e^{j2\pi f_o t} dt = 0$$

The final Fourier transform formulation includes the time limits on the integral.

$$\boxed{X(f) = \frac{1}{T} \int_{Tstart}^{Tstart+T} x(t) \cdot e^{-j2\pi ft} dt}$$

Let us work a few examples using the continuous version of the Fourier transform.

Example 2.8: *The Complex Sinusoid*

Here we compute the continuous Fourier transform of a complex sinusoid, *exp(j2πf₀t)*, which extends from negative to positive infinity.

$$X(f) = \frac{1}{T} \int_{Tstart}^{Tstart+T} e^{j2\pi f_o t} \cdot e^{-j2\pi ft} dt$$

$$= \frac{1}{T} \int_{Tstart}^{Tstart+T} e^{j2\pi (f_o - f)t} dt$$

Note that the mean value, which is in fact what we are calculating above, of the complex exponential *exp(j2π(f₀-f)t) = cos(2π(f₀-f)t) + j·sin(2π(f₀-f)t)* is zero, except for the case when *f₀=f*. In that case, the complex exponential collapses to *exp(j2π(0)t) = cos(0) + j·sin(0) = 1*. For the case when *f₀=f*, we integrate 1.0 over the time period *T* and scale the result by *1/T* to get back to unity.

$$X(f) = \begin{vmatrix} 1 & if & f_o = f \\ 0 & otherwise \end{vmatrix}$$

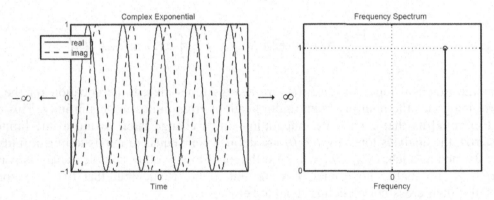

Figure 2-14: CTFT of Complex Exponential Function *x(t) = exp(j2πf₀t)*

Example 2.9: *The Gate or Rectangular Window Function*

Now compute the continuous Fourier transform for the rectangular window function defined as unity over the time interval *T*.

$$x(t) = \begin{vmatrix} 1 & for & -T/2 \leq t \leq T/2 \\ 0 & otherwise \end{vmatrix}$$

Therefore,

$$X(f) = \frac{1}{T} \int_{-T/2}^{T/2} 1 \cdot e^{-j2\pi ft} dt = \frac{1}{T} \left(\frac{e^{-j2\pi ft}}{-j2\pi f} \right) \Big|_{-T/2}^{T/2}$$

$$X(f) = \frac{1}{T}\left(\frac{e^{-j2\pi f(T/2)}}{-j2\pi f} - \frac{e^{-j2\pi f(-T/2)}}{-j2\pi f}\right)$$

$$= -\frac{1}{T}\frac{1}{2j}\left(\frac{e^{-j2\pi f(T/2)}}{\pi f} - \frac{e^{j2\pi f(T/2)}}{\pi f}\right)$$

$$= \frac{1}{T}\frac{1}{2j}\left(\frac{e^{j2\pi f(T/2)}}{\pi f} - \frac{e^{-j2\pi f(T/2)}}{\pi f}\right)$$

$$= \frac{1}{T}\frac{\sin(\pi fT)}{\pi f}$$

$$= \frac{\sin(\pi fT)}{\pi fT} = sinc(\pi fT)$$

The *sinc* function is something that we see again and again in signal processing since it indicates how the frequency spectrum of a signal is affected by time truncation.

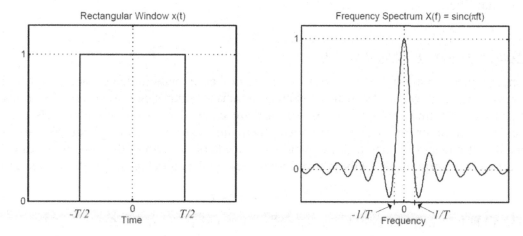

Figure 2-15: CTFT of Gate Function with Time Limit T

Example 2.10: The Unit Impulse Function

The unit impulse function, $\delta(t)$, is the ideal stimulus for the excitation and characterization of linear systems such as filters. Although the unit impulse is more a mathematical quantity rather than a real signal that we could generate in practice, its properties and use reveal invaluable information to us as engineers and scientists. Let us define this interesting function and evaluate its Fourier transform.

→ The unit impulse function, $\delta(t)$, is equal to zero everywhere except at $t = 0$.
→ The unit impulse function, $\delta(t)$, integrates to a value of 1.0.

Whereas the continuous version of the impulse lends itself best to mathematical integration operation as done in the Fourier transform that follows, the discrete version of the impulse $\delta[n]$ is perfectly applicable in practical situations, as we have seen in the section on convolution.

$$X(f) = \frac{1}{T} \int_{-\infty}^{\infty} \delta(t) \cdot e^{-j2\pi ft} dt$$

$$= \frac{1}{T} \int_{-\infty}^{\infty} \delta(t) \cdot (\cos(2\pi f 0) - j \sin(2\pi f 0)) dt$$

$$= \frac{1}{T} \int_{-\infty}^{\infty} \delta(t) dt$$

$$= \frac{1}{T}$$

The Fourier transform of the unit impulse is a constant, which indicates that complex sinusoids, *exp(j2πft)*, at all frequencies are equally represented in the function. Taking the Fourier transform of the impulse response of a filter demonstrates how all frequencies are affected by the transit through that filter.

The Fourier Transform of Real Signals

The spectrum of a real input signal, *x(t)*, is unique in that its magnitude components will always be an even function versus frequency, whereas its phase components is an odd functions. The Fourier transform finds complex sinusoids that are embedded in our input signals at certain frequencies *f*. In order for the signal to be real, there must exist an additional complex sinusoid at the negative frequency *–f* such that the imaginary portion of the sum of the two sinusoids cancels. The Fourier transform results at these two frequencies are the complex conjugates of one another.

$$X(f) = X(-f)^*$$

The most obvious real function that demonstrates this fact is the $cos(2\pi f_o t) = \frac{1}{2} \cdot e^{j2\pi f_o t} + \frac{1}{2} \cdot e^{-j2\pi f_o t}$ expression. The two sinusoids rotate in opposite directions and are guaranteed to be complex conjugates of one another since they both start at an angle of zero. For a slightly more challenging real function, let us look at *sin(2πf_o t)*, which features two complex sinusoids rotating in opposite directions but starting at different phases.

$$\sin(2\pi f_o t) = \frac{1}{2j}(e^{j2\pi f_o t} - e^{-j2\pi f_o t})$$

$$= -j\frac{1}{2}(e^{j2\pi f_o t} + e^{-j2\pi f_o t}e^{j\pi})$$

$$= e^{-j\frac{\pi}{2}}\frac{1}{2}(e^{j2\pi f_o t} + e^{-j2\pi f_o t}e^{j\pi})$$

$$= \frac{1}{2}(e^{j2\pi f_o t} \cdot e^{-j\frac{\pi}{2}} + e^{-j2\pi f_o t} \cdot e^{j\frac{\pi}{2}})$$

In the next figure, we can see two complex positions that are changing with time. The first one, *½·exp(j2πf_o t)·exp(-jπ/2)*, rotates in the positive frequency or counter clockwise direction and starts out at an angle of -90 degrees. The other, *½·exp(j2πf_o t)·exp(-jπ/2)*, rotates in the negative frequency or clockwise direction and starts at an angle of 90 degrees. At any instant in time these sinusoids sum to a value located on the real axis. Real signals are composed of many such pairs of complex sinusoids, each rotating at different frequencies, featuring different magnitudes and starting at different angles.

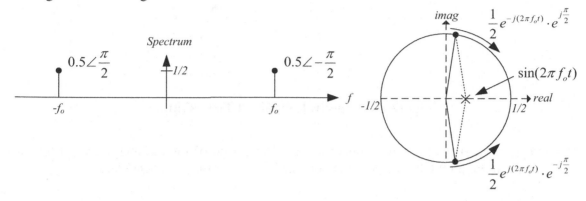

Figure 2-16: The Decomposition of *sin(2πf₀t)* into a Pair of Complex Sinusoids

The Time-shifting Property

The time-shifting property of the Fourier transform illustrates the change in the frequency spectrum of a signal as it is time-shifted from $x(t)$ to $x(t-t_o)$. Given that

$$FT(x(t)) = \frac{1}{T} \int_{-\infty}^{\infty} x(t) \cdot e^{-j2\pi ft} dt = X(f)$$

find

$$FT(x(t-t_0)) = \frac{1}{T} \int_{-\infty}^{\infty} x(t-t_0) \cdot e^{-j2\pi ft} dt = X_{new}(f)$$

By introducing a simple change in variables $z = t-t_o$, the equation changes as follows.

$$X_{new}(f) = \frac{1}{T} \int_{-\infty-t_o}^{\infty-t_o} x(z) \cdot e^{-j2\pi f(z+t_o)} dz$$

$$= e^{-j2\pi ft_o} \frac{1}{T} \int_{-\infty}^{\infty} x(z) \cdot e^{-j2\pi fz} dz$$

$$= e^{-j2\pi ft_o} X(f)$$

Therefore, delaying or time advancing a time domain signal causes a phase shift, which changes linearly as a function of frequency.

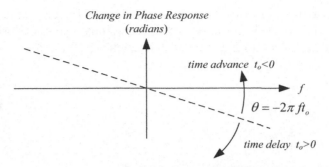

Figure 2-17: Spectral Impact of Time-shifting

This simple property is the key to understanding the design of digital interpolation filters and timing acquisition in OFDM modems, both of which we will discuss in detail later.

The Frequency-shifting Property

The frequency-shifting property of the Fourier transform illustrates the change in frequency spectrum of an input signal $x(t)$ after it has been multiplied by a complex sinusoid $exp(j2\pi f_o t)$.

Given that

$$FT(x(t)) = \frac{1}{T}\int_{-\infty}^{\infty} x(t) \cdot e^{-j2\pi ft} dt = X(f)$$

then

$$FT(x(t) \cdot e^{j2\pi f_o t}) = \frac{1}{T}\int_{-\infty}^{\infty} (x(t) \cdot e^{j2\pi f_o t}) \cdot e^{-j2\pi ft} dt$$

$$= \frac{1}{T}\int_{-\infty}^{\infty} x(t) \cdot e^{-j2\pi (f-f_o)t} dt$$

$$= X(f - f_o)$$

A positive f_o value pushes the spectrum of the input signal $x(t)$ to the right side while a negative value moves it left.

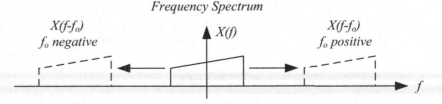

Figure 2-18: Spectral Impact of Frequency Shifting

Example 2.11*: The Frequency-shifted Gate Function*

We now take the gate function of Example 2.9 and multiply it by a complex tone $exp(j2\pi f_o t)$. According to the frequency-shifting property of the Fourier transform, the *sinc* response we saw in the example will simply be moved in frequency by f_o.

$$X(f) = \frac{1}{T} \int_{-T/2}^{T/2} (1 \cdot e^{j2\pi f_o t}) \cdot e^{-j2\pi ft} dt$$

$$= \frac{1}{T} \int_{-T/2}^{T/2} e^{-j2\pi(f-f_o)t} dt$$

This integration is virtually identical to that seen in Example 2.9, with the exception of the frequency f, which now has been replaced by the quantity f-f_o. Therefore, instead of the *sinc* function being centered at zero Hertz, it now appears at f_o.

$$X(f) = sinc(\pi(f - f_o)T)$$

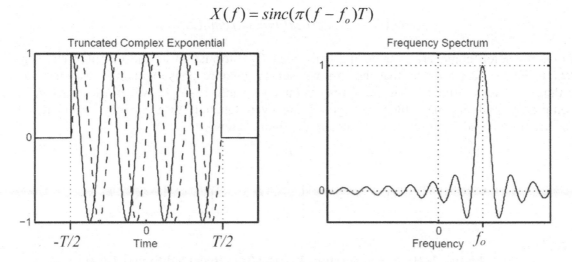

Figure 2-19: CTFT of a Time-truncated Complex Exponential Function with Time Limit *T*

The figure above shows us the frequency-shifted gate function, which looks a lot like a time-limited sinusoid. They are one and the same. Time domain signals, which are truncated or limited in time, will have spectra that extend from positive to negative infinity in the frequency domain.

Example 2.12*: Frequency-shifting in a Zero-IF Receiver*

Using our understanding of real signals and the frequency-shifting property of the Fourier transform, we can now examine the spectral changes of an RF signal as it works its way through a communication receiver.

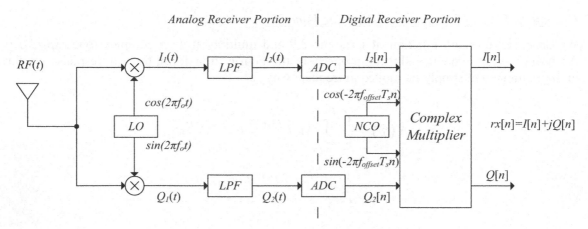

Figure 2-20: Typical Zero-IF Receiver Chain

Even though the received signal $RF(t)$ carries I and Q information, it is a completely real valued waveform as should be clear given the lack of the operator j.

$$RF(t) = I(t) \cdot cos(2\pi f_c t) + Q(t) \cdot sin(2\pi f_c t)$$

Communication engineers refer to the embedded IQ information as the complex envelope of $RF(t)$, but the fact remains that the in-phase and quadrature phase streams are riding on two orthogonal carrier signals that are simply summed up into a single waveform. Since $RF(t)$ is completely real, its magnitude spectrum is an even function of frequency, while its phase spectrum is an odd function, as illustrated in the figure below.

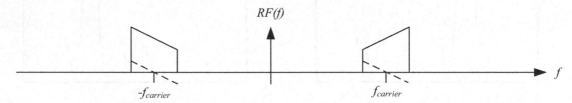

Figure 2-21: The Spectrum, $X_{RF}(f)$, of the Received Signal $RF(t)$

Given the expression and spectrum of $RF(t)$ shown above, we now formulate the respective signal equations at nodes $I_1(t)$ and $Q_1(t)$.

$$I_1(t) = RF(t) \cdot cos(2\pi f_o t) = RF(t) \cdot \frac{1}{2}(e^{j2\pi f_o t} + e^{-j2\pi f_o t})$$

$$Q_1(t) = RF(t) \cdot sin(2\pi f_o t) = RF(t) \cdot \frac{1}{2}(e^{j2\pi f_o t} \cdot e^{-j\frac{\pi}{2}} + e^{-j2\pi f_o t} \cdot e^{j\frac{\pi}{2}})$$

According to the frequency-shifting property of the Fourier transform, the frequency spectrum of $RF(t)$ is translated by the cosine and sine functions in the following way.

$$I_1(f) = \frac{1}{2}RF(f - f_o) + \frac{1}{2}RF(f + f_o)$$

$$Q_1(f) = \frac{1}{2}RF(f - f_o)e^{-j\frac{\pi}{2}} + \frac{1}{2}RF(f + f_o)e^{j\frac{\pi}{2}}$$

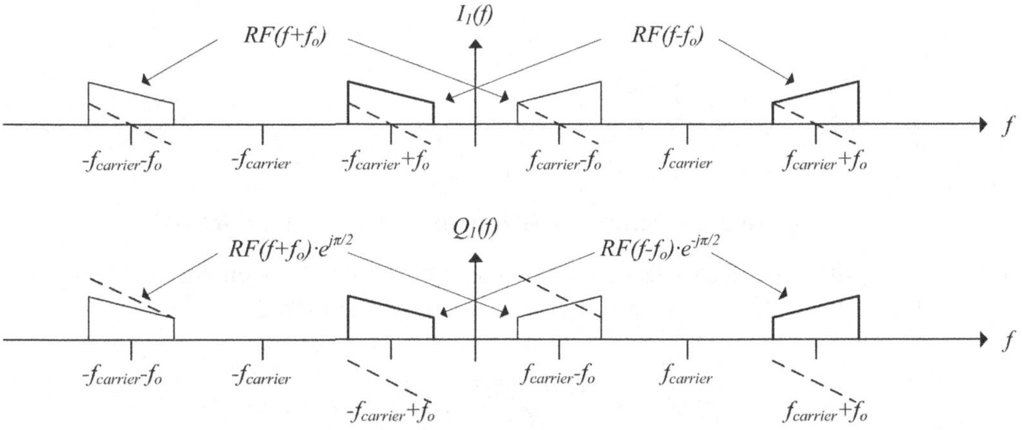

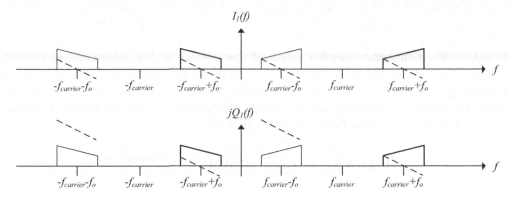

Figure 2-22: Spectra of $I_1(t)$ and $Q_1(t)$

Notice the phase shifts experienced by the signal $Q_1(f)$ which result from the multiplication by the sine wave. We think of the $I_1(t)$ and $Q_1(t)$ waveforms as the real and imaginary components of a complex signal, $C_1(t) = I_1(t) + jQ_1(t)$. To appreciate the spectra of this complex signal we must take into account the rotation by 90 degrees due to j term in front of the $Q_1(t)$ term.

Figure 2-23: Spectra of $I_1(t)$ and $jQ_1(t)$

$$jQ_1(f) = e^{j\frac{\pi}{2}} \cdot (\frac{1}{2}RF(f - f_o)e^{-j\frac{\pi}{2}} + \frac{1}{2}RF(f + f_o)e^{j\frac{\pi}{2}})$$

$$= \frac{1}{2}RF(f - f_o) + \frac{1}{2}RF(f + f_o)e^{j\pi}$$

$$= \frac{1}{2}RF(f - f_o) - \frac{1}{2}RF(f + f_o)$$

Summing the two spectra yields a result of $RF(f-f_o)$, which we might have expected from the frequency shifting property had the LO signal be the complex waveform $exp(j2\pi f_o t)=$

$\cos(2\pi f_o t) + j\sin(2\pi f_o t)$. The RF down-conversion structure of Figure 4.20 is such that it can be modeled as a complex frequency translator.

$$C_1(f) = I_1(f) + jQ_1(f) = RF(f - f_o)$$

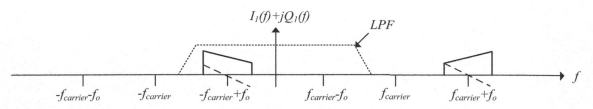

Figure 2-24: Summed Spectrum of $C_1(t) = I_1(t) + jQ_1(t)$

We now low pass filter the higher frequency image and digitize the signal to arrive at the quantity $C_2[n] = I_2[n]$ and $Q_2[n]$. The image at $f_{carrier} + f_o$ is thereby eliminated.

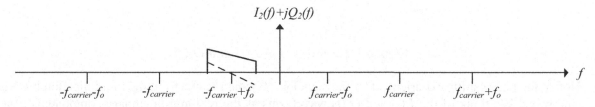

Figure 2-25: Summed Spectrum of $C_2(t) = I_2(t) + jQ_2(t)$

The figure above clearly shows that the spectrum of our complex signal $I_2[n] + jQ_2[n]$ has not been centered correctly, as the transmit LO frequency, $f_{carrier}$, and that of the receiver, f_o, were not identical. This remaining frequency offset, $-f_{carrier} + f_o$, is eliminated by multiplying the signal $I_2[n] + jQ_2[n]$ by a discrete complex sinusoid that rotates at the frequency offset but in the negative direction. All being well, we'll get it right this time. Actually, the base band processor goes through quite a bit of effort to figure out what this final frequency offset is. The process of computing the final frequency offset is part of the hardware's synchronization effort. Let us assume here that we know what that frequency offset is.

$$I[n] + jQ[n] = (I_2[n] + jQ_2[n]) \cdot e^{j2\pi \cdot (-FreqOffset) \cdot n \cdot Ts}$$

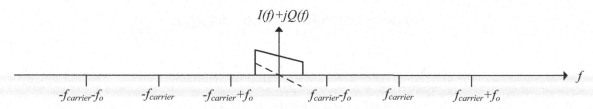

Figure 2-26: Spectrum after Final Digital Frequency Shift

Whereas the receiver *LO* signal provided the initial frequency translation, it is now the job of the numerically controlled oscillator, *NCO*, to shift the spectrum such that it is located squarely at the frequency origin. The *NCO* resides in the digital portion of the receiver and produces two signals

NCO_{real} and NCO_{imag} where T_s and n refer to the sample period and the discrete time index, respectively.

$$NCO_{real}[n] = \cos(2\pi \cdot (-FreqOffset) \cdot n \cdot T_s)$$
$$NCO_{imag}[n] = \sin(2\pi \cdot (-FreqOffset) \cdot n \cdot T_s)$$

The NCO is a simple device consisting mainly of cosine and sine ROM tables, whose content is being repeatedly accessed to form the sinusoid at the correct frequency. The final product $rx[n] = I[n] + jQ[n]$ is formed as follows.

$$rx[n] = I[n] + jQ[n] = (I2[n] + j \cdot Q_2[n]) \cdot (NCO_{real}[n] + j \cdot NCO_{imag}[n])$$

$$I[n] = I2[n] \cdot NCO_{real}[n] - Q_2[n] \cdot NCO_{imag}[n]$$
$$Q[n] = I2[n] \cdot NCO_{imag}[n] + Q_2[n] \cdot NCO_{real}[n].$$

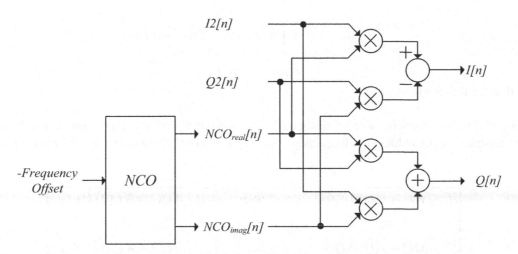

Figure 2-27: Complex Multiplier as Implemented in a Baseband Processor

The Convolution Property of the Fourier Transform

The convolution property of the Fourier transform directly points to our first and simplest method of digital filter design. Given that the two time domain waveforms $x(t)$ and $h(t)$ transform into $X(f)$ and $H(f)$, the convolution property of the Fourier transform makes the following claim.

$$If \quad y(t) = x(t) \otimes h(t) \qquad then \qquad Y(f) = X(f) \cdot H(f)$$

The property shows that if an input signal $x(t)$ is convolved with or processed by an LTI system with impulse response $h(t)$, then the spectrum, $Y(f)$, of the output signal $y(t)$ is the product of $X(f)$ and $H(f)$. The property is suggesting that if we knew how to get back from $H(f)$ to $h(t)$, we could influence the signal spectrum of our signal $x(t)$ at will. The technique of moving back from $H(f)$

to $h(t)$ is called the inverse Fourier transform, which we will meet in the section on the DFT. Proving the convolution property is actually not as difficult as it may look.

$$FT(x(t) \otimes h(t)) = \int_{-\infty}^{\infty} [x(t) \otimes h(t)] \cdot e^{-j2\pi ft} dt$$

$$= \int_{-\infty}^{\infty} [\int_{-\infty}^{\infty} x(\tau) \cdot h(t-\tau) d\tau] e^{-j2\pi ft} dt$$

We now rearrange the integrals and use the time-shifting property of the Fourier transform to finish the derivation.

$$FT(x(t) \otimes h(t)) = \int_{-\infty}^{\infty} x(\tau) [\int_{-\infty}^{\infty} h(t-\tau) e^{-j2\pi ft} dt] d\tau$$

$$= \int_{-\infty}^{\infty} x(\tau) H(f) \cdot e^{-j2\pi f\tau} d\tau$$

$$= \int_{-\infty}^{\infty} x(\tau) \cdot e^{-j2\pi f\tau} d\tau \cdot H(f)$$

$$= X(f) \cdot H(f)$$

And that is all there is to it.

The second part of the convolution property of the Fourier transform states that multiplication in time translates to convolution in frequency. The two parts of the property are summarized below.

if $y(t) = x(t) \otimes h(t)$		*then*	$Y(f) = X(f) \cdot H(f)$
if $y(t) = x(t) \cdot h(t)$		*then*	$Y(f) = X(f) \otimes H(f)$

2.3.2   The Discrete Time Fourier Transform, or DTFT

The discrete time Fourier transform, or DTFT, computes the correlation between a sampled, or discrete, test waveform, $x[n]$, and a sampled complex sinusoid, $exp(j2\pi nf)$, which serves as the analysis waveform.

$$X(f) = \frac{1}{N} \sum_{n=0}^{N-1} x[n] \cdot e^{-j2\pi n f}$$

$$= \frac{1}{N} \sum_{n=0}^{N-1} x[n] \cdot (\cos(2\pi n f) - j\sin(2\pi n f))$$

$X(f)$ → is a complex scalar representing the DTFT output at frequency f.
f → is a continuous variable representing normalized frequencies between -0.5 and 0.5 Hz.
$x[n]$ → is the discrete input sequence.
n → is the time index (causes the limit assignment / usually $n = 0,1 \ldots N$-1).
N → is the length of the input sequence $x[n]$.

Note
→ The input, $x[n]$, of the DFT must be a sequence of evenly spaced time domain samples.
→ The output $X(f)$ is a complex scalar that yields the orientation (phase) and extent (magnitude) to which the analysis waveform $exp(j2\pi nf)$ is correlated with or similar to $x[n]$.
→ Whereas $x[n]$ is restricted to a time domain range of 0 through N-1, the complex analysis sinusoid $exp(j2\pi nf)$ extends from negative to positive infinity. The limit assignment on the summation symbol is due only to the finite length of sequence $x[n]$.
→   The DTFT is the method of choice to compute the magnitude spectrum of a signal, which previously we could see only by using a spectrum analyzer. The spectrum is plotted by graphing $|X(f)|$ or $20 \cdot log_{10}(|X(f)|)$.

From the CTFT to the DTFT

Deriving the DTFT equation from that of the CTFT follows a simple progression of just a few steps. The first and most obvious difference between the operations is the change from continuous to discrete time domain. The change reduces the integration to the simpler summation operation.

Change 1 → The continuous variable t is replaced by $n \cdot Ts$ where $n = 0, 1 \ldots N$-1. We've arrived at the first intermediate expression of the DTFT. Because of the sampled nature of the waveform, the frequency range is restricted to the Nyquist bounds of $\pm Fs/2$.

Change 2 → Change $x(n \cdot Ts)$ to $x[n]$. More than anything else, this change in nomenclature nudges us into thinking in terms of number sequences rather than sampled waveforms.

TRANSFORM	EQUATION	FREQUENCY DOMAIN	TIME DOMAIN
CTFT	$$X_{CTFT}(f_{actual}) = \frac{1}{T}\int_{-\infty}^{\infty} x(t)\cdot e^{-j2\pi f_{actual}}\,dt$$	$-\infty < f_{actual} < \infty$	continuous time (t)
1st Intermediate DTFT	$$X_{DTFT}(f_{actual}) = \frac{1}{N}\sum_{n=-\infty}^{\infty} x(n\cdot T_s)\cdot e^{-j2\pi(n\cdot T_s)f_{actual}}$$	$-\frac{Fs}{2} < f_{actual} < \frac{Fs}{2}$	discrete time (n·Ts)
2nd Intermediate DTFT	$$X_{DTFT}(f_{actual}) = \frac{1}{N}\sum_{n=0}^{N-1} x[n]\cdot e^{-j2\pi n\cdot T_s f_{actual}}$$	$-\frac{Fs}{2} < f_{actual} < \frac{Fs}{2}$	discrete index n=0,1...N-1
DTFT	$$X_{DTFT}(f) = \frac{1}{N}\sum_{n=-\infty}^{\infty} x[n]\cdot e^{-j2\pi nf}$$	$-0.5 < f < .5$	discrete index n=0,1...N-1

Figure 2-28: Progression from the CTFT to the DTFT

Although either intermediate DTFT expression above is perfectly valid, textbooks, DSP software and engineers alike prefer to work with a variable, f, representing normalized frequency with a valid range of -0.5Hz < f < 0.5Hz. The variable f_{actual} can be expressed in terms of normalized frequency as $f_{actual} = f\cdot Fs$.

Change 3 → Replace the variable f_{actual} in the 2nd intermediate DTFT expression with $f\cdot Fs$.

The term $exp(-j2\pi nTs\cdot f_{actual})$ changes to $exp(-j2\pi nTs\cdot f\cdot Fs)$ and finally simplifies to $exp(-j2\pi nf)$, which brings us to the final expression of the DTFT. Working with normalized frequency, f, rather than with the actual frequency may seem awkward at first but becomes second nature with a little practice. Just evaluate the DTFT over the range of -0.5Hz and 0.5Hz and then scale the frequency axis by the sampling rate, Fs, when you need to plot the output as a function of actual frequency.

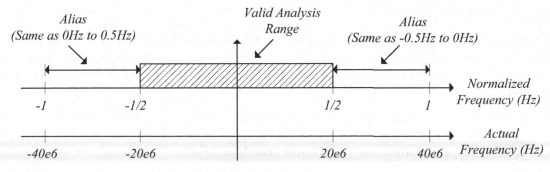

Figure 2-29: Comparison of Normalized and Actual Frequency Ranges for a Sample Rate, Fs, of 40MHz

Example 2.13: *Evaluating Complex Sinusoids*

In the following example, we compute the discrete time Fourier transform for three waveforms of length $N = 20$ sampled at 1Hz. We evaluate the DTFT and plot its magnitude, $|X(f)|$, between frequencies -0.5Hz to 0.5Hz in 0.01Hz steps.

→ $x_1[n] = cos(2\pi n \ 0.1) = 0.5 \ exp(j2\pi n \ 0.1) + 0.5 \ exp(j2\pi n \ (-0.1))$
→ $x_2[n] = exp(j2\pi n \ 0.315)$
→ $x_3[n] = exp(j2\pi n \ 0.1) + 0.01 \cdot exp(j2\pi n \ (-0.25))$

As expected, for test waveform 1, the DTFT has produces two peaks with magnitude 0.5 at ±0.1Hz. The magnitude spectrum does not consist of two lonely peaks, however, but indicates spectrum content at every frequency. Some textbooks call this unexpected content *frequency leakage* of the DTFT, as if some amount of content had leaked away from the two peaks. But there is nothing wrong with the DTFT itself. What you see below really is the magnitude spectrum of a cosine waveform that has been limited in time (see Example 2.11).

Test waveform 2 shows that this leakage is a blessing in disguise. Note that the complex sinusoid of test sequence 2 is located at $f = 0.315$Hz, and given that we ran the DTFT at a frequency resolution of 0.01Hz, the closest places at which we calculated the transform were 0.31 and 0.32Hz (but not 0.315Hz). This frequency leakage effect caused by the time-limiting process has spread frequency content to places where we did in fact run the analysis.

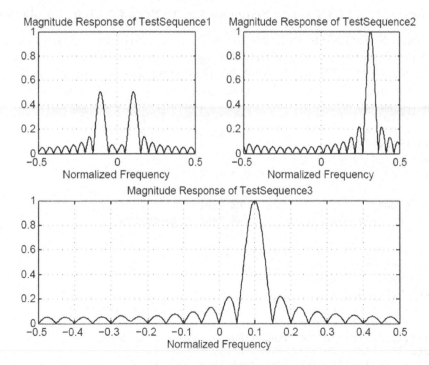

Figure 2-30: Magnitude Results for Test Sequences 1, 2, and 3

The third test sequence exposes a serious problem when analyzing time-limited sequences as the plot below shows. The tone at 0.1Hz is clearly visible, while the one at -.25Hz, which features a magnitude of 0.01, is drowned out by the leakage of the bigger tone. To counteract this problem, we increase the number of samples of test sequence 3 from 20 to 100 and plot both DTFT magnitude spectra in *dB* as $20 \cdot log_{10}(|X(f)|)$. In the second plot, which originated from the 100 sample variant, the tone at -0.25Hz becomes just barely visible, courtesy of a modest improvement in frequency leakage.

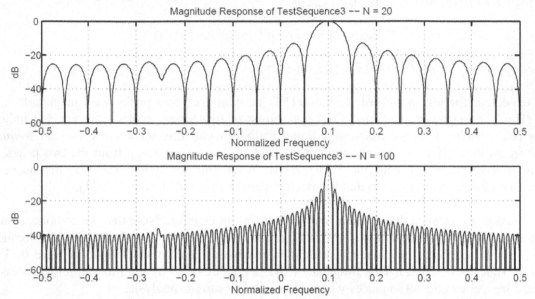

Figure 2-31: DTFT for Test Sequence 3 at *N*=20 and *N*=100 Sample Points

Clearly, we must do something different in order to properly see the magnitude spectrum of signals that have content at vastly different magnitude levels. We could simply increase the number of samples to $N = 200$ and beyond, but the computational load of our DTFT calculation would increase too much. The code below computes the DTFT of all three test waveforms.

```
N            = 20;                      %% Sequence length
n            = 0:1:N-1;                 %% Discrete time index
TestSequence1 = cos(2*pi*0.1*n);
TestSequence2 = exp(j*2*pi*0.315*n);
TestSequence3 = exp(j*2*pi*0.1*n) + 0.02*exp(j*2*pi*(-0.25)*n);


FResolution = 0.001;                    %% Hertz
Frequencies = -0.5:FResolution:0.5;     %% Test frequencies


DTFT_Out1    = zeros(1,length(Frequencies));
DTFT_Out2    = zeros(1,length(Frequencies));
DTFT_Out3    = zeros(1,length(Frequencies));
for i = 1:length(Frequencies)
    f            = Frequencies(1,i); %% Current analysis frequency
    AnalysisTone = exp(j*2*pi*n*f);
    DTFT_Out1(1,i)= (1/N)*TestSequence1*AnalysisTone'; %% Summation
    DTFT_Out2(1,i)= (1/N)*TestSequence2*AnalysisTone';
    DTFT_Out3(1,i)= (1/N)*TestSequence3*AnalysisTone';
end
```

Adding a Window

Applying one of many possible windows has the effect of reducing the frequency leakage caused by time-limiting operation. The next section takes a close look at what types of windows are available and explains how they reduce frequency leakage. In the example below, we will apply the popular Hanning window, $0.5-0.5 \cdot \cos(2\pi(n+1)/(N+1))$, to a long sample record, $s[n]$, to yield $x[n]$.

$$x[n] = s[n] \cdot HanningWindow[n]$$

The performance differences between the Hanning windowed sequence, $x[n]$, and its counterpart that was simply truncated become very clear. Note that simply time-truncating $s[n]$ to $x[n]$ is equivalent to applying a rectangular window, as the next section will show.

Example 2.14: *Advantages of the Hanning over the Default Rectangular Window*

Using the discrete time Fourier transform, we compute the spectrum from -0.5 to 0.5Hz in 0.002Hz steps of the test sequence $x[n] = exp(j2\pi \cdot n \cdot 0.1) + 0.01 \cdot exp(-j2\pi \cdot n \cdot 2.5)$. To truncate $x[n]$ to a length of $N = 40$, we use both rectangular and Hanning windows.

```
N            = 40;                          %% Sequence length
n            = 0:1:N-1;
Rectangular  = ones(1,N);                   %% Rectangular window
Hanning      = (40/20.5)*(0.5- 0.5*cos(2*pi*(n+1)/(N+1)));
                                            %% Scaled Hanning window
TestSequence = exp(j*2*pi*n*0.1) + 0.01*exp(-j*2*pi*n*0.25);
FResolution  = 0.002;                       %% Hertz
Frequencies  = -0.5:FResolution:0.5;        %% Test frequencies
DTFT_OutRect = zeros(1,length(Frequencies));
DTFT_OutHann = zeros(1,length(Frequencies));

for i = 1:length(Frequencies)               %% DTFT calculation
    f            = Frequencies(1,i);        %% Current analysis frequency
    AnalysisTone = exp(j*2*pi*n*f);
    DTFT_OutRect(1,i)=(1/N)*(TestSequence.*Rectangular)*AnalysisTone';
    DTFT_OutHann(1,i)=(1/N)*(TestSequence.*Hanning)*AnalysisTone';
end
MagSpectrumRect = abs(DTFT_OutRect);
MagSpectrumHann = abs(DTFT_OutHann);
```

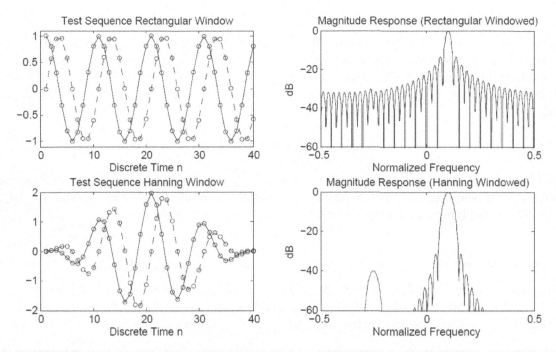

Figure 2-32: Magnitude Response of $x[n] = exp(j2\pi \cdot n \cdot 0.1) + 0.01 \cdot exp(-j2\pi \cdot n \cdot 0.25)$

The results clearly show the advantage that the Hanning window provides to the DTFT calculation. Although the width of the main frequency lobe of the Hanning window is slightly broader in the immediate vicinity of the detected sinusoid, it falls off very rapidly thereafter. The smaller tone at $f=-2.5f_o$ was completely undetectable owing to the excessive spectral splatter of the rectangular window, whereas the pre-multiplication of the Hanning window revealed its existence. The fact that the main lobe of the Hanning windowed sequence is broader and appears to degrade the frequency resolution slightly can be mitigated by simply choosing a larger length N. The figure below illustrates two spectrum calculations for an EDGE (Enhanced Data Rate for GSM Evolution) signal. The EDGE technology upgrades GSM by enabling higher transmission rates. Calculating the spectrum using the rectangular window would indicate a very large occupied bandwidth and it is not until the Hanning window is used that the true spectrum emerges.

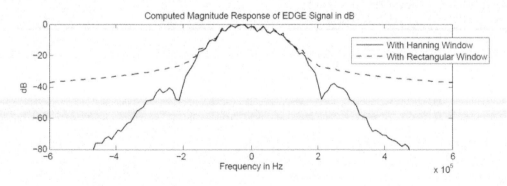

Figure 2-33: Averaged Magnitude DTFT Result of EDGE Signal

2.3.3   The Effect of Windows

What Are Windows?

Windows are functions that effectively time-limit a long sample record, $s[n]$, which we wish to analyze via the DTFT. This time-limitation process extracts N consecutive samples from the overall test record to yield the vector $x[n]$.

Multiplication in Time Domain

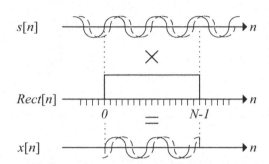

Figure 2-34: Time Limitation using the Rectangular Window

Clearly, applying the rectangular window is akin to simply discarding samples prior to $n = 0$ and after N-1. Note, however, that the resulting input vector, $x[n]$, features discontinuous or abrupt changes at the beginning and end, thus producing high frequency content in the form of our well-known $sinc(x)$ function. See Examples 2.09 and 2.11 in our section on the CTFT. We therefore consider window functions that smoothly approach a value of zero at the beginning and end of the time-limited region in order to avoid the high frequency behavior of the discontinuities. The following windows are defined over N samples and feature a discrete time index $n = 0, 1 \ldots N$-1.

WINDOW FUNCTION	*EQUATION*
Rectangular	1
Hanning	$0.50 - 0.50 \cos(\ 2\pi\ (n+1)/(N+1)\ )$
Hamming	$0.54 - 0.46 \cos(\ 2\pi\ (n+1)/(N+1)\ )$
Custom	$(Hanning)^b$ or $(Hamming)^b$

Figure 2-35: Equations for Different Window Functions

There are additional window functions that could be used for the purpose of time limitation [10], but the ones shown in the table above suffice for now. Note that the definition for these windows differs minutely from text to text, and the way they have been formulated here insures that none of the sequences features a value of zero as the start or end sample. The next figure plots some of these window functions and indicates that the side lobe suppression of each window is directly proportional to the smoothness with which they approach zero. As an exercise, you could write some MatLab code that plots the time sequence and associated frequency response for the custom Hanning and Hamming windows. By choosing different values for the exponent *b*, we may set the manner in which zero is approached at the beginning and end or the window. In this way, we may steer the trade-off between the thickness of the main lobe at lower frequencies and the amount of suppression of the side lobes at higher frequencies. The custom Hanning window will be used later in this chapter during the design of FIR filter coefficients.

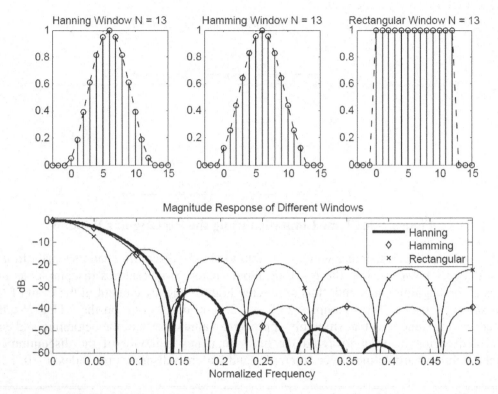

Figure 2-36: Hanning, Hamming, and Rectangular Windows of Length N=13

The Hanning window features the least amount of selectivity (the widest main lobe), but proves to be the best candidate for side lobe suppression.

Scaling the Window

We have already seen that the rectangular window is the most basic way of time-limiting a large sample record, *s[n]*. Since all coefficients of the rectangular window of length *N* are equal to one, the sum of all coefficients adds to $N \cdot 1 = N$. To avoid changing the mean of different window functions, we scale them such that their sum of their coefficients is equal to *N* as well. Note that this scaling operation was already implemented in the last example's code snippet, which compared the magnitude response of the rectangular and Hanning windows. The scaling factor for a 40 tap Hanning window was 40/20.5.

Convolution in the Frequency Domain

Let us summarize a few important insights we have made in these last few sections and provide a mathematical foundation that we may use to better understand how windows affect the spectrum of very long test waveforms. First of all, we have come to understand the spectrum of infinitely extending test sequences such as the complex sinusoid $exp(j2\pi nf)$ to be either a single peak or a collection of many peaks in the frequency domain. We have also become familiar with the magnitude spectrum of several popular windows that we apply for purposes of time-limiting long test waveforms prior to the DTFT operation. So, how does the spectrum of the long test waveform actually change as we overlay (via multiplication) the desired window? The answer is simple. Based on the convolution property of the Fourier transform, multiplication in the time domain results in convolution in the frequency domain, as can be seen in the figure below. This fact is presented to augment your overall understanding of the windowing process, but convolution itself does not have to be undertaken in your search for the perfect window. Just apply your window of choice to the long test waveform, $s[n]$, and plot the dB magnitude response as $20 \cdot log10(|X(f)|)$.

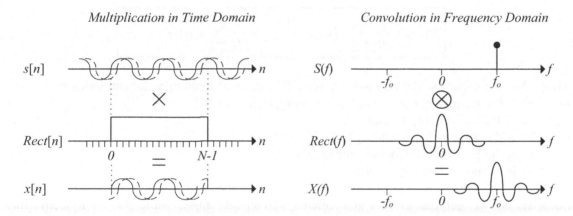

Figure 2-37: Multiplication in Time Results in Convolution in Frequency

When to Use Windows

Windows are used when evaluating the spectrum of a long waveform, as discussed in detail above. It is also common practice to apply a window function in order to enhance filter coefficients that were calculated via the IDFT, as we will see in later sections. When using the DTFT to analyze the frequency response of a set of FIR filter coefficients, we do *not* enhance the DTFT calculation via a window since we must see its correct and natural response and not one that has been artificially embellished via a window.

2.3.4   The Discrete Fourier Transform, or DFT, and Its Inverse

The discrete Fourier transform, or DFT, is like a gentlemen's agreement between you, the user, and the DTFT. For an input sequence, $x[n]$, of length N, you agree to evaluate the DTFT N times at certain unique frequencies, which of course lie in our valid range of -0.5 to 0.5 Hertz. The DFT result, $X[m]$, is a vector of N DTFT outputs, where the range of the index m is from 0 to N-1. The N unique frequencies at which we agree to evaluate the DTFT are m/N Hertz.

$$ValidFrequencies = \frac{m}{N} = \begin{bmatrix} \dfrac{0}{N} & \dfrac{1}{N} & \cdots & \dfrac{N-1}{N} \end{bmatrix} Hz$$

Wait a minute, you may say; clearly a frequency of $(N-1)/N$ is larger than 0.5Hz and therefore outside our legal range. Those frequencies, f, that exceed 0.5Hz are in fact the negative frequencies equal to f-1 Hz. We allow these aliased frequencies into our equation so that the index m neatly extends from 0 to N-1. The equation for the DFT is seen below.

$$X[m] = \frac{1}{N} \sum_{n=-\infty}^{\infty} x[n] \cdot e^{-j2\pi n \frac{m}{N}} = \frac{1}{N} \sum_{n=0}^{N-1} x[n] \cdot e^{-j2\pi n \frac{m}{N}}$$

$X[m]$ → is a complex scalar representing the DFT output at frequency domain index m.
m → is the frequency domain index extending from 0 to N-1.
$x[n]$ → is the discrete input sequence.
n → is the time index extending from 0 to N-1.
N → is the length of the input sequence $x[n]$.
m/N → is the normalized frequency.

So what do we get out of this gentlemen's agreement? First, while the DFT seems like a somewhat restrictive way of calculating the frequency spectrum of an input sequence $x[n]$, this frequency arrangement leads to a highly efficient implementation called the FFT, which we will meet in the next section. Without the FFT, spectrum calculation in digital baseband processors would be almost unthinkable. Second, given a set of N DFT results, we can recreate the original time domain input sequence, $x[n]$, by calculating the inverse discrete Fourier transform, or IDFT.

$$x[n] \overset{DFT}{\to} X[m] \overset{IDFT}{\to} x[n]$$

Therefore, in addition to being a tool for spectral calculation, the DFT is a method of parameterizing the input sequence, $x[n]$, (as a linear combination of weighted complex sinusoids) such that none of the original input time domain information, $x[n]$, is lost.

Sample and Frequency Indices

Once a sample rate, F_S, is other than 1Hz, the sample time instances and analysis frequencies assume the following actual values.

$$SampleInstances_n = \frac{n}{F_s} = n \cdot T_s \quad seconds$$

$$AnalysisFrequency_m = (\frac{m}{N}) \cdot F_s \quad Hertz$$

The variable N modifies the analysis frequency, not the sample instant. The quantity m/N is also called the normalized frequency.

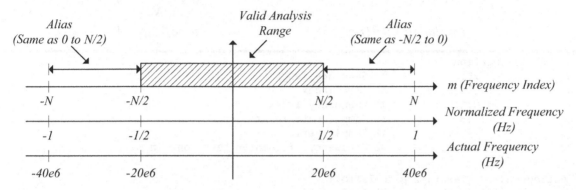

Figure 2-38: Valid Frequency Range of the DFT for F_S = 40MHz

Example 2.15: DFT Indices

Assume that we are presented with an input sequence, $x[n]$, with length $N=8$ and a sample rate of 40MHz. What would be the time and analysis frequency indices?

$$TimeInstances = n \cdot \frac{1}{40e6Hz} = \begin{bmatrix} 0 & 25nsec & 50nsec & 75nsec & \cdots & 175nsec \end{bmatrix}$$

$$FrequencyInstances = m \cdot \frac{40e6Hz}{N} = \begin{bmatrix} 0 & 5e6 & 10e6 & 15e6 & 20e6 & 25e6 & 30e6 & 35e6 \end{bmatrix}$$

$$= \begin{bmatrix} 0 & 5e6 & 10e6 & 15e6 & -20e6 & -15e6 & -10e6 & -5e6 \end{bmatrix}$$

Example 2.16: DFT versus DTFT

Assume that $x[n]$ is a sampled complex sinusoid oscillating at 100 KHz. Given a sample rate of 1MSPS and an input sample length of $N=15$, compare the output of the high resolution DTFT to that of the DFT.

$$x[n] = e^{j2\pi n(100e3/1e6)}$$

In the case of the DTFT, we will assume that the analysis frequencies are spaced in steps of 0.01Hz within our legal limit of -0.5Hz to 0.5Hz. For the DFT, the analysis frequencies will assume 15 unique values from 0 to 14/15 in steps of 1/15.

$$DTFT$$

$$X(f) = \frac{1}{N}\sum_{n=0}^{N-1} x[n] \cdot e^{-j2\pi nf}$$

$$AnalysisFreq = f$$

$$-0.5 < f < 0.5$$

$$resolution = 0.01Hz$$

$$DFT$$

$$X[m] = \frac{1}{N}\sum_{n=0}^{N-1} x[n] \cdot e^{-j2\pi n\frac{m}{N}}$$

$$AnalysisFreq = \frac{m}{N}$$

$$-0.5 < \frac{m}{N} < 0.5$$

$$m = 0,1...N-1$$

```
%% DFT Calculation
N           = 15;           %% Sequence length
n           = 0:1:N-1;      %% Time index
m           = 0:1:N-1;      %% Frequency index
Fs          = 1e6;          %% Sample rate in Hz
Ts          = 1/Fs;         %% Sample period
fo          = 100e3;        %% Frequency of complex test tone in Hz


TestSequence  = exp(j*2*pi*n*Ts*fo);
FrequenciesDFT = m/N;       %% Analysis frequencies
DFT_Out       = zeros(1,N);


for i = 1:N                 %% DFT calculation
    f           = FrequenciesDFT(1,i);
    AnalysisTone = exp(j*2*pi*n*f);
    DFT_Out(1,i) = (1/N)*(TestSequence.*Rectangular)*AnalysisTone';
end
MagSpectrumDFT   = abs(DFT_Out);
PhaseSpectrumDFT = angle(DFT_Out);
```

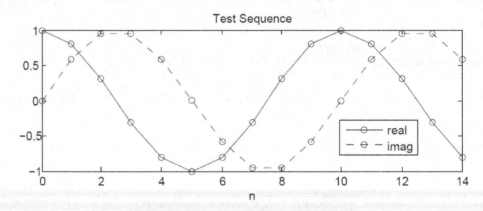

Figure 2-39: The Complex Test Sequence

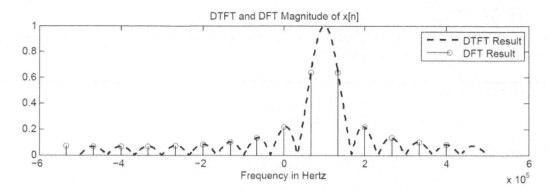

Figure 2-40: DTFT and DFT Results Compared

The DTFT shows us the true quasi-continuous spectrum (frequency step = 0.01Hz) within the Nyquist bounds, whereas the DFT produces a subset that clearly did not give us a single non-zero peak at 100 KHz. However, at the points set by the analysis frequencies m/N we did in fact compute the correct results. Note that the x axis was scaled by Fs to reflect the actual, unnormalized frequency scale.

> The fact that the DTFT can be evaluated at any frequency makes it the tool of choice when we need to characterize the full frequency behavior of short sequences or filter impulse responses where the length, N, is limited. The DFT together with its efficient FFT implementation shines when the spectra of signals for which larger amounts of samples are available must be found.

The IDFT – Inverse Discrete Fourier Transform

In a sense, the IDFT is the recomposition of the time waveform $x[n]$ from the linear combination of the complex, scaled analysis tones calculated by the DFT. The formula looks very similar to the DFT. In this case, the size of the input vector, $X[m]$, determines the length N of the IFFT.

$$x[n] = \sum_{m=0}^{N-1} X[m] \cdot e^{j2\pi nm/N}$$

Let us use our previous example to expand the equation above and show that in fact the IDFT is nothing but a superposition of scaled analysis tones. Each vector $exp(j2\pi nm/15)$ of length 15 is multiplied by the complex scalar $X[m]$ and summed on a point-by-point basis. Just like the time index, n, the frequency index, m, extends from 0 to $N-1$.

$$x[n] = X[0] \cdot e^{j2\pi n0/15} + \dots$$
$$X[1] \cdot e^{j2\pi n1/15} + \dots$$
$$X[2] \cdot e^{j2\pi n2/15} + \dots$$
$$\vdots$$
$$X[14] \cdot e^{j2\pi n14/15}$$

Example 2.17: IDFT

In this example, we take the DFT result of the previous example and inverse Fourier transform it to recreate the original test waveform.

```
%% IDFT Calculation
N           = 15;               %% Sequence length
n           = 0:1:N-1;          %% Time index

IDFT_Out = zeros(1,length(n));
for i = 1:N                     %% IDFT calculation
    m               = i - 1;
    f               = m/N;
    AnalysisTone    = exp(j*2*pi*n*f);
    IDFT_Out        = IDFT_Out + DFT_Out(1,i)*AnalysisTone;
end
```

The IDFT_Out vector contains the exact values of the initial test sequence.

$$IDFT\_Out = [1.00 + j0 \quad 0.809 + j.588 \quad 0.309 + j.951 \quad -.309 + j.951 \quad -.809 + j.588$$
$$-1.00 + j0 \quad -.809 - j.588 \quad -.309 - j.951 \quad 0.309 - j.951 \quad 0.809 - j.588$$
$$1.00 + j0 \quad 0.809 + j.588 \quad 0.309 + j.951 \quad -.309 + j.951 \quad -.809 + j.588]$$

Example 2.18: The All-important Spectrum Function

You have seen it in text-books, on spectrum analyzers, on brochures, and in any variety of other communication paraphernalia: The average magnitude spectrum of a communication signal. How is the magnitude spectrum of the EDGE signal, which we saw in the last section, really computed? The function we used for this purpose is called Spectrum and takes in an input signal, the DFT length, N, and the sample rate. It will split up the input waveform into 200 separate sections and compute the magnitude spectrum of an N_ point DFT for each one of them. The separate magnitude results, |DFT|, are averaged and normalized to take into account the number of sections and the magnitude gain introduced by the Hanning window.

The parameter 'step' is computed as a starting point for each one of these sections. Next, the function defines a Hanning window and the vector that will hold the DFT output. Note that the 'SampleRate' parameter does not impact any of the calculations. It exists merely to scale the output frequency vector returned from the function, exactly as explained above.

The function computes the DFT and may require quite some time to execute for large values of N such as 1024 and beyond. The code can be trivially retrofitted to use the built-in MatLab function called fft(), which massively accelerates the spectrum calculation. The spectrum function returns two vectors that are conveniently formatted to be used via the MatLab function call *plot(f, Mag,* '*k*').

```
function [Mag f] = Spectrum(Input, N, SampleRate)

if(N < 2)
    msgbox('N must be larger than 1','Error Message');
    Mag = []; f = [];
    return;
end

if(length(Input) < N)
    msgbox('Length of Input must be larger than N','Error Message');
    Mag = []; f = [];
    return;
end

Step  = floor((length(Input)- N) / 200);
if(Step < 1)
    msgbox('Please increase Input vector length','Error Message');
    Mag = []; f = [];
    return;
end

%% Setup parameters for the DFT calculation
n         = 0:N-1;                          %% Discrete time index
Hanning   = 0.5 - 0.5*cos(2*pi*(n+1)/(N+1));%% Basic Hanning definition
H         = (N/sum(Hanning)) * Hanning;     %% Normalized Hanning window
DFTOutput = zeros(1,N);                     %% X[m]

%% Here we calculate the average magnitude response
for Repetitions = 0:199
    Current_Range = (Repetitions*Step + 1):(Repetitions*Step + N);
    x             = Input(1,Current_Range);
    for m = 0:N-1                           %% The DTFT Function
        DFTOutput(1,m+1) = DFTOutput(1,m+1) + ...
                        (1/N)*abs( (x.*H) * exp(-j*2*pi*n*m/N).' );
    end
end

%% Move the aliased region above f >= 0.5 to the negative frequencies
if(mod(N,2) == 0)   %% even number
    NegStart = (N/2) + 1;
    PosStop  = (N/2);
else
    NegStart = ceil(N/2);
    PosStop  = floor(N/2);
end

DFTOutput  = [ DFTOutput(1, NegStart:end) DFTOutput(1,1:PosStop) ];
F_Norm     = 0:1/N:(N-1)/N;                 %% Normalized frequency

%% Function Outputs
Mag        = 20*log10(DFTOutput/200);
f          = [ F_Norm(1, NegStart:end)-1  F_Norm(1,1:PosStop) ]*SampleRate;
                                            %% Actual reordered frequencies
```

2.3.5 The Fast Fourier Transform (Radix-2)

The fast Fourier transform, or FFT, is a family of efficient algorithms used to compute the discrete Fourier transform (DFT). While the most popular FFT method was proposed by J.W. Cooley and John Tukey in 1965 [5], the inception of the FFT can be traced back all the way to Carl Friedrich Gauss [4]. The Cooley-Tukey approach subdivides a size N DFT into two DFTs of length $N/2$. These two new DFTs are then themselves divided again and the division process continues until we reach the basic 2-point DFT. This methodology allows only DFTs that feature a size $N = 2^b$, where b is any non-zero positive integer, to be computed. The method of continually dividing an N-point DFT into two equal $N/2$-point DFT modules is called the radix-2 implementation. Other implementations, like the radix-4 and split radix techniques, feature further gains in computational efficiency. Computing the DFT of a time sequence is one of the most important operations in the field of digital signal processing. It is an algorithm that is heavily used in many different engineering disciplines, including in the area of electronic communication, where it estimates the frequency content of signals. Further, the DFT and its inverse, the IDFT, are the core processing blocks in all OFDM modems. Efficient implementations via the FFT are paramount when designing low power communication transceivers of this type. First, let us review some basic DFT/FFT definitions and constraints.

The **DFT** requires an evenly sampled time sequence $x[n]$ of any length N. The DFT is computed at N normalized frequencies, m/N, where m ranges from 0 to N-1. Due to the effects of aliasing, normalized frequencies, m/N, that are larger than 0.5 in fact represent negative oscillations at a rate of (m/N)-1.

The **FFT** (radix-2) is based on the decomposition of an N-point DFT into two $N/2$-point DFT modules. The length N is restricted to values of $N = 2^b$, where b is a non-zero positive integer (N = 2, 4, 8, 16 … etc.). Otherwise, the same restrictions for the DFT apply to the radix-2 FFT.

The **FFT** (radix-4) is based on the decomposition of an N-point DFT into four $N/4$-point DFT modules. The length N is restricted to values of $N = 4^b$, where b is a non-zero positive integer (N = 4, 16, 64 … etc.). Compared to the radix-2 structure, some further digital hardware can be saved at the expense of slightly increased abstraction.

The **FFT** (split radix [6], [7], [8]) tries to get around some of the restrictions on the length N that are placed on higher radix FFT operations. A 32-point FFT could for example start with a radix-2 operation to split the task into two 16-point DFT blocks, and then decompose them into 4-point DFT elements via the radix-4 approach.

Hardware Efficiency

Minimizing the number of multiplication in both hardware and software implementations of the DFT operation is of paramount importance for large values of N. The following formulas compare the number of complex multiplications required for a DFT and the radix-2/4 FFT counterparts.

Multiplication ➔ $DFT = N^2$ *radix*-2 $FFT = (N/2) \cdot log_2(N)$ *radix*-4 $FFT = (3N/8) \cdot log_2(N)$

There are plenty of instances where complex multiplication by certain convenient values can be implemented without the use of a full multiplier, as for example multiplications by 1, -1, j and $-j$. Keep this in mind in case you start counting multiplication operations at the end of your own FFT custom design. Note that the radix-4 implementation of the FFT yields an additional multiplier saving of 25% over the radix-2 implementation but the improvement is nowhere near as great as that experienced during the initial migration from the DFT to the radix-2 FFT.

DFT LENGTH N	DFT	RADIX-2 FFT	RADIX-4 FFT
8	64	12	
16	256	32	24
32	1024	80	
64	4096	192	144
128	16384	448	
256	65536	1024	768

Figure 2-41: Number of Complex Multiplications for the DFT, the Radix-2 FFT, and the Radix-4 FFT

Although the values above are a reasonable reflection of the number of complex multiplications that must be computed, they are not a reasonable reflection of how many actual hardware multipliers are needed in a digital ASIC/FPGA implementation of an FFT. The FFT computation does not need to happen all at once and usually has a set number of clock periods within which it must finish. This fact allows us to reuse complex hardware multipliers in the FFT implementation, giving rise to further gate savings.

Before we show how the DFT decomposes into the FFT, let us quickly discuss which FFT algorithm is right for us. In communication systems, the FFT is primarily used in embedded DSP and dedicated digital base band processors. Many years ago, when the feature size of CMOS devices was larger, minimizing the transistor count in base band processors was a virtue, and extensive research was done to design FFT algorithms of radix-8 [9] or higher. With today's high performance CMOS technology, transistors can be used somewhat more liberally, and engineers should consider the radix-2 and radix-4 topologies first. Designers should know that higher radix FFT implementations also increase routing complexity in digital ASICs.

From the DFT to the Radix-2 FFT

There are three distinct topics regarding the transition from the DFT to the radix-2 FFT algorithm that must be well understood before to any programming or design is undertaken.

1. The first topic is the decomposition process of an N-point DFT into two separate $N/2$ DFT modules.
2. The second feature is the radix-2 butterfly structure that forms the basic calculation element of the FFT.
3. Finally, there is an input index reversal phenomenon that will become clear after working through topic one. This index reversal of any size FFT must be well understood.

Decomposing an N Point DFT into Two N/2 Point DFT Elements

We start off with our well-known DFT equation, which for this application is restricted to lengths of $N = 2^b$, where b is an integer greater than zero. For convenience and ease of notation, we will skip the normalization term 1/N for now and call the equation below the un-normalized DFT.

$$X[m] = \sum_{n=0}^{N-1} x[n] \cdot e^{-j2\pi nm/N}$$

We now break the above expression into two parts, one for even, $x[2n]$ and one for odd $x[2n+1]$ input samples. Take a moment to convince yourself that the equations above and below are in fact equal.

$$X[m] = \sum_{n=0}^{\frac{N}{2}-1} x[2n] \cdot e^{-j2\pi(2n)m/N} + \sum_{n=0}^{\frac{N}{2}-1} x[2n+1] \cdot e^{-j2\pi(2n+1)m/N}$$

$$= \sum_{n=0}^{\frac{N}{2}-1} x[2n] \cdot e^{-j2\pi(2n)m/N} + e^{-j2\pi m/N} \sum_{n=0}^{\frac{N}{2}-1} x[2n+1] \cdot e^{-j2\pi(2n)m/N}$$

$$= \sum_{n=0}^{\frac{N}{2}-1} x[2n] \cdot e^{-j2\pi nm/(N/2)} + e^{-j2\pi m/N} \sum_{n=0}^{\frac{N}{2}-1} x[2n+1] \cdot e^{-j2\pi nm/(N/2)}$$

If, for just a moment, we ignore the term $e^{-j2\pi m/N}$ to the left of the second summation, we realize that the two summation expressions are completely normal DFT operations of length $N/2$. Their input sequences are different but they are DFT operations nevertheless. The expression $e^{-j2\pi m/N}$ is called a *twiddle factor*, for which we define the following common notation.

$$W_N^x = e^{-j2\pi x/N} = W_{N/x}^1 = e^{-j2\pi/(N/x)}$$

Therefore,

$$W_N^m = e^{-j2\pi m/N}$$

and we may rewrite the original DFT equation yielding $X[m]$ as follows.

$$X[m] = \sum_{n=0}^{N-1} x[n] \cdot e^{-j2\pi nm/N} = \underline{\sum_{n=0}^{\frac{N}{2}-1} x[2n] \cdot W_{N/2}^{nm}} + W_N^m \underline{\sum_{n=0}^{\frac{N}{2}-1} x[2n+1] \cdot W_{N/2}^{nm}}$$

The underlined expressions are the separate DFT operations with the lengths of $N/2$. One DFT processes the even samples $x[2n]$, while the other processes the odd ones $x[2n+1]$. Each one of these smaller DFT operations can now be repeatedly broken down until we reach the most basic structure, the DFT of length 2. The figure below illustrates how an 8-point DFT is decomposed into a combination of two 4-point DFT modules and the associated twiddle elements. Each 4-point DFT is itself decomposed into two 2-point DFT modules with their own twiddle logic.

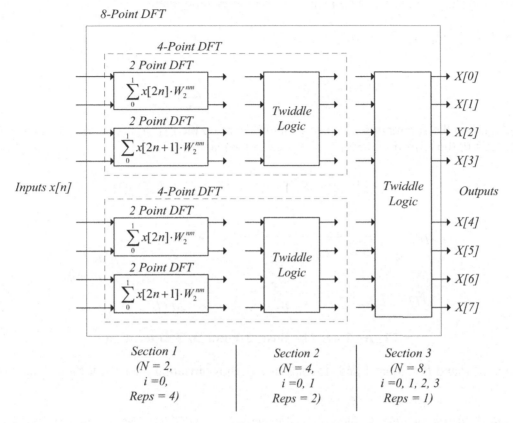

Figure 2-42: Repeated Decomposition of 8-Point DFT into 4-Point and 2-Point DFT Elements

The diagram above is subdivided into different sections that delineate the different levels of computation. Section one contains the most basic DFT elements of length $N = 2$. The concepts behind the variable $i = 0$ will become obvious once we discuss the fundamental radix-2 butterfly and Reps, or repetitions, indicates the number of N-point DFT elements that are computed. The FFT MatLab code we will present later will use these three variables extensively.

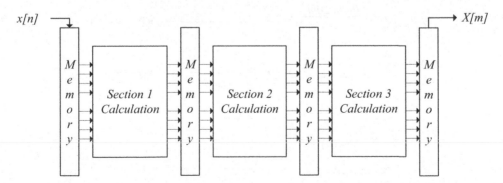

Figure 2-43: Hardware FFT Computation Flow with Intermediate Section Storage

Real-life hardware implementations will feature SRAM or register banks to store the intermediate results of each section calculation. The algorithm consecutively executes the calculation of each section until the result is ready for pick in the output memory bank.

Example 2.19: Illustrate the Basic 2-Point DFT Pictorially

$$X[m] = \sum_{n=0}^{1} x[n] \cdot e^{-j2\pi nm/2} = \sum_{n=0}^{1} x[n] \cdot W_2^{nm}$$

Let us remove the summation sign and explicitly sum up the two terms for $n = 0$ and $n = 1$. Note that for $n = 0$, the term W_2^{nm} defaults to W_2^0, which equals 1.

$$X[m] = x[0] \cdot W_2^{0 \cdot m} + x[1] \cdot W_2^{1 \cdot m} = x[0] + W_2^m \cdot x[1]$$

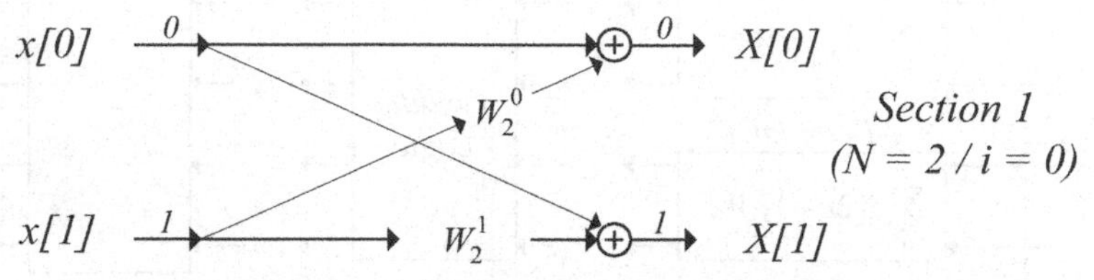

Figure 2-44: The Basic 2-Point DFT Operation

Note that W_2^0 and W_2^1 equal 1 and -1, respectively, thus obviating the need for multiplication.

Example 2.20: Pictorially Decompose a Generic 4-Point DFT into Two 2-Point DFTs

$$X[m] = \sum_{n=0}^{3} x[n] \cdot W_4^{nm} = \sum_{n=0}^{1} x[2n] \cdot W_2^{nm} + W_4^m \sum_{n=0}^{1} x[2n+1] \cdot W_2^{nm}$$

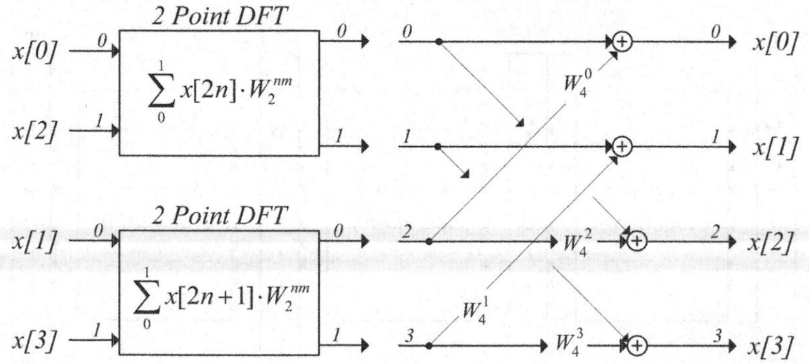

Figure 2-45: Diagram of a 4-Point DFT Broken into Two 2-Point DFT Elements

Example 2.21: *Decomposition of a Generic 8-Point DFT into Two 4-Point DFTs*

$$X[m] = \sum_{n=0}^{7} x[n] \cdot W_8^{nm} = \sum_{n=0}^{3} x[2n] \cdot W_4^{nm} + W_8^{m} \sum_{n=0}^{3} x[2n+1] \cdot W_4^{nm}$$

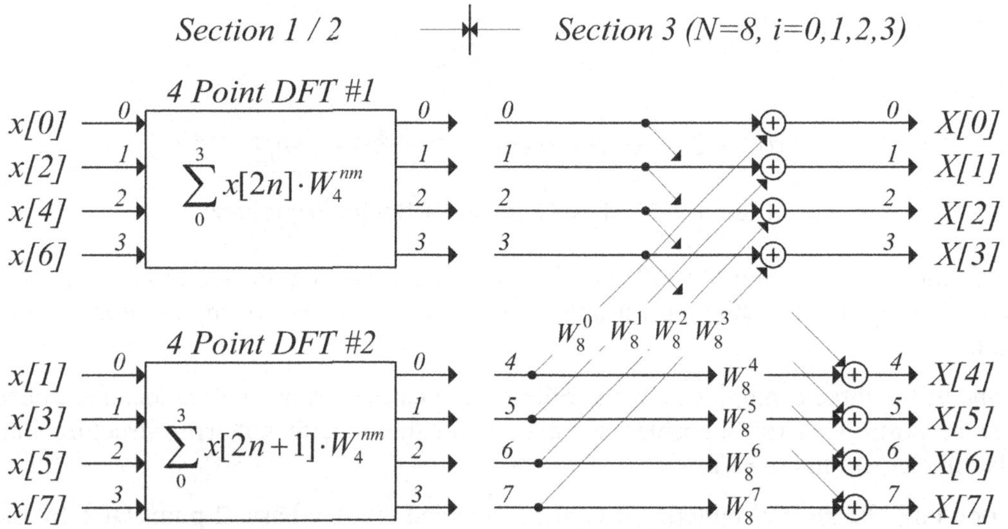

Figure 2-46: Diagram of an 8 Point DFT Broken into two 4 Point DFT

Note that the input / output port numbers in the figure of Example 2.20 range from 0 to 1 for the 2-point DFT modules (section 1) and 0 to 3 for the twiddle elements, combining them into the four final outputs (section 2). The input/output port numbers for Examples 2.19 through 2.21 are helpful when looking at the generalized radix-2 butterfly structure.

The last three examples are presented consecutively to allow us a certain insight into the calculation flow. First, take a look at the figure of Example 2.19, which shows the inputs of a 2-point DFT block being stitched together by some adders and two multiplications by twiddle factors. Note sub- and superscripts of the W term and that the multiplications emanate from the bottom input.

Second, review the structure in Example 2.20 that stitches together the two 2-point DFT blocks into four output values. Again we note the sub- and superscripts of the twiddle element W and can see that the twiddle factor multiplications emanate from the bottom input ports.

Finally, the twiddle elements merge two 4-point DFT modules into the final 8-point FFT output, as seen in Example 2.21. These examples, including the twiddle factor sub- and superscripts progression, point to a generalized radix structure, which we now introduce.

The Radix-2 Butterfly

The radix-2 butterfly is the computational structure that is repeated throughout the entire FFT flow. In its most basic form, it appears as the 2-point-DFT. The structure below represents the generalized radix-2 butterfly, which describes the computational pattern that we have seen in the last three examples.

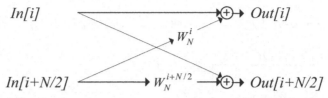

Figure 2-47: The General Butterfly Structure

Note that the indices 'i' and '$i + N/2$' are the input/output port numbers that form the boundaries of each twiddle section, while the variable N refers to the number of input/output ports of each element.

Each one of the three examples shown previously indicates sections with associated indices i and number of ports N. These variables serve to mold the general butterfly structure into every calculation element in the FFT.

In section one, the general butterfly structure is molded into the basic 2-point DFT by fixing the index $i = 0$ and $N = 2$, resulting in the following setup.

➔ $i = 0$

$In[i] = In[0]$	$Out[i] = Out[0]$	$W_N{}^i = W_2{}^0$
$In[i+N/2] = In[1]$	$Out[i+N/2] = Out[1]$	$W_N{}^{i+N/2} = W_2{}^1$

In section two, the general butterfly structure is molded into a twiddle network that combines the outputs of two 2-point DFT elements (index $i = 0$ and 1, while $N = 4$). Given that the index i now assumes two numbers, 0 and 1, we must implement two butterfly elements.

➔ $i = 0$

$In[i] = In[0]$	$Out[i] = Out[0]$	$W_N{}^i = W_2{}^0$
$In[i+N/2] = In[2]$	$Out[i+N/2] = Out[2]$	$W_N{}^{i+N/2} = W_4{}^2$

➔ $i = 1$

$In[i] = In[1]$	$Out[i] = Out[1]$	$W_N{}^i = W_4{}^1$
$In[i+N/2] = In[3]$	$Out[i+N/2] = Out[3]$	$W_N{}^{i+N/2} = W_4{}^3$

In section three, the general butterfly structure is molded into a twiddle network that combines the outputs of two 4-point DFT elements (index $i = 0$, 1, 2, and 3, while $N = 8$). Given that the index i now assumes four numbers, we must implement four butterfly elements. You may verify the input/output port order and twiddle factors to be those of Example 2.20.

The primary lesson that we take away from this exercise is that we may follow the idea of sections and its associated parameters i and N to easily build a radix-2 FFT of any size.

Input Index Reversal

As you have surely noticed, the index of our input sequence $x[n]$ becomes a bit scrambled as can plainly be seen by the last few examples. In Example 2.21, the input sequence was scrambled into $x[0]$, $x[2]$, $x[4]$, $x[6]$, $x[1]$, $x[3]$, $x[5]$, and $x[7]$, where the first four inputs fed the first 4-point DFT and the latter four inputs fed the second 4-point DFT. Now, the 4-point DFT units shown below receive four numbers at four input ports and they have their own method of fetching these inputs and connecting them to their 2-point DFT blocks (Example 2.19). This scenario is seen on the left portion of the figure below.

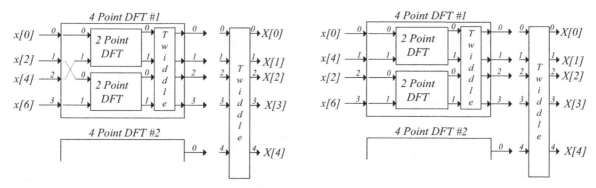

Figure 2-48: Radix-2 8-Point FFT with Streamlined Input Index Ordering

It requires additional scrambling to arrange the inputs such that the diagram and the computational flow become streamlined, as seen on the right side of the figure above. The streamlined input sequence is as follows: $x[0]$, $x[4]$, $x[2]$, $x[6]$, $x[1]$, $x[5]$, $x[3]$, and $x[7]$. As the FFT length N increases, this issue becomes cumbersome and must be addressed in more general terms.

Luckily, the index scrambling problem is solved trivially by reversing the bit order of the input index, as shown below for a length 8 radix-2 FFT. The bit reversal technique works for any radix-2 FFT of legal length N.

NORMAL INDEX	BIT REPRESENTATION	REVERSED BIT REPRESENTATION	SCRAMBLED INDEX
0	000	000	0
1	001	100	4
2	010	010	2
3	011	110	6
4	100	001	1
5	101	101	5
6	110	011	3
7	111	111	7

Figure 2-49: Input Index Reversal Process for 8-point Radix-2 FFT

Example 2.22: *MatLab Code for 16-Point DFT Implemented as Radix-2 FFT*

The figure below illustrates the calculation flow typically found in a FFT hardware implementation. Section one of the algorithm handles the basic 2-point DFT elements whose outputs are merged into 4-point, 8-point, and 16-point DFT outputs in sections two, three and four respectively. The input sequence as well as the result of each section are stored in memory element or register banks. This allows us to compute the results of each section in order, starting with section one and finishing with section four.

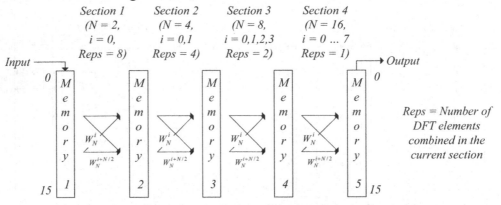

Figure 2-50: Calculation Progression of Radix-2 16-Point FFT

```
Memory        = zeros(16, 5);
Length        = 16;
Order         = 1 + [0 8 4 12 2 10 6 14   1 9 5 13 3 11 7 15];
%% We add a 1 because MatLab arrays must start at index 1 not 0
%% Input sequence cos(2 pi Freq *n/SampleRate)
n             = 0:1:Length-1;
Freq          = .05;
Input         = cos(2*pi*Freq*n/1);   %% Sample rate is 1
Memory(Order,1) = Input.';


%% Butterfly Calculations for Every Section
for Section = 1:1:4
    N       = 2^Section; %% Current section processes N-point DFT modules
    Reps    = 16/N; %% Number of N-point DFT combinations in current section
    I       = 2^(Section-1); %% Limit for index i
    Offset  = 1;
    for a = 0:Reps-1
        for i = 0:I-1  %% Start of butterfly
            Ini     = Memory( Offset + i      , Section);
            IniN_2  = Memory( Offset + i + N/2, Section);
            Wi      = exp(-j*2*pi*i/N);
            WiN_2   = exp(-j*2*pi*(i+N/2)/N);
            Memory( Offset + i      , Section + 1) = Ini + Wi    * IniN_2;
            Memory( Offset + i + N/2, Section + 1) = Ini + WiN_2 * IniN_2;
        end             %% End of butterfly
        Offset = Offset + N;
    end
end
Offset = Memory(:, 5).';
```

2.3.6 The Fast Fourier Transform (Radix-4)

From the DFT to the Radix-4 FFT

Let us once again look at our un-normalized DFT equation below.

$$X[m] = \sum_{n=0}^{N-1} x[n] \cdot e^{-j2\pi nm/N} = \sum_{n=0}^{N-1} x[n] \cdot W_N^{nm}$$

This time, $N = 4^b$, where b is any integer greater than 0 and we decompose the un-normalized length N DFT into four smaller DFT modules of length $N/4$. The new input sequences are $x[4n]$, $x[4n+1]$, $x[4n+2]$, and $x[4n+3]$.

$$
\begin{aligned}
X[m] = &\sum_{n=0}^{\frac{N}{4}-1} x[4n] \cdot e^{-j2\pi(4n)m/N} + \sum_{n=0}^{\frac{N}{4}-1} x[4n+1] \cdot e^{-j2\pi(4n+1)m/N} \\
&+ \sum_{n=0}^{\frac{N}{4}-1} x[4n+2] \cdot e^{-j2\pi(4n+2)m/N} + \sum_{n=0}^{\frac{N}{4}-1} x[4n+3] \cdot e^{-j2\pi(4n+3)m/N} \\
= &\sum_{n=0}^{\frac{N}{4}-1} x[4n] \cdot e^{-j2\pi nm/(N/4)} + \sum_{n=0}^{\frac{N}{4}-1} x[4n+1] \cdot e^{-j2\pi nm/(N/4)} \cdot e^{-j2\pi m/N} \\
&+ \sum_{n=0}^{\frac{N}{4}-1} x[4n+2] \cdot e^{-j2\pi nm/(N/4)} \cdot e^{-j2\pi 2m/N} + \sum_{n=0}^{\frac{N}{4}-1} x[4n+3] \cdot e^{-j2\pi nm/(N/4)} \cdot e^{-j2\pi 3m/N}
\end{aligned}
$$

We reintroduce our twiddle factor notation via the following rule.

$$W_N^x = e^{-j2\pi x/N} = W_{N/x}^1 = e^{-j2\pi/(N/x)}$$

The result is the following expression.

$$
\begin{aligned}
X[m] \quad = &\sum_{n=0}^{\frac{N}{4}-1} x[4n] \cdot W_{N/4}^{nm} + W_N^m \cdot \sum_{n=0}^{\frac{N}{4}-1} x[4n+1] \cdot W_{N/4}^{nm} \\
&+ W_N^{2m} \cdot \sum_{n=0}^{\frac{N}{4}-1} x[4n+2] \cdot W_{N/4}^{nm} + W_N^{3m} \cdot \sum_{n=0}^{\frac{N}{4}-1} x[4n+3] \cdot W_{N/4}^{nm}
\end{aligned}
$$

The equation above is analogous to the expression for the radix-2 case, except that in this case we have four separate and equally sized DFT blocks of length $N/4$ and three twiddle factor W_N^m, W_N^{2m}, and W_N^{3m}.

Example 2.23: *Illustrate the 4-Point DFT Pictorially*

The 4-point DFT equation states the following.

$$X[m] = \sum_{n=0}^{3} x[n] \cdot e^{-j2\pi nm/4} = \sum_{n=0}^{3} x[n] \cdot W_4^{nm}$$

Let us remove the summation sign and explicitly sum up the terms for $n = 0$, 1, 2, and 3. Note that for $n = 0$, the term W_4^{nm} defaults to W_4^0, which equals 1.

$$X[m] = x[0] \cdot W_4^{0 \cdot m} + x[1] \cdot W_4^{1 \cdot m} + x[2] \cdot W_4^{2 \cdot m} + x[3] \cdot W_4^{3 \cdot m}$$

$$= x[0] + W_4^{1 \cdot m} x[1] + W_4^{2 \cdot m} x[2] + W_4^{3 \cdot m} x[3]$$

Note that the factors, $W_4^{nm} = e^{-j2\pi nm/4}$, will cause a rotation that features a resolution of $2\pi/4 = \pi/2$ radians. Therefore, they can be equal only to 1, -1, j and $-j$ as seen in the figure below.

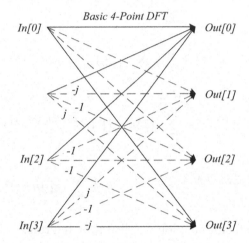

Figure 2-51: Basic 4-Point DFT Structure

We compute each output to verify that the W^{nm}_4 terms correctly contribute to the structure above.

$$X[0] = x[0] + x[1] \cdot W_4^0 + x[2] \cdot W_4^0 + x[3] \cdot W_4^0 = x[0] + x[1] + x[2] + x[3]$$

$$X[1] = x[0] + x[1] \cdot W_4^1 + x[2] \cdot W_4^2 + x[3] \cdot W_4^3 = x[0] + x[1] \cdot e^{-j\pi/2} + x[2] \cdot e^{-j\pi} + x[3] \cdot e^{-j\pi3/2}$$
$$= x[0] - j \cdot x[1] - x[2] + j \cdot x[3]$$

$$X[2] = x[0] + x[1] \cdot W_4^2 + x[2] \cdot W_4^4 + x[3] \cdot W_4^6 = x[0] + x[1] \cdot e^{-j\pi} + x[2] \cdot e^{-j2\pi} + x[3] \cdot e^{-j3\pi}$$
$$= x[0] - x[1] + x[2] - x[3]$$

$$X[3] = x[0] + x[1] \cdot W_4^3 + x[2] \cdot W_4^6 + x[3] \cdot W_4^9 = x[0] + x[1] \cdot e^{-3j\pi/2} + x[2] \cdot e^{-j3\pi} + x[3] \cdot e^{-j9\pi/2}$$
$$= x[0] + j \cdot x[1] - x[2] - j \cdot x[3]$$

The Radix-4 Butterfly

The radix-4 butterfly is the computational structure that is repeated throughout the entire FFT flow. In its most basic form, it appears as the 4-point DFT, as should readily become apparent when comparing the figure below with that of the last example. The structure below represents the generalized radix-4 butterfly.

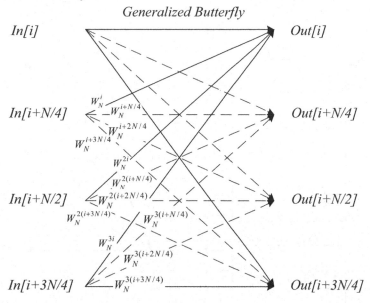

Figure 2-52: Generalized Butterfly Structure for Radix-4 Implementation

Note that the indices 'i', '$i + N/4$', '$i + N/2$', and '$i + 3N/4$', are the input/output port numbers that form the boundaries of each twiddle section, while the variable N refers to the number of input/output ports of each element. The significance of the index i is the same as that in the radix-2 example. As seen in the example below, the index $i = 0 / N = 4$ for section one and $i = 0, 1, 2$, and $3 / N = 4$ for section two. You can convince yourself that the figure above reduces to the basic 4-point DFT structure by choosing $i = 0 / N = 4$.

Example 2.24: Pictorially Break a Generic 16 Point FFT into Four 4-Point DFTs

Let us first take our general equation for the radix-4 FFT and adjust it for the $N=16$ variant.

$$X[m] = \sum_{n=0}^{3} x[4n] \cdot W_4^{nm} + W_{16}^{m} \cdot \sum_{n=0}^{3} x[4n+1] \cdot W_4^{nm}$$
$$+ W_{16}^{2m} \cdot \sum_{n=0}^{3} x[4n+2] \cdot W_4^{nm} + W_{16}^{3m} \cdot \sum_{n=0}^{3} x[4n+3] \cdot W_4^{nm}$$

The 16-point FFT breaks into four 4-point DFT blocks connected by four additional butterfly structures, of which we show only one in the next figure.

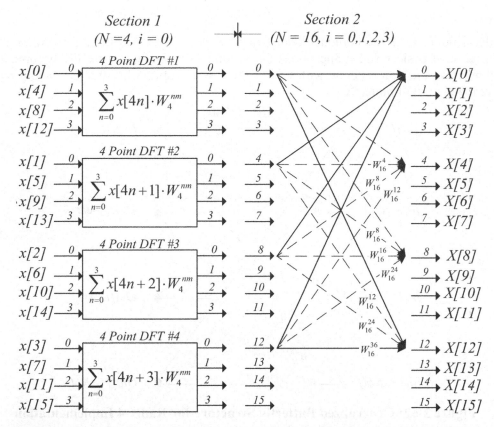

Figure 2-53: Diagram of a 16-Point DFT Broken into Four 4-Point DFTs

As is painfully obvious, the diagram is no longer quite as simple as the one for the radix-2 implementation. Because of space restriction, not all the twiddle factor multiplications and paths are shown. The figure above illustrates one of four butterfly structures belonging to section two.

Input index reversal is, at first glance, slightly more confusing than in the radix-2 case. Note that we use two pairs of bits, where each pair actually represents one base 4 number. It is these base 4 numbers that are reversed, leading to the bit exchange pattern shown here.

NORMAL INDEX	BIT REPRESENTATION	REVERSED BIT REPRESENTATION	SCRAMBLED INDEX
0	00 00	00 00	0
1	00 01	01 00	4
2	00 10	10 00	8
3	00 11	11 00	12
⋮	⋮	⋮	⋮
13	11 01	01 11	7
14	11 10	10 11	11
15	11 11	11 11	15

Figure 2-54: Reversed Bit Representation for Radix-4 16-Point FFT

Example 2.25: <u>MatLab Code for 16-Point DFT Implemented as Radix-4 FFT</u>

```
Memory      = zeros(16, 3);
Length      = 16;

%% Input Sequence cos(2 pi Freq *n/SampleRate)
n             = 0:1:Length-1;
Freq          = .05;
SampleRate    = 1;
Input         = cos(2*pi*Freq*n/SampleRate);
Order         = 1 + [0 4 8 12 1 5 9 13   2 6 10 14 3 7 11 15];
            %% we add a 1 because MatLab arrays must start at index 1 not 0
Memory(Order,1) = Input.';
for Section = 1:1:2
    N     = 4^Section; %% Current section contains N-point DFT modules
    Reps  = 16/N;        %% Number of N-point DFT modules in current section
    I     = 4^(Section-1); %% Limit for index i
    Offset = 1;
    for a = 0:Reps-1
        for i = 0:I-1       %% Start of radix-4 butterfly
            Ini     = Memory( Offset + i        , Section);
            IniN_4  = Memory( Offset + i + N/4  , Section);
            IniN_2  = Memory( Offset + i + N/2  , Section);
            Ini3N_4 = Memory( Offset + i + 3*N/4, Section);

            Wi      = exp(-j*2*pi*i/N);
            WiN_4   = exp(-j*2*pi*(i+N/4)/N);
            WiN_2   = exp(-j*2*pi*(i+N/2)/N);
            Wi3N_4  = exp(-j*2*pi*(i+3*N/4)/N);

            W2i     = exp(-j*2*pi*2*i/N);
            W2iN_4  = exp(-j*2*pi*2*(i+N/4)/N);
            W2iN_2  = exp(-j*2*pi*2*(i+N/2)/N);
            W2i3N_4 = exp(-j*2*pi*2*(i+3*N/4)/N);

            W3i     = exp(-j*2*pi*3*i/N);
            W3iN_4  = exp(-j*2*pi*3*(i+N/4)/N);
            W3iN_2  = exp(-j*2*pi*3*(i+N/2)/N);
            W3i3N_4 = exp(-j*2*pi*3*(i+3*N/4)/N);

            Memory( Offset + i        , Section + 1) = ...
                Ini + Wi*IniN_4     + W2i*IniN_2    + W3i*Ini3N_4;
            Memory( Offset + i + N/4  , Section + 1) = ...
                Ini + WiN_4*IniN_4  + W2iN_4*IniN_2 + W3iN_4 *Ini3N_4;
            Memory( Offset + i + N/2  , Section + 1) = ...
                Ini + WiN_2*IniN_4  + W2iN_2*IniN_2 + W3iN_2 *Ini3N_4;
            Memory( Offset + i + 3*N/4, Section + 1) = ...
                Ini + Wi3N_4*IniN_4 + W2i3N_4*IniN_2 + W3i3N_4*Ini3N_4;
        end                 %% End of radix-4 butterfly
    Offset = Offset + N;
    end
end
Output = Memory(:, 3).';
```

Side Note: The computation flow that calls for the index, *n*, scrambling at the input of the FFT is called the decimation-in-time (DIT) flow. This flow will result in an output array $X[m]$ with a monotonically increasing frequency index *m*. Literature on the FFT also mentions a decimation-in-frequency (DIF) flow, which is a simple reformulation of the FFT algorithm that maintains a monotonically increasing time index *n*, for the input samples, and ends with a reordered frequency index *m*, for the output samples. There is no difference in computational efficiency, so we will not discuss the DIF flow in this text.

Example 2.26: *Radix-4 64-Point FFT*

Now let us do a somewhat more complicated radix-4 64-point FFT/IFFT example, which is used as the main demodulator / modulator blocks in the 802.11a OFDM transceiver of chapter 6. In the scheme below, we show three base 4 numbers, which are order reversed to form the correct input index, *n*.

NORMAL INPUT INDEX	BIT REPRESENTATION	REVERSED BIT REPRESENTATION	SCRAMBLED INPUT INDEX
0	00 00 00	00 00 00	0
1	00 00 01	01 00 00	16
2	00 00 10	10 00 00	32
3	00 00 11	11 00 00	48
4	00 01 00	00 01 00	4
5	00 01 01	01 01 00	20
:	:	:	:
60	11 11 00	00 11 11	15
61	11 11 01	01 11 11	31
62	11 11 10	10 11 11	47
63	11 11 11	11 11 11	63

Figure 2-55: Reversed Bit Representation for Radix-4 64-Point FFT

The complex input index is shown in the sequence below in compact form.

$$\begin{array}{llll}
[\,0 & 16 & 32 & 48 & \quad 4 & 20 & 36 & 52 & \quad 8 & 24 & 40 & 56 & \quad 12 & 28 & 44 & 60 & \dots \\
1 & 17 & 33 & 49 & \quad 5 & 21 & 37 & 53 & \quad 9 & 25 & 41 & 57 & \quad 13 & 29 & 45 & 61 & \dots \\
2 & 18 & 34 & 50 & \quad 6 & 22 & 38 & 54 & \quad 10 & 26 & 42 & 58 & \quad 14 & 30 & 46 & 62 & \dots \\
3 & 19 & 35 & 51 & \quad 7 & 23 & 39 & 55 & \quad 11 & 27 & 43 & 59 & \quad 15 & 31 & 47 & 63\,]
\end{array}$$

The figure below illustrates a similar calculation flow as the one shown in the radix-2 16-point FFT of Example 2-21. Although the FFT length of 64 is significantly larger this time around, we actually use fewer memory/register banks to hold the intermediate results. As before, we use several variables to quantify the number and type of butterfly operations that are executed in each section. The variable *N* indicates the length of the DFT operation that is being handled in each section, while '*Reps*' (meaning repetitions) represents the number of *N*-point DFT operations. As before, the variable *i* represents a loop index indicating the number of times a butterfly operation is performed to calculate the output of the current N-point DFT element.

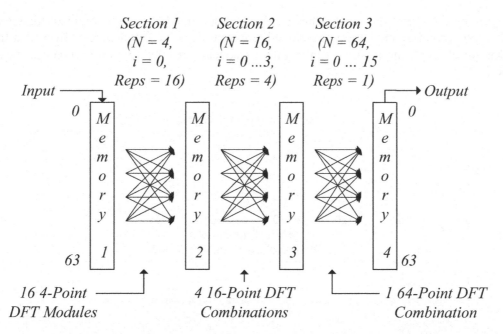

Figure 2-56: Calculation Progression of Radix-4 64-Point FFT

```
Memory     = zeros(64, 4);
Length     = 64;


%% Input Sequence cos(2 pi Freq *n/SampleRate)
n              = 0:1:Length-1;
Freq           = .05;
SampleRate     = 1;
Input          = cos(2*pi*Freq*n/SampleRate);
Order  = 1 + [0 16 32 48    4 20 36 52    8 24 40 56    12 28 44 60 ...
              1 17 33 49    5 21 37 53    9 25 41 57    13 29 45 61 ...
              2 18 34 50    6 22 38 54   10 26 42 58    14 30 46 62 ...
              3 19 35 51    7 23 39 55   11 27 43 59    15 31 47 63];
          %% We add a 1 because MatLab arrays must start at index 1 not 0
Memory(Order,1) = Input.';


for Section = 1:1:3
    N     = 4^Section; %% Current section contains N-point DFT modules
    Reps  = 64/N;       %% Number of N-point DFT modules in current section
    I     = 4^(Section-1); %% Limit for index i
    Offset = 1;
    for a = 0:Reps-1
        for i = 0:I-1       %% Start of radix-4 butterfly
            %%  Same Butterfly code as last example
        end                 %% End of radix-4 butterfly
    Offset = Offset + N;
    end
end
Output     = Memory(:, 4).';
```

The figure below shows the input sequence $x[n] = \cos(2\pi \cdot n \cdot 0.05)$ with its real and imaginary FFT outputs. Indices m beyond 31 are aliased and represent negative frequencies. Given that the input signal is purely real, the positive and negative frequency results are complex conjugates of one another.

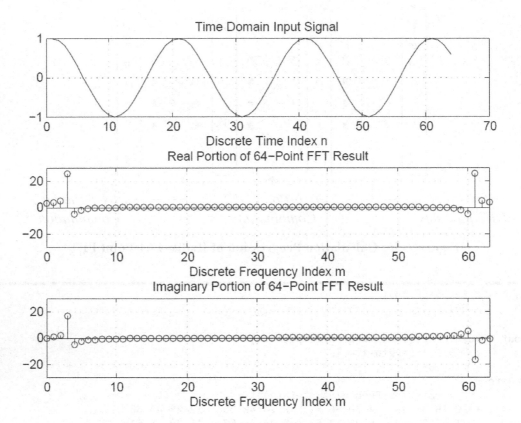

Figure 2-57: Input and Output Sequences for 64-Point Radix-4 FFT Example

2.3.7   Using the FFT Structure to Calculate the IFFT

Orthogonal frequency division multiplexing, or OFDM, is a new modulation technique that has found great popularity in the WLAN (802.11a/g/n) arena and now in the cell phone industry via the new LTE standard. OFDM uses an IFFT structure to modulate the signal and an FFT structure to demodulate it. In WLAN systems, modems communicate in half duplex mode, meaning they either transmit or receive but don't do both simultaneously. To minimize hardware resources, a single FFT hardware implementation may assume a dual-purpose role, of the modulator (in IFFT mode) and the demodulator (in FFT mode). In this section, we review the manner in which an un-normalized FFT/DFT structure may be used to calculate the IFFT/IDFT result. The equations below review the DFT and IDFT computation. We use the indices a and b to break with our tendency to think in terms of time and frequency indices.

$$Output[b] = \frac{1}{N}\sum_{a=0}^{N-1} Input[a] \cdot e^{-j2\pi ab/N} \quad \leftarrow DFT$$

$$Output[b] = \sum_{a=0}^{N-1} Input[a] \cdot e^{j2\pi ab/N} \quad \leftarrow IDFT$$

The core of both expressions is the summation of the point-by-point products of an input sequence and a complex, sinusoidal analysis function. The real difference is that the complex sinusoid for the DFT/FFT rotates in the clockwise or negative direction, while the IDFT/IFFT version has that same complex sinusoid rotating in the counter-clockwise or positive direction. To use the FFT block in the IFFT mode, we conjugate the input as well as the result. Let us show the equation for the un-normalized DFT as we have been implementing it in the last couple of examples. *Input[a]* is a vector of time domain samples, but, to the algorithm, it's just a number sequence.

$$Output[b] = \sum_{a=0}^{N-1} Input[a] \cdot e^{-j2\pi ab/N}$$

We now assume the vector *Input[a]* represents a vector of DFT outputs representing frequency samples. We now conjugate that input vector and process it through the DFT equation.

$$Output[b] = \sum_{a=0}^{N-1} conj(Input[a]) \cdot e^{-j2\pi ab/N}$$

$$= \sum_{a=0}^{N-1} conj(Input[a] \cdot e^{j2\pi ab/N})$$

$$= conj(\sum_{a=0}^{N-1} Input[a] \cdot e^{j2\pi ab/N})$$

$$= conj(IDFT)$$

Thus, if we input a conjugated frequency vector into the FFT engine, it will yield the conjugate of the IFFT result. The figure below shows the final dual-purpose implementation with the FFT path located at the top and the IFFT path at the bottom.

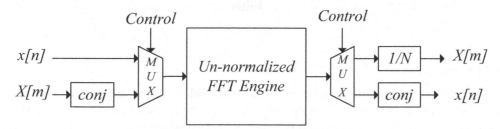

Figure 2-58: Dual Use I/FFT Engine

2.4   The Finite Impulse Response or FIR Filter

Finite impulse response filters are immensely important tools in the realm of digital signal processing and communication systems. Their ability to change the frequency spectrum of a signal makes them indispensable in several locations inside the digital portion of a communication receiver. First and foremost, digital filters are used to remove unwanted signal content such as noise and interference from our discrete waveform. Second, FIR filters are heavily used to reshape the frequency content of a signal that has suffered from linear distortion during its transmission. Such filters are commonly referred to as equalizers.

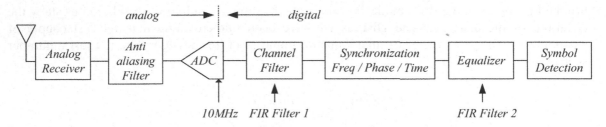

Figure 2-59: FIR Filter in a Typical Receiver Chain

The figure above shows three separate filter modules, one analog and two discrete types. To better illustrate the functionality of the different filters, we will assume that the sample rate of the system is set to $Fs = 10MHz$. It is the responsibility of the analog filter then to attenuate frequency content above the Nyquist rate of $Fs/2 = 5MHz$, which would otherwise fold or alias into the pass band of our waveform. It is for this reason that the module is called an anti-aliasing filter during the analog to digital conversion step.

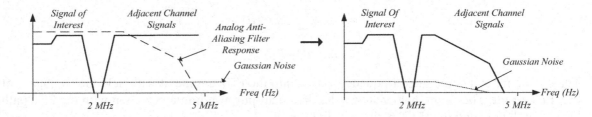

Figure 2-60: Frequency Response and Signal Spectra Before and After the Anti-aliasing Filter

Channel Filtering

If the signal we are attempting to receive has a single sided bandwidth of 2MHz, then it falls to the first FIR filter to remove all frequency content above that threshold from the sampled waveform. Since it isolates our channel of interest, it is commonly referred to as a channel filter. The frequency content beyond 2MHz usually consists of other radio transmissions on adjacent channels and Gaussian thermal noise that was originally created in the RF receiver section. You may ask why the anti-aliasing filter does not limit the signal to 2MHz right away. Why distribute this function to two separate filter modules? FIR filters have two very important advantages over their analog counterparts. First and foremost, a FIR filter can reject unwanted content far more efficiently than an analog filter. Let us say that we wanted to pass all frequency content below

2MHz and reject content above 2.5MHz by at least 50dB. The hardware resource (capacitors, operational amplifier, resistors…) of an analog filter with such specifications would be substantial. A FIR filter, consisting of nothing more than delays, multipliers, and adders, would require comparatively few resources to accomplish the same task. Second, analog filters will always introduce some small amount of phase distortion into the signal. The FIR filter, however, can be designed to be completely phase linear over its bandwidth. A filter with a linear phase response causes the input signal to experience nothing more than a simple time delay (see the time-shifting property of the Fourier transform).

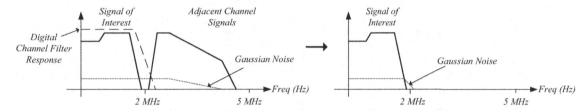

Figure 2-61: The Frequency Response of the Channel Filter

Note that even if there was no adjacent channel signal present, the filter would still maximize the signal to noise ratio by removing noise content above 2MHz.

Equalization

In communication systems, the pristine signal leaving the transmitter is often linearly distorted by the transmission channel. These phenomena, which are usually referred to as multi-path or echo effects, may alter the frequency spectrum of the original transmitted signal in very dramatic ways. In general, the equalizer is a complex FIR filter that approximately assumes the inverse frequency response of the linear distortion. If we assume that the originally transmitted signal had a flat frequency spectrum, then according to the figures in this section, the lower frequency portions of the received signal of interest have indeed suffered from some amount of unwanted attenuation. As can be seen below, the equalizer would boost these frequencies to flatten the spectrum prior to the symbol detection module. We will discuss equalizers in chapter 4.

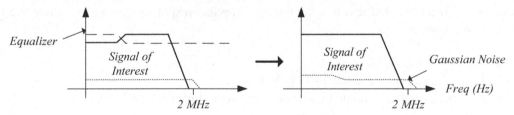

Figure 2-62: Impact of the Equalizer on the Frequency Response of Our Signal of Interest

All in all, FIR filters are flexible tools that allow us to manipulate the spectrum of signals almost arbitrarily. The design effort is more reasonable than that of its more hardware-efficient infinite impulse response (IIR) counterpart, making the FIR a very popular filter implementation.

The FIR Filter Structure

The structure of a FIR filter is that of a tapped delay line whose tap values are multiplied by a set of coefficients with the resulting products added to form one output $y[n]$. In fact, the figure below is the same figure that illustrated the linear time-invariant, or LTI, system in our earlier section on convolution. In the digital hardware realm, the implementation of such a structure is trivial. The delay elements are in fact registers that propagate the data according to the clock rate at which they are being run. Therefore, for a frequency of 10MHz, the delay through each register is 100 nanoseconds.

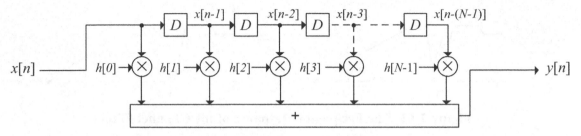

Figure 2-63: FIR Filter Structure

The mathematical operation relating the input and output of such a system is of course our well known principle of convolution.

$$y[n] = x[n] \otimes h[n] = \sum_{k=0}^{N-1} x[n-k] \cdot h[k]$$

$$= x[n] \cdot h[0] + x[n-1] \cdot h[1] + ... + x[n-(N-1)] \cdot h[N-1]$$

From sections 2.1.1., an LTI system is characterizing by its impulse response. Given that we can directly implement the above structure in digital hardware, we can arbitrarily choose the impulse response by simply assigning values to coefficients $h[k]$. As we mentioned before, filters are used to modify the frequency response of a time domain waveform. It is therefore our goal to specify the frequency response of the filter that will alter the content of our time waveform and then use some mathematical operation to find the corresponding impulse response, $h[k]$. There are three methods we will use in this text to calculate the impulse response that will produce the desired frequency behavior.

1 → The frequency sampling method.
2 → The least squares error method of FIR design.
3 → The Minimax method leading to an equiripple type filter response.

The frequency sampling method is the focus of this section, and the least squares error and Minimax techniques will be discussed in chapter 3 on optimization and approximation. The latter two methods produce an FIR filter that is optimal in the least squares and Minimax sense. Although these three methods may differ in complexity, the fundamental concept underlying them all is the convolution property of the Fourier transform, which we repeat here for convenience.

If
$$y[n] = x[n] \otimes h[n]$$
then
$$Y(f) = X(f) \cdot H(f)$$

Based on the property above, we specify the filter's frequency response, $H(f)$ or $H[m]$, and take the inverse Fourier transform to arrive at the corresponding impulse response coefficients, $h[n]$.

2.4.1   The Frequency Sampling Method of FIR Filter Design

The frequency sampling method [10, 11] of FIR filter design consists of the following steps.

1 → Specify the desired frequency response, $H[m]$.
2 → Take the inverse discrete Fourier transform that yields the impulse response, $h[n]$.
3 → Multiply this impulse response by one of many possible windowing functions.
4 → Normalize the magnitude of the impulse response to the overall desired DC gain.
5 → Compare the results with the specified frequency response using the discrete time FT and if desired pass a test signal through a filter with the final coefficients using convolution.

Defining the Frequency Response of the FIR Filter

In order to correctly define the frequency response, there are three separate vectors that we must set. We need a vector indicating the independent variable, which is the normalized frequency. Further, we require a vector specifying the magnitude response at the given frequencies and another specifying the phase response. Clearly, the vector specifying the frequencies at which we will define our response must be formatted such that the IDFT can correctly compute the impulse response. The vector defining normalized frequency obeys the format m/N where $m=0, 1 \ldots N-1$, and N represents the tap length of the FIR filter.

$$F_{Normalized}[m] = \frac{m}{N} = \left[0 \quad \frac{1}{N} \quad \frac{2}{N} \quad \frac{3}{N} \quad \cdots \quad \frac{N-1}{N} \right]$$

Example 2.27*: (Part 1) Defining the Frequency Response, H[m], of a Low Pass FIR Filter*

Assume a low pass FIR filter of tap length $N = 13$ running at a sample rate of 20MHz. It is our goal to force a response that passes frequency content below 3.5MHz (normalized frequency = 3.5/20 MHz = 0.175Hz) and blocks it everywhere else. According to the rule regarding frequency assignments above, we set the frequency vector as follows.

$$F_{Normalized}[m] = \frac{m}{N} = \begin{bmatrix} 0 & \dfrac{1}{13} & \dfrac{2}{13} & \dfrac{3}{13} & \dfrac{4}{13} & \dfrac{5}{13} & \dfrac{6}{13} & \dfrac{7}{13} & \dfrac{8}{13} & \dfrac{9}{13} & \dfrac{10}{13} & \dfrac{11}{13} & \dfrac{12}{13} \end{bmatrix}$$

Remember that the frequencies larger than 0.5 are in fact aliased to negative frequencies and appear at f-1. At our convenience, we may therefore redefine the vector as follows.

$$F_{Normalized}[m] = \frac{m}{N} = \begin{bmatrix} 0 & \dfrac{1}{13} & \dfrac{2}{13} & \dfrac{3}{13} & \dfrac{4}{13} & \dfrac{5}{13} & \dfrac{6}{13} & \dfrac{-6}{13} & \dfrac{-5}{13} & \dfrac{-4}{13} & \dfrac{-3}{13} & \dfrac{-2}{13} & \dfrac{-1}{13} \end{bmatrix}$$

As mentioned before, the actual frequencies are equal to the normalized frequencies times the sample rate.

$$F_{Actual} = SampleRate \cdot F_{Normalized}$$

We now define the magnitude and phase response of the filter. We force a magnitude response of unity at the normalized frequencies -2/13 ≈ -3.1MHz, -1/13, 0/13, 1/13, and 2/13 ≈ 3.1MHz while leaving all others at zero. This even-functioned magnitude behavior is the first step in creating a filter impulse response, or coefficient vector that is completely real valued. The second step is to choose a phase response that is an odd function of frequency. Here we set the phase response to zero everywhere.

$$MagnitudeResponse = Mag[m] = [1\ 1\ 1\ 0\ 0\ 0\ 0\ 0\ 0\ 0\ 0\ 1\ 1]$$

$$PhaseResponse = Phase[m] = [0\ 0\ 0\ 0\ 0\ 0\ 0\ 0\ 0\ 0\ 0\ 0\ 0]$$

In order to get a meaningful result from the inverse Fourier transform, we must first convert the frequency response to its rectangular form.

$$H[m] = Mag[m] \cdot cos(Phase[m]) + j \cdot Mag[m] \cdot sin(Phase[m])$$

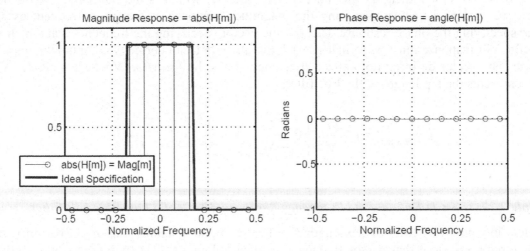

Figure 2-64: Magnitude and Phase Response Specifications for Low Pass Filter

Example 2.27: (Part 2) Computing the IDFT

Once $H[m]$ is known, we use the *IDFT* to compute the impulse response, which may or may not be complex. Since we have set the magnitude response to be even and the phase response to be an odd function of frequency, the impulse response will be completely real.

$$h[n] = \sum_{m=0}^{12} H[m] \cdot e^{j2\pi mn/13} \quad \leftarrow \textit{The IDFT calculation}$$

```
%% Step 1: Defining the Frequency Response
N              = 13;                        %% Sequence Length
MagResponse    = [1 1 1 0 0 0 0 0 0 0 0 1 1];
PhaseResponse  = [0 0 0 0 0 0 0 0 0 0 0 0 0];
H              = MagResponse.*cos(PhaseResponse) + ...
                 j*MagResponse.*sin(PhaseResponse);

%% Step 2: Taking the Inverse Discrete Fourier Transform
n              = (0:N-1) - floor(N/2); %% Time index at which to evaluate
                                       %% the IDFT
h = zeros(1, N);                       %% Impulse response h[n]
for m = 0:N-1                          %% IDFT calculation
    f          = m/N;                  %% Normalized frequencies
    h          = h + H(1,m+1)*exp(j*2*pi*n*f);
end
```

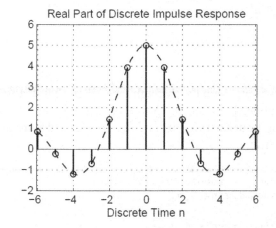
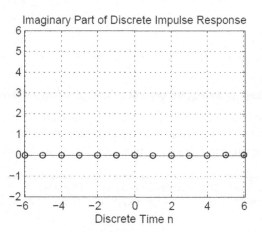

Figure 2-65: The Discrete Impulse Response $h[n]$

Note that we chose to evaluate the inverse DFT at $n = (0:N\text{-}1) - \text{floor}(N/2)$ resulting in $n = $ -6, -5, … 5, 6 rather than $n = 0, 1 … N\text{-}1$ (I knew in advance that the impulse response would be centered at time equal to zero because of the specification that the phase response should be zero everywhere). Requesting an impulse response value at a time of $n < 0$ is a dubious thing to do since it would appear as if we were creating a non-causal filter. If this makes you uneasy, then read the next two bullets; if not, you may want to skip them. There are two ways to accommodate our feeling of foreboding regarding this seemingly illegal concept.

1. We may adjust the desired phase response to introduce a time delay equal to or greater than $N/2$. If you review the time-shifting property of the Fourier transform, you will know exactly how to force the delay. We discuss an example (in chapter 3) on interpolation using FIR filters, where that same delay technique is used.

2. Specifying the all-zero phase response was done for convenience and easy visualization. Imagine wanting to tweak the phase response slightly to undo some type of phase distortion in the input signal. Always having to include the time delay feature would make plotting the desire phase response much less of a visual aid. So, what we do instead is request a phase response that excludes the time delay, and then simply assign the IDFT values at $n = -6, -5 \ldots 6$ to FIR tap multipliers at $n = 0, 1 \ldots 12$. This simple act implements the delay, which, for convenience, we avoided specifying.

Normalizing the Impulse Response

The purpose of digital filters is to pass some frequencies unhindered and reject others that may feature noise content or interference. Assuming that we want neither to increase nor to decrease the magnitude of those frequencies that we wish to pass, we need to scale the impulse response to assure that the gain in the pass band is in fact unity. In the case of the low pass filter, we fix the DC gain at unity by making sure the sum of all filter coefficients features a magnitude of 1.0. Therefore, the un-normalized filter impulse response, $h[n]$, is rescaled as follows.

$$h_{norm}[n] = \frac{1}{|\sum h[n]|} \cdot h[n]$$

Checking the Result via the DTFT

We have computed the impulse response given the magnitude and phase response at orthogonal frequencies m/N. Clearly, however, we will be passing signals through our filter that feature content at all frequencies. What does the actual response between the orthogonal frequencies look like? To find that out, we compute the discrete time Fourier transform of the normalized impulse response at a high frequency resolution, as seen below.

```
%% Step 2b: Normalize h[n] and check the frequency response H(f) via DTFT
h_norm          = (1/abs(sum(h)))*h;    %% Normalizing h[n]


Frequencies  = -0.5:.002:0.5;            %% Resolution = 0.002Hz
DTFT_Output  = zeros(1,length(Frequencies));


for i=1:length(Frequencies)              %% The DTFT operation
    f                = Frequencies(1,i);
    AnalysisTone     = exp(j*2*pi*n*f);
    DTFT_Output(1,i) = h_norm*AnalysisTone';
end
MagnitudeResponse     = abs(DTFT_Output);
MagnitudeResponse_dB  = 20*log10(abs(DTFT_Output));
```

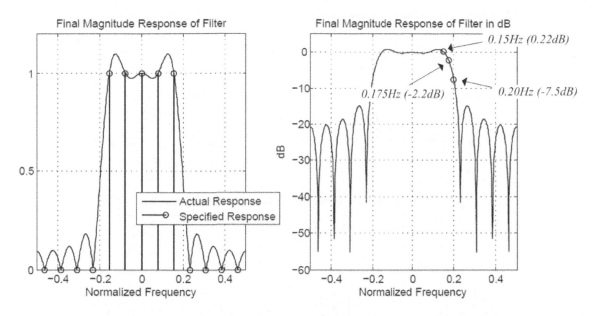

Figure 2-66: Magnitude Response H(*f*) (Linear and in dB)

Examining the magnitude response of our filter more closely, we find that rejections of 0.22 dB, -2.2 dB, and -7.5 dB are achieved at 0.15Hz (or 3MHz scaled), 0.175Hz (or 3.5MHz scaled), and 0.2Hz (or 4MHz scaled), respectively. The initial pass/stop band transition point requirement of 3.5MHz is met reasonably well. However, upon closer examination, the amount of rejection in the stop band (beyond 3.5MHz) is somewhere in the neighborhood of 20dB, which is hardly something to write home about. To improve the rejection, we can either increase the number of taps in the filter or, more importantly, multiply a window function on top of the calculated impulse response.

Example 2.27: *(Part 3) Adding a Window*

In section 3.3 of this chapter, we discussed the importance of multiplying windows on top of test sequences both to limit their time span to $n = 0, 1 \ldots N\text{-}1$ and to reduce the frequency leakage effects associated with the limiting operation. The excessive side lobes due to the time truncation of our calculated filter impulse response, $h[n]$, causes deterioration in the filters stop band rejection, as we can clearly see in the last figure. As was the case for the DTFT calculation before, here too, the only options are to increase the number of taps of the filter or enhance the impulse response via a window. In part three of this example, we will employ the customized Hanning window to optimize the frequency response of our filter to our liking.

$$Custom\_Hanning = (0.5 - 0.5\ cos(2\pi\ (n\text{+}1)/(N\text{+}1)\ ))^b$$

After overlaying the custom Hanning window on top of $h[n]$, the final impulse response should again be scaled such that the overall DC gain of the FIR filter becomes unity.

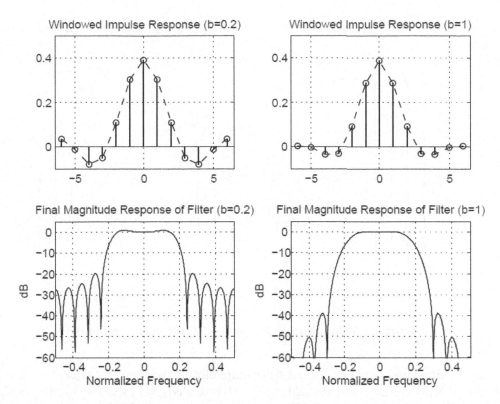

Figure 2-67: Performance Comparison Between Different Custom Hanning Windows

The figure above illustrates the impact of a custom Hanning window with index $b = 0.2$ and that of the default version with index $b = 1.0$. The impulse response modified via the default Hanning window features a less aggressive pass to stop band skirt but better stop band rejection. After the design steps are finished, and the frequency response is available for inspection, our last verification step should be to exercise the FIR with a time domain signal and observe the output.

```
%% Step 3: Adding a custom Hanning Window
n        = 0:1:N-1;
b        = 1;                   %% Exponent of customizabled Hanning window
CustHann = (0.5 - 0.5*cos(2*pi*(n+1)/(N+1))).^b;
New_h    = (1/abs(sum(h.*CustHann)))*h.*CustHann; %% Renormalized windowed
                                                  %% Impulse response
%% Step 3: Computing the new Frequency Response via the DTFT
Frequencies = -0.5:.002:0.5;
DTFT_Output  = zeros(1,length(Frequencies));

for i=1:length(Frequencies)
    f            = Frequencies(1,i);
    AnalysisTone = exp(j*2*pi*n*f);
    DTFT_Output(1,i) = New_h*AnalysisTone';
end
MagnitudeResponse    = abs(DTFT_Output);
MagnitudeResponse_dB = 20*log10(abs(DTFT_Output));
```

Example 2.27: *(Part 4) A Quick Reality Check Using Convolution*

Once the design effort is completely finished, there is nothing quite as satisfying as seeing the FIR filter in action. Does it really perform according to the magnitude response that was indicated by final DTFT verification? Why not find out by actually convolving a test waveform with the calculated impulse response? The test waveform below features components at a frequency of 20MHz·1/13 = 1.54MHz and 20MHz·4/13 = 6.15MHz.

$$TestWaveform[n] = cos(\ 2\pi n\ 1/13\ ) + cos(2\pi n\ 4/13)$$

$$= e^{\ j2\pi n\ 1/13} + e^{-j2\pi n\ 1/13} + \ e^{\ j2\pi n\ 4/13} + e^{-j2\pi n\ 4/13}$$

The normalized impulse response was derived in part three of this example using the default Hanning window (*b*=1). The coefficients are normalized to feature a unity DC gain and empower the filter to elegantly suppress the complex tones at 20MHz·4/13 = 6.15MHz.

$$h_{normalized}[n] = [0.0032 \quad -0.0037 \quad -0.0363 \quad -0.0331 \quad 0.0895 \quad 0.2871 \quad 0.3866 \quad 0.2871$$
$$0.0895 \quad -0.0331 \quad -0.0363 \quad -0.0037 \quad 0.0032\ ]$$

```
%% Defining the Composite Input Sequence
n           = 0:200;
TestWaveform = cos(2*pi*n*1/13) + cos(2*pi*n*4/13);
Output      = zeros(1,length(TestWaveform));
DelayLine   = zeros(1,13);
h           = [0.0032 -0.0037 -0.0363 -0.0331  0.0895 0.2871  0.3866  ...
              0.2871  0.0895 -0.0331 -0.0363 -0.0037  0.0032];
%% The Convolution Operation
for n = 1:length(TestWaveform)
    DelayLine(1,2:13) = DelayLine(1,1:12);   %% Time advance
    DelayLine(1,1)    = TestWaveform(1,n);   %% New sample enters
    Output(1,n)       = DelayLine*h.';       %% Calculating the output
end
```

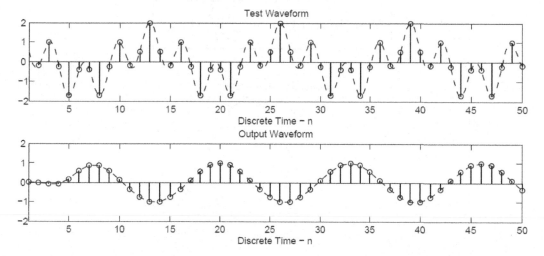

Figure 2-68: Testing Low Pass FIR filter via Multi-tone Input Test Waveform

2.4.2 Understanding the Phase and Group Delay

The phase and group delays are two common metrics that illustrate phase distortion in linear systems such as filters. These metrics calculate the transit time that a sinusoid at a particular frequency or groups of sinusoids at different but close frequencies experience as they traverse a filter. In communication systems, a great number of filters are tasked to attenuate interference and noise outside the bandwidth of the signal of interest. The time that it takes each frequency to traverse these filters should be the same. If this is not the case, then the different frequency components of the signal in the pass band will reassemble out of phase at the output, causing linear distortion. The phase and group delay function are tools that allow us to easily visualize the presence of phase distortion. The phase response of a linear system that does not introduce phase distortion and thus delays all frequency components equally is a straight line as is suggested by the time shifting property of the Fourier transform (see Section 3.1 of this chapter.)

$$x(t) \quad \xrightarrow{FT} X(f)$$

$$x(t - t_0) \xrightarrow{FT} X(f) \cdot e^{-j2\pi f t_o}$$

Phase and group delay are functions of frequency and can be calculated directly from the phase response of the linear time invariant system.

Phase Delay

The *phase delay* is a measure of transit time, t_o, experienced by a complex sinusoid, $exp(j2\pi ft)$, as it travels through a linear time-invariant system such as a filter. The transit time is calculated by comparing the input and output phases of the complex sinusoid.

$$\theta(f) = \angle Output(f) - \angle Input(f) = 2\pi f(t - t_o) - 2\pi ft$$

$$\theta(f) = -2\pi ft_o$$

$$\boxed{PhaseDelay(f) = t_o = -\frac{\theta(f)}{2\pi f} \, seconds}$$

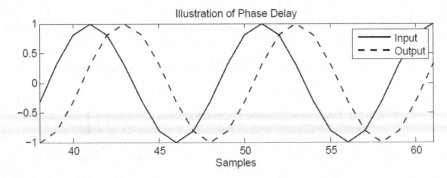

Figure 2-69: Input and Output Sinusoid with a Phase Delay of Two Samples

The Group Delay

The *group delay* represent the transit time through a filter of the amplitude envelope of a modulated signal. For a signal to feature an amplitude envelope at least two sinusoids must be present as is seen in the figure below.

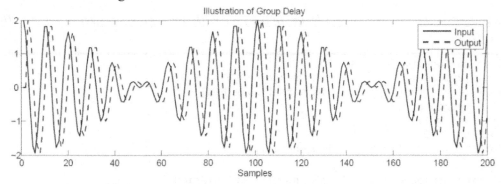

Figure 2-70: Input and Output Sinusoid Pairs Featuring a Group Delay of Two Samples

Consider the input of a linear time invariant system to be the sum of two complex sinusoids, one at frequency $f + 0.5 \cdot \Delta f$ and the other at frequency $f - 0.5 \cdot \Delta f$. The two sinusoids are passed through a filter, which delays the input by a time equal to t_o.

$$Input[n] = e^{j2\pi(f+0.5\Delta f)t} + e^{j2\pi(f-0.5\Delta f)t}$$

$$Output[n] = Input[t - t_o] = e^{j2\pi(f+0.5\Delta f)(t-t_o)} + e^{j2\pi(f-0.5\Delta f)(t-t_o)}$$

The phase responses, $\theta(f)$, of the filter at the two frequencies, $f + \Delta f$ and $f - \Delta f$, is shown next.

$$\theta(f + 0.5\Delta f) = \angle Output(f + 0.5\Delta f) \quad - \angle Input(f + 0.5\Delta f)$$
$$= 2\pi(f + 0.5\Delta f)(t - t_o) - 2\pi(f + 0.5\Delta f)(t) = -2\pi(f + 0.5\Delta f)t_o$$

$$\theta(f - 0.5\Delta f) = \angle Output(f - 0.5\Delta f) \quad - \angle Input(f - 0.5\Delta f)$$
$$= 2\pi(f - 0.5\Delta f)(t - t_o) - 2\pi(f - 0.5\Delta f)(t) = -2\pi(f - 0.5\Delta f)t_o$$

Let us now see what happens as we subtract the phase response at $f - 0.5 \cdot \Delta f$ from that at $f + 0.5 \cdot \Delta f$ and finally allow the value of Δf to approach zero.

$$\theta(f + 0.5\Delta f) - \theta(f - 0.5\Delta f) = -2\pi(f + 0.5\Delta f)t_o + 2\pi(f - 0.5\Delta f)t_o$$
$$= -2\pi 0.5\Delta f t_o + 2\pi 0.5(-\Delta f)t_o$$
$$= -2\pi \Delta f t_o$$

$$\boxed{GroupDelay(f) = t_o = -\frac{1}{2\pi}\frac{\theta(f + 0.5\Delta f) - \theta(f - 0.5\Delta f)}{\Delta f} = -\frac{1}{2\pi}\frac{d\theta(f)}{df} seconds}$$

The group delay is proportional to the slope of the phase response, $\theta(f)$, while the phase delay is proportional to the ratio of phase response and frequency. If no phase distortion exists in the phase response of a filter, then the equations for group and phase delay produce the identical result, t_o. If phase distortion does exist, the two equations will produce slightly different results. The example below calculates the group delay for two different impulse responses.

_Example 2.28__: Compute the Group Delay of a Filter with an Impulse Response for h = [0 1 1 0]_ _and h = [0 .5 1 0]_

```
%% Define an Impulse Response
h       = [0 1 1 0];              %% Or h[n] = [0 .5 1 0]
h_norm  = (1/abs(sum(h)))*h;      %% Normalizing h[n]
N       = length(h_norm);
n       = 0:1:N-1;
Frequencies  = -0.5:.002:0.5;
DTFT_Output  = zeros(1,length(Frequencies));

for i=1:length(Frequencies)       %% The DTFT calculation
    f               = Frequencies(1,i);
    AnalysisTone    = exp(j*2*pi*n*f);
    DTFT_Output(1,i) = h_norm*AnalysisTone';
end


MagnitudeResponse   = abs(DTFT_Output);
PhaseResponse       = unwrap(angle(DTFT_Output));

%% The Group Delay Calculation
GroupDelay          = zeros(1, length(Frequencies));
for i = 2:length(Frequencies)-1
    FrequencyDelta      = Frequencies(1, i+1) - Frequencies(1, i-1);
    PhaseResponseDelta = PhaseResponse(1,i+1) - PhaseResponse(1, i-1);
    GroupDelay(1,i)     = (-1/(2*pi))* PhaseResponseDelta/FrequencyDelta;
end
GroupDelay(1,1) = GroupDelay(1,2);
GroupDelay(1,length(Frequencies)) = GroupDelay(1,length(Frequencies)-1);
```

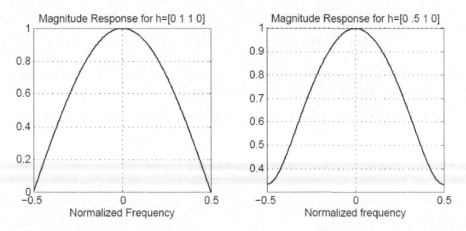

Figure 2-71: Magnitude Response for Filters with h = [0 1 1 0] and h = [0 0.5 1 0]

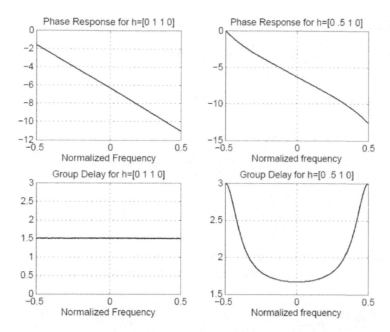

Figure 2-72: Phase Response and Group Delay Calculation for Filters with Impulse Responses h = [0 1 1 0] and h = [0 0.5 1 0]

2.5 Correlation and Filtering in Software Defined Radios

In this section, we will take another look at correlation and digital filtering in the time domain and explore how we could use the FFT to execute these same processing steps. The immediate question that everybody should ask themselves is why we would want to use a frequency domain algorithm to compute something that can easily be done in the time domain. The simple answer is that in software (defined radios) the use of the FFT in both filtering and correlation is far more efficient than the time domain equivalent. For guidance, let us take a look at the generic time domain structure below, which allows us to both convolve $x[n]$ with $h[n]$ and execute a sliding correlation between $x[n]$ a $h[n]$.

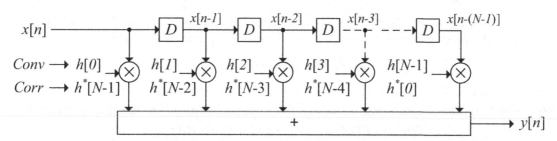

Figure 2-73: Simple Time Domain Transversal Filter Structure

In section 2.4, we find this very same diagram used as a FIR filter structure, which convolves an input sequence $x[n]$ with the discrete impulse response $h[k]$ to produce a filtered output $y[n]$.

The FIR Filtering (Convolution) Expession

$$y[n] = x[n] \otimes h[n] = \sum_{k=0}^{N-1} x[n-k] \cdot h[k] \tag{1}$$

$$= x[n] \cdot h[0] + x[n-1] \cdot h[1] + \ldots + x[n-(N-1)] \cdot h[N-1]$$

The Discrete Correlation Expression

Rememer that the simple correlation between two sequences $x[n]$ and $h[n]$ is provided by the expression below and discussed in section 2.2.

$$R_{xh} = \frac{1}{N} \sum_{n=0}^{N-1} x[n] \cdot h^*[n]$$

However, when we are trying to detect a known sequence that is buried in an incoming signal $x[n]$, we execute a sliding correlation and mathematical expression changes as follows.

$$R_{xh}[n] = \frac{1}{N} \sum_{k=0}^{N-1} x[n+k-Delay] \cdot h^*[k] \tag{2}$$

In software base radio applications, were the entire input sequence has been captured before we compute the correlation, the delay in the the formula is equal to zero. In the case of a real-time hardware sliding correlation, as shown in the last figure, the delay is N-1. This sliding correlation process is commonly used to detect synchronization signals in communication waveforms.

From our previous discussions on filtering, we know that steeper filtering skirts can only be achieved if we allow for plenty of coefficients and thus lots of complex multiplications. In fact, computing a single filtered output sample, $y[n]$, or correlation result, $R_{xh}[n]$, requires N complex and N-1 complex additions. With modern sub-micron CMOS processes, these multipliers can all be instantiated as digial hardware to execute in parallel, and an output $y[n]$ can be computed for every clock cycle. Implementing the time domain expressions above in software is straight forward but as we don't have access to parallel complex multipliers, the effort and time required to filter an input waveform is enormous. A better way must be found.

2.5.1 Digital Filtering using the FFT

To understand how to use the FFT in the digital filtering process we recall the convolution property of the Fourier transform which states the following well known fact.

$$y[n] = x[n] \otimes h[n] \rightarrow FFT \rightarrow Y[m] = X[m] \cdot H[m]$$

Review section 2.3.1 where we had derived this result for the continuous time case stating that $Y(f) = X(f) \cdot H(f)$. The discrete time version of this expression appears to suggest the following:

$$Y[m] = FFT(x[n]) \cdot FFT(h[n]) = X[m] \cdot H[m]$$

Followed by

$$y[n] = IFFT(Y[m])$$

Note that the input sequence $x[n]$ is likely very long requiring that we break it up into separate sections that feature a convenient length for our FFT operation. Furthermore, as the discrete impulse doesn't change over time, we only take its FFT once when the software is booting up and simply apply the result $H[m]$ afterwards. Assuming a section index, k, let's rewrite the expression as follows.

$$y_k[n] = IFFT(FFT(x_k[n]) \cdot H[m])$$

The equations above leads us to a straight forward, but somewhat premature conclusion, which indicates the following course of action.

→ Zero-pad the coefficient vector, $h[n]$, until its length is equal to that of $x[n]$, and then take the FFT to produce $H[m]$. As mentioned above, this is only done only once.
→ Take the FFT of the input section $x_k[n]$ to produce $X_k[m]$.
→ Take the IFFT of product of $X_k[m]$ and $H[m]$ to arrive at the filtered output section, $y_k[n]$.
→ Repeat the last two steps for each section k of the total input signal $x[n]$.

The steps above is essentially correct, but we must take care of how we select the sections of the input waveform. In *method 1*, which is suboptimal, we break up the input waveform $x[n]$ into different contiguous section $x_k[n]$ each featuring 2048 samples as follows.

$$x_k[n] = x[n + k \cdot 2048] \quad n = 0, 1...2047$$

The output waveform $y[n]$ is generated by simply concatenating the individual output sections $y_k[n]$. The problem with this approach is that it can produce discontinuities at the boundaries of the output sections. In part *a*) of the figure below, observe the real continuous waveform, $x[n]$, as it progresses from section 1 through section 3. Once we isolate section 2 to form a vector $x_2[n]$, we take its Fourier transform, remove some of the frequency information by multiplying via $H[m]$ and then reassemble the time domain waveform via the IFFT. You can see the result of a hypothetical filtering operation in part *b*) of the figure below. Notice that each section $x_k[n]$ has been separately low pass filtered. During the filtering operation of section 2, the algorithm was oblivious to what the samples of section 1 and 3 looked like. As a matter of fact, once we take the Fourier transform of $x_2[n]$, the reverse process via the inverse Fourier transform would produce a periodic waveform if we were to evaluate outside of its target sample range of 0 to 2047.

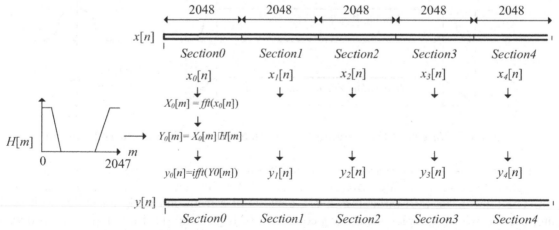

Figure 2-74: Filtering Method Processing Contiguous Section of the Input Waveform

Thus samples $y_2[2048, 2049, ...]$ would be identical to samples $y_2[0, 1, ...]$. The more signal content that is filtered away, the larger this problem becomes

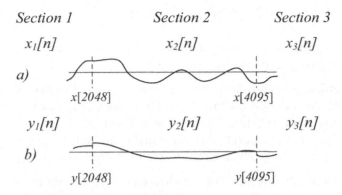

Figure 2-75:

In a time-domain filter implementation, the situation is very different, as filtering any sample in $x[n]$ requires that both past samples ($x[n-1]$, $x[n-2]$, ..) and future samples ($x[n+1]$, $x[n+2]$, ...) are accessible in the FIR filter's shift register.

Method 2: Selecting Sections Properly

The next figures illustrate the proper way of solving the discontinuity problem associated with frequency domain filtering. This method still divides the input waveform $x[n]$ into sections of length 2048, but this time there are overlapping regions of length 256 samples. Sections $x_1[n]$, $x_2[n]$, and $x_3[n]$ span samples $x[0 ... 2047]$, $x[1792 ... 3839]$, $x[3584 ... 5631]$ and so forth.

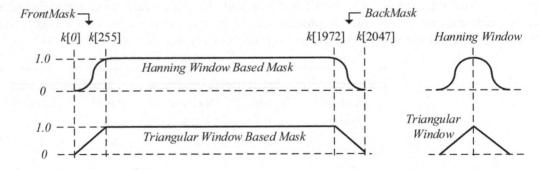

Figure 2-76: Using Ovelapping Section to Solve the Discontinuity Problem

We multiply a mask onto each section to smoothly force the sample values at the start and end of each section to zero. There are different possibilities for this mask and we showed just two of them. The total mask, $k[0]$ through $k[2047]$, consists of a front and back portion as well as the center section whose samples are always equal to 1.0. If you add the front and back portions of the mask, all samples must also add to 1.0. The front and back masks can be the first and seconds halves of a *Hanning* or triangular window as shown in the figure above. The equation for the Hanning window is provided below. The total length of the window would be $N = 512$, which splits up into two halves each with a length of 256 samples.

$$Hanning[n] = 0.5 - 0.5cos(\frac{2\pi(n+1)}{N+1})$$

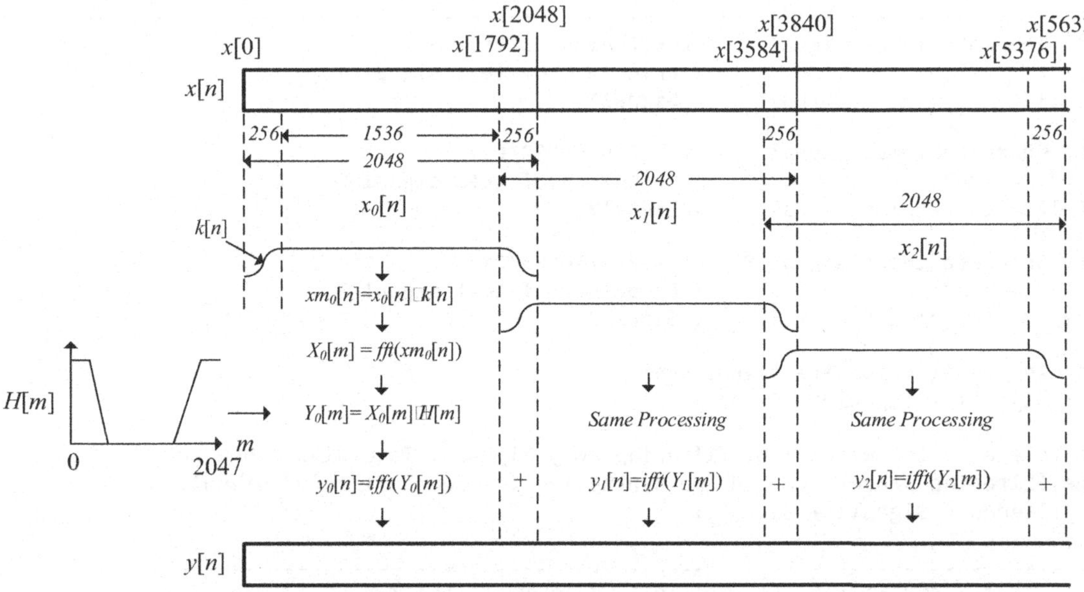

Figure 2-77: The Filtering Process Using Overlapping Sections

The figure above illustrates the processing steps for sections $x_1[n]$, $x_2[n]$, and $x_3[n]$, which produce output sections $y_1[n]$, $y_2[n]$, and $y_3[n]$. The big difference to the first method is that that each section now has knowledge of the final samples of the last section and initial samples of the next section as sections now overlap. Take care to overlap these output sections as well to yield the final output waveform $y[n]$.

***Example**: FFT Based Filtering using Both Non-Overlapping and Overlapping Sections*

In the following example, we will write MatLab code that executes both methods of frequency domain filtering. The sample rate of the signal is set to Fs = 1 MHz, and the input signal, $x[n]$, and filter magnitude response are defined as follows. Note further, that in this example the sections are 4096 samples long, and the filter will eliminate the tone at 75KHz.

$$x[n] = \cos\left(\frac{2\pi \cdot n \cdot 3000}{Fs}\right) + \cos\left(\frac{2\pi \cdot n \cdot 10000}{Fs}\right) + \cos\left(\frac{2\pi \cdot n \cdot 75000}{Fs}\right)$$

$$H(f) = \begin{bmatrix} 1 \ from - 39KHz \ to + 39KHz \\ 0 \qquad\qquad otherwise \end{bmatrix}$$

```
% --------------------------------------------------------------
% 0. Simulation Setup
Fs  = 1e6;                  % Sample Rate in Hz
```

Digital Signal Processing in Modern Communication Systems (Edition 2)

```
Len = 4*4096;                % Number of samples in test waveform
n   = 0:Len-1;               % The sample indices
disp(['Nyquist range spans: ' num2str(-Fs/2) ' to ' num2str(Fs/2) ' Hz.']);


% --------------------------------------------------------------------------
% 1a. Generate input signal 1 -> cos(2*pi*n*F1/Fs)
F1      = 3e3;                % Frequency of test signal 1
Signal1 = cos(2*pi*n*F1/Fs); % Signal1

% 1b. Generate input signal 2 -> cos(2*pi*n*F2/Fs)
F2      = 10e3;              % Frequency of test signal 2
Signal2 = cos(2*pi*n*F2/Fs); % Signal2

% 1c. Generate input signal 3 -> cos(2*pi*n*F3/Fs)
F3      = 75e3;              % Frequency of test signal 3
Signal3 = cos(2*pi*n*F3/Fs); % Signal3

% 1d. The final composite signal x[n]
x_n = Signal1 + Signal2 + Signal3;

% In this exercise we will be filtering away Signal3. To evaluate how well
% the filtering process worked, we synthesize the ideal filtered signal.
x_n_Filtered = Signal1 + Signal2;


% --------------------------------------------------------------------------
% 2. Defining the Filter frequency Response. Note, the frequency response will
% be entirely real valued.
% Our goal is the reject the signal with Frequency F3.
% Remember, that the frequency step is equal to Fs/SectionLength
NumSections         = 4;
SectionLength       = 4096;
FStep               = Fs/SectionLength;
disp(['Frequency Step: ' num2str(FStep)]);
H_m                 = zeros(1, SectionLength);
H_m(1,1:160)        = ones(1, 160);   % Define passband for positive frequencies
H_m(1,end-158:end)  = ones(1, 159);   % Define passband for the negative frequencies


% ----------------------------------------------------
% 3a. Method 1: Non-Overlapping sections
y1_n = zeros(1, length(x_n));
for SectionIndex  = 0:(NumSections-1)
    Range          = (1:SectionLength) + SectionIndex*SectionLength;
    Section        = x_n(1, Range);
    DFT_Section    = fft(Section);
    Y_m            = H_m .* DFT_Section;
    y1_n(1, Range) = ifft(Y_m);
end


% ----------------------------------------------------
% 3b. Method 2: Overlapping Sections
HannLength    = 256;                 % 256 samples long - try 512 or 128 if you like
HanningWindow = hann(HannLength)';
FrontMask     = HanningWindow(1, 1:HannLength/2);              % 128 samples long
BackMask      = HanningWindow(1, (1+HannLength/2):HannLength); % 128 samples long
TotalMask     = [FrontMask, ones(1,SectionLength-HannLength), BackMask]; % 4096 samples
y2_n          = zeros(1, length(x_n));
for SectionIndex  = 0:(NumSections-1)
```

```
   Range           = (1:SectionLength) + SectionIndex*(SectionLength - HannLength/2);
   Section         = x_n(1, Range).*TotalMask;
   DFT_Section     = fft(Section);
   Y_m             = DFT_Section .* H_m ;
   OutputSection   = ifft(Y_m);
   y2_n(1, Range)  = y2_n(1, Range) + OutputSection;
end
```

Observe the first subplot of the figure below, which illustrates a small section of what would otherwise be a very busy plot. We see a certain number of samples of the original waveform $x[n]$ as well as what we would be expecting the filtered waveform to look like. The sample range was chosen to span the boundary between input sections 1 and 2. Subplot 2 illustrates the output of filtering method 1 and 2, and the discontinuity at the section boundary around sample 4086 is clearly visible for the filtered waveform produced by method 1. The third subplot illustrates the very small error between the ideally filtered signal, that only consists of sinusoids at 3KHz and 10KHz, and the filtered signal of method 2, thus proving its effectiveness. Clearly the error is tiny compared to the one from method 1. Notice that however small the error is between the ideal filtered waveform and the actual counterpart, it does reach a maximum just before the end of the section and thus inside the overlapping region.

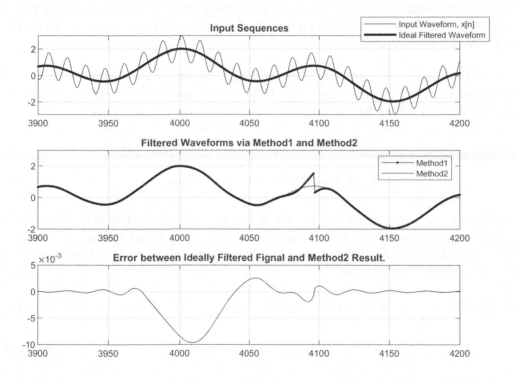

Figure 2-78: Filtering Performance for Method 1 (Non-Overlapping Sections) and Method 2 (Overlapping Sections)

2.5.2   Discrete Correlation using the FFT

As mentioned at the start of section 2.5, software based radios will significantly benefit from the use of the FFT to filter incoming signals and to detect embedded synchronization sequences within them via sliding correlation. If we assume that a receiver fully captures the input sequence $x[n]$ and passes it off to the software routines for further processing then the delay in equation (2) of section 2.5 for sliding correlation becomes zero and simplifies as follows.

$$R_{xy}[n] = \frac{1}{N}\sum_{k=0}^{N-1} x[n+k] \cdot h^*[k]$$

Just as we did in the last section, we now need to find the frequency domain equivalent to the above expression. For the convolution of an input signal $x[n]$ and the impulse response $h[n]$, this frequency expression was $Y[m] = X[m] \cdot H[m]$. For the sliding correlation process, the frequency expression is similarly simple
$R_{xy}[m] = X[m] \cdot H[m]^*$

At the end of this section, we will show how the time-domain sliding correlation expression reduces to the frequency domain equivalent. But for now, let us first discuss how to use it properly. Just as we did in the last section, we assume that the input sequence $x[n]$ is very long and needs to be divided into different sections of a length that is convenient for the FFT, such as $N = 2^n$. We further assume that the length, L, of the synchronization signal, $h[n]$, is smaller than the FFT size N. Our course of action is now as follows.

→ Zero pad the synchronization sequence, $h[n]$, until its length is equal to N. We now take the Fourier transform to arrive at the expression $H[m]$. As was the case in the last section, $H[m]$ needs to be computed only once when the software first boots up as $h[n]$ never changes.
→ From the input signal, $x[n]$, separate out the first section of N samples. Just as in the last section we will use the section index k resulting in sections called $x_k[n]$.

$$R_{xh0}[n] = IFFT(FFT(x_0[n]) \cdot H[m]^*)$$

Notice that the last L samples in the correlation result $R_{xh0}[n]$ are useless to us due to the fact that the correlation is circular. Assume that the length of a section and that of the synchronization sequence $h[n]$ are equal to $N = 2048$ and $L = 128$ respectively. The correlation result at sample $n = 2046$ will actually look as follows.

$$R_{xh0}[2046] = \frac{1}{N}(x[2046] \cdot y[0] + x[2047] \cdot y[1] + x[0] \cdot y[2] + \dots + x[125] \cdot y[127])$$

As there are no samples beyond $x[2047]$, the computation reuses values from the start of the section $x_0[n]$. For our application this is an error and we therefore discard that last L result samples. The procedure of dividing up the different section and processing them properly is described in the figure below. Once all outpur sections are computed, they are simply concatenated to form the final sliding correlation result $R_{xh}[n]$.

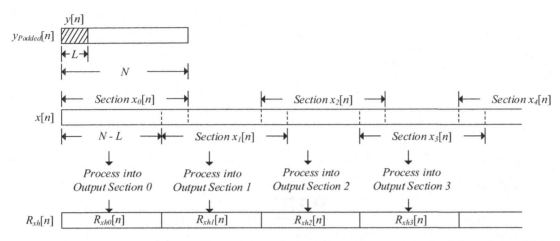

Figure 2-79: Procedure for Section Processing during FFT Based Sliding Correlation

References

[1] IEEE 802.11: Wireless LAN Medium Access Control (MAC) and Physical Layer (PHY) Specifications (2007 revision), 12 June 2007

[2] Lee J. and Miller L. (1998), *CDMA Systems Engineering Handbook*, Artech House Publishers, 425-478

[3] Fourier, Baron J. B. J. (trans. A. Freeman) (1878), *The Analytical Theory of Heat*, Cambridge University Press, Cambridge, UK

[4] Gauss, C. F. (1866), *Theoria interpolationis methodo nova tractata*, Werke, Königliche Gesellschaft der Wissenschaften, Göttingen, Band 3, 265–327

[5] Cooley, J. W. and Tukey, J. W. (1965). *An algorithm for the machine calculation of complex Fourier series, Math. Comput.* 19, (90): 297-301.

[6] Yavne R. (1968), *An economical method for calculating the discrete Fourier transform, Proc. AFIPS Fall Joint Computer Conf.*, 33: 115–125

[7] P. Duhamel and H. Hollmann, *"Split-Radix FFT Algorithm,"* Electronics Letters, Vol. 20, No. 1, pp. 14-16, 1984.

[8] Vetterli M., Nussbaumer H. J. (1984), *Simple FFT and DCT algorithms with reduced number of operations*, Signal Processing, 6 (4): 267–278

[9] Widhe T., Melander J., and Wanhammar L. (1997), *Design of efficient radix-8 butterfly PEs for VLSI, IEEE Intern. Symp. on Circuits and Systems* (ISCAS), 3: 2084 -2087

[10] Oppenheim A. V. and Schafer R. W. (1975), *Digital Signal Processing*, Prentice-Hall, Englewood Cliffs, NJ, 251-255

PART 2

Advanced Topics in Digital Signal Processing

3   <u>Optimization and Approximation</u>

There comes a moment in the career of any DSP engineer when the time is ripe for their next big step in theoretical knowledge. You've seen the discrete Fourier transform (DFT) at work, designed your own simple FIR filters, mastered the basic concepts of probability and random processes, and have, in general, worn down the pages of your introductory DSP textbook. The time has arrived to step up to the big leagues, where the pros use linear algebra to optimally solve many problems in the area of signal processing. It's the world of equalizers, maximum ratio combiners, channel estimators, and adaptive signal processing algorithms. At the core of most of these advanced signal processing concepts lies the least squares error, or LSE, technique [1], a concept that was pioneered by the great German mathematician Karl Friedrich Gauss (1777–1855) and used by him to estimate planetary motion. The least squares error concept, itself, is actually a criterion of quality or goodness used to find optimal solutions to engineering problems that can be formulated as overdetermined systems of linear equations. Its derivation, shown in this chapter, is one of the most elegant treatments in the area of mathematics, and I hope very much that you, the reader, will take your time to completely understand it. The derivation is also the perfect opportunity to put many of the linear algebra concepts mentioned in chapter 1 to good use. A second criterion of quality or goodness discussed in this chapter is the Minimax concept. This technique attempts to minimize the maximum absolute error of the solution and forms the core algorithm in the design of equiripple filters.

Let's summarize the distinct topics that we will cover in this chapter.

→ Primarily, we will be focusing on the least squares error concept and how we may use it to find optimal solutions to engineering problems that can be expressed as linear systems of equations. To make this introduction as visually intuitive as possible, we cover the classic application of curve fitting first.

→ Next, we will examine a popular variant of the LSE concept called the weighted least squares technique, which allows the user to customize the optimal solution.

→ We then extend the LSE technique to the design of optimum finite impulse response, or FIR, filters.

→ Next, we introduce the technique of Minimax approximation, which we will use both for curve fitting and equiripple FIR filter design. This technique uses an iterative procedure called Remez exchange to find a different type of optimum solution to the same engineering problems addressed by the least squares error method.

→ To finish off the chapter, we show how waveforms may be approximated for the purpose of interpolation and waveform resampling. We will use techniques that are based on FIR filter design, as well as polynomial approximation. Finally, the ideas of up and downsampling signals is introduced.

So, without further delay, let's introduce one of the most powerful tools available in the area of signal processing and numerical methods – the method of least squares.

3.1 The Method of Least Squares

The method of least squares, also known as the LSE (least squares error) technique, finds the optimal solution of an overdetermined system of linear equations. To develop a scenario that allows its use, consider the figure below in which we attempt to fit a straight line to a set of data points.

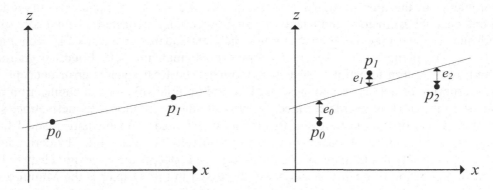

Figure 3-1: Straight Line Fit

In the figure to the left, we are given two points, $p_0 = [x_0, z_0]$ and $p_1 = [x_1, z_1]$, whereas the one on the right features three points, $p_0 = [x_0, z_0]$, $p_1 = [x_1, z_1]$, and $p_2 = [x_2, z_2]$. Remember that the equation of a straight line uses intercept and slope coefficients, which we will call a_0 and a_1, respectively.

$$z = Intercept + Slope \cdot x$$
$$or$$
$$z = a_0 + a_1 \cdot x$$

Our system of equation for the two-point scenario changes into matrix form and yields the following easy solution:

$$z_0 = a_0 + a_1 \cdot x_0$$
$$z_1 = a_0 + a_1 \cdot x_1$$

$$\begin{bmatrix} z_0 \\ z_1 \end{bmatrix} = \begin{bmatrix} 1 & x_0 \\ 1 & x_1 \end{bmatrix} \cdot \begin{bmatrix} a_0 \\ a_1 \end{bmatrix}$$

$$Z = F \cdot A$$
$$\boxed{A = inv(F) \cdot Z}$$

As long as the x coordinates of the two points differ, there will always exist the inverse of the matrix F, and its multiplication by the column vector $Z = [z_0 \ z_1]^T$ yields the solution vector A. Remember, the inverse is only defined for square matrices. However, the graph on the right side of the figure above presents a dilemma because we cannot draw a line that passes neatly through three different coordinates.

Setting up this system of equations takes a similar path until reaching the point where the inverse of F cannot be computed because the system is overdetermined. The system of equations is referred to as overdetermined if the number of rows (constraints) of matrix F is larger than the number of columns (degrees of freedom). A system of equations is underdetermined if the number of rows is smaller than the number of columns.

$$z_0 = a_0 + a_1 \cdot x_0$$

$$z_1 = a_0 + a_1 \cdot x_1$$

$$z_2 = a_0 + a_1 \cdot x_2$$

$$\begin{bmatrix} z_0 \\ z_1 \\ z_2 \end{bmatrix} = \begin{bmatrix} 1 & x_0 \\ 1 & x_1 \\ 1 & x_2 \end{bmatrix} \cdot \begin{bmatrix} a_0 \\ a_1 \end{bmatrix}$$

$$Z = F \cdot A$$

Because F has no inverse, we cannot find an exact solution. However, could we construct a line that passes by the three points as closely as possible? This idea is generally called curve fitting, and the least squares criterion defines the goodness or quality of such a fit.

> According to the least squares criterion, the optimum solution vector, A, is the one that minimizes the sum of the square errors between the given points and the fit.

We, therefore, relax the requirement of exactness and allow for the inclusion of the error vector, E.

$$\begin{bmatrix} z_0 \\ z_1 \\ z_2 \end{bmatrix} = \begin{bmatrix} 1 & x_0 \\ 1 & x_1 \\ 1 & x_2 \end{bmatrix} \cdot \begin{bmatrix} a_0 \\ a_1 \end{bmatrix} + \begin{bmatrix} e_0 \\ e_1 \\ e_2 \end{bmatrix}$$

$$\begin{bmatrix} e_0 \\ e_1 \\ e_2 \end{bmatrix} = \begin{bmatrix} z_0 \\ z_1 \\ z_2 \end{bmatrix} - \begin{bmatrix} 1 & x_0 \\ 1 & x_1 \\ 1 & x_2 \end{bmatrix} \cdot \begin{bmatrix} a_0 \\ a_1 \end{bmatrix}$$

$$E = Z - F \cdot A$$

The purpose of the optimization is to find the solution vector $A = [a_0 \ a_1]^T$ that will minimize the square error (SE), defined as $E^T E = e_0 \cdot e_0 + e_1 \cdot e_1 + e_2 \cdot e_2 + \dots$. The procedure that accomplishes this task is an extension of the first semester calculus task of finding the minimum of a curve. Assume that we are given a curve that represents the error as a function of some independent parameter. Finding the minimum involves taking the derivative of the error function, setting that expression to zero, and solving for the independent variable. The only real difference now is that the error is a function of not one, but several independent variables, which reside in vector A. Therefore, it is a multidimensional problem, which requires the use of the gradient operation rather than the simple one-dimensional derivative. The three steps that follow spell out our plan of action.

Finding the Least Squares Solution

1. Define the cost function (the expression for the square error, *SE*) by taking the inner product of column vector E.

$$SE = E^T E = (Z - FA)^T \cdot (Z - FA)$$

Note that the *SE* quantity reduces to a 1 by 1 matrix (a simple scalar), which is a function of the solution vector, A.

$$SE = \begin{bmatrix} e_0^* & e_1^* & e_2^* \end{bmatrix} \cdot \begin{bmatrix} e_0 \\ e_1 \\ e_2 \end{bmatrix} = e_0^* \cdot e_0 + e_1^* \cdot e_1 + e_2^* \cdot e_2$$

Because we deal with entirely real quantities in our example, the conjugation that occurs during the transpose disappears, and we are left with a real square error (*SE*).

$$SE = e_0^2 + e_1^2 + e_2^2$$

2. The second step requires that we compute the gradient (multi-dimensional derivative) of the cost, or *SE* function, with respect to solution vector A and set the result equal to zero.

$$\nabla_A (E^T E) = 0$$

3. Finally, we solve the previous equation to find the vector A, which is the optional solution in the least squares sense.

$$LSE = min(SE)$$

Understanding Error Functions of Multiple Independent Variables

The aim of the following example is to help us visualize square errors that are a function of several independent variables. These error functions take the shape of surfaces, in the case of two independent variables, or collections of surfaces, in cases where the number of independent variables is even higher. A secondary aim is to show why the least squares error criterion (*LSE* = $min[e_0 \cdot e_0 + e_1 \cdot e_1 + e_2 \cdot e_2 + \ldots]$) is preferable as a metric of goodness more than other possible metrics, such as the minimum sum of the absolute errors = $min[|e_0| + |e_1| + |e_2| + \ldots]$.

Example 3.1: *Least Squares Optimization*

Therefore, we consider a slightly different scenario that allows us to properly see the error surfaces that require optimization. Consider the equation of a plane in three-dimensional space.

$$z = Intercept + Slope1 \cdot x + Slope2 \cdot y$$

Assume further that we are given four *XYZ* coordinates ([1, 1, 1], [1, -1, .95], [-1, -1, 1], and [-1, 1, 1]), for which we must find the coefficients for a plane that passes by these coordinates as closely as possible.

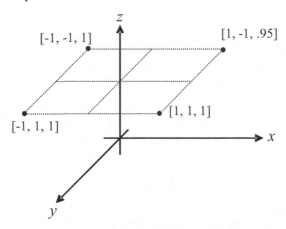

Figure 3-2: Finding Optimal Coefficients for a Plane

The three coefficients that must be calculated are the Intercept (a_0), Slope1 (a_1), and Slope2 (a_2), and as the reader may have guessed already, it takes only three *XYZ* coordinates (or constraints) to precisely specify the plane. The fact that we must consider four coordinates causes our system of equations to be overdetermined.

$$z_1 = a_0 + a_1 x_1 + a_2 y_1 + e_1$$
$$z_2 = a_0 + a_1 x_2 + a_2 y_2 + e_2$$
$$z_3 = a_0 + a_1 x_3 + a_2 y_3 + e_3$$
$$z_4 = a_0 + a_1 x_4 + a_2 y_4 + e_4$$

$$\begin{bmatrix} e_0 \\ e_1 \\ e_2 \\ e_3 \end{bmatrix} = \begin{bmatrix} z_0 \\ z_1 \\ z_2 \\ z_3 \end{bmatrix} - \begin{bmatrix} 1 & x_0 & y_0 \\ 1 & x_1 & y_1 \\ 1 & x_2 & y_2 \\ 1 & x_3 & y_3 \end{bmatrix} \cdot \begin{bmatrix} a_0 \\ a_1 \\ a_2 \end{bmatrix}$$

Because we know what the coordinates $x_0 y_0 z_0$ through $x_3 y_3 z_3$ are, we select a range of reasonable intercept points and slopes and compute the squares error ($SE = e_0^2 + e_1^1 + e_2^2 + e_3^2$). In the code segment that follows, we initially hold the *z* intercept, a_o, at a constant value of 1.0 and compute the square error values for a range of different *x* and *y* slopes. This allows us to see the square error as a function of only two variables, thus displaying a surface, rather than a graphic of higher dimension.

```
P1 = [1 1 1]; P2 = [1 -1 .95];  P3 = [-1 -1 1]; P4 = [-1 1 1];
InterceptGuess = 1;
Slope1Range    = -1:.1:1;
Slope2Range    = -1:.1:1;

[Slope1,Slope2]  = meshgrid(Slope1Range, Slope2Range);
LSE              = zeros(size(Slope1));
AbsoluteError    = zeros(size(Slope1));
S                = size(Slope1);

for a = 1:S(1,1)
    for b = 1:S(1,2)
        S1   = Slope1(a,b);
        S2   = Slope2(a,b);
        e1   = P1(1,3) - (InterceptGuess + S1*P1(1,1) + S2*P1(1,2));
        e2   = P2(1,3) - (InterceptGuess + S1*P2(1,1) + S2*P2(1,2));
        e3   = P3(1,3) - (InterceptGuess + S1*P3(1,1) + S2*P3(1,2));
        e4   = P4(1,3) - (InterceptGuess + S1*P4(1,1) + S2*P4(1,2));
        LSE(a,b)          = e1*e1 + e2*e2 + e3*e3 + e4*e4;
        AbsoluteError(a,b)= abs(e1) + abs(e2) + abs(e3) + abs(e4);
    end
end

figure(1);
mesh(Slope1, Slope2, LSE);
xlabel('Intercept'); ylabel('Slope'); zlabel('LSE');
title('Least Squares Error Surface'); axis([-1 1 -1 1 0 10]);
colormap('gray'); caxis([.00 20]);
```

The range of the *x* and *y* slopes is used in the *meshgrid* statement, which sets up an *XY* grid used for plotting the mesh graph below.

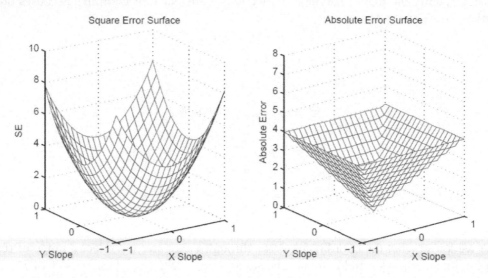

Figure 3-3: Square Error and Absolute Error Surfaces (*Z* Intercept = 1.0)

The graph on the left of Figure 3-3 shows the square error surface as a paraboloid, which is a shape that features well-defined derivatives and one unique minimum. The graph on the right represents the error surface according to the absolute error value criterion. This criterion also features a clear minimum, but the derivatives are not well behaved at that minimum.

$$AbsoluteError = abs(e_1) + abs(e_2) + abs(e_3) + abs(e_4)$$

The intercept, which was fixed at 1.0, must, unfortunately, also be optimized. Therefore, in the figure below, we recomputed the two surfaces with a z intercept of 1.5, rather than 1.0. The square error surface simply moves upward but maintains its other geometric properties. However, the absolute error surface has flattened at its bottom and now features a large set of minima, which basically render that criterion useless.

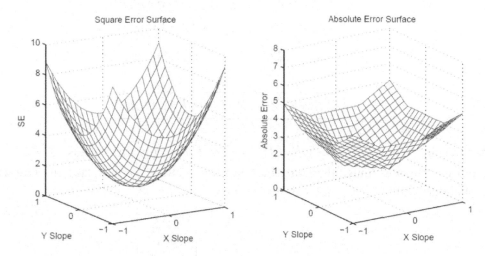

Figure 3-4: Square Error and Absolute Error Surfaces (Z Intercept = 1.5)

The least squares error criterion is useful because it allows us to find the optimal solution in a mathematically reasonable fashion. It has well-defined derivatives everywhere and features one unique minimum for which we can solve. Other criteria of optimality, like the absolute value criterion, simply don't feature the geometric properties that allow for a reasonable mathematical solution. You may use the least squares error criterion and its accompanying optimal solution approach for engineering situations where data is approximated as a linear combination of differently scaled basis functions. Only those formulations feature a unique minimum.

$$z[n] = a_0 \cdot basis_0[n] + a_1 \cdot basis_1[n] + a_2 \cdot basis_2[n] + \cdots$$

The basis functions, or basis sequences, change from application to application and may be quite non-linear with respect to n, as for example $basis_0[n] = \cos[2\pi n]$ or $basis_0[n] = n^2$. The following are two examples of approximations that allow the use of the method of least squares.

$$z[n] = a_0 + a_1 \cdot x[n] + a_2 \cdot y[n] \qquad \leftarrow Example 3.1$$
$$z[n] = a_0 \cdot \cos[n] + a_1 \cdot n^2 + a_2 \cdot \exp(2\pi n) + a_3 \cdot [n-1] + \cdots$$

However, there are plenty of processes in nature that are mathematically approximated using equations that are non-linear with respect to their coefficients. The equation below illustrates such an approximation, which cannot be solved using the method of least squares.

$$z[n] = a_0 \cdot n + a_1 \cdot e^{a_2 n}$$

Finding the Optimal Solution

Now that we understand the reasoning behind using the least squares criterion, let's proceed with the topic of finding the optimum solution vector A. Because this derivation involves a fair bit of linear algebra, an occasional look back at chapter 1 is definitely warranted.

$$E^T E = (Z - FA)^T \cdot (Z - FA)$$
$$= (Z^T - (FA)^T) \cdot (Z - FA)$$
$$= Z^T Z - Z^T FA - (FA)^T Z + (FA)^T (FA)$$
$$= Z^T Z - Z^T FA - (Z^T FA)^T + (FA)^T (FA)$$
$$= Z^T Z - 2Z^T FA + (FA)^T (FA)$$

$$\boxed{E^T E = Z^T Z - 2Z^T FA + (FA)^T (FA)}$$

Note that the expression $-Z^T FA - (Z^T FA)^T = -2Z^T FA$ is only true because $Z^T FA$ is a real 1x1 matrix and its transpose must, therefore, be that same matrix. We now take the gradient (multi-dimensional derivative) $\nabla_A (E^T E)$ with respect to A.

$$\nabla_A (E^T E) = \frac{\partial (E^T E)}{\partial A} = \begin{bmatrix} \dfrac{\partial (E^T E)}{\partial a_0} \\[2mm] \dfrac{\partial (E^T E)}{\partial a_1} \\[2mm] \vdots \\[2mm] \dfrac{\partial (E^T E)}{\partial a_{n-1}} \end{bmatrix}$$

We define the gradient of the cost function $E^T E$, with respect to the matrix A, as a column vector in which $E^T E$ is partially differentiated versus each scalar entry of A. The gradient looks daunting, but is really not an involved concept, as the examples in chapter 1 showed.

$$\nabla_A (E^T E) = \nabla_A (Z^T Z) - \nabla_A (2Z^T FA) + \nabla_A ((FA)^T (FA))$$

Before we take the gradients of each one of these three expressions, let's remember that $E^T E$ is a scalar, and the expressions $Z^T Z$, $2Z^T FA$, and $(FA)^T (FA)$ are, thus, also scalars. However, the gradient is by definition a column vector. Take a moment to review Examples 1.11 and 1.12 of chapter 1 to familiarize yourself with the gradient operation.

Step 1

$$\nabla_A(Z^T Z) = 0$$

The gradient is zero because $Z^T Z$ is not a function of A.

Step 2

$$\nabla_A(2Z^T F A) = 2\nabla_A(\ (Z^T F)\cdot A)$$
$$= 2\nabla_A(\ P\cdot A)$$
$$= 2\nabla_A(\ \begin{bmatrix} p_0 & p_1 & .. & p_{n-1} \end{bmatrix}\cdot \begin{bmatrix} a_0 \\ a_1 \\ .. \\ a_{n-1} \end{bmatrix})$$
$$= 2\nabla_A(p_0\cdot a_0 + p_1\cdot a_1 + .. + p_{n-1}\cdot a_{n-1})$$

For readability, we let $P = Z^T F$, which turns out to be a row vector that when multiplied by A, results in a scalar value (1x1 matrix). We now take the gradient to arrive at the following result.

$$\nabla_A(2Z^T F A) = 2\begin{bmatrix} \dfrac{\partial(p_0\cdot a_0 + .. + p_{n-1}\cdot a_{n-1})}{\partial a_0} \\ \vdots \\ \dfrac{\partial(p_0\cdot a_0 + .. + p_{n-1}\cdot a_{n-1})}{\partial a_{n-1}} \end{bmatrix} = 2\begin{bmatrix} p_0 \\ \vdots \\ p_{n-1} \end{bmatrix} = 2P^T = 2(Z^T F)^T = 2F^T Z$$

Step 3

$$\nabla_A((F A)^T F A) = \nabla_A(A^T F^T F A) = 2F^T F A$$

Thinking back to scalar calculus, the previous result makes sense because the vector A seems to be squared.

$$\frac{d(a\cdot C\cdot a)}{da} = \frac{C\cdot d(a^2)}{da} = 2Ca$$

From the scalar calculus version, we may be led to believe that $\nabla_A(A^T T A) = 2T A$, where T is some matrix with appropriate dimensions. It actually turns out that this conclusion is true only if T is symmetric. Because $F^T F$ is guaranteed to be a symmetric matrix, the equation holds. A quick example should illustrate this point.

$$F = \begin{bmatrix} 1 & 1 \\ 2 & 3 \end{bmatrix} \text{ then}$$

$$A^T F^T FA = \begin{bmatrix} a_0 & a_1 \end{bmatrix} \begin{bmatrix} 1 & 2 \\ 1 & 3 \end{bmatrix} \begin{bmatrix} 1 & 1 \\ 2 & 3 \end{bmatrix} \begin{bmatrix} a_0 \\ a_1 \end{bmatrix}$$

$$A^T F^T FA = \begin{bmatrix} a_0 & a_1 \end{bmatrix} \begin{bmatrix} 5 & 7 \\ 7 & 10 \end{bmatrix} \begin{bmatrix} a_0 \\ a_1 \end{bmatrix} = 5a_0{}^2 + 10a_1{}^2 + 14a_0 a_1$$

We now apply the gradient with respect to A, which yields the following result.

$$\nabla_A (A^T F^T FA) = \begin{bmatrix} \dfrac{\partial(A^T F^T FA)}{\partial a_0} \\[2mm] \dfrac{\partial(A^T F^T FA)}{\partial a_1} \end{bmatrix} = \begin{bmatrix} 10a_0 + 14a_1 \\ 14a_1 + 20a_2 \end{bmatrix} = 2 \begin{bmatrix} 5 & 7 \\ 7 & 10 \end{bmatrix} \begin{bmatrix} a_0 \\ a_1 \end{bmatrix} = 2F^T FA$$

As an exercise, use a non-symmetric square matrix instead of $F^T F$ to prove that the conclusion does not hold.

Final Step

Adding up these three separate results yields the final gradient of the squares error.

$$\nabla_A (E^T E) = -2F^T Z + 2F^T FA$$

If we now set the gradient to zero and solve for A, we arrive at the final solution.

$$2F^T FA = 2F^T Z$$

$$[F^T F]^{-1} F^T FA = [F^T F]^{-1} F^T Z$$

$$\boxed{A = [F^T F]^{-1} F^T Z}$$

This is the moment we have all been waiting for. The previous equation is one of the most useful and elegant results in the area of mathematics and forms the basic concept underlying optimal and adaptive signal processing, which we explore in this and several of the subsequent chapters in this text.

3.1.1   Curve Fitting Using the Least Squares Method

To cement our understanding of the results at which we have just arrived, let's introduce the concept of curve fitting. Curve fitting is a useful technique allowing engineers to better understand the results of test data. The example below illustrates a typical experiment designed to measure the input/output power relationship of an RF amplifier. In the experiment, a test engineer transmits different amounts of input power to an RF amplifier and measures the resulting output power. In an ideal world, the gain of such an amplifier would be a constant and, thus, be independent of the input power level. The real world, of course, dictates that a RF amplifier can only output a finite amount of power, regardless of how far the input signal increases in magnitude. Curve fitting allows us to express the input-output relationship as a continuous function.

Example 3.2: *Polynomial Regression*

Below, we are given a set of data points in dBm and are tasked to calculate a function that expresses the output power over a continuous range of input powers.

$$In = \begin{bmatrix} -10 & -5 & 0 & 5 & 10 & 15 & 20 & 25 \end{bmatrix}^T \quad \leftarrow InputPower(dBm)$$

$$Out = \begin{bmatrix} 0 & 5.5 & 10 & 14 & 18.5 & 21 & 21 & 21 \end{bmatrix}^T \leftarrow OutputPower(dBm)$$

The purpose is to express the relationship given by the test data using a fourth order polynomial curve expressed in the equation that follows. This type of curve fit is also called *polynomial regression* because of the type of curve used to approximate our tabulated data. Had the equation below been reduced to that of a line and, therefore, only included the a_0 and a_1 terms, we would have called the curve fit a *linear regression*.

$$Out = a_0 + a_1 \cdot In + a_2 \cdot In^2 + a_3 \cdot In^3 + a_4 \cdot In^4$$

Therefore,

$$Out_1 = a_0 + a_1 \cdot In_1 + a_2 \cdot In_1^2 + a_3 \cdot In_1^3 + a_4 \cdot In_1^4 + e_1$$
$$Out_2 = a_0 + a_1 \cdot In_2 + a_2 \cdot In_2^2 + a_3 \cdot In_2^3 + a_4 \cdot In_2^4 + e_2$$
$$\vdots$$
$$Out_8 = a_0 + a_1 \cdot In_8 + a_2 \cdot In_8^2 + a_3 \cdot In_8^3 + a_4 \cdot In_8^4 + e_8$$

$$\begin{bmatrix} 0 \\ \vdots \\ 21 \end{bmatrix} = \begin{bmatrix} 1 & -10 & (-10)^2 & (-10)^3 & (-10)^4 \\ \vdots & \vdots & \vdots & \vdots & \vdots \\ 1 & 25 & (25)^2 & (25)^3 & (25)^4 \end{bmatrix} \cdot \begin{bmatrix} a_0 \\ \vdots \\ a_4 \end{bmatrix} + \begin{bmatrix} e_1 \\ \vdots \\ e_8 \end{bmatrix}$$

Replacing the matrices with variables, we reformulate the last expression as follows.

$$Out = F \cdot A + E$$
$$E = Out - F \cdot A$$

Using our least squares formula produces the following solution vector and graph.

$$A = \left[F^T F \right]^{-1} F^T Z = \begin{bmatrix} 10.094156 \\ 0.952597 \\ -0.00992 \\ -0.00052 \\ 0.000003 \end{bmatrix}$$

$$Out = 10.09 + 0.9526 \cdot In - 0.00992 \cdot In^2 - 0.00052 \cdot In^2 + 0.000003 \cdot In^4$$

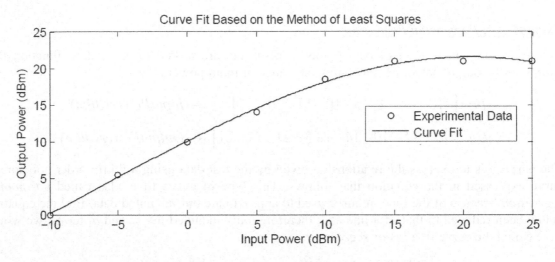

Figure 3-5: Curve Fit Using Polynomial Regression

```
InputPower   = [-10 -5    0    5   10    15   20   25];
OutputPower  = [  0   5.5 10   14   18.5  21   21   21];
X            = InputPower';
Z            = OutputPower';        %% Output power as a column vector


F = [ones(size(Z))  X X.^2 X.^3 X.^4]; %% Defining matrix F
A = inv(F'*F)*F'*Z;                    %% Here we solve for A


InRange  = -10:.1:25;
CurveFit = A(1,1)    + A(2,1)*InRange  + A(3,1)*InRange.^2  ...
                     + A(4,1)*InRange.^3    + A(5,1)*InRange.^4;
```

3.1.2   The Weighted Least Squares Method

The least squares algorithm, as we have come to know it in the last section, uses the degrees of freedom (the polynomial coefficients) to optimally satisfy the constraints (test points) given. What happens if we are more confident about the accuracy of some test points than others and wish to include this knowledge to improve our curve fit? For example, we might be very confident that the test points at input powers of 15dBm, 20dBm, and 25dBm are accurate and may wish to bias the curve fit to more faithfully represent them. We, therefore, weight the errors between the curve fit and the last three data points more heavily, thereby increasing their contribution to the final cumulative square error. This extension to the LSE technique is called the weighted least squares method [2]. Recall that we had previously defined the error vector as follows.

$$E = Z - F \cdot A = \begin{bmatrix} e_{OfTestPoint1} \\ \vdots \\ e_{OfTestPoint5} \\ e_{OfTestPoint6} \\ e_{OfTestPoint7} \\ e_{OfTestPoint8} \end{bmatrix}$$

We now exaggerate the error for each data point by a certain amount by multiplying E with a diagonal matrix C to arrive at the weighted error vector E_w.

$$E_W = C \cdot (Z - F \cdot A) = \begin{bmatrix} 1 & \cdots & 0 & 0 & 0 & 0 \\ \vdots & \vdots & \vdots & \vdots & \vdots & \vdots \\ 0 & \cdots & 1 & 0 & 0 & 0 \\ 0 & \cdots & 0 & 10 & 0 & 0 \\ 0 & \cdots & 0 & 0 & 10 & 0 \\ 0 & \cdots & 0 & 0 & 0 & 10 \end{bmatrix} \cdot \begin{bmatrix} e_{OfTestPoint1} \\ \vdots \\ e_{OfTestPoint5} \\ e_{OfTestPoint6} \\ e_{OfTestPoint7} \\ e_{OfTestPoint8} \end{bmatrix} = \begin{bmatrix} 1 \cdot e_{OfTestPoint1} \\ \vdots \\ 1 \cdot e_{OfTestPoint5} \\ 10 \cdot e_{OfTestPoint6} \\ 10 \cdot e_{OfTestPoint7} \\ 10 \cdot e_{OfTestPoint8} \end{bmatrix}$$

In the previous expression, we exaggerated the curve fitting errors of the last three test points by a factor of 10. Now, we will work this diagonal weighting matrix C into our squared error or cost function, as follows.

$$E_w{}^T \cdot E_w = [C(Z - FA)]^T \cdot C(Z - FA)$$

$$= (Z - FA)^T C^T \cdot C(Z - FA)$$

In most texts, the term W is used instead of the product C^TC, which is a diagonal matrix with the squared weighting terms of C. Given that W is a matrix composed of constants, simplifying the new weighted cost function, $E_w{}^TE_w$, is completely analogous to the procedure used in Section 3.1 for the straight LSE method.

$$E_W^T E_W = (Z - FA)^T \cdot W \cdot (Z - FA)$$
$$= (Z^T W - (FA)^T W) \cdot (Z - FA)$$
$$= Z^T WZ - Z^T WFA - (FA)^T WZ - (FA)^T WFA$$
$$= Z^T WZ - Z^T WFA - ((WZ)^T FA)^T - (FA)^T WFA$$
$$= Z^T WZ - Z^T WFA - (Z^T W^T FA)^T - (FA)^T WFA$$
$$= Z^T WZ - 2Z^T WFA - A^T F^T WFA$$

We again apply the gradient to the cost function $E_w^T E_w$, set the result to zero, and solve for A.

$$\nabla_A (E^T E) = 2Z^T WF - 2F^T WFA = 0$$

and

$$F^T WFA = Z^T WF$$

thus

$$\boxed{A = inv(F^T WF) \cdot F^T WZ}$$

```
Weight = [ 1 1 1 1 1 100 100 100];      %% Add this line to previous code
W      = diag(Weight);                  %% Add this line to previous code
AW     = inv(F'*W*F)*F'*W*Z;            %% Draw Addition Polynomial based on AW
```

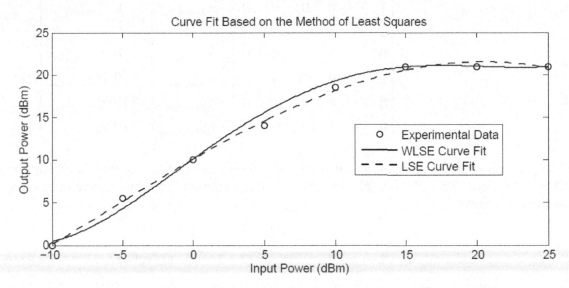

Figure 3-6: Polynomial Regression Using LSE and WLSE Techniques

The figure above illustrates the higher accuracy that the weighted curve fit provides for the last three samples. It clearly does this at the expense of the other samples at lower input powers.

3.1.3 Optimal - Least Squares - Filter Design

From the frequency sampling method covered at the end of chapter 2, we saw that the simplest way to compute the coefficients of a FIR filter is to take the inverse DFT (IDFT) of the desired frequency response. Take a moment to review the method in chapter 2. Let's assume, then, that we need to calculate FIR coefficients of a filter with length $N = 17$ for the following magnitude response H.

$$Frequencies = \begin{bmatrix} -8 & -7 & -6 & -5 & -4 & -3 & -2 & -1 & 0 & 1 & 2 & 3 & 4 & 5 & 6 & 7 & 8 \end{bmatrix} \cdot \frac{SampleRate}{N}$$

$$H_M = \begin{bmatrix} .1 & .1 & .1 & .1 & .5 & 1 & 1 & 1 & 1 & 1 & 1 & 1 & 1 & .5 & .1 & .1 & .1 & .1 \end{bmatrix}$$

$$H_{Ph} = \begin{bmatrix} 0 & 0 & 0 & 0 & 0 & 0 & 0 & 0 & 0 & 0 & 0 & 0 & 0 & 0 & 0 & 0 & 0 \end{bmatrix}$$

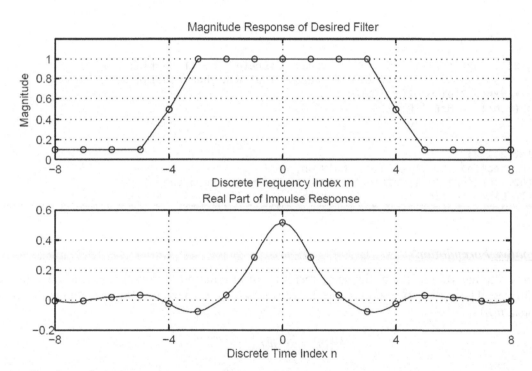

Figure 3-7: Magnitude Response and Corresponding Impulse Response of Desired FIR Filter

After recasting the desired magnitude and phase response into rectangular form $H[m]$, we use the inverse discrete Fourier transform to determine the impulse response for the time indices of $n = -8, -7 \dots 8$.

$$h[n] = \sum_{m=-8}^{8} H[m] \cdot e^{j2\pi nm/N}$$

The following MatLab code finds the discrete impulse response as well as the quasi continuous waveform, from which it is sampled.

```
N = 17;                    %% Number of FIR filter taps
H = [ .1 .1 .1 .1 .5 1 1 1 1 1 1 1 .5 .1 .1 .1 .1];


figure(1); subplot(2,1,1);
plot(-8:1:8, abs(H),'k-o'); grid on;
title('Magnitude Response of Desired Filter');
xlabel('Discrete Frequency Index m'); ylabel('Magnitude');
axis([-8.0 8.0 0 1.2]); set(gca, 'XTick', [-8 -4 0 4 8]);


n = -8:1:8; m = -8:1:8;
h = zeros(1,length(n));
for d=0:1:16               %% Take IDFT to find impulse response
    E = exp(j*2*pi*m*n(1,d+1)/N);
    h(1,d+1) = H*E.'/N;
end


n2 = -8:.25:8;
h2 = zeros(1,length(n2));
for d=0:1:length(n2)-1     %% Take inverse discrete Time FT to find the
                           %% impulse response at a higher resolution
    E = exp(j*2*pi*m*n2(1,d+1)/N);
    h2(1,d+1) = H*E.'/N;
end


subplot(2,1,2);
plot(n, real(h),'ko'); grid on; hold on;
plot(n2, real(h2),'k'); title('Real Part of Impulse Response ');
xlabel('Discrete Time Index n'); set(gca, 'XTick', [-8 -4 0 4 8]);
```

Alternative Formulation

We may alternatively arrive at the same result by reformulating the un-normalized DFT in terms of matrices. The vector H contains the desired frequency response of the yet unknown impulse response $h[n]$.

$$H[m] = \sum_{n=-8}^{8} h[n] \cdot e^{-j2\pi nm/N}$$

$$
\begin{bmatrix} H_{-8} \\ H_{-7} \\ \vdots \\ H_0 \\ \vdots \\ H_7 \\ H_8 \end{bmatrix}
=
\begin{bmatrix}
e^{-j2\pi(-8)(-8)/17} & e^{-j2\pi(-7)(-8)/17} & e^{-j2\pi(-6)(-8)/17} & \cdots & e^{-j2\pi(8)(-8)/17} \\
e^{-j2\pi(-8)(-7)/17} & e^{-j2\pi(-7)(-7)/17} & e^{-j2\pi(-6)(-7)/17} & \cdots & e^{-j2\pi(8)(-7)/17} \\
\vdots & \vdots & \vdots & \ddots & \vdots \\
e^{-j2\pi(-8)(0)/17} & e^{-j2\pi(-7)(0)/17} & e^{-j2\pi(-6)(0)/17} & \cdots & e^{-j2\pi(8)(0)/17} \\
\vdots & \vdots & \vdots & \ddots & \vdots \\
e^{-j2\pi(-8)(7)/17} & e^{-j2\pi(-7)(7)/17} & e^{-j2\pi(-6)(7)/17} & \cdots & e^{-j2\pi(8)(7)/17} \\
e^{-j2\pi(-8)(8)/17} & e^{-j2\pi(-7)(8)/17} & e^{-j2\pi(-6)(8)/17} & \cdots & e^{-j2\pi(8)(8)/17}
\end{bmatrix}
\cdot
\begin{bmatrix} h_{-8} \\ h_{-7} \\ h_{-6} \\ \vdots \\ h_8 \end{bmatrix}
$$

$$H = 17\text{x}1 \qquad\qquad\qquad F = 17\text{x}17 \qquad\qquad\qquad h = 17\text{x}1$$

$$H = F \cdot h$$

$$\boxed{h = inv(F) \cdot H}$$

```
N                    = 17;
n                    = -8:1:8;
FrequencyResponse    = [.1 .1 .1 .1 .5 1 1 1 1 1 1 1 .5 .1 .1 .1 .1];
m                    = -8:1:8;

%% Building the H and F matrices

H = FrequencyResponse.';
F = zeros(length(FrequencyResponse), N);
for a=1:length(FrequencyResponse)
    for b = 1:N
        F(a,b) = exp(-j*2*pi*n(1,b)*m(1,a)/17);
    end
end
h = inv(F)*H;        %% Same Result with h = inv(F'*F)*F'*H
```

The result of the direct IDFT calculation and the matrix solution are the same.

$$h = [-.0077 \quad -.0085 \quad .0189 \quad .0332 \quad -.0265 \quad -.0806 \quad .0309 \quad .2815 \quad .5176 \ldots$$
$$.2815 \quad .0309 \quad -.0806 \quad -.0265 \quad .0332 \quad .0189 \quad -.0085 \quad -.0077]$$

In the case where we have 17 constraints and 17 degrees of freedom ($N = 17$ taps), the least squares solution $h = inv(F^T \cdot F) \cdot F^T \cdot H$ is able to satisfy all constraints and produces the same result as $h = inv(F) \cdot H$.

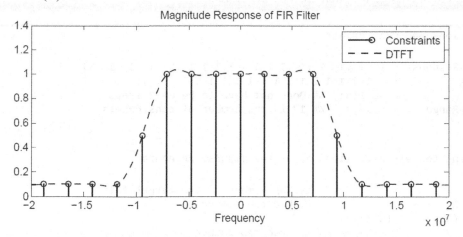

Figure 3-8: FIR Filter Magnitude Response for a Sample Rate of 40MHz

As we can see, the continuous magnitude response of the FIR filter passes through all constraints, a fact that we will not be able to achieve when we decrease the number of tabs.

Using the Least Squares Technique

Let's assume that we have 17 constraints (frequency response specifications) but fewer than 17 degrees of freedom (coefficients) available for our FIR filter. We may use the LSE technique to find the set of coefficients that best produces the desired frequency response. If, for example, we only had 9 coefficients available, then the linear system of equations becomes overdetermined as follows.

$$
\begin{bmatrix} H_{-8} \\ H_{-7} \\ \vdots \\ H_0 \\ \vdots \\ H_7 \\ H_8 \end{bmatrix} = \begin{bmatrix} e^{-j2\pi(-4)(-8)/17} & e^{-j2\pi(-3)(-8)/17} & e^{-j2\pi(-2)(-8)/17} & \cdots & e^{-j2\pi(4)(-8)/17} \\ e^{-j2\pi(-4)(-7)/17} & e^{-j2\pi(-3)(-7)/17} & e^{-j2\pi(-2)(-7)/17} & \cdots & e^{-j2\pi(4)(-7)/17} \\ \vdots & \vdots & \vdots & \ddots & \vdots \\ e^{-j2\pi(-4)(0)/17} & e^{-j2\pi(-3)(0)/17} & e^{-j2\pi(-2)(0)/17} & \cdots & e^{-j2\pi(4)(0)/17} \\ \vdots & \vdots & \vdots & \ddots & \vdots \\ e^{-j2\pi(-4)(7)/17} & e^{-j2\pi(-3)(7)/17} & e^{-j2\pi(-2)(7)/17} & \cdots & e^{-j2\pi(4)(7)/17} \\ e^{-j2\pi(-4)(8)/17} & e^{-j2\pi(-3)(8)/17} & e^{-j2\pi(-2)(8)/17} & \cdots & e^{-j2\pi(4)(8)/17} \end{bmatrix} \cdot \begin{bmatrix} h_{-4} \\ h_{-3} \\ h_{-2} \\ \vdots \\ h_4 \end{bmatrix}
$$

$$H = 17\text{x}1 \qquad\qquad F = 17\text{x}9 \qquad\qquad h = 9\text{x}1$$

$$H = F \cdot h$$

$$\boxed{h = inv(F^T \cdot F)F^T H}$$

In the following code, we will go even a step further and build weighting into the least squares equation, just as we did with the curve fitting example. Initially, we'll keep the weighting vector at unity and then change it to demonstrate the effect.

```
%% Conditions and Constraints

FrequencyResponse = [ .1 .1 .1 .1 .5 1 1 1 1 1 1 1 .5 .1 .1 .1 .1];
Weighting         = [ 1 1 1 1 1 1 1 1 1 1 1 1 1 1 1 1 1 ];
m                 = -8:1:8;  %% Does not have to be unit steps
FrequencyRange    = m/17;    %% 17 is the number of constraints


%% Defining the discrete time, n, of the impulse response

N = 9;                           %% Tap Number of the target filter
if(mod(N,2) == 1)                %% Odd tap number
    n = -(N-1)/2:1:(N-1)/2;
else                             %% Even tap number
    n = -((N/2)-.5):1:(N/2)-.5;
end

%% Execute least squares calculation
H = FrequencyResponse.';
F = zeros(length(FrequencyResponse), N);
```

```
for a=1:length(FrequencyResponse)
    for b = 1:N
        F(a,b) = exp(-j*2*pi*n(1,b)*m(1,a)/17);
    end
end

W = diag(Weighting);
h = inv(F'*W*F)*F'*W*H; %% Impulse response as column vector

%% Check result using the DTFT
Frequencies = -0.5:.01:0.5;   %% Defining the analysis frequencies
DTFT_Out = zeros(1,length(Frequencies));
for i = 1:length(Frequencies)
    f           = Frequencies(1,i); %% Current analysis frequency
    AnalysisTone = exp(j*2*pi*n*f);  %% Current analysis tone
    DTFT_Out(1,i) = h.'*AnalysisTone';
end

figure(1);
subplot(2,1,1); stem(FrequencyRange, FrequencyResponse,'k'); hold on;
xlabel('Frequency');
plot(Frequencies, abs(DTFT_Out), 'k:'); legend('Constraints','DTFT');
title('Magnitude Response of FIR Filter');
subplot(2,1,2); stem(n, real(h), 'k');
title('Impulse Response of Optimal FIR Filter');
```

Although still doing a reasonable job, the optimal 9 tap FIR filter cannot meet all 17 criteria (just as we expected).

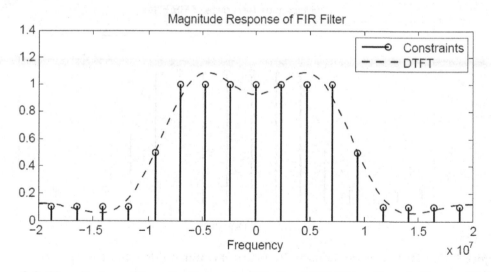

Figure 3-9: Magnitude and Impulse Response of Optimal FIR Filter (Sample Rate = 40MHz)

$$h_{optimal}[n] = [-0.265 \ -0.806 \ -0.309 \ 0.218 \ 0.5176 \ 0.2815 \ 0.0309 \ -0.0806 \ -0.0265]$$

Things to note:

→ As can be seen in all previous figures, the frequency response constraints are even functions. As we assign constraints that are not even (or symmetric around 0Hz), the impulse response will become complex.

→ The constraints do not have to be placed at evenly spaced frequencies for the algorithm to perform properly.

→ We may resort to special weighting vectors, as shown in the following code segment, where we ask the algorithm to pay special attention to only the first four constraints.

To demonstrate the LSE algorithm's flexibility, let's redefine the task as follows.

```
%% Conditions and Constraints

FrequencyResponse = [ .2 .1 .1 .1 .5 1 1 1 1 1 1 .5 .1 .1 .1 .1];
Weighting         = [ 100 100 100 100 1 1 1 1 1 1 1 1 1 1 1 1 ];
m                 = [-7.5 -7:1:8];   %% Does not have to be unit steps
FrequencyRange    = m/17;            %% 17 is the number of constraints
```

The algorithm solves for the impulse response, $h_{optimal}[n]$, which becomes complex because of the uneven frequency constraints. We can clearly see that the offbeat constraint of 0.2 at $m = -7.5$ caused no problems whatsoever, and the added emphasis on the four most-negative frequencies is clearly visible.

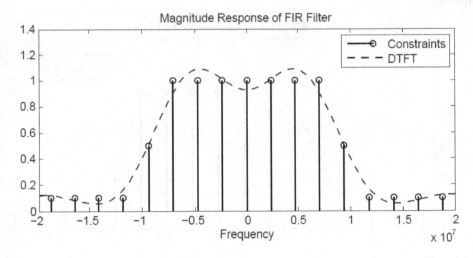

Figure 3-10: Magnitude and Impulse Response of Optimal FIR Filter (Sample Rate =40MHz)

$$h_{optimal}[n] = [-0.029 + j0.0307 \quad -0.092 + j0.0161 \quad 0.0677 - j0.0041 \quad 0.2595 - j0.0154 \quad ...$$
$$0.522 \quad 0.2595 + j0.0154 \quad 0.0677 + j0.0041 \quad -0.092 - j0.0161 \quad -0.029 - j0.0307]$$

3.2   Minimax Approximation

In the previous sections, we constructed polynomial curves that optimally approximated a given function or data set in the least squares error (LSE) sense. The LSE method computed the optimal polynomial approximation by minimizing the sum of the square errors. The Minimax approximation [3, 4], on the other hand, derives a polynomial curve of degree N such that the maximum absolute error between the function that is to be fitted and the polynomial approximation is minimized within a given interval. Chebyshev showed that the resulting optimized error function exhibits an equiripple like oscillatory behavior. The expressions below summarize the goals of the respective criteria, where $F(x)$ is the original function and $P(x) = c_0 + c_1 x + ... + c_N x^N$ is the polynomial approximation that is to be computed.

$$LSE \rightarrow Minimize \sum (P(x) - F(x))^2$$

$$Minimax \rightarrow Minimize(MaxError(|P(x) - F(x)|))$$

3.2.1   Polynomial Approximation Using the Minimax Criterion

To illustrate the difference in error behavior between the two optimization methods, let's fit two polynomial curves of fifth order, one using the LSE and one using the Minimax criterion, to the following function.

$$F(x) = \frac{1}{1 + 25x^2}$$

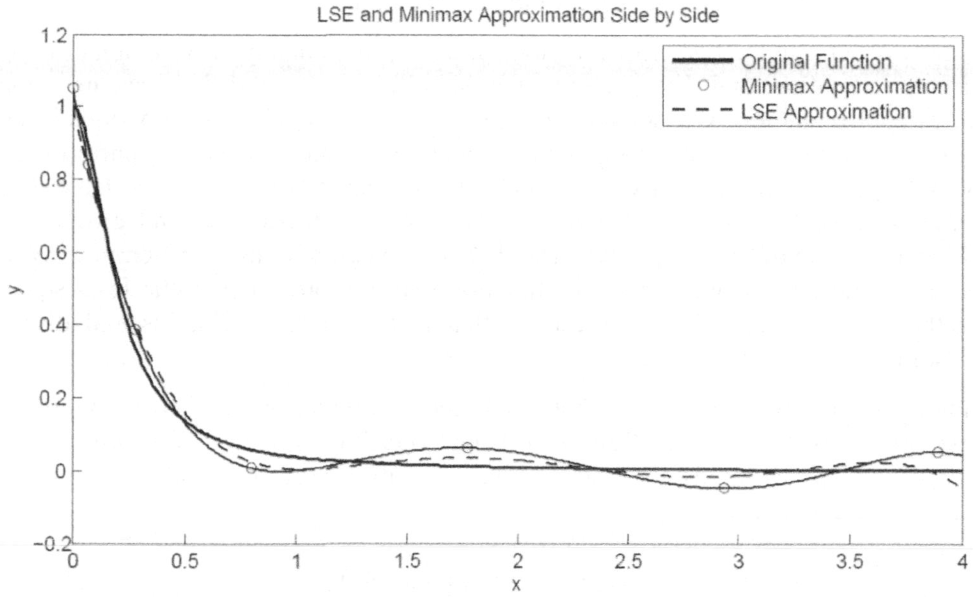

Figure 3-11: LSE and Minimax Approximations Compared

It can quickly be observed that both criteria yield approximations that waver around the original function, and it is not until we plot their respective error curves that the difference between the two techniques becomes apparent.

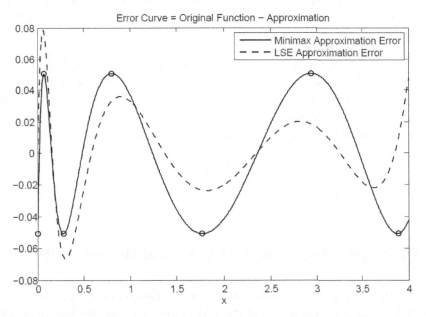

Figure 3-12: Error = $F(x)$ - $P(x)$ Curves for LSE and Minimax Approximation

Chebyshev showed that the optimal error curve (in the Minimax sense) will feature $N + 2$ positions (shown as circles in the plot above) along the x-axis, where the absolute error reaches its maximum value. N is the order, or degree, of the polynomial.

After casually observing the two error functions, it is not immediately clear which criterion yields the better approximation, and their performances should be compared for every new application that requires approximation. We can say, however, that polynomial approximations based on the Minimax criteria are best suited for applications where the data set needing approximation takes the shape of a quasi-continuous curve or function rather than a collection of isolated data points. This is true because the Minimax technique is an iterative procedure, and errors need to be calculable at any value of the independent variable at the beginning of each iteration. Whereas the Minimax technique does produce an error function that is more orderly, the least squares error (LSE) method can be applied to applications that feature both continuous and discrete data, making it somewhat more flexible as a tool.

Computing the polynomial coefficients that satisfy the Minimax criterion is usually performed via an iterative procedure based on the Remez exchange algorithm. Given that we know the solution will feature $N + 2$ points along the x-axis, where the absolute error reaches its maximum, we define the following conditions.

$$a \le x_0 < x_1 < ... < x_{N+1} \le b$$
$$P(x_i) - F(x_i) = (-1)^i E$$
$$i = 0,1...N+1$$

The values a and b constitute the range over which we intend to compute the curve fit. The optimum $N + 2$ positions, x_i, are as yet unknown, but they exist over the range delineated by a and b. The second equation is at the heart of the computational procedure that is about to follow.

It forces the equiripple behavior onto the coefficient solution. Starting the iterative process requires picking initial positions, x_i, at which we force a constant absolute error, E. For simplicity, we select positions that are evenly spaced over the interval from a to b.

$$P(x_i) + (-1)^i E = F(x_i)$$

The previous expression expands to the following system of equations and corresponding matrix formulation.

$$
\begin{aligned}
c_0 + c_1 x_0 + c_2 x_0{}^2 + \ldots + c_N x_0{}^N \qquad\quad + (-1)^0 E &= F(x_o) \\
c_0 + c_1 x_1 + c_2 x_1{}^2 + \ldots + c_N x_1{}^N \qquad\quad + (-1)^1 E &= F(x_1) \\
\vdots \qquad\qquad\qquad\qquad\qquad & \\
c_0 + c_1 x_{N+1} + c_2 x_{N+1}{}^2 + \ldots + c_N x_{N+1}{}^N + (-1)^{N+1} E &= F(x_{N+1})
\end{aligned}
$$

$$
\begin{bmatrix}
1 & x_0 & \cdots & x_0{}^N & 1 \\
1 & x_1 & \cdots & x_1{}^N & -1 \\
\vdots & \vdots & \ddots & \vdots & \vdots \\
1 & x_{N+1} & \cdots & x_{N+1}{}^N & (-1)^{N+1}
\end{bmatrix}
\begin{bmatrix}
c_0 \\ c_1 \\ \vdots \\ c_N \\ E
\end{bmatrix}
=
\begin{bmatrix}
F(x_0) \\ F(x_1) \\ \vdots \\ F(x_N) \\ F(x_{N+1})
\end{bmatrix}
$$

Using variables, we may express the previous formulation, as well as, its solution in a more compact form.

$$X \cdot CE = F$$

$$CE = inv(X) \cdot F$$

The solution vector, CE, contains both the coefficients of the polynomial, as well as the Minimax error, E.

Example 3.3: *Minimax Optimization*

In this example, we will find the coefficients for the example curve that was given earlier in the section and then use the Remez exchange method to iteratively improve on the results. The example function that was shown earlier is evaluated from $x = 0$ to 4 ($a = 0$ and $b = 4$) in steps of 0.1. The accompanying MatLab text shows the set up for the function.

$$F(x) = \frac{1}{1 + 25x^2}$$

```
x    = 0:.01:4;                    %% independent variable
F    = 1./(1 + 25*x.^2);           %% Test Function
```

To initially solve for the coefficients, we pick evenly spaced x coordinates (x_guess in the code below) at which we want the Minimax condition to be solved. The function values at these initial guess positions are referred to as F_guess in the following code.

```
Ind     = [1 68 134 201 268 334 401];%% Starting indices for equal spacing
x_guess = x(1, Ind)';            %% = [0 0.67 1.33 2.0 2.67 3.3 4.0]'
F_guess = F(1, Ind)';            %% Test function values at x_guess
```

In this next code section, we build the matrix X and solve for solution vector CE, which contains both the desired polynomial coefficients and the resulting error, E.

```
Order       = 5;
Sign        = ((-1).^(0:1:Order+1))';  %% The needed sign vector
X = ones(Order+2,1);                   %% Building the matrix X
for n = 1:Order
   X = [X x_guess.^n];
end
X = [X Sign];                          %% Adding the sign vector

CE = inv(X)*F_guess                    %% Calculating solution vector
C  = CE(1:Order+1,1)                   %% Extracting the coefficients
```

We now use the calculated polynomial coefficients, C, to evaluate the polynomial approximation, $P(x)$, at the index values of $x = 0$, 0.01 … 4 and compare it to the original curve, $F(x)$. The following graph shows the error function $E(x) = F(x)-P(x)$, as well as the positions at which we forced the absolute error to be constant.

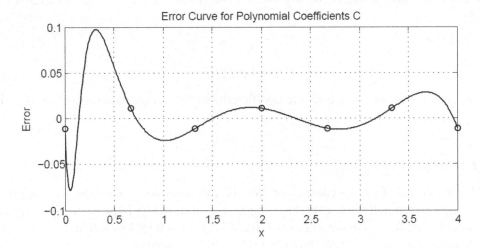

Figure 3-13: Error Curve P(x) – F(x) after First Evaluation

The solution forced a constant absolute error, E (0.0109), at the desired positions, x_i. These points, however, are not at the extrema of the error curve and, therefore, do not satisfy the Minimax criterion. We need a procedure that allows us to fine tune the original guess positions, x_guess, such that the accompanying errors settle at the absolute maxima of the error curve.

The Remez Exchange

There exists an algorithm that examines the error function shown in the last figure and iteratively improves the guess of the x_i vector until the Minimax criterion is met. This technique is called the Remez exchange algorithm. The Remez exchange algorithm moves each circular marker (the present guess) to the nearest local maximum/minimum. Let's restate the original guess positions.

$$x_i = \begin{bmatrix} 0.00 & 0.67 & 1.33 & 2.00 & 2.67 & 3.33 & 4.00 \end{bmatrix}$$

Just from pure visual inspection of the last figure, we select the closest maxima/minima and refine our guess to the following vector.

$$x_i = \begin{bmatrix} 0.1 & 0.3 & 1.0 & 1.9 & 2.7 & 3.6 & 4.00 \end{bmatrix}$$

The more formal procedure for the Remez exchange states the following:

1. Find all $N+1$ roots (or zero crossings) of the current error functions and label them r_1 through r_{N+1}. There are a myriad of simple techniques that numerically find the zero crossings. In our code example, the error function is defined as a discrete vector of 401 entries. We may simply search through the vector to see where the sign changes.

2. Once the roots are found, we define the following segments: $[a \ r_1] \ [r_1 \ r_2] \ \ldots\ldots \ [r_5 \ r_6] \ [r_6 \ b]$. Note that a is defined as the start point at $x=0$, whereas b is defined as the end point at $x=4$.

3. Each segment will have one minimum or one maximum, which can be found by taking the derivative of the error function, setting it to zero, and solving for x. We may just as well look for the maximum and minimum value in each interval by brute force. The new set of extrema constitutes the refined guess vector, x_i.

4. Using the new guess vector, build the matrix equations again and resolve for the polynomial coefficients.

Executing the Remez exchange algorithm once produces the following refined guess vector, closely resembles our visual inspection above.

$$x_i = \begin{bmatrix} 0.05 & 0.32 & 1.02 & 1.89 & 2.79 & 3.68 & 4.00 \end{bmatrix}$$

```
%% Start Minimax Curve Fit - Defining the Curve

x              = 0:.01:4;          %% independent variable
F              = 1./(1 + 25*x.^2); %% Test Function
Order          = 5;                %% Polynomial order of fit
RemezIterations = 5;               %% Let's try just 5 for now
a = x(1);    b = x(length(F));
Sign   = ((-1).^(0:1:Order+1))';   %% Column sign vector
NumberPoints = Order + 2;
Ind    = [1 68 134 201 268 334 401]; %% Let's pick starting indices
F_guess = F(1, Ind)';              %% Points of test function
x_guess = x(1, Ind)';              %% Accompanying x coordinates


Ind = [];
```

```
for n = 1:RemezIterations
    X = ones(NumberPoints,1); %% Build system of equations for the
    for z = 1:Order           %% current Remez Iteration
        X = [X x_guess.^z];
    end
    X  = [X Sign];
    CE = inv(X)*F_guess;      %% Here we solve for the coefficients
    C  = CE(1:Order+1,1);     %% Extracting polynomial coefficients
    E  = CE(Order+2,1);       %% Extracting the error

    P1 = polyval(flipud(C),x_guess); %% Eval only at select points
    P2 = polyval(flipud(C),x);       %% Eval for high res index
    Error1 = F_guess - P1;           %% Error only at select points
    Error2 = F - P2;                 %% Error for high res index

    Ind = Remez(Error2, Order, Ind); %% Calling Remez function
    x_guess  = x(1,Ind)';
    F_guess  = F(1,Ind)';
end
```

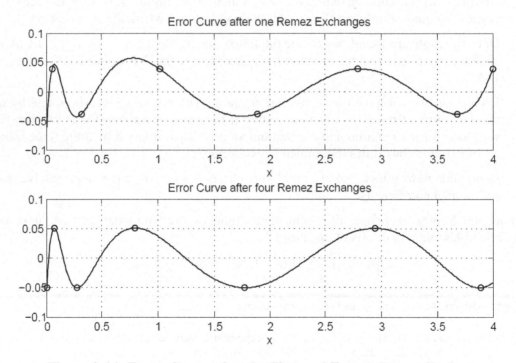

Figure 3-14: Error Curve after the First and Fourth Remez Exchange

The last code segment computes the final polynomial coefficients that produced the Minimax fit shown in Figure 3-14. The figure shows the vector, *Error2* (see the code above), at iterations *n*=2 and *n*=5, which correspond to the conditions after one and four Remez operations, respectively. The final coefficients are stored in vector *C*.

```
function [Output] = Remez(ErrorFunction, Order, PreviousPointsIndex)

NumberOfRoots  = Order + 1;
NumberOfPoints = Order + 2;

%% Finding roots (via brute search)/Store segment boundaries in Range

Range   = zeros(NumberOfPoints, 2);
Counter = 1;
for r = 1:NumberOfRoots
    Range(r,1) = Counter;
    CurrentSign = sign(ErrorFunction(Counter));
    while(sign(ErrorFunction(Counter)) == CurrentSign ...
                    && Counter < length(ErrorFunction))
        Counter = Counter + 1;
    end
    Range(r,2) = Counter - 1;     %% Change in sign was found
end
Range(NumberOfPoints,1) = Range(NumberOfPoints-1,2) + 1;
Range(NumberOfPoints,2) = length(ErrorFunction);

if( Range(NumberOfPoints-1,1) == length(ErrorFunction) )
        %% Houston, we have a problem, we don't have enough roots
    Output = PreviousPointsIndex;
    return;
end

%% Let's now look through each segment to find the min max values

StartIndex = 0;
for r=1:NumberOfPoints
    CurrentSection = ErrorFunction(Range(r,1):Range(r,2));
    [Cmax Imax] = max(CurrentSection);
    [Cmin Imin] = min(CurrentSection);

    if(Cmax > abs(Cmin))
        Output(1,r) = Imax + StartIndex;
    else
        Output(1,r) = Imin + StartIndex;
    end
    StartIndex = Range(r,2);
end
```

Only a few exchanges are required to converge on the Minimax criterion, although that number may increase with larger order polynomials.

3.2.2   Optimal - Equiripple - Filter Design

The effort of applying Chebyshev approximation and the Remez exchange algorithm to optimal FIR filter design was undertaken by James McClellan and Thomas Park in their seminal paper published in 1972 [5]. The resulting equiripple filter design process [6] is completely analogous to that of polynomial approximation using the Minimax criterion discussed in the last section. There, we had built a matrix expression that, when solved, yielded coefficients that would produce a unique polynomial. When comparing this polynomial function, $p(x)$, to the one that we were trying to approximate, $f(x)$, we ended up with error functions whose maximum absolute error was minimized. The following formulation was initially established to solve the polynomial approximation problem.

$$X \cdot CE = F$$

$$
\begin{bmatrix}
1 & x_0 & \cdots & x_0{}^N & 1 \\
1 & x_1 & \cdots & x_1{}^N & -1 \\
\vdots & \vdots & \ddots & \vdots & \vdots \\
1 & x_{N+1} & \cdots & x_{N+1}{}^N & (-1)^{N+1}
\end{bmatrix}
\begin{bmatrix}
c_0 \\
c_1 \\
\vdots \\
c_N \\
E
\end{bmatrix}
=
\begin{bmatrix}
F(x_0) \\
F(x_1) \\
\vdots \\
F(x_N) \\
F(x_{N+1})
\end{bmatrix}
$$

$$N+2 \; by \; N+2 \qquad N+2 \; by \; 1 \qquad N+2 \; by \; 1$$

$F(x_0)$ through $F(x_{N+1})$ were the desired y coordinates that the calculated polynomial was meant to approximate at the x coordinates x_0 through x_{N+1}. The calculated polynomial didn't pass through the desired y coordinates exactly because we allowed for an error to exist between the desired and actual calculated polynomial. This error would be opposite in sign for each successive constraint (desired $F(x_i)$ coordinate).

Regarding the filter, the only aspect of our mathematic expression that changes is the type of information stored in the matrices. To understand what we need to insert into the matrices, let's recall the expression for the discrete time Fourier transform (DTFT) of an impulse response, c_n, featuring an even and odd number of coefficients.

$$c_0 \cdot e^{-j2\pi f_D(-4.5)} + c_1 \cdot e^{-j2\pi f_D(-3.5)} + \cdots + c_8 \cdot e^{-j2\pi f_D(3.5)} + c_9 \cdot e^{-j2\pi f_D(4.5)} = H(f_D) \quad \leftarrow even$$

$$c_0 \cdot e^{-j2\pi f_D(-5)} + c_1 \cdot e^{-j2\pi f_D(-4)} + \cdots + c_5 \cdot e^{-j2\pi f_D(0)} + \cdots + c_{10} \cdot e^{-j2\pi f_D(5)} = H(f_D) \quad \leftarrow odd$$

In the previous equations, we calculate the frequency response, $H(f)$, at a particular normalized frequency, f_D, that lies between -0.5 and 0.5 Hz. To make the notation less general and confusing, we show examples of the DTFT equation above using impulse responses with 10 and 11 taps, respectively. Note further that for the sake of convenience, we allow the time instance, n, to start at the negative indices of -4.5 and -5. The type of filter we will focus on in this section is one whose frequency response is $H(f) = conj(H(-f))$. Therefore, the magnitude and phase response will respectively be an even and odd function of frequency. To make our task easier yet, we will set the phase response of $H(f)$ to zero everywhere. The resulting impulse response, c_n, will be an even function of time and feature completely real coefficients. Forcing the impulse response into an even function gives the coefficients the following symmetry: $c_0 = c_9$, $c_1 = c_8$ ... (for $N = 10$) and $c_0 = c_{10}$, $c_1 = c_9$ ... (for $N = 11$).

For the $N = 10$ scenario, the DTFT simplifies in the following manner.

$$c_0 \cdot e^{-j2\pi f_D(-4.5)} + c_1 \cdot e^{-j2\pi f_D(-3.5)} + \cdots + c_1 \cdot e^{-j2\pi f_D(3.5)} + c_0 \cdot e^{-j2\pi f_D(4.5)} = H(f_D)$$

$$c_0 \cdot (e^{j2\pi f_D(4.5)} + e^{-j2\pi f_D(4.5)}) + c_1 \cdot (e^{j2\pi f_D(3.5)} + e^{-j2\pi f_D(3.5)}) + \cdots c_4 \cdot (e^{j2\pi f_D(0.5)} + e^{-j2\pi f_D(0.5)}) = H(f_D)$$

$$2 \cdot [c_0 \cdot \cos(2\pi f_d(4.5)) + c_1 \cdot \cos(2\pi f_d(3.5)) + \cdots + c_4 \cdot \cos(2\pi f_d(0.5))] = H(f_d)$$

The odd tap length version differs only in the fact that it features a lone tap at time equal to zero, which gives the equation a slightly different form.

$$c_0 \cdot e^{-j2\pi f_D(-5)} + c_1 \cdot e^{-j2\pi f_D(-4)} + \cdots + c_5 \cdot e^{-j2\pi f_D(0)} + \cdots + c_{10} \cdot e^{-j2\pi f_D(5)} = H(f_D)$$

$$2 \cdot [c_0 \cdot \cos(2\pi f_d 5) + c_1 \cdot \cos(2\pi f_d 4) + \cdots + c_4 \cdot \cos(2\pi f_d 1)] + c_5 = H(f_d)$$

The new matrix expression $X \cdot CE = F$ can now be formulated as seen below. The vectors X, CE, and F hold the analysis functions of the Fourier transform, the unknown impulse response plus the error scalar, and the desired frequency response, respectively.

$$\begin{bmatrix} 2 \cdot \cos(2\pi f_0 4.5) & \cdots & 2 \cdot \cos(2\pi f_0 \cdot 1.5) & 2 \cdot \cos(2\pi f_0 \cdot 0.5) & (-1)^0 \\ 2 \cdot \cos(2\pi f_1 4.5) & \cdots & 2 \cdot \cos(2\pi f_1 \cdot 1.5) & 2 \cdot \cos(2\pi f_1 \cdot 0.5) & (-1)^1 \\ \vdots & \vdots & \ddots & \vdots & \vdots \\ 2 \cdot \cos(2\pi f_5 4.5) & \cdots & 2 \cdot \cos(2\pi f_5 \cdot 1.5) & 2 \cdot \cos(2\pi f_5 \cdot 0.5) & (-1)^5 \end{bmatrix} \begin{bmatrix} c_0 \\ \vdots \\ c_3 \\ c_4 \\ E \end{bmatrix} = \begin{bmatrix} H(f_0) \\ H(f_1) \\ \vdots \\ H(f_4) \\ H(f_5) \end{bmatrix} \leftarrow even$$

$$X = (N/2)+1 \ by \ (N/2)+1 \qquad CE = (N/2)+1 \ by \ 1 \qquad F = (N/2)+1 \ by \ 1$$

$$\begin{bmatrix} 2 \cdot \cos(2\pi f_0 5) & \cdots & 2 \cdot \cos(2\pi f_0 1) & 1 & (-1)^0 \\ 2 \cdot \cos(2\pi f_1 5) & \cdots & 2 \cdot \cos(2\pi f_1 1) & 1 & (-1)^1 \\ \vdots & \vdots & \ddots & \vdots & \vdots \\ 2 \cdot \cos(2\pi f_6 5) & \cdots & 2 \cdot \cos(2\pi f_6 1) & 1 & (-1)^6 \end{bmatrix} \begin{bmatrix} c_0 \\ \vdots \\ c_4 \\ c_5 \\ E \end{bmatrix} = \begin{bmatrix} H(f_0) \\ H(f_1) \\ \vdots \\ H(f_5) \\ H(f_6) \end{bmatrix} \leftarrow odd$$

$$X = (N+1)/2+1 \ by \ (N+1)/2+1 \qquad CE = (N+1)/2+1 \ by \ 1 \qquad F = (N+1)/2+1 \ by \ 1$$

Ok, now that we have gone through this arithmetic stretching exercise, let's see how we set up our constraints for a typical filter response.

Specifying the Desired Frequency Response

Given the fact that the coefficients for the previous matrix expressions are real and symmetric around zero, there is no need to specify the filter's response at negative frequencies. The magnitude response of the filter will be symmetric, and the phase response will be zero for all frequencies. Figure 3-15 shows a typical scenario, where we wish to design a low-pass filter given pass and stop band frequencies (F_pass and F_stop) as well as the number of taps in the filter. Note that for an even number of taps, N, there are $N/2$ coefficients plus an error, E, for which we solve directly via $N/2+1$ constraints, $H(f)$. For an odd number of taps, we solve for $(N+1)/2$ coefficients plus the error using $(N+1)/2 +1$ constraints.

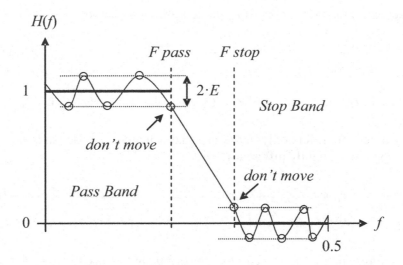

Figure 3-15: Typical Frequency Response Constraint for Equiripple Filter

If we define the error as $E(f) = Desired\_H(f) - Actual\_H(f)$, then at the *F_pass* and *F_stop* frequencies, the error should reach a positive and negative maximum, respectively, just as it is shown in the previous figure. To guarantee that the maximum absolute errors occur at these specified frequencies, we will not move them during the Remez exchange iterations. This yields a slightly inferior ripple behavior but is specified here nevertheless. The area between the pass and stop bands is somewhat of a no-man's land and does not need to be included in the optimization process. So, without further delay, let's jump right in and work an example.

Example 3.4: Equiripple Filter

Design a 15-tap equiripple filter featuring pass and stop band frequencies of 0.2Hz and 0.3Hz, respectively. The number of coefficients for which we directly solve is $(N+1)/2 = 8$, whereas the number of constraints is equal to 9. We will distribute the frequencies where we want the maximum errors to occur somewhat evenly.

$$f = [0 \quad .1 \quad .15 \quad .2 \quad \quad .3 \quad .35 \quad .4 \quad .45 \quad .5]$$
$$F = [1 \quad 1 \quad 1 \quad 1 \quad \quad 0 \quad 0 \quad 0 \quad 0 \quad 0]$$

Below, we show how the matrix needs to be set up before calculating the solution vector.

$$\begin{bmatrix} 2 \cdot \cos(2\pi f_0 7) & \cdots & 2 \cdot \cos(2\pi f_0 1) & 1 & (-1)^0 \\ 2 \cdot \cos(2\pi f_1 7) & \cdots & 2 \cdot \cos(2\pi f_1 1) & 1 & (-1)^1 \\ \vdots & \ddots & \vdots & \vdots & \vdots \\ 2 \cdot \cos(2\pi f_8 7) & \cdots & 2 \cdot \cos(2\pi f_8 1) & 1 & (-1)^8 \end{bmatrix} \begin{bmatrix} c_0 \\ \vdots \\ c_6 \\ c_7 \\ E \end{bmatrix} = \begin{bmatrix} H(f_0) \\ H(f_1) \\ \vdots \\ H(f_7) \\ H(f_8) \end{bmatrix} \leftarrow odd$$

$$9 \, by \, 9 \qquad\qquad 9 \, by \, 1 \qquad 9 \, by \, 1$$

```
%% Defining the Response --> Tap length is N = 15 (odd number)
%% We control (N+1)/2 coefficients via (N+1)/2 +1 constraints


F = [0.0 .1 .15 .2  .3 .35 .4 .45  0.5].'; %% 9 constraints
H = [1 1 1 1   0 0 0 0 0].';              %% Frequency response
A = [1 -1 1 -1  1 -1 1 -1 1].';           %% Alternating +/- 1


%% Initial Solution
X = [2*cos(2*pi.*F*7) 2*cos(2*pi.*F*6) 2*cos(2*pi.*F*5) ...
     2*cos(2*pi.*F*4) 2*cos(2*pi.*F*3) 2*cos(2*pi.*F*2) ...
     2*cos(2*pi.*F*1) ones(9,1) A];   %% Building matrix X
CE = inv(X)*H;                        %% Solve for the CE vector


Coefficients = [CE(1:8,1); flipud(CE(1:7,1))].'; %% Extract C
MinimaxError = CE(9,1);                          %% Extract E
```

The previous code yields the first stab at the optimal coefficients in the Minimax sense. The following figure shows the initial impulse response, $c[n]$, and frequency response, of which we only plot the real part (the imaginary part is zero). The goal now is to compute the error between the current and desired frequency responses inside the pass and stop bands.

$$c[n]=[-.023 \quad 0 \quad .049 \quad 0 \quad -.097 \quad 0 \quad .315 \quad .5 \quad .315 \quad 0 \quad -.097 \quad 0 \quad .049 \quad 0 \quad -.023]$$

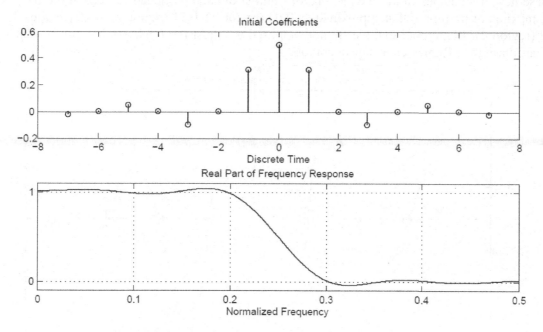

Figure 3-16: Initial Impulse and Frequency Response

```
%% Compute and Plot the Magnitude Response
f       = 0:.001:0.5;
N       = length(Coefficients);
n       = -7:7;
Actual  = zeros(1,length(f));
```

```
for i = 1:length(f)
    AnalysisTone  = exp(j*2*pi*n*f(1,i));
    Actual(1,i)   = Coefficients*AnalysisTone'; %% Summation
end
figure(1);
plot(f, real(Actual), 'r'); grid on; hold on;
plot(f, imag(Actual), 'b'); title('Frequency Response');
```

The following figure shows us the error between the desired and actual (current) frequency responses. The algorithm did indeed assure that the absolute error was equal at the frequencies [0 0.1 0.15 0.2 0.3 0.35 0.40 0.45 0.5], as we requested in the specification. During the Remez exchange iteration, we take the same steps as we did for the polynomial curve fitting problem.

1 → Divide the error curve into regions whose boundaries are the end points of the curve, or its roots. The regions in the pass and stop bands would be approximately the following: pass band regions = [0 0.09] [0.09 0.14] [0.14 0.19] [0.19 0.20]; stop band regions = [0.3 0.31] [0.31 0.36] [0.36 0.42] [0.42 0.48] [0.48 0.5].

2 → Select new frequencies in the pass and stop bands at which to generate new constraints. These new frequencies fall at the maxima or minima of each particular region. In the pass band, the maxima or minima fall at approximately 0.05 (region 1), 0.11 (region 2), 0.18 (regions 3), and 0.2 (region 4). Remember that we are not supposed to adjust the $F\_pass = 0.2$ and $F\_stop = 0.3$ values during the Remez exchange operation.

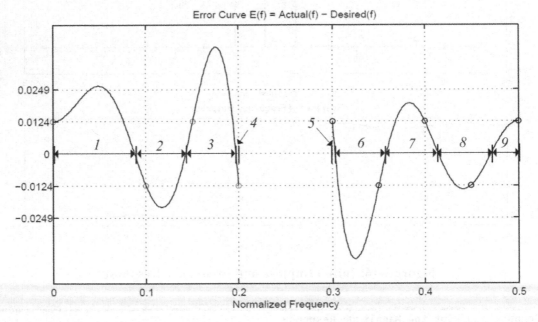

Figure 3-17: The Error Curve after the First Solution

In the stop band, we will try the following new frequencies: 0.3 (region 5), 0.32 (region 6), 0.38 (region 7), 0.44 (region 8), and 0.5 (region 9).

$$F = [0.05 \quad 0.11 \quad 0.18 \quad 0.2 \quad 0.3 \quad 0.32 \quad 0.38 \quad 0.44 \quad 0.5]$$
$$A = [1 \quad\quad 1 \quad\quad 1 \quad\quad 1 \quad\quad 0 \quad\quad 0 \quad\quad 0 \quad\quad 0 \quad\quad 0]$$

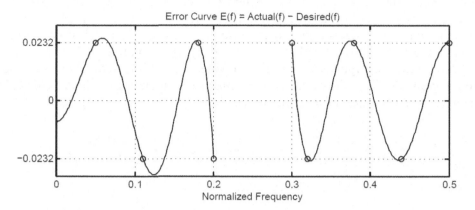

Figure 3-18: The Error Curve after the First Remez Exchange

It becomes immediately clear, that after just one Remez exchange we are moving toward the equiripple error condition. The logarithmic magnitude response for the coefficients computed this time around can be seen below.

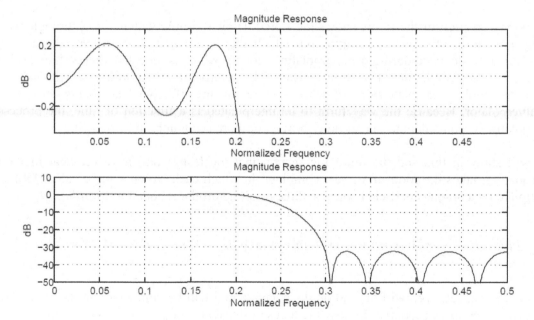

Figure 3-19: Magnitude Response of Second Set of Coefficient

The magnitude response in dB confirms that the pass and stop bands gain is approaching the equiripple condition. The stop band behavior in the preceding figure is typically what is seen when we choose the equiripple design method in MatLab's filter design tool "fdatool".

3.3   Interpolation

In communication systems, we are often presented a sequence of evenly spaced, discrete samples that are derived from a continuous waveform. There are times when we must deduce values of that original continuous waveform other than those provided by the set of discrete samples. The proper engineering terminology for the act of deducing or estimating values in between given samples is the term "interpolation". The following figure illustrates how continuous-time information is converted into the discrete time domain via an analog-to-digital converter (ADC) and then stored in a shift register for further processing.

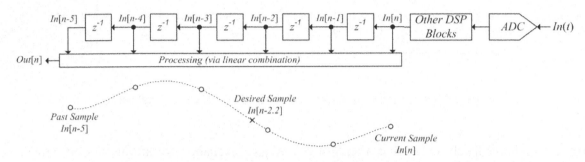

Figure 3-20: Sample Storage in Digital Portion of a Receiver

The bottom portion of the figure shows the stored, known sample values, $In[n]$ through $In[n-5]$, along with the continuous waveform from which they originated. In-between values, such as $In[n-2.2]$, are calculated during the interpolation process, which is repeated after a clock cycle has moved a new sample into the shift register. The interpolation algorithm should use every available sample to calculate $In[n-d]$, where d represents the effective delay that is introduced by the interpolator. Because the waveform to be interpolated is a function of time, the process of interpolation is equivalent to that of time shifting, or more accurately, time delaying.

We will show in this and the following section that the fastest and most efficient method of interpolation takes the form of a linear combiner, in which each stored sample value, $In[n-k]$, is multiplied by a unique coefficient, $a[k]$, and the resulting products added to form $In[n-d]$.

$$In[n-d] = \sum_{k=0}^{N} a[k] \cdot In[n-k] = a[0] \cdot In[n] + a[1] \cdot In[n-1] + \cdots + a[N-1] \cdot In[x - (N-1)]$$

The coefficient set, $a[k]$, will depend on the delay, d, and will be computed via the methods that are reviewed in these sections. Because the focus of this text is communication systems, the set of values used in the interpolation process is almost always derived from an evenly sampled, time-domain waveform, $In(t)$, as shown in the previous figure. The sampled values residing in the shift register and the linear combiner, which merges them into a final result, form the well-known FIR filter structure introduced in chapter 2.

Of the four methods discussed in this section, two will lend themselves to the high-speed FIR filter implementations, whereas to others require more general-purpose computational hardware. The table below presents an overview of these four techniques.

INTERPOLATION METHOD	BASIS FUNCTIONS	IMPLEMENTATION
Polynomial Approximation (algebraic form)	monomials of the form x^i	Restricted to general-purpose microprocessors
Polynomial Approximation (Lagrange form)	polynomials	High speed FIR implementation
Time Shifting	complex sinusoids	High speed FIR implementation
DFT / IDFT method	complex sinusoids	Restricted to general-purpose microprocessors

Figure 3-21: A List of Interpolation Techniques

Basis Functions

To find values positioned in-between known sample instances, an approximation of the unknown, continuous waveform, from which the samples were taken, must be conceived. Each algorithm listed in the previous table approximates the original, continuous waveform as the sum of scaled basis functions.

$$In(t) = \sum_{k=0}^{N-1} a_k \cdot basis_k(t)$$

In the case of polynomial approximation, these basis functions take the form $basis_k(t) = t^k$ for the algebraic form, and specialized polynomial functions for the Lagrange form. In the other two cases, the basis functions take the form of a complex sinusoid, $basis_k(t) = exp(j2\pi kt)$. The introduction to chapter 2 discusses the idea of basis functions in more detail.

Interpolation Performance

Note that on the last page, we chose to estimate (interpolate) the input stream, $In[n]$, at a delay of 2.2 samples, which placed the final value $In[n-2.2]$ between known sample values $In[n-2]$ and $In[n-3]$. At delays of 2.0 to 3.0, the interpolation process reaches its highest performance because it has a similar number of future and past samples to work with. The more close-by samples are available, the more we know about the curvature and behavior of the original waveform between sample values $In[n-2]$ and $In[n-3]$.

3.3.1   Interpolation Using Polynomial Approximation (Algebraic Form)

Given N consecutive values of a sampled waveform, $In(t)$, we may construct a polynomial function, $y(t)$, of degree N-1 that passes through each one of the N points. The basic technique of computing the coefficients of the polynomial were already discussed in Example 1.10, and the more sophisticated method of least squares extended the use of polynomials approximation in cases where the number of samples to be fitted was larger than the number of coefficients available. The following figure shows the equation of a polynomial of degree N-1.

$$y(t) = \sum_{i=0}^{N-1} a_i \cdot t^i = a_0 + a_1 t + a_2 t^2 + a_3 t^3 + \cdots a_{N-1} t^{N-1}$$

For N waveform samples, a system of linear equations may be constructed and solved via simple matrix inversion as follows.

$$\begin{bmatrix} a_0 + a_1 t_0 + a_2 t_0^2 + \cdots + a_{N-1} t_0^{N-1} \\ a_0 + a_1 t_1 + a_2 t_1^2 + \cdots + a_{N-1} t_1^{N-1} \\ \vdots \\ a_0 + a_1 t_{N-1} + a_2 t_{N-1}^2 + \cdots + a_{N-1} t_{N-1}^{N-1} \end{bmatrix} = \begin{bmatrix} y(t_0) \\ y(t_1) \\ \vdots \\ y(t_{N-1}) \end{bmatrix}$$

$$\underbrace{\begin{bmatrix} 1 & t_0 & t_0^2 & \cdots & t_0^{N-1} \\ 1 & t_1 & t_1^2 & \cdots & t_1^{N-1} \\ & & \vdots & & \\ 1 & t_{N-1} & t_{N-1}^2 & \cdots & t_{N-1}^{N-1} \end{bmatrix}}_{F} \cdot \underbrace{\begin{bmatrix} a_0 \\ a_1 \\ \vdots \\ a_{N-1} \end{bmatrix}}_{A} = \underbrace{\begin{bmatrix} y(t_0) \\ y(t_1) \\ \vdots \\ y(t_{N-1}) \end{bmatrix}}_{Y}$$

Finally,

$$\boxed{A = inv(F) \cdot Y}$$

After computing the coefficients of solution vector A, we still needed to generate the basis functions, t^i, multiply them by those coefficients, and sum the resulting products to arrive at the final answer, $y(t)$. To ease the computational burden, it is common practice to precompute the basis functions for a certain number of tightly spaced time instances, t, that lie within a range that interests us. The values are stored in read-only memory, or ROM. However, this pre-computation step does not alleviate the initial effort involved in finding the coefficients, a_i, and this approach is practical in dedicated digital hardware only for polynomial approximations of degree 3 or less. To improve the accuracy of the polynomial approximation process, we need to increase the number of sample points, which requires the inverse operation of larger matrices. A far more efficient approach to polynomial approximation is named after the French / Italian mathematician Joseph Louis Lagrange (1736–1813). This approach allows for real-time approximation because may be computed by a FIR filter.

3.3.2 Interpolation using Polynomial Approximation (Lagrange Form)

The Lagrange formulation differs from the algebraic version previously discussed only in form, and both lead to identical approximations. The Lagrange [7, 8] formulation approximates the continuous curve between $N+1$ coordinates (x_0, y_0), (x_1, y_1), (x_2, y_2) ... (x_N, y_N) as a linear combination of scaled basis functions, b_i. The basis functions for the algebraic formulation previously computed were x^0, x^1, x^2, ... etc. Whereas the ones belonging to the Lagrange formulation seem much more cumbersome, they provide significant advantages for real-time implementation.

$$y(x) = \sum_{i=0}^{N} y_i \cdot b_i(x) = y_0 \cdot b_0(x) + y_1 \cdot b_1(x) + y_2 \cdot b_2(x) + \cdots + y_N \cdot b_N(x) \quad \leftarrow Lagrange$$

$$y(x) = \sum_{i=0}^{N} a_i \cdot x^i = a_0 + a_1 x + a_2 x^2 + \cdots + a_N x^N \qquad \leftarrow Algebraic$$

In digital hardware, the basis functions, x^i or $b_i(x)$, are precomputed for all required values of x and stored in ROM. With this effort out of the way, note that the variables scaling the basis functions of the Lagrange formulations are the y coordinates of the sample points, not a collection of coefficients, a_i, that have to be cumbersomely computed via matrix inversion. The basis functions of the Lagrange formulation are expressed as follows.

$$b_i(x) = \prod_{\substack{0 \leq m \leq N \\ m \neq i}} \frac{x - x_m}{x_i - x_m}$$

To demystify the awkward formulation of the previous expression, we will calculate the Lagrange basis functions for a third order polynomial in the following example.

__Example 3.5:__ Lagrange Basis Functions for a Third-Order Polynomial

Let's assume that we are to fit a third-order polynomial to four xy coordinates: $(0, y_0)$, $(1, y_1)$, $(2, y_2)$, and $(3, y_3)$. The four basis functions, $b_i(x)$, reduce to the following expressions.

$$b_0(x) = \frac{x - x_1}{x_0 - x_1} \cdot \frac{x - x_2}{x_0 - x_2} \cdot \frac{x - x_3}{x_0 - x_3} = \frac{x-1}{-1} \cdot \frac{x-2}{-2} \cdot \frac{x-3}{-3}$$

$$= -\frac{1}{6}(x-1) \cdot (x-2) \cdot (x-3)$$

$$b_1(x) = \frac{x - x_0}{x_1 - x_0} \cdot \frac{x - x_2}{x_1 - x_2} \cdot \frac{x - x_3}{x_1 - x_3} = \frac{x}{1} \cdot \frac{x-2}{-1} \cdot \frac{x-3}{-2}$$

$$= \frac{1}{2} x \cdot (x-2) \cdot (x-3)$$

$$b_2(x) = \frac{x - x_0}{x_2 - x_0} \cdot \frac{x - x_1}{x_2 - x_1} \cdot \frac{x - x_3}{x_2 - x_3} = \frac{x}{2} \cdot \frac{x-1}{1} \cdot \frac{x-3}{-1}$$

$$= -\frac{1}{2} x \cdot (x-1) \cdot (x-3)$$

$$b_4(x) = \frac{x - x_0}{x_3 - x_0} \cdot \frac{x - x_1}{x_3 - x_1} \cdot \frac{x - x_2}{x_3 - x_2} = \frac{x}{3} \cdot \frac{x-1}{2} \cdot \frac{x-2}{1}$$

$$= \frac{1}{6} x \cdot (x-1) \cdot (x-2)$$

Example 3.6: *Interpolation via Both Polynomial Forms*

The following code illustrates the interpolation process via both polynomial formulations.

```
%% The Coordinates
X = [0 1 2 3].';  %% X and Y coordinates of the four points
Y = [1 3 2 2].';  %% as column vectors

%% The Desired x coordinate at which we interpolate
x = 1.5;


%% Solution via the algebraic form of the polynomial
F           = [ones(4,1) X  X.^2 X.^3];
A           = inv(F)*Y;
y_Algebraic = [1 x x^2 x^3]*A


%% Solution via the Lagrange form of the polynomial
b0          = -(1/6)*(x-1)*(x-2)*(x-3);
b1          =  (1/2)*x*(x-2)*(x-3);
b2          = -(1/2)*x*(x-1)*(x-3);
b3          =  (1/6)*x*(x-1)*(x-2);
y_Lagrange  = Y.' * [b0; b1; b2; b3]
```

In the last example, we computed $b0$, $b1$, $b2$, and $b3$ for every desired x coordinate that was within our range of 0 to 3.0. As mentioned before, in a hardware implementation, these coefficients would be precomputed and stored in ROM tables for access on a clock-by-clock basis at some desired step size.

3.3.3 Interpolation Using All-Pass FIR Filters

An all-pass filter is a special FIR implementation that is frequently used for interpolation and phase-equalization. The all-pass filter leaves the magnitude of the input signal's frequency content unchanged and, instead, modifies only its phase characteristics. Phase equalization is often undertaken to undo distortion caused by preceding IIR or analog filters. In this section, however, the interpolation application is our main focus. Because in our time-sampled system, interpolation is equivalent to time shifting, let's recall the time-shifting property of the Fourier transform.

$$x[n-n_o] \overset{DFT}{\leftrightarrow} X[m] \cdot e^{-j2\pi n_o m/N}$$

The property suggests that to delay or advance the signal $x[n]$ by a time offset, n_o, we pass the signal $x[n]$ through a filter with the following frequency response.

$$H[m] = 1 \cdot e^{-j2\pi n_o m/N}$$

The code below uses the frequency sampling method to set up a 13-tap all-pass FIR filter that causes a time shift of $n_o = 6.25$ sample periods. Delaying or advancing the waveform by fractional sample periods reveals waveform values that are in between the given samples.

```
N     = 13;
m     = [-6 -5 -4 -3 -2 -1 0 1 2 3 4 5 6];   %% Force Response via the
Mag   = [1 1 1 1 1 1 1 1 1 1 1 1 1];         %% Frequency Sampling Method
no    = 6.75;                                %% Sample Delay
Phase = -2*pi*m*no/N;
H     = Mag.*cos(Phase) + j*Mag.*sin(Phase); %% Rectangular Freq Response
```

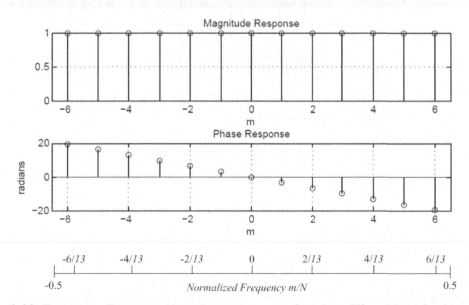

Figure 3-22: Frequency Response *H[m]* (Polar Format) of All Pass Filter Setup for Interpolation

After the inverse Fourier transform (IFT) operation (as was done in Example 2.27), we arrive at the discrete impulse response of our all-pass filter.

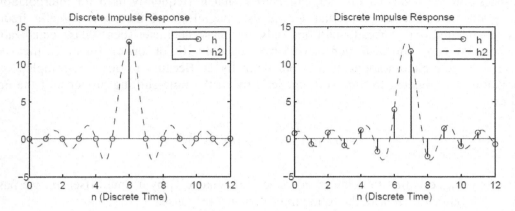

Figure 3-23: Discrete Impulse Responses for All-Pass Filters with Delay of 6.00 and 6.75 Samples

Taking the inverse Fourier transform of a frequency response that features constant magnitude yields a *sinc* function in the time domain. The peak of this *sinc* function is positioned to reflect the delay of 6.75 samples, as seen on the right side of the previous figure. Taking the DTFT of the impulse response provides us with the frequency characteristics of the filter, which features a phase response that may be correct at the discrete frequency indices *m* (-6, -5...6) but is not straight in between. Whereas this is not easily seen in the following plot, it becomes painfully obvious when we calculate the corresponding group delay, which we would expect to be constant at 6.75 samples across the normalized frequency range of $\pm 6.5/N = \pm 0.5$Hz.

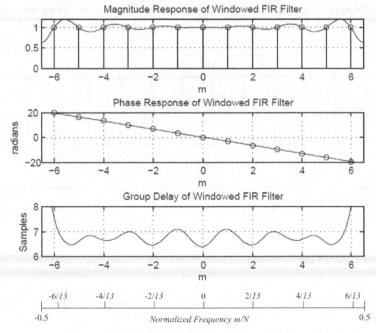

Figure 3-24: Frequency Response Filter Coefficients (Rectangular Window)

Many articles on digital signal processing lead us to believe that FIR filters always feature linear phase response. The reality is that we may make the phase response as linear or non-linear as we wish. The first step in crafting a linear phase response is to force the phase into a straight line at those positions that we can directly control via our frequency sampling technique. To further straighten the phase response, we multiply the impulse response by a Hanning window.

$$Hanning[n] = 0.5 - 0.5\cos(2\pi\frac{n+1}{N+1})$$

After recalculating the frequency response via the DTFT and computing the group delay, we find that the filter's performance is far superior to that of the un-windowed version.

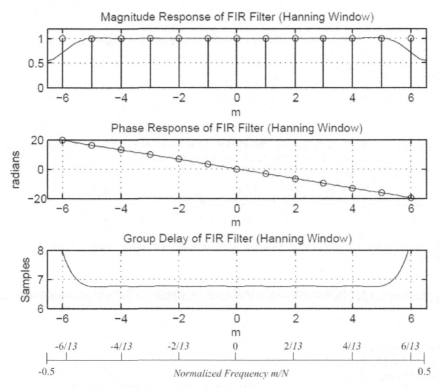

Figure 3-25: Discrete Time Fourier Transform of Windowed Impulse Response

Notes

Note that the impulse response of this interpolating filter could have just as easily been computed via the method of least squares. Via the LSE technique, we are able to force the phase response into a straight line at far more positions than those N locations that were available to us via the frequency sampling method.

Note further that we chose to delay the input by approximately half the filter length, such that the structure had approximately as many past and future samples to work with. You may verify yourself that choosing delays of less than 1 sample or more than 11 will produce an ill-behaved group delay because of the lack of future or past samples available for estimation (interpolation).

```matlab
%% Computing the inverse discrete Fourier transform
DiscreteTime = 0:N-1;
h = zeros(1,N);
for i=1:N
    n       = DiscreteTime(1,i);
    h(1,i) = X*exp(j*2*pi*m*n/N).';
end


%% Hanning Overlay
n                = 0:N-1;
Hanning          = (ones(1,N) - cos(2*pi*(n+1)/(N+1)));
h_windowed       = h.*Hanning/sum(Hanning.*h); % Windowed impulse response
                                               % scaled for DC gain = 1.0


%% Calculated frequency response via DTFT - done separately
DiscreteFrequency = -6.5:.1:0; %% Negative Fractionally Valued Frequencies
NegResult         = zeros(1,length(DiscreteFrequency));
for i=1:length(DiscreteFrequency)
    m             = DiscreteFrequency(1,i);
    NegResult(1,i) = h_windowed*exp(-j*2*pi*n*m/N).';
end


DiscreteFrequency = 0.1:.1:6.5; %% Positive fractionally valued frequencies
PosResult         = zeros(1,length(DiscreteFrequency));
for i=1:length(DiscreteFrequency)
    m             = DiscreteFrequency(1,i);
    PosResult(1,i) = h_windowed*exp(-j*2*pi*n*m/N).';
end


%% Required operation to properly unwrap the phase response for group delay
%% calculation
MagNew       = abs([NegResult PosResult]);
PhaseNew     = [(fliplr(unwrap(fliplr(angle(NegResult))))) ...
unwrap(angle(PosResult))];
DiscreteFrequency = -6.5:.1:6.5;


%% Calculate group delay
GroupDelay = zeros(1,length(PhaseNew));
for i = 2:length(PhaseNew) - 1
    FrequencyDelta = (DiscreteFrequency(1, i+1)-DiscreteFrequency(1, i))/N;
    PhaseResponseDelta = PhaseNew(1, i+1) - PhaseNew(1, i);
    GroupDelay(1,i)    = (-1/(2*pi))*PhaseResponseDelta/FrequencyDelta;
end


GroupDelay(1,1) =  GroupDelay(1,2);
GroupDelay(1,length(PhaseNew)) =  GroupDelay(1,length(PhaseNew)-1);


figure(3);
subplot(3,1,1);
plot(DiscreteFrequency, MagNew,'k'); grid on; hold on;
stem(-6:6, Mag,'k'); xlabel('m');
title('Magnitude Response of Windowed FIR Filter'); axis([-6.5 6.5 0 1.2]);
subplot(3,1,2);
plot(-6:6, Phase,'ko'); grid on; hold on;
```

```
plot(DiscreteFrequency, PhaseNew,'k');
title('Phase Response of Windowed FIR Filter'); ylabel('radians'); xlabel('m');
axis([-6.5 6.5 -20 20]);
subplot(3,1,3);
plot(DiscreteFrequency, GroupDelay,'k'); grid on; hold on;
title('Group Delay of Windowed FIR Filter'); ylabel('Samples'); xlabel('m');
axis([-6.5 6.5 6 8]);
```

3.3.4   Interpolation Using the Fourier Transform/Inverse Fourier Transform

The discrete time Fourier transform, or DTFT, is an algorithm that computes the correlation or similarity of a sampled input sequence, $x[n]$, and a complex analysis sinusoid, $exp(j2\pi fnT_s)$, of a certain frequency, f. The DTFT may validly be computed at any frequency f that is valued within the Nyquist sampling bounds of $\pm Fs/2$.

The discrete Fourier transform, or DFT, is a collection of DTFT results that are computed at evenly spaced normalized frequencies, m/N.

$$X[m] = \frac{1}{N}\sum_{n=0}^{N-1} x[n] \cdot e^{-j2\pi nm/N} \qquad m = 0,1,...N-1$$

This collection, stored in the vector $X[m]$, not only provides information regarding the spectrum of the input sequence, $x[n]$, but may also be used to reconstruct it. The sampled input sequence, $x[n]$, is reconstructed by taking the inverse discrete Fourier transform, IDFT, of $X[m]$. In fact, the DFT approximates the original continuous waveform, $x(t)$, from which its input samples, $x[n]$, are derived as a collection of complex sinusoids of different magnitude, phase, and frequency. All we really have to do is evaluate each one of the complex sinusoids at the desired time, n, and add up the results. And that is, in fact, exactly what the inverse Fourier transform does. Let's take another look at the equation for the IDFT.

$$x[n] = \sum_{m=0}^{N-1} X[m] \cdot e^{j2\pi nm/N} \qquad n = 0,1,...N-1$$

Let's suppose, then, that we have an input sequence of length $N = 8$ and an accompanying discrete time index $n = 0, 1 ... 7$. We know that we may recreate $x[n]$ from $X[m]$ at any one of the valid time indices n. But what about a fractional value, such as $n = 1.5$? Could we reformulate the IDFT to evaluate $x(t)$ where $0 \le t < 8$?

$$x(t) = \sum_{m=0}^{N-1} X[m] \cdot e^{j2\pi tm/N} \qquad 0 \le t < N$$

The simple answer is "no". The DFT, processing an input sequence $x[n]$ of length $N = 8$, will produce an output sequence $X[m]$ of length $N = 8$. The discrete, normalized frequencies at which the DFT evaluates the input sequence, $x[n]$, are equal to m/N, where $m = 0, 1 \ldots N\text{-}1$. Those frequencies, m/N, that are greater than or equal to 0.5 are aliased to negative frequencies equal to (m/N) -1. For our case of $N = 8$, the discrete frequencies can be stated in two ways.

$$discreteFrequencies \quad \rightarrow dFreq[m] = \begin{bmatrix} \dfrac{0}{8} & \dfrac{1}{8} & \dfrac{2}{8} & \dfrac{3}{8} & \dfrac{4}{8} & \dfrac{5}{8} & \dfrac{6}{8} & \dfrac{7}{8} \end{bmatrix}$$

or

$$discreteFrequencies \quad \rightarrow dFreq[m] = \begin{bmatrix} \dfrac{0}{8} & \dfrac{1}{8} & \dfrac{2}{8} & \dfrac{3}{8} & \dfrac{-4}{8} & \dfrac{-3}{8} & \dfrac{-2}{8} & \dfrac{-1}{8} \end{bmatrix}$$

If we modify the inverse discrete Fourier transform using the second discrete frequency vector, then the original continuous waveform, $x(t)$, may be uniquely evaluated at any time between 0 and N.

$$x(t) = \sum_{m=0}^{N-1} X[m] \cdot e^{j2\pi t \cdot dFreq[m]} \quad 0 \le t < N$$

Example 3.7: Interpolation Using the DFT/IDFT Method

In this example, we consider an input sequence $x[n]$ that obeys the following equation and features a length of $N = 17$. The sample rate is assumed to be 1Hz.

$$x[n] = \cos(2\pi n \frac{2}{N} + \frac{\pi}{8}) + \sin(2\pi n \frac{4}{N}) + 0.5\cos(2\pi n \frac{.2}{N}) - n/16$$

While we evaluate the previous equation for a discrete time index $n = 0, 1 \ldots N\text{-}1$, we will also precompute a higher-resolution version of the waveform for times t = 0, .1 … 24 seconds. This higher-resolution version will be compared to the interpolated output from the modified inverse discrete Fourier transform.

```
%% Input Waveform Definition (Sample Rate is 1Hz)
N     = 17;           %% Length N of input sequence x[n]
n     = 0:1:N-1;      %% normal discrete time index n
x_n   = cos((2*pi*n*2/N)+pi/8) +      sin(2*pi*n*4/N) + ...
        0.5*cos((2*pi*n*0.2/N))     -  n/16;  %% x[n] = Input Waveform


t     = 0:.1:24;      %% high resolution time index
x_t   = cos((2*pi*t*2/N)+pi/8) +      sin(2*pi*t*4/N) + ...
        0.5*cos((2*pi*t*.2/N))      - .5* t/8;   %% x(t) = high res version
```

The next code segment computes the DFT and allows us to either select the correct (non-aliased) frequencies or the aliased versions. Choose the aliased versions, and watch the interpolation fail.

```
%% Computing the DFT of x[n]
m            = 0:1:N-1;      %% Frequency index m
UseCorrect   = 1;           %% 0/1 = aliased/correct norm. frequencies
dFreq        = zeros(1,N);  %% holds normalized frequencies
for i = 1:N
    if(m(1,i)/N >= 0.5 && UseCorrect == 1); dFreq(1,i) = m(1,i)/N -1;
    else                                    dFreq(1,i) = m(1,i)/N;
    end
end


DFT_Out      = zeros(1,N);
for i = 1:N                     %% DFT Calculation
    f            = dFreq(1,i);
    AnalysisTone = exp(j*2*pi*n*f);
    DFT_Out(1,i) = (1/N)*x_n*AnalysisTone';
end

%% Interpolation via the modified IDFT at any time instance 0 <= t < N
IDFT_Out = zeros(1,length(t));
for i = 1:N                     %% modified IDFT calculation
    f            = dFreq(1,i);
    AnalysisTone = exp(j*2*pi*t*f);
    IDFT_Out     = IDFT_Out + DFT_Out(1,i)*AnalysisTone;
end


figure(1);
plot(t, real(IDFT_Out), 'k'); hold on;
plot(t, x_t, 'k:'); plot(n, x_n, 'ko');
legend('Interpolator Result', 'Original High Resolution Waveform', 'Input Samples');
title('Results'); xlabel('Time in seconds'); axis([0 24 -3 3]);
```

The code both interpolates and attempts to extrapolate or guess at the waveform at times larger than $N - 1$ where it does not have information.

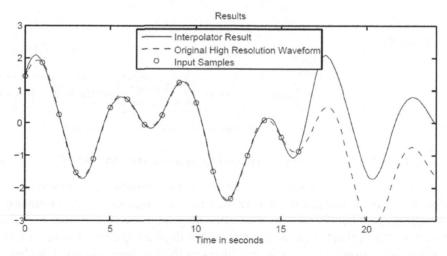

Figure 3-26: Comparing the Interpolation and the Original High Resolution Waveform

From the resulting plot, we may draw several conclusions. The first and most important result is that the low-resolution input waveform $x[n]$ was perfectly recreated, as can be seen by circular markers in the graph. Comparing the original high-resolution waveform and the interpolated results, we conclude that interpolated values are most accurate towards the center of our collection of samples. Given that our algorithm lacks information at times less than 0 and larger than N -1, it cannot accurately estimate interpolated values close to the beginning and end of the waveform. Whereas the original high-resolution waveform continues to become more negative with increasing time, the interpolation via the modified IDFT assumes that the original waveform is periodic. Note that by definition, the IDFT returns N sample values at $n = 0, 1 \ldots N$-1. The modified version of the IDFT, used for interpolation, returns just one unique value at $0 \leq t < N$.

3.4   Multirate Signal Processing

There are many applications in the realm of signal processing that require us to change the sample rate of a signal of interest. In general, we define the term "upsampling" as increasing the sample rate of our waveform by the integer n ($F_{Snew} = F_S \cdot n$), whereas "downsampling" is the process of reducing the original rate by the integer n ($F_{Snew} = F_S/n$.) We reserve the term "resampling" as the change of the sample rate by any integer or fractional number.

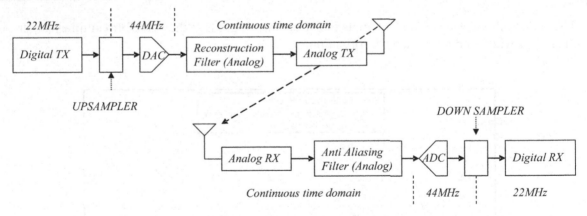

Figure 3-27: Multirate Signal Processing in a Modern WLAN Transceiver

A popular application of multirate signal processing in the domain of communication systems is the upsampling of transmit waveforms before they pass through the digital-to-analog converters and the downsampling of receive waveforms after they have been sampled. Running the DAC and ADC modules at a higher rate than the rest of the digital transceiver eases the requirements on the analog reconstruction and anti-aliasing filters in the transmit and receive chains. The next subsections illustrate these concepts using the hypothetical 802.11b transceiver shown here.

3.4.1 The Upsampling Process

Increasing the sample rate of a signal is a valuable exercise in many different signal processing applications. In this section, we will show off a popular use of this processing algorithm in the transmitter section of a digital modem. In the 802.11b WLAN modem, internal signal processing occurs at the minimum sample rate of 22MHz. The complex signal occupies a bandwidth of -9 to 9 MHz, which is close to the Nyquist limits of ±11MHz. Passing the output I and Q signals out of the modem via the DAC modules at 22MHz is a bad idea for several reasons.

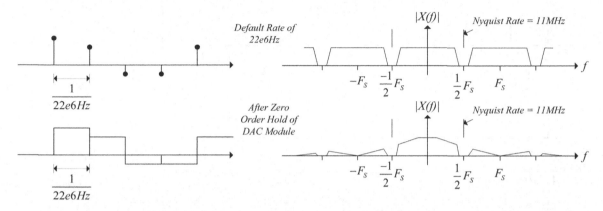

Figure 3-28: Zero-Order Hold Impact on Signal Spectrum

1. As we will show in detail in chapter five, the zero-order hold mechanism of the DAC modules causes the magnitude spectrum of our signal to be multiplied by a *sinc* function. The overlay causes magnitude distortion in the signal's pass band, as seen by the spectral drooping in the previous figure.
2. Whereas the *sinc* function has attenuated the sampling images beyond the Nyquist rate of 11 MHz, they must be further suppressed by an external analog reconstruction filter. The filter has the hard task of passing content below 9MHz and attenuating it above 13MHz. Aggressive requirements of this type translate to resource-hungry filter implementations.

It is far easier to first upsample the signal to 44MHz before sending it out the DAC module. The current consumption of the DAC modules does not appreciably increase, and the analog reconstruction filter requirements can be relaxed because there will be no content between 11 and 33MHz. Further, the wider *sinc* overlay reduces the dropping in the magnitude spectrum. The upsampling process consists of a zero-stuffing step that increases the sample rate to 44MHz and a filtering step that removes the first sampling image between 11 and 33MHz.

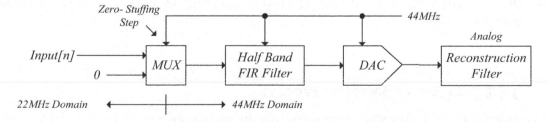

Figure 3-29: The Zero-Stuffing and Filtering Process

The previous and next figures illustrate the multistep upsampling process in block diagram and waveform formats. The zero-stuffing operation, itself, is as easy as switching between the 22MSPS sample stream and a zero value at a rate of 44MHz. The operation does not cause any change in the signal spectrum. We added a zero in a position that was already zero by default. We simply make it official now and include the zero in our waveform. The one thing that does change, however, is the sampling and associated Nyquist rates, which increase to 44MHz and 22MHz, respectively.

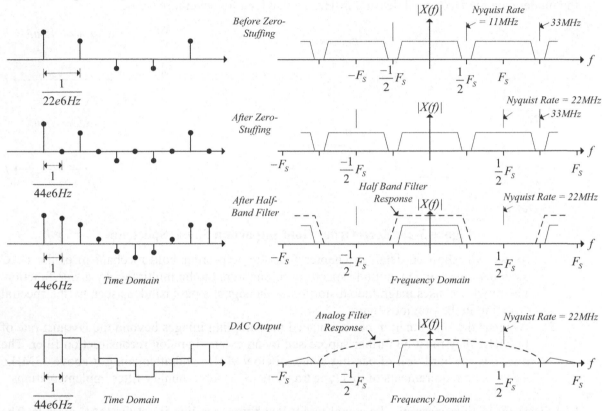

Figure 3-30: The Two Step Upsampling Operation in the Time and Frequency Domains

The low-pass FIR filter removes the frequency content between ¼ F_S and ¾ F_S and, thereby, simplifies the subsequent analog reconstruction filter that now must reject content at 33MHz rather than 11MHz. The hardware-efficient implementation of the FIR structure is called a half-band filter and will be discussed shortly. The final analog filter smoothens the step like sequence into a continuous waveform.

3.4.2 The Down Sampling Process

Whereas upsampling is frequently used before the digital-to-analog conversion in the transmitter, downsampling is a common feature that follows the analog-to-digital conversion at the input of the digital receiver section. Like upsampling, downsampling is a two-step process that includes a filtering and sample selection (decimation) step.

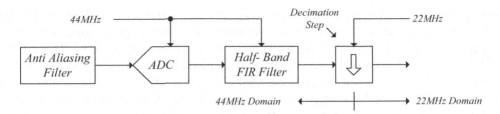

Figure 3-31: The Downsampling Process

Given our 802.11b example signal from before, we know that the frequency content of the desired waveform stays within ±9MHz and features a default processing rate in the baseband processor of 22MSPS. Sampling the signal directly at 22MSPS would require an anti-aliasing filter that passes signal content up to around 9MHz and rejects adjacent channels that reside close by. Given the complexity of such an anti-aliasing filter, we are better off increasing the ADC sampling rate from 22MHz to 44MHz. This gives our baseband process access to frequency content up to an increased Nyquist rate of 22MHz.

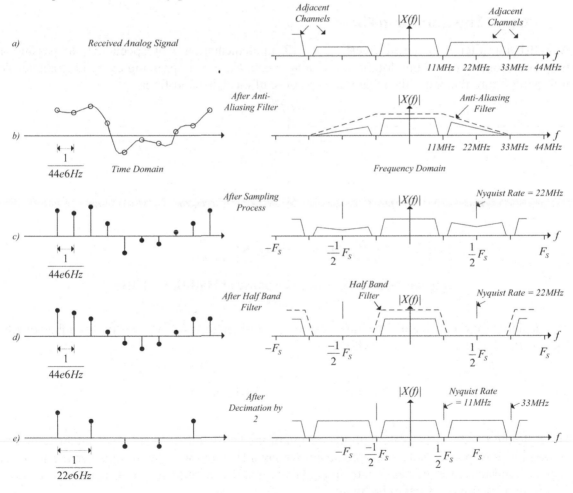

Figure 3-32: The Downsampling Process

The last figure shows the entire downsampling process quite nicely. Part *a*) illustrates the received signal content at the input of the anti-aliasing filter that removes content beyond 33MHz resulting in the situation of part *b*). As we sample the waveform at 44MHz, the content greater that 22MHz is aliased into our Nyquist range, as can clearly be seen in part *c*). The half-band filer now eliminates whatever part of the adjacent channel signals still remaining. We move from part *d*) to part *e*) of the figure by simply skipping every second sample, thus moving the Nyquist boundary to 11MHz. Had there been no traffic on adjacent channels immediately next to our pass band, we could have avoided the digital half-band filtering step and skipped samples directly. So, there we are. The sample rate of the signal is now at its minimum required level, thus maintaining a low current consumption for the digital receiver part of our modem.

3.4.3   The Half-Band Filter

A half-band filter is a hardware-efficient FIR implementation that specializes at passing all frequency content below one fourth of the sample rate, F_S, and suppressing everything above. The following figure illustrates the magnitude response of the half-band filter.

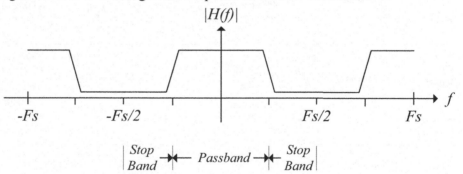

Figure 3-33: Magnitude Response of Half Band Filter

The design of the half-band filter turns out to be straight forward. For an odd number of filter taps, *N*, the design equation is as follows.

$$h[n] = sinc[\frac{n}{2} - \frac{N-1}{4}] \cdot [Window(n)]$$

The following code takes us through the design of the half-band filter step by step and computes its frequency response. Note, we once again overlay a Hanning window to improve the stop band rejection and decrease the pass band ripple. Given that the impulse response is an even function, the phase response turns out to be linear.

```
N    = 15;                              % Number of taps
n    = 0:N-1;
Arg  = n/2 - (N-1)/4;                   % Argument inside sinc function
Hann = (ones(1,N) - cos(2*pi*(n+1)/(N+1))); % The Hanning window as a row
                                        % vector
%% Half Band Filter impulse response
h    = sinc(Arg).*Hann;
h    = h/sum(h);                        % normalize to unity DC gain


%% Computing the frequency response
FResolution = 0.002;
Frequencies = -0.5:FResolution:0.5;
FrequencyResponse = zeros(1,length(Frequencies));


for i = 1:length(Frequencies)           % The DTFT operation
    f                   = Frequencies(1,i);
    AnalysisTone        = exp(j*2*pi*n*f);
    FrequencyResponse(1,i) = h*AnalysisTone';
end
LogResponse = 20*log10(abs(FrequencyResponse));
```

The next figure shows the impulse response and normalized magnitude response for a 15-tap half-band filter based on the previous equation.

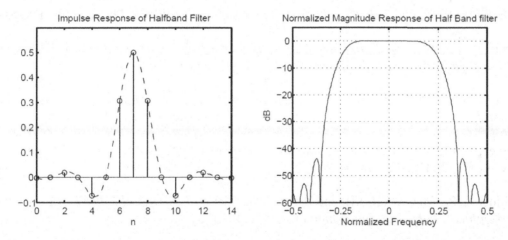

Figure 3-34: Impulse and Normalized Magnitude Response of a Typical 15 Tap Half Band Filter

The important thing to note is that aside from the center tap, every second [n=1, 3, 5, 9, 11, 13] coefficient is zero. Therefore, to implement a 15-tap half-band filter, we, in fact, would only require nine hardware multipliers. Note further that the impulse response is an even function. Thus, the coefficients at taps 0 and 14, 2 and 12, 4 and 10, and 6 and 8 are equal. This allows us to first add the values at those location pairs and then execute one multiplication instead of two. In fact, a 15-tap half band filter requires, at most, five hardware multiplications and potentially four if the center tap is unity. We may increase or decrease the number of taps, N, to make the filter skirt at $\pm\frac{1}{4}Fs$ more or less aggressive. The next figure illustrates a 15-tap half band filter implemented with only four multipliers. In digital hardware we can create half-band filters with large numbers of taps and very steep skirts around $\pm\frac{1}{4}Fs$ with surprisingly few resources.

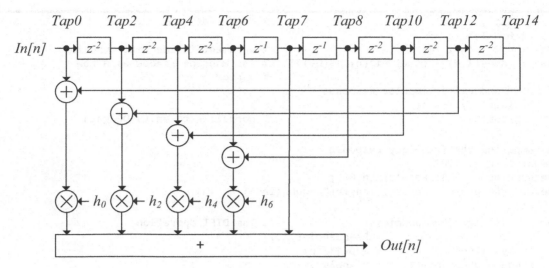

Figure 3-35: Hardware Implementation of 15 Tap Half Band Filter

References

[1] Björck, Å. (1996) *Numerical Methods for Least Squares Problems*, Philadelphia, PA, Siam

[2] Strutz, T. (2010) *Data Fitting and Uncertainty (A Practical Introduction to Weighted Least Squares and Beyond)*, Wiesbaden, Germany, Vieweg+Teubner

[3] Fraser, W. (1965) *A Survey of Methods of Computing Minimax and Near-Minimax Polynomial Approximations for Functions of a Single Independent Variable*, J. ACM 12

[4] Remez, E. (1934) *Sur la Détermination des Polynômes D'approximation de Degré Donnée*, Comm. Soc. Math. Kharkov 10, 41

[5] Rabiner, L.R., McClellan, J.H. and Parks, T.W. (1975) *FIR Digital Filter Design Techniques Using Weighted Chebyshev Approximations*, Proc. IEEE 63

[6] Cetin, A. E., Gerek, O.N., Yardimci, Y. (1997) *Equiripple FIR Filter Design by the FFT Algorithm*, IEEE Signal Processing Magazine, March, 60-64

[7] Chapra, S. C., Canale, R. P. (2002) *Numerical Methods for Engineer*, McGraw-Hill Higher Education, New York, NY, 486-493

[8] Meijering, Erik (2002), *A Chronology of Interpolation: from Ancient Astronomy to Modern Signal and Image Processing*, Proceedings of the IEEE 90 (3): 319–342

4 Estimation, Equalization and Adaptive Signal Processing

Communication channels are afflicted by various calamities including different noise phenomena and offsets, as well as linear and nonlinear distortions. We will discuss each of these deleterious effects in detail later in this text, but it is appropriate to examine the issue of linear distortion here and now, since the techniques to remedy this problem are a natural extension to the topics of the last chapter.

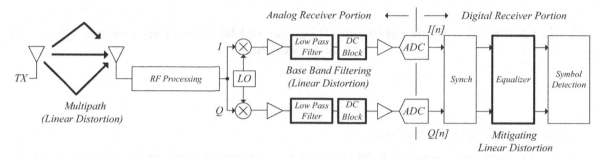

Figure 4-1: Contributors to Linear Distortion in Receivers

What Is Linear Distortion?

Linear distortion is caused by processes that scale the frequency and phase portion of the signal spectrum in an undesirable manner. Although there are a host of processing modules that may introduce linear distortion, the chief culprits in communication systems are the over-the-air multipath channel and overzealous filtering in the analog domain, as seen in the figure above. Frequency responses of processes with and without linear distortion are shown below.

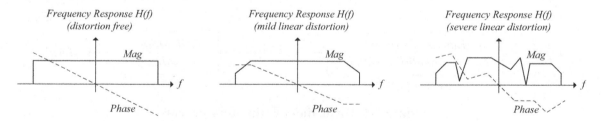

Figure 4-2: Frequency Responses for Channels with No/Mild/Severe Linear Distortion

How Do We Model Linear Distortion?

In this chapter, we will model processes that introduce linear distortion as complex valued FIR filters with time-variant and time-invariant characteristics. These characteristics are represented by the filters' impulse response, $h[n]$, and associated frequency response, $H(f)$. The next figure illustrates the real portion of potential FIR model impulse responses for no, mild, and severe linear distortion. The imaginary portion of the impulse response would look similar. For lack of a better term, we refer to the collection of distorting processes we are modeling as the *communication channel* or just plain *channel*.

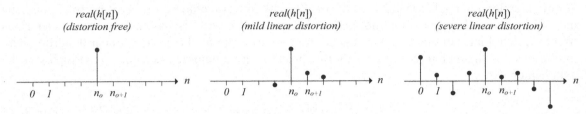

Figure 4-3: The Real Portion of Impulse Responses of FIR Filter Models for No/Mild/Severe Linear Distortion

How to Deal with Linear Distortion

Linear distortion is a serious problem in wireless communication links. Understanding the mechanisms that allow us to characterize and remedy this degrading effect is paramount if optimal detection of received information is to be achieved. The technically appropriate term for guessing at data that has been linearly distorted is *estimation*. In communication systems we also refer to it as equalization. The figure below illustrates a simplified problem statement of the linear distortion and equalization process. Pristine transmit information, $In[n]$, enters the FIR distortion model and is observed as $y[n]$ at its output. The FIR coefficients of the model are chosen for the scenario at hand. Clearly, the multipath effects will differ depending on the location and environment within which the communication link is established (see chapter 5). The distortion is mitigated by the equalizer and estimates, $\hat{In}[n]$, of the original input information are produced.

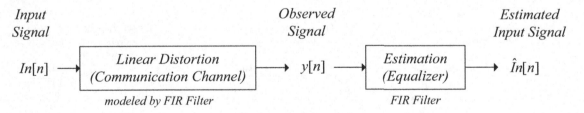

Figure 4-4: Distortion / Estimation Process

Linear Estimator (Equalizer) Structure

The FIR distortion model will usually feature an impulse response with several scaled unit pulses at different time delays, thus causing the information of a transmitted sample to appear in several observations at the receiver. The estimators we will discuss in this chapter take the form of a linear combiner that scales and sums the observations to guess at the original transmit values. Since the observed values exist at evenly spaced sample periods and are stored in a shift register, the linear combiner takes the shape of a FIR filter. More advanced equalizer configurations include feedback paths and nonlinear decision elements, as we will see later in this chapter.

The Bottom Line

The figure below illustrates the basic action of the linear estimator or equalizer. It attempts to reestablish an overall channel frequency response that is flat in magnitude and a straight line in phase. More sophisticated equalizers will actually strike a balance between the task of straightening the overall channel frequency response and maximizing the overall signal to noise ratio.

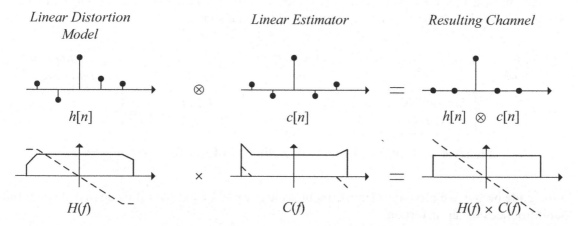

Figure 4-5: Characteristics of the Communication Channel and Estimator Combination

In this chapter, we will cover the following algorithms involved in equalization and estimation.

1. *Zero forcing (ZF) equalizers*
   This equalizer (FIR) straightens the overall channel response without considering noise. The process of finding its coefficients is crude but simple.
2. *Least squares error (LSE) equalizers and channel estimators*
   The FIR LSE equalizer also straightens the overall channel response without considering noise, but the coefficient calculation process is more elegant and thereby provides additional flexibilities.
3. *Minimum mean squares error (MMSE) equalizers*
   The MMSE equalizer is a FIR structure but uses a highly sophisticated method of coefficient calculation that strikes the optimal balance between straightening the overall channel response and mitigating the impact of noise in the system.
4. *Least mean squares (LMS) adaptive algorithm*
   The LMS algorithm leads to the MMSE optimal coefficients of a FIR equalizer via an iterative process. This adaptive scheme is very popular in base band processors.
5. *Decision feedback (DFE) equalizers*
   The decision feedback equalizer augments the FIR structure by adding certain feedback elements. This feature allows better equalizer performance, which would be difficult to achieve for a traditional FIR structure without increasing the number of taps significantly.

4.1 Zero Forcing Equalizers

The figure below introduces us to the manner and notation that we will use to represent and express the different estimation scenarios throughout this chapter. The input signal, $In[n]$, is the high quality waveform produced at the transmitter, while $y[n]$ represents the distorted waveform observed at the input of the receiver's base band processor. The distortion is modeled via a FIR filter with an impulse response we will name either '$Channel[n]$' or '$h[n]$'. Our goal is to find the coefficients of a FIR estimator, which will undo this distortion and reproduce the input signal. The output, $\hat{In}[n]$, is referred to as the estimate of the input signal.

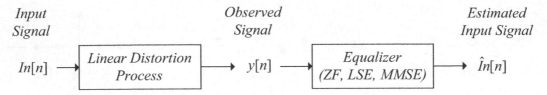

Input Signal $In[n]$ → [Linear Distortion Process] → *Observed Signal* $y[n]$ → [Equalizer (ZF, LSE, MMSE)] → *Estimated Input Signal* $\hat{In}[n]$

Figure 4-6: Standard Linear Distortion and Equalizer Arrangement

In the code below, we produce a simple input sequence of ±1s and pass it through a channel filter that introduces linear distortion.

```
%% Input symbol and Distorting Channel definitions
In      = [1 1 -1 -1 -1 1 -1 -1 1 -1 1 1 1 -1 1 -1 -1 -1 1 1 -1 1 1 -1 1 ...
          -1 -1 -1 1 -1 -1 1 1 -1 1 1 1 1 -1 1 1 -1]; %% Random +/-1 samples
Channel = [.1 -.1 .05 1 .05];        %% The coefficients of our
                                     %% Distorting FIR filter
y       = filter(Channel, 1, In);    %% The observed sequence
```

Different amounts of each sample arrive at the output at different times. While each input sample exits the distorting FIR filter in all its glory at a delay of three samples (coefficient = 1), they also leak through to the output early (coefficients = .1, -.1 and 0.05) and one sample late (coefficient = 0.05). The figure below shows how the distortion has altered the input signal.

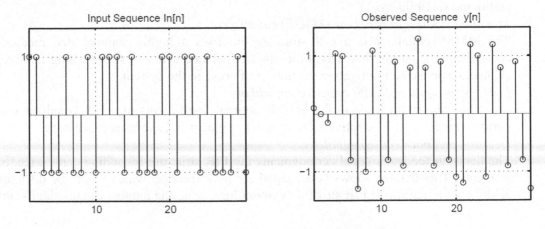

Figure 4-7: Effect of Linear Distortion

The observations, $y[n]$, serve as the input sequence to the equalizer, which ideally reproduces the undistorted transmit information, $In[n]$. We now write the convolution equations given the two sequences above, as well as the equalizer's impulse response $c[n]$.

$$In[n] = y[n + (N-1)] \otimes c[n]$$
$$= c[n] \otimes y[n + (N-1)]$$
$$= \sum_{k=0}^{N-1} c[k] \cdot y[(n + (N-1)) - k]$$

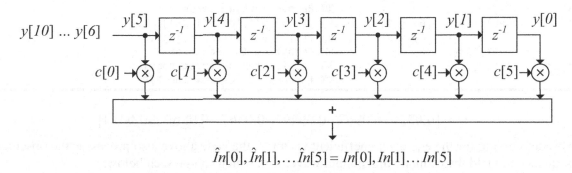

$$\hat{In}[0], \hat{In}[1], \dots \hat{In}[5] = In[0], In[1] \dots In[5]$$

Figure 4-8: The FIR Equalizer (Linear Estimator)

The figure above shows the FIR equalizer with the first non-zero observed sample $y[0]$ shifted all the way to the end of the filter. Clearly, the estimate of $In[0]$ does not happen at the discrete time instant $n = 0$ since we need to draw upon observations $y[1]$ through $y[5]$, which are not available yet. Once all observed samples that contain some amount of $In[0]$ are inside the shift register, the first estimate, $\hat{In}[0]$, may be calculated.

$$\begin{bmatrix} y[5] & y[4] & y[3] & y[2] & y[1] & y[0] \\ y[6] & y[5] & y[4] & y[3] & y[2] & y[1] \\ y[7] & y[6] & y[5] & y[4] & y[3] & y[2] \\ y[8] & y[7] & y[6] & y[5] & y[4] & y[3] \\ y[9] & y[8] & y[7] & y[6] & y[5] & y[4] \\ y[10] & y[9] & y[8] & y[7] & y[6] & y[5] \end{bmatrix} \cdot \begin{bmatrix} c[0] \\ c[1] \\ c[2] \\ c[3] \\ c[4] \\ c[5] \end{bmatrix} = \begin{bmatrix} In[0] \\ In[1] \\ In[2] \\ In[3] \\ In[4] \\ In[5] \end{bmatrix}$$

$$Y \cdot C = In$$

$$C = inv(Y) \cdot In$$

Although the output of the equalizer is only an estimate of $In[n]$, we may in fact program the FIR filter to produce exactly $In[0]$ to $In[5]$, as should be clear from the matrix expression above. This type of estimator is called a zero-forcing equalizer because the error between the estimate of $In[n]$ and $In[n]$ is forced to zero. The disadvantage of this estimator is that the number of observations used to craft our equalizer coefficients is limited to the number of equalizer taps, *N*. The equalizer works perfectly for those observations used during the coefficient computation, but sub optimally for the rest of the observed samples.

From the matrix expression of the last page, we can see that the matrix *Y* has a unique structure in which all diagonals that move from the top left to bottom right feature identical values. This unique arrangement is called a Toeplitz matrix and appears frequently when we analyze FIR filters. The code below sets up the equation and solves for the FIR coefficients of the zero-forcing equalizer.

```
%% Computing the Zero-Forcing Solution
y      = y(1,3:40);                     %% Remove initial zeros
Y      = toeplitz([y(1,6) y(1,7) y(1,8) y(1,9) y(1,10) y(1,11)], ...
                  [y(1,6) y(1,5) y(1,4) y(1,3) y(1,2) y(1,1)] ) ;
C      = inv(Y)*In(1,1:6)'              %% Computing the solution
E_In   = filter(C, 1, y);               %% Passing the observed samples
                                        %% through our estimation FIR
```

$$C = [0.035 \quad -0.0967 \quad 0.0949 \quad -0.0567 \quad 0.9864 \quad -0.0414]$$

Besides solving for the equalizer coefficient vector *C*, the code above also processes the observed sequence to yield the estimate of the input waveform, which can be seen below.

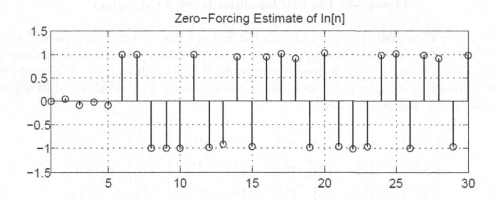

Figure 4-9: The Output of Our 6 Tap FIR Equalizer

Unsurprisingly, the first six useful estimates are exactly equal to *In[0]* through *In[5]*. We did, after all, train our equalizer to estimate these input samples perfectly. However, the rest of the estimates – as mentioned earlier – cannot be perfect.

An additional reason for the suboptimal estimates is that we have only a limited number of available taps, which restricts our ability to invert the frequency response of the distorting filter. In our case, we are limited to $N = 6$ taps, which allows us precise control over the equalizer's magnitude and phase response at only six unique frequencies. Only with an infinite number of taps could we perfectly recover all possible combinations of distorted input samples.

4.2 The Least Squares Error Equalizer

We pointed out in the previous section that one of the major drawbacks of our zero-forcing equalizer was that we were unable to include a larger number of observations in our equalizer coefficient calculation. As our matrix equation showed, only $(2 \cdot N)-1 = 11$ observations, $y[0]$ through $y[10]$, were used to solve for an equalizer with six taps. At this point, we will relax the requirement that the equalizer estimate the first six input samples perfectly. The least squares technique allows us to include a larger number of observations when crafting the N equalizer coefficients. In equation form then, the situation looks as follows.

$$
\begin{bmatrix}
y[5] & y[4] & y[3] & y[2] & y[1] & y[0] \\
y[6] & y[5] & y[4] & y[3] & y[2] & y[1] \\
y[7] & y[6] & y[5] & y[4] & y[3] & y[2] \\
y[8] & y[7] & y[6] & y[5] & y[4] & y[3] \\
\vdots & \vdots & \vdots & \vdots & \vdots & \vdots \\
y[L+4] & y[L+3] & y[L+2] & y[L+1] & y[L] & y[L-1]
\end{bmatrix}
\cdot
\begin{bmatrix}
c_0 \\
c_1 \\
c_2 \\
c_3 \\
c_4 \\
c_5
\end{bmatrix}
=
\begin{bmatrix}
In[0] \\
In[1] \\
In[2] \\
In[3] \\
\vdots \\
In[L-1]
\end{bmatrix}
$$

$$Y \cdot C = In$$

Given that the matrix Y is no longer square, we must take the pseudo inverse (least squares solution) for overdetermined systems to solve for the coefficient vector C. (See section 3.1)

$$C = inv(Y^T \cdot Y) \cdot Y^T \cdot In$$

Below we will use 30 observations to solve for the six coefficients of our equalizer.

```
%% Here we compute the LSE Solution
Y      = toeplitz(y(1,6:30), [y(1,6) y(1,5) y(1,4) y(1,3) y(1,2) y(1,1)]);
C      = inv(Y'*Y)*Y'*In(1,1:25)'
E_In   = filter(C, 1, y);          %% Filtering the observed samples
```

$$C = [0.0291 \quad -0.1063 \quad 0.1104 \quad -0.0637 \quad 1.0108 \quad -0.0426]$$

This time around, the coefficients are slightly different from those of the zero-forcing scenario, and to properly determine which set of coefficients is superior, we calculate the RMS error produced by both methods. The root mean square metric was defined in chapter 1 and is shown again below for reference.

$$Error[n] = \hat{I}n[n] - In[n]$$

$$Error_{RMS} = \sqrt{\frac{1}{N} \sum_{n=0}^{n-1} Error[n] \cdot conj(Error[n])}$$

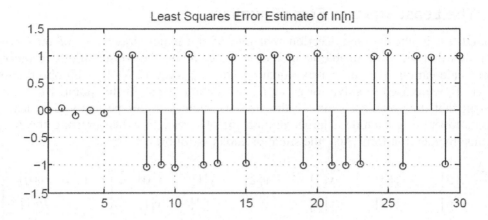

Figure 4-10: Estimate of *In*[*n*] Using a Least Squares Error Equalizer

Comparing the estimate plots of the zero-forcing and least squares equalizers may not conclusively show which of the coefficient sets is superior, but the statistical RMS metric of the error unequivocally tips the balance in favor of the least squares error derived taps.

$$Error_{RMS} = 0.0390 \quad \leftarrow Zero-Forcing$$

$$Error_{RMS} = 0.0258 \quad \leftarrow Least\ Squares$$

It should be apparent at this point that the zero-forcing and least squares error equalizers differ only in the method of computing their FIR filter coefficients. They are not different physical structures. We can also conclude that the equalizer performance could be further improved by considering a yet larger set of observations. This obvious fact brings us to our next class of equalizers, the minimum mean squares error estimator, which indirectly considers all observation through knowledge of the channel.

But before we dive into this topic, let us clear up a few issues that may be nagging engineers new to the topic of equalization. We have learned that we need to consider as many observations as possible to compute good equalizer coefficients, and to that end we also need to know an equally large set of source symbols, *In*[*n*]. Clearly, there is not much sense in communicating a large amount of information to a receiver if that receiver must already know all the input information to compute proper equalizer coefficients. There is no information exchange if the receiver already knows what is coming. So, how do we transmit useful data and also train our equalizers?

In almost all data communication schemes, the transmitter will regularly send information sequences that are known to the receiver. These 'a priori' known sequences, also known as training sequences, are usually placed at the beginning of a transmission stream and may be repeated regularly during that stream if the transmission duration is long. Compared to the overall number of transmitted symbols, the training sequences are usually very short. The receiver uses the initial training sequence to program the equalizer and the later ones to update it. Equalizers may also employ a technique called decision feedback to continually adjust their coefficients. A detector guesses at the correct value of the transmitted sample, and the equalizer coefficient computation uses the guess rather than a priori known input samples. We will see this technique in later sections.

Channel Estimation

In addition to computing proper equalizer coefficients, there is another closely related task that is frequently undertaken in modern digital modems. This task, called channel estimation, finds the coefficients of a FIR filter that models the linearly distorting channel. Here we must make clear that the distorting channel is not a FIR filter. The channel is a combination of imperfections in the analog hardware of our transceivers, as well as the before-mentioned multipath effect. The FIR filter is simply a structure that adequately models the linear distortion introduced by the channel. The task of channel estimation can thus be summed up as follows. Find the coefficients set, $h[n]$, of a hypothetical FIR (tap length N) filter that distorts the input sequence $In[n]$, such that the observations $y[n]$ are produced.

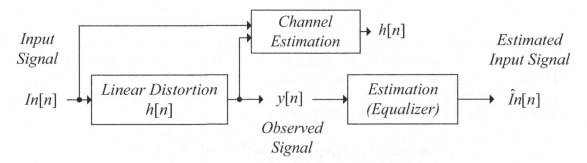

Figure 4-11: The Channel Estimation Problem

The mathematical formulation is analogous to that of the equalization problem, except that, this time, it is the input samples that enter the FIR model and we solve for a coefficient vector H that would produce the observed samples. The matrix formulation that follows uses the least squares technique to solve for a set of $N = 6$ filter coefficients. Note, of course, that the larger the number of FIR coefficients for which we solve, the better our model will emulate the linear distortion of the channel.

$$y[n] = In[n] \otimes h[n] = h[n] \otimes In[n]$$
$$= \sum_{k=0}^{N-1} h[n] \cdot In[n-k]$$

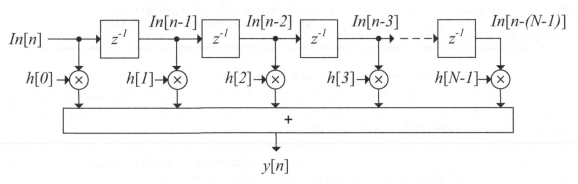

Figure 4-12: FIR Filter Model of the Distorted Channel

$$\begin{bmatrix} In[0] & 0 & 0 & 0 & 0 & 0 \\ In[1] & In[0] & 0 & 0 & 0 & 0 \\ In[2] & In[1] & In[0] & 0 & 0 & 0 \\ In[3] & In[2] & In[1] & In[0] & 0 & 0 \\ \vdots & \vdots & \vdots & \vdots & \vdots & \vdots \\ In[L-1] & In[L-2] & In[L-3] & In[L-4] & In[L-5] & In[L-6] \end{bmatrix} \cdot \begin{bmatrix} h_0 \\ h_1 \\ h_2 \\ h_3 \\ h_4 \\ h_5 \end{bmatrix} = \begin{bmatrix} y[0] \\ y[1] \\ y[2] \\ y[3] \\ \vdots \\ y[L-1] \end{bmatrix}$$

$$L \times 6 \qquad\qquad\qquad\qquad 6 \times 1 \qquad L \times 1$$

$$In \cdot H = Y$$

Since the matrix In is not square, we use the pseudo inverse to solve for the coefficient set H based on $L > N$ observations. We could obviously have used the zero-forcing formulation as well, and it may come as a surprise that this simpler and cruder estimate may sometimes be preferred in dedicated baseband processors for the very reason that it requires less computational effort. However, here we use the least squares error method.

$$H = inv(In^T \cdot In) \cdot In^T \cdot Y \qquad H = inv(In^T \cdot In) \cdot In^T \cdot Y$$

Example 4.1: *Channel Estimation Using the Least Squares Error Method*

In the code below, we consider the same input stream and distorting FIR channel that was used previously. We build our Toeplitz matrix exactly as shown in the expression above and solve to get the least squares channel estimate.

```
%% 4. The Channel Estimation Task
In      = [1 1 -1 -1 -1 1 -1 -1 1 -1 1 1 1 -1 1 -1 -1 -1 1 1 -1 1 1 -1 1 ...
           -1 -1 -1 1 -1 -1 1 1 -1 1 1 1 1 -1 1 1 -1]; %% Random +/-1 samples
Channel    = [.1 -.1 .05 1 .05];        %% The coefficients of our
                                         %% distorting FIR filter
y          = filter(Channel, 1, In);    %% The Observed sequence
L          = 30;                         %% Number of observations used
IN         = toeplitz(In(1,1:L), [In(1,1) 0 0 0 0 0]); %% Toeplitz matrix
Y          = y(1,1:L).';                 %% Column vector of observations
H_Estimate = inv(IN'*IN)*IN'*Y;
```

Unsurprisingly, the solution H is identical to the distorted FIR model $H = [0.1 \quad -0.1 \quad 0.05 \quad 1.0 \quad 0.05 \quad 0.0]$. At this point, we should experiment a little with the setup of the Toeplitz matrix and make some of the distortion channel taps complex. Try the following commands and examine the impact of each of them on the solution vector H.

```
→ Channel = [.1 -.1 .05 1+j .05 0]; %% Complex Coefficients
→ IN      = toeplitz(In(1,2:31)     , [In(1,2) In(1,1) 0 0 0 0]);
→ IN      = toeplitz([0 In(1,1:29)], [0 0 0 0 0 0]);
```

4.3 The Minimum Mean Squares Error Based Equalizer

Rather than finding N equalizer coefficients to estimate only N input samples perfectly (zero-forcing scheme) or finding N coefficients to optimally estimate a few more than N input samples (least squares technique), we now wish to calculate N optimal FIR tap weights for all possible inputs and associated observations. This scheme, called minimum mean squares error (MMSE) equalization, appears impractical, since, as we mentioned before, the original transmit information is available to the receiver only during the short training sequences that appear intermittently throughout our transmit stream. However, the coefficient calculation based on the MMSE criterion of optimality is not just possible but also forms the basis of a great number of equalizer implementations in modern communication systems. There are two practical methods of arriving at the MMSE optimal equalizer coefficients:

1. In this section, we present equations that allow for the calculation of the MMSE optimal equalizer coefficients, given knowledge of the channel impulse response, $h[n]$.
2. In the next section, we have the equalizer start out with suboptimal coefficients and allow an adaptive algorithm called LMS to iteratively fine-tune them until they have reached MMSE optimal values.

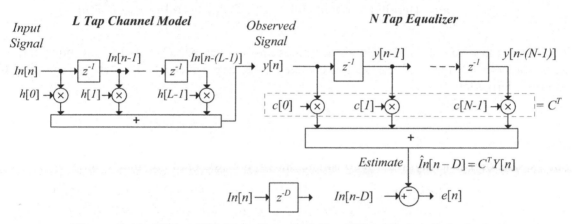

Figure 4-13: N Tap FIR Equalizer with Real Time Error Calculation

Developing the MMSE Optimal Solution (the Wiener–Hopf Equation)

The figure above shows an L tap channel model as well as the FIR equalizer of length N. The goal is to compute equalizer coefficients, residing in row vector C^T, which optimally estimate input samples that entered the channel model D samples ago. Here we define some of the quantities we will be using during the derivation of the MMSE optimal solution vector C^T.

$$C^T = \begin{bmatrix} c[0] & \cdots & c[N-1] \end{bmatrix} \quad C = \begin{bmatrix} c[0] \\ c[1] \\ \vdots \\ c[N-1] \end{bmatrix} \quad Y[n] = \begin{bmatrix} y[n] \\ y[n-1] \\ \vdots \\ y[n-(N-1)] \end{bmatrix} \quad \hat{I}n[n-D] = C^T Y[n]$$

Note that $y[n]$ represents the first observation of sample $In[n]$. Since our distorting channel is modeled as a FIR filter with tap length L, we can easily picture that $In[n]$, traveling down the FIR shift register, also influences future observations $y[n+1]$, $y[n+2]$ all the way to $y[n+(L-1)]$. If we are to estimate the input sample $In[n]$ competently, then it will be prudent to have most observations that contain some of amount of $In[n]$ present in the equalizer's shift register ready for processing. Ideally, therefore, the equalizer tap length N should be equal to or greater than the tap length L of the channel FIR model. To allow the observations to properly enter the equalizer's shift register, we wait for D samples. Therefore, at any one moment, n, we are actually estimating the input sample $In[n - D]$ transmitted D sample periods ago. The delay D, which should be larger than or equal to L but smaller than $N + L$, will influence the effectiveness of the final equalizer solution. The error between the estimate, $\hat{In}[n-D]$, and the actual input sample, $In[n-D]$, is shown below.

$$e[n] = In[n - D] - \hat{In}[n - D] = In[n - D] - C^T Y[n]$$

In a similar spirit to the LSE technique, it is our objective to compute equalizer coefficients that minimize the mean squares error produced by the comparison block of the previous figure. The equation below summarizes the objective of the MMSE criterion.

$$\boxed{\min_c E(|e[n]|^2) = \min_c E(\, (In[n - D] - C^T Y[n]) \cdot (In[n - D] - C^T Y[n])^T \,)}$$

The derivation below assumes that all quantities are real valued. The final result, however, will work for both the real and the complex valued case. See Sayed [5] for details regarding the complex derivation.

Step 1: Simplify $E(|e[n]|^2)$

For the sake of readability, we will do without the $[n]$ notation until the final result.

$$|e|^2 = e \cdot e^T = (In - C^T Y) \cdot (In - C^T Y)^T = (In - C^T Y) \cdot (In^T - (C^T Y)^T)$$
$$= In \cdot In^T - C^T Y \cdot In^T - In \cdot (C^T Y)^T + C^T Y \cdot (C^T Y)^T$$

Note that $C^T Y \cdot In^T = (In \cdot (C^T Y)^T)^T$, which reduces to $In \cdot (C^T Y)^T$, simply because these products are all real scalar quantities. Note also that $(C^T Y)^T = Y^T C$.

$$|e|^2 = e \cdot e^T = In \cdot In^T - 2In \cdot (C^T Y)^T + C^T Y \cdot (C^T Y)^T$$

$$= In \cdot In^T - 2In \cdot Y^T C + C^T Y \cdot Y^T C$$

We now take the mean of that error by applying the expectation operator, E, to those quantities that vary over time, ($In[n - D]$ and $Y[n]$). Remember that the squared errors are scalars and, consequently, each of their expected values on the right side of the equation below is also a scalar.

$$E(e \cdot e^T) = E(In \cdot In^T) - 2E(In \cdot Y^T)C + C^T E(Y \cdot Y^T)C$$

Step 2: _Compute the Gradient with Respect to the Vector C_

$$\nabla_c E(e \cdot e^T) = \nabla_c E(In \cdot In^T) - \nabla_c(2E(In \cdot Y^T)C) + \nabla_c(C^T E(Y \cdot Y^T)C)$$

Clearly, the gradient of $E(In \cdot In^T)$ will reduce to zero, since it is not a function of C. We arrive at the expressions below via the same process used during the LSE derivation in the last chapter, except that, this time, we define the gradient as a row vector.

$$\nabla_c E(e \cdot e^T) = 0 - 2E(In \cdot Y^T) + 2C^T E(Y \cdot Y^T)$$

$$= -E(In \cdot Y^T) + C^T E(Y \cdot Y^T)$$

Step 2: _Minimize by Setting the Gradient to Zero_

We now set the gradient expression on the right of the equals sign to zero and solve for the equalizer coefficient vector.

$$0 = -E(In \cdot Y^T) + C^T E(Y \cdot Y^T)$$

$$C^T E(Y \cdot Y^T) = E(In \cdot Y^T)$$

$$C^T = E(In \cdot Y^T) \cdot inv(E(Y \cdot Y^T))$$

Given that we are nearing the final answer, we now reintroduce the notation $[n]$.

$$C^T = E(In[n-D] \cdot Y[n]^T) \cdot inv(E(Y[n] \cdot Y[n]^T))$$

$$= E(In[n-D] \cdot Y[n]^T) \cdot inv(R_y)$$

So here we are. This is the grand minimum mean squares error solution for our equalizer coefficients. (The expression above is called the Wiener-Hopf equation named after mathematicians Norbert Wiener [1, 2] and Eberhard Hopf.)

$R_y = E(Y[n] \cdot Y[n]^T)$ is the autocorrelation matrix of the observed sample stream, $y[n]$. Remember that the autocorrelation matrix of a vector containing random variables measures the similarity of any one member variable, $y[n]$, to itself as well as to its neighbors, $y[n-2]$, $y[n-1]$, $y[n+1]$, $y[n+2]$ … etc. Note further that R_y is a square matrix, whereas $E[In[n-D] \cdot (Y[n]^T)]$ is a row vector.

$$R_y = E\begin{bmatrix} y[n] \cdot y[n]^* & y[n] \cdot y[n-1]^* & \cdots & y[n] \cdot y[n-(N-1)]^* \\ y[n-1] \cdot y[n]^* & y[n-1] \cdot y[n-1]^* & \cdots & y[n-1] \cdot y[n-(N-1)]^* \\ \vdots & \vdots & \ddots & \vdots \\ y[n-(N-1)] \cdot y[n]^* & y[n-(N-1)] \cdot y[n-1]^* & \cdots & y[n-(N-1)] \cdot y[n-(N-1)]^* \end{bmatrix}$$

$$E(In[n-D] \cdot Y[n]^T) = E(In[n-D] \cdot [y[n]^* \cdots y[n-(N-1)]^*])$$

Example 4.2: _Minimum Mean Squares Error Equalization_

To clear up the confusion that almost certainly exists after a derivation like that, let us step through an example featuring a simple scenario that consists of a 2 tap channel filter distorting the input samples, $In[n]$, followed by a 2 tap equalizer attempting to estimate them. The coefficients of each FIR structure are column vectors of the following form.

$$H = \begin{bmatrix} h_0 \\ h_1 \end{bmatrix} = \begin{bmatrix} 0.25 \\ 1 \end{bmatrix} \qquad Y[n] = \begin{bmatrix} y[n] \\ y[n-1] \end{bmatrix} \qquad C = \begin{bmatrix} c[0] \\ c[1] \end{bmatrix}$$

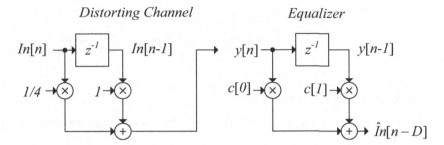

Figure 4-14: Simple MMSE Equalizer

In order to use our MMSE solution, we must first compute the autocorrelation matrix $Ry = E(Y[n] \cdot Y[n]^T)$.

$$Ry = E\left(\begin{bmatrix} y[n] \\ y[n-1] \end{bmatrix} \cdot \begin{bmatrix} y[n]^* & y[n-1]^* \end{bmatrix} \right) = \begin{bmatrix} E(y[n] \cdot y[n]^*) & E(y[n] \cdot y[n-1]^*) \\ E(y[n-1] \cdot y[n]^*) & E(y[n-1] \cdot y[n-1]^*) \end{bmatrix}$$

At this point, we can either choose to record thousands of observations and compute the matrix above or use the known impulse response H of the channel. One might ask whether the impulse response would be known in practice. The receiver would first have to estimate the impulse response before the equalizer coefficient calculation could proceed. In some cases, such as wired communication modems, the transmitter would simply send an impulse across the channel (cable), which the receiver would sample and, voilà, the channel impulse response is known with no further calculation required. Other communication scenarios may require more effort to find the impulse response. If we assume that successive input symbols are statistically independent, then we may proceed, knowing that $y[n] = \frac{1}{4} In[n] + In[n-1]$ and $y[n-1] = \frac{1}{4} In[n-1] + In[n-2]$.

$$E(y[n] \cdot y[n]^*) = E\left((\frac{1}{4} In[n] + In[n-1]) \cdot (\frac{1}{4} In[n] + In[n-1])^*\right) = \frac{1}{16}\sigma_{In}^2 + \sigma_{In}^2$$

$$E(y[n] \cdot y[n-1]^*) = E\left((\frac{1}{4} In[n] + In[n-1]) \cdot (\frac{1}{4} In[n-1] + In[n-2])^*\right) = \frac{1}{4}\sigma_{In}^2$$

$$E(y[n-1] \cdot y[n]^*) = E\left((\frac{1}{4} In[n-1] + In[n-2]) \cdot (\frac{1}{4} In[n] + In[n-1])^*\right) = \frac{1}{4}\sigma_{In}^2$$

$$E(y[n-1] \cdot y[n-1]^*) = E\left((\frac{1}{4} In[n-1] + In[n-2]) \cdot (\frac{1}{4} In[n-1] + In[n-2])^*\right) = \frac{1}{16}\sigma_{In}^2 + \sigma_{In}^2$$

$$Ry = \begin{bmatrix} \frac{17}{16}\sigma_{In}^2 & \frac{1}{4}\sigma_{In}^2 \\ \frac{1}{4}\sigma_{In}^2 & \frac{17}{16}\sigma_{In}^2 \end{bmatrix}$$

Next we compute the expression $E(In[n-D]\cdot Y[n]^T)$, which is, as we have mentioned before, a row vector. (We will work it out for a delay of both $D = 1$ and $D = 2$.)

$$E(In[n-1]\cdot\begin{bmatrix} y[n]^* & y[n-1]^* \end{bmatrix}) = \ldots$$

$$E(In[n-1]\cdot\begin{bmatrix} \frac{1}{4}In[n]^* + In[n-1]^* & \frac{1}{4}In[n-1]^* + In[n-2]^* \end{bmatrix}) = \begin{bmatrix} \sigma_{In}^2 & \frac{1}{4}\sigma_{In}^2 \end{bmatrix}$$

$$E(In[n-2]\cdot\begin{bmatrix} y[n]^* & y[n-1]^* \end{bmatrix}) = \ldots$$

$$E(In[n-2]\cdot\begin{bmatrix} \frac{1}{4}In[n]^* + In[n-1]^* & \frac{1}{4}In[n-1]^* + In[n-2]^* \end{bmatrix}) = \begin{bmatrix} 0 & \sigma_{In}^2 \end{bmatrix}$$

Combining these results, we form our solution for $D = 2$.

$$C^T = E(In[n-D])\cdot Y[n]^T)\cdot inv(R_y) = \begin{bmatrix} 0 & \sigma_{In}^2 \end{bmatrix}\cdot inv(\begin{bmatrix} \frac{17}{16}\sigma_{In}^2 & \frac{1}{4}\sigma_{In}^2 \\ \frac{1}{4}\sigma_{In}^2 & \frac{17}{16}\sigma_{In}^2 \end{bmatrix})$$

Assuming that the signal variance σ_{In}^2 is unity, the N equalizer coefficients are as follows.

$$C^T = [-0.2344 \quad 0.9963]$$

The figure below shows the magnitude response of the original channel H, the equalizer with coefficient C^T, and the convolution of the two, which represents the corrected channel. Clearly, the combined magnitude response is significantly better behaved and would improve further as the length N of the equalizer increases. Repeat the last two steps to find the coefficients for the $D = 1$ case and verify the overall inferior performance.

$$Combined[n] = [0.25 \quad 1] \otimes [-0.2344 \quad 0.9963] = [-0.586 \quad 0.0147 \quad 0.9963]$$

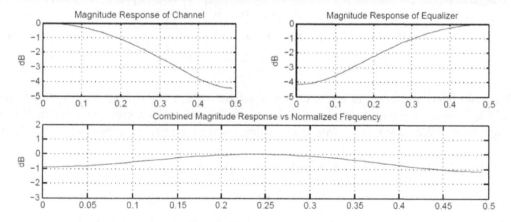

Figure 4-15: Magnitude Responses of the Channel, the Equalizer, and their Convolution

Introducing Noise and the Channel Matrix H into the Model

The previous expression for the MMSE optimal equalizer coefficients was cumbersome because it did not provide a more direct way of using the impulse response of the distorting channel. We will remedy this situation by introducing the channel matrix, H, as well as Gaussian noise, $v[n]$, which invariably forms part of all communication systems.

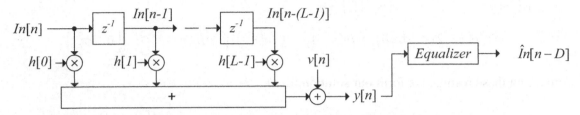

Figure 4-16: Channel FIR Model (Length L) with Equalizer (Length N) and Noise

White Gaussian noise $v[n]$ is mainly being produced by the resistive portion of our receiving antenna and subsequent RF amplifiers. We follow the model in the figure above to modify the Wiener-Hopf equation.

$$Y[n] = H \cdot IN[n] + V[n]$$

The channel matrix H features the structure shown in the example expression below.

$$Y[n] \to N \times 1 \qquad H \to N \times N + L - 1 \qquad IN[n] \to N + L - 1 \times 1 \qquad V[n] \to N \times 1$$

$$
\begin{bmatrix} y[n] \\ y[n+1] \\ y[n+2] \\ y[n+3] \\ y[n+4] \end{bmatrix}
=
\begin{bmatrix}
h_0 & h_1 & h_2 & h_3 & 0 & 0 & 0 & 0 \\
0 & h_0 & h_1 & h_2 & h_3 & 0 & 0 & 0 \\
0 & 0 & h_0 & h_1 & h_2 & h_3 & 0 & 0 \\
0 & 0 & 0 & h_0 & h_1 & h_2 & h_3 & 0 \\
0 & 0 & 0 & 0 & h_0 & h_1 & h_2 & h_3
\end{bmatrix}
\cdot
\begin{bmatrix} In[n] \\ In[n-1] \\ In[n-2] \\ In[n-3] \\ In[n-4] \\ In[n-5] \\ In[n-6] \\ In[n-7] \end{bmatrix}
+
\begin{bmatrix} v[n] \\ v[n+1] \\ v[n+2] \\ v[n+3] \\ v[n+4] \end{bmatrix}
$$

The dimensions are indicated in terms of the equalizer $[c_0\ c_1\ c_2\ c_3\ c_4]$ tap length N and distortion channel $[h_0\ h_1\ h_2\ h_3]$ tap length L, which in the example expression above work out to $N = 5$ and $L = 4$, respectively. We now reformulate the solution for the equalizer coefficients C as follows.

$$R_y = E(Y[n] \cdot Y[n]^T) = E[(H \cdot IN[n] + V[n]) \cdot (H \cdot IN[n] + V[n])^T]$$

$$= E[(H \cdot IN[n] + V[n]) \cdot ((H \cdot IN[n])^T + V[n]^T)]$$

$$= E[(H \cdot IN[n] + V[n]) \cdot (IN[n]^T \cdot H^T + V[n]^T)]$$

$$= H \cdot E(IN[n] \cdot IN[n]^T) \cdot H^T + E(V[n] \cdot V[n]^T)$$

$$= H \cdot R_{In} \cdot H^T + R_v \qquad \leftarrow NxN \quad Square\ Matrix$$

$$E(In[n-D] \cdot Y[n]^T) = E(In[n-D] \cdot (H \cdot IN[n] + V[n])^T)$$
$$= E(In[n-D] \cdot (H \cdot IN[n])^T) + E(In[n-D] \cdot V[n]^T)$$
$$= E(In[n-D] \cdot (IN[n]^T \cdot H^T)) + E(In[n-D] \cdot V[n]^T)$$
$$= E(In[n-D] \cdot IN[n]^T) \cdot H^T \qquad \leftarrow 1xN \quad Row \ Vector$$

After introducing the channel matrix H and Gaussian noise into the model, the Wiener–Hopf equation changes to the following expression.

$$\boxed{C^T = E(In[n-D] \cdot IN[n]^T)H^T \cdot inv(H \cdot R_{In} \cdot H^T + R_v)}$$

At first glance, this expression looks even more complicated than the earlier Wiener-Hopf equation, but a closer look shows that the individual components are simpler. First of all, we know that the input autocorrelation matrix R_{In} is available without any type of observation at the receiver. It is the transmitter specification of the communication system that will specify how transmit symbols are sent and whether they are in any way correlated. We assume here that the neighboring samples of our input symbol stream are uncorrelated, making R_{In} an identity matrix, I, scaled by the variance of $In[n]$. Assuming that we are dealing with white noise, the autocorrelation matrix, R_v, reduces to an identity matrix, I, scaled by the variance of $v[n]$.

Example 4.3: *Minimum Mean Squares Error Equalization in the Presence of Noise*

In this example, we start with a complex channel impulse response of length $L = 5$, small amounts of Gaussian white noise $v[n]$ ($SNR = 40dB$), and an equalizer of tap length $N = 5$. The delay through the cascaded filter structures will be set at $D = L + N - 4$ samples.

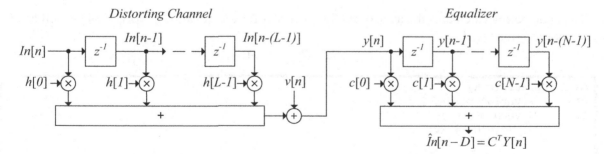

Figure 4-17: System Setup (Distortion Model L=5 / Equalizer N=5)

After evaluating the initial performance of the solution coefficients, we lengthen the equalizer to $N = 12$ taps and finally reduce the signal to noise ratio to 12.2 dB in order to see the impact of the added noise on the final equalizer tap values.

$$Channel = [h_0 \ h_1 \ h_2 \ h_3 \ h_4] = [0.1 \ -0.1 \ 0.05 \ 0.9+j0.1 \ 0.5]$$

```
%% 1. Definitions
VarIn    = 1;                      %% Variance of input signal
VarN     = 1e-4;                   %% Noise variance
Channel = [.1 -.1 .05 .9+j*.1 .05]; %% L = 5
N        = 5;                      %% Number of taps in equalizer
L        = length(Channel);        %% Number of taps in channel
D        = N + L - 4;              %% D = 6 / Delay must be less than N+L


%% 2. Calculate H
row      = zeros(1,N+L-1); row(1,1:L) = Channel;
col      = zeros(1,N);      col(1,1)   = Channel(1,1);
H        = toeplitz(col,row);
```

$$N \ x \ N+L\text{-}1$$

$$H = \begin{bmatrix} .1 & -.1 & .05 & .9+j.01 & .05 & 0 & 0 & 0 & 0 \\ 0 & .1 & -.1 & .05 & .9+j.01 & .05 & 0 & 0 & 0 \\ 0 & 0 & .1 & -.1 & .05 & .9+j.01 & .05 & 0 & 0 \\ 0 & 0 & 0 & .1 & -.1 & .05 & .9+j.01 & .05 & 0 \\ 0 & 0 & 0 & 0 & .1 & -.1 & .05 & .9+j.01 & .05 \end{bmatrix}$$

The dimensions of the column vectors $IN[n]$ and $V[n]$ are $N+L$-1 x 1 and N x 1 respectively, giving rise to the input and noise autocorrelation matrices shown below.

$$N+L-1 \ x \ N+L-1 \qquad\qquad N \ x \ N$$

$$R_{In} = \sigma_{In}^2 \cdot I = \begin{bmatrix} 1 & 0 & \cdots & 0 \\ 0 & 1 & \cdots & 0 \\ \vdots & \vdots & \ddots & \vdots \\ 0 & 0 & \cdots & 1 \end{bmatrix} \qquad R_V = \sigma_v^2 \cdot I = \begin{bmatrix} 1e-4 & 0 & \cdots & 0 \\ 0 & 1e-4 & \cdots & 0 \\ \vdots & \vdots & \ddots & \vdots \\ 0 & 0 & \cdots & 1e-4 \end{bmatrix}$$

The code below calculates the autocorrelation matrices for the quantities $IN[n]$, $V[n]$ as well as of the observations $y[n]$.

$$R_Y = H \cdot R_{In} \cdot H^T + R_V$$

```
%% 3. Calculate Ry
RIn      = VarIn*eye(N+L-1);
Rv       = VarN*eye(N);
Ry       = H*RIn*H' + Rv;


%% 4. Calculate E(In(n-D)IN(n)')H'
E_InIN            = zeros(1,N+L-1);
E_InIN(1,1+D)     = VarIn;              %% 1+D due to MatLab array indexing
E_InIN_H          = E_InIN*H';


EC                = E_InIN_H*inv(Ry)    %% Equalizer coefficients EC = C'
Corrected_Channel = conv(Channel, EC); %% The corrected channel
```

To correctly compute $E(In[n-D] \cdot IN[n]^T) \cdot H^T$, the expression $E(In[n-D] \cdot IN[n]^T)$ must obey the dimension $1 \times N+L-1$. Assume below that $D = N+L-4$.

$$E(In[n-D] \cdot IN[n]^T) \cdot H^T$$
$$= E\left[In[n-D] \cdot In[n]^* \quad \cdots \quad In[n-D] \cdot In[n-(N+L-4)]^* \quad \cdots \quad In[n-D] \cdot In[n-(N+L-2)]^* \right] \cdot H^T$$
$$= E\left[0 \quad \cdots \quad \sigma_{In}^2 \quad 0 \quad 0 \right] \cdot H^T \quad \leftarrow \ 1 \times N+L-1$$

We finally calculate the row vector C^T, which in the code we refer to as the equalizer coefficients vector, *EC*.

$$C^T = EC = E(In[n-D] \cdot IN[n]^T)H^T \cdot inv(R_Y)$$
$$= E(In[n-D] \cdot IN[n]^T)H^T \cdot inv(H \cdot R_{In} \cdot H^T + R_v)$$

$$EC = [-0.127+j0.03 \quad 0.13-j0.031 \quad -0.069+j0.017 \quad 1.102-j0.124 \quad -0.058+j0.0134]$$

We convolve the channel impulse response with the just calculated equalizer coefficients, *EC*, to get our first glimpse at the performance of the equalized (corrected) channel. Note that the equalized channel impulse response features a lone, large peak and thus an approximate average group delay of $D = N + L - 4 = 6$.

$$EqualizedChannel = [h_0 \ h_1 \ h_2 \ h_3 \ h_4] \otimes [c_0 \ c_1 \ c_2 \ c_3 \ c_4]$$

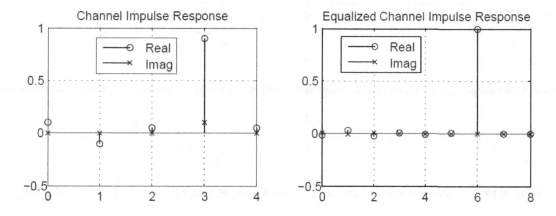

Figure 4-18: Complex Impulse Responses of the Original and Corrected Channel

Now that we have figured out the estimator coefficients, we pass a complex input stream $In[n]$ through the channel and watch the equalizer at work. The distorted observation clearly shows the damage done by the channel filter, which is then mitigated by our equalizer. The estimated constellation is significantly tighter than the observed version, but we can do better yet by increasing the number of taps of our equalizer. The bottom part of the next figure illustrates the magnitude response of the distorting channel, as well as the cascaded response of the channel and equalizer. It is clear that the equalizer is trying to flatten the overall magnitude response, and, although not seen, will linearize the phase response as well.

```
%% 5. Generate input sequence and filter it through the channel
SimL    = 2000;                      %% Number of Symbols in Simulation
Input   = sqrt(VarIn)*(sign(randn(1,SimL)) + ...
                   i*sign(randn(1,SimL)))/sqrt(2);
                                     %% QPSK input sequence (variance=VarS)
y       = filter(Channel,1,Input);   %% Variance has changed after filter
Noise   = sqrt(VarN)*(((randn(1,SimL)) + i*(randn(1,SimL))))/sqrt(2);
                                     %% Complex Noise (variance VarN)
Observations = y + Noise;
```

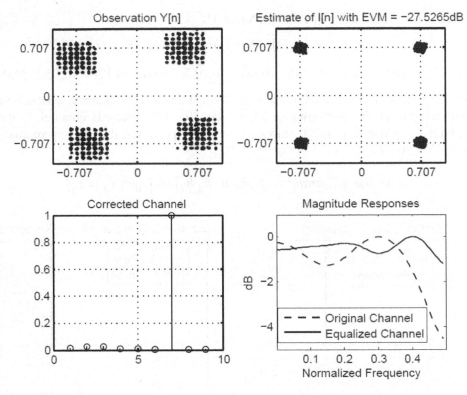

Figure 4-19: Equalizer Performance for N=5, SNR = 40dB, and h[n] = [.1 -.1 .05 .9+*j*0.1 .05]

The equalized constellation features an *EVM* or error vector magnitude of -27.5dB, which may be computed according to the following equations. Both the decibel and percent RMS versions compare the mean square of the estimator error with the mean square of the ideal input or reference samples. As mentioned in chapter 1, we use the terms mean square and total average power interchangeably.

$$EVM_{dB} = 10 \cdot log_{10}(\frac{AverageErrorVectorPower}{ReferencePower})$$

$$EVM_{\%rms} = \sqrt{\frac{AverageErrorVectorPower}{ReferencePower}} \cdot 100$$

The code below estimates the input sequence, calculates the error between the estimate and the ideal input vectors, and establishes both the percent RMS and decibel based error vector magnitudes.

```
%% 6. The Equalization and Analysis Process
Estimation       = filter(EC,1,Observations); %% Executing the equalizer
EstimationVector = Estimation(1,D+1:SimL);    %% Estimated samples
InputVector      = Input(1,1:SimL-D);         %% Reference samples


Error            = EstimationVector - InputVector;
EVM_Percent_RMS  = sqrt(var(Error)/var(InputVector))*100
EVM_dB           = 10*log10(var(Error)/var(InputVector))
SNR              = 10*log10(var(y)/VarN)
```

In the experiment below, we raise the estimator length, N, to 12 while maintaining a noise variance of 1e-4. The EVM improves to -38.4dB and the magnitude response of the total channel straightens almost completely.

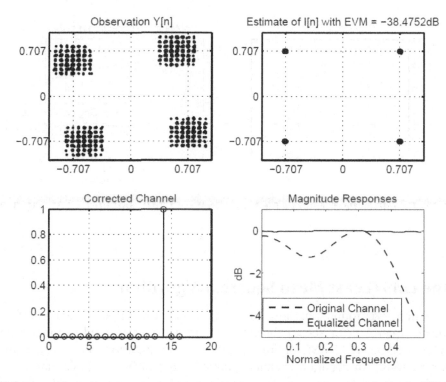

Figure 4-20: Equalizer Performance N=12, SNR = 40dB, and h[n] = [.1 -.1 .05 .9+j0.1 .05]

In our final test, we add a significant amount of noise to our observation and examine the performance of the equalizer. We stay with our 12 tap equalizer but add white Gaussian noise with a variance of 5e-2, yielding a signal to noise ratio of 12.2dB, and adjust the distorting channels impulse response to [.01 -.01 0.3 0.9+j0.1 0.05]. Even though the equalizer is perfectly programmed, it can obviously not remove the noise, which is now a much larger problem than the channel distortion itself. Note, however, that the equalizer has decided not to completely straighten the magnitude response at the upper frequency range. Given the fact that the noise has a flat power spectrum, the signal to noise ratio of the observed sample stream becomes quite a bit worse as frequency increases. The equalizer has determined that it is beneficial to forgo straightening the magnitude response completely to keep the signal to noise ratio in check. By including the noise component in our channel model, we have given the equalizer the opportunity to strike the optimal balance (in the minimum mean squares error sense) between mitigating linear distortion and maximizing the signal to noise ratio.

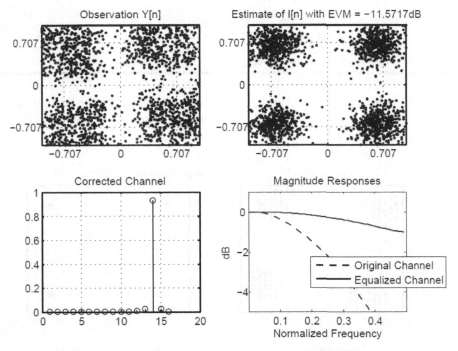

Figure 4-21: Equalizer Performance for N=12, SNR = 12.2dB, and h[n] = [.01 -.01 .3 .9+j0.1 .05]

4.4 The LMS (Least Mean Squares) Algorithm

In the last section, we found out how to compute the optimal equalizer coefficients (in the minimum mean squares error sense) for cases where we have a good channel estimate readily available. This method of MMSE optimal coefficient computation can become impractical for two common reasons. An accurate channel estimate may not be available, and even if it is, the computational resources to compute the sophisticated linear algebra involved may be absent. Linear algebra lends itself well to digital signal processors for which flexible programming languages like *C++* are available. Teaching a dedicated digital ASIC, such as a baseband processor, the fine art of linear algebra is a daunting task. Its strength lies in signal processing applications needing large numbers of parallel operations, as in digital filters. Some DSP solutions now provide both the dedicated digital hardware resources found in ASIC modules and multipurpose DSP processing cores to handle complex mathematical operations like channel estimation and equalizer coefficient calculations, which require matrix inverse operations. Wireless LAN transceivers are a good example of communication systems that omit the traditional digital signal processing chip and rely exclusively on base band processors. This is done for both cost and current consumption reasons and forces us into a new technique to compute the MMSE optimal equalizer coefficients. The technique in question is the least mean squares or LMS algorithm [3, 4], co-invented by Bernard Widrow and Edward Hoff in 1960, which iteratively adjusts imperfect equalizer coefficients until they have reached their MMSE optimal values. There are techniques that allow even a dedicated digital ASIC to calculate a crude channel estimate and associated equalizer coefficient set. These coefficients, though far from being optimal, are iteratively improved via the adaptive LMS technique. For proper operation, the algorithm must have access to training information. Consider the now familiar picture below.

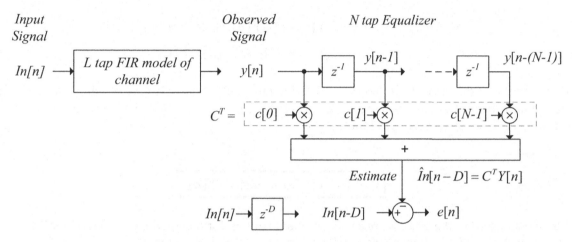

Figure 4-22: Our Generic Equalizer Setup

The LMS algorithm uses the instantaneous error $e[n]$, shown above, to update the equalizer coefficients on a clock-by-clock basis. In anticipation of the mathematical derivation for the real valued LMS scenario, let us review the main variables and their dimensions.

$$C^T = \begin{bmatrix} c[0] & \cdots & c[N-1] \end{bmatrix} \quad C = \begin{bmatrix} c[0] \\ c[1] \\ \vdots \\ c[N-1] \end{bmatrix} \quad Y[n] = \begin{bmatrix} y[n] \\ y[n-1] \\ \vdots \\ y[n-(N-1)] \end{bmatrix} \quad \hat{I}n[n-D] = C^T Y[n]$$

Deriving the LMS Update Equations

To gain insight into the operation of the LMS technique, we now trace through the derivation of the MMSE optimal equalizer discussed in the last section. We assume that all quantities are real valued and pick up the derivation at step two, where we calculate the gradient of the squared error.

$$\nabla_C E(e \cdot e^T) = \nabla_C E(In \cdot In^T) - \nabla_C (2E(In \cdot Y^T)C) + \nabla_C [C^T E(Y \cdot Y^T)C]$$

Clearly, the gradient of $E(In \cdot In^T)$ will reduce to zero, as it is not a function of C. Review section 4.3 to see how the remaining two expressions, $2E(In \cdot Y^T)C$ and $C^T E(Y \cdot Y^T)C$, respond to the gradient operation.

$$\nabla_C E(e \cdot e^T) = 0 - 2E(In \cdot Y^T) + 2C^T E(Y \cdot Y^T)$$

For the *LMS* derivation, however, we now forgo the expectation operation, since it would require us to know the exact channel impulse response. We thus stay with the instantaneous squared error and proceed with the gradient calculation. We also reintroduce the notation $[n]$, since the expressions are reasonably short.

$$\nabla_C(e[n] \cdot e[n]^T) = -2In[n-D] \cdot Y[n]^T + 2C^T \cdot Y[n] \cdot Y[n]^T$$
$$= -2(In[n-D] - C^T Y[n]) \cdot Y[n]^T$$

Remember that the error itself was defined as the ideal output value, $In[n-D]$, minus the estimate. In equation form, the error was expressed as $e[n] = In[n-D] - C^T Y[n]$. The final gradient of the instantaneous squared error is as follows.

$$\boxed{\nabla_C(e[n] \cdot e[n]^T) = -2e[n] \cdot Y[n]^T}$$

Rather than setting the gradient result to zero and solving for the equalizer coefficients, $EC = C^T$, we instead use the instantaneous gradient as a guide to improving the solution vector in an iterative fashion. The gradient itself is a row vector with the same dimension as the observation vector $Y[n]^T$. It indicates the change in error for each of the coefficients.

$$\nabla_C(e[n] \cdot e[n]^T) = -2e[n] \cdot \begin{bmatrix} y[n] \\ y[n-1] \\ y[n-2] \\ \vdots \\ y[n-(N-1)] \end{bmatrix}^T = \begin{matrix} \textit{error slope for coefficient } c[0] \\ \textit{error slope for coefficient } c[1] \\ \textit{error slope for coefficient } c[2] \\ \vdots \\ \textit{error slope for coefficient } c[N\text{-}1] \end{matrix}$$

Now, how does the knowledge of the slope of the error or cost function for any one equalizer coefficient show us how to improve that coefficient's value?

Take a look at the next figure, which draws the error curve as a function of just one equalizer coefficient. No matter what the values of the other coefficients are, the cost function as a function of just one coefficient will be quadratic in nature. As can be seen in the figure, if the coefficient is too large, the slope will be positive, whereas a negative slope indicates that the coefficient value is too small and must increase.

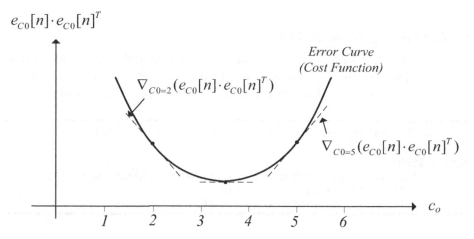

$$e_{C0}[n] \cdot e_{C0}[n]^T$$

$$\nabla_{C0=2}(e_{C0}[n] \cdot e_{C0}[n]^T)$$

Error Curve (Cost Function)

$$\nabla_{C0=5}(e_{C0}[n] \cdot e_{C0}[n]^T)$$

c_o

1   2   3   4   5   6

Figure 4-23: Error Gradient as a Function of Just One of the Equalizer Coefficients

Therefore, to arrive at a new set of coefficients, we multiply the instantaneous gradient by a small factor, μ, and *subtract* it from the original coefficient set. We adjust the equalizer coefficient by only a fraction, μ, of the instantaneous gradient, since it is at best a crude approximation of the gradient computed during the MMSE derivation, where the expectation was still present. This iterative adjustment achieves a low pass filtering effect, which allows the equalizer coefficients to approach their MMSE optimal values smoothly. The variable μ, where $0 < \mu \ll 1$, controls the rate at which the coefficients approach their MMSE optimal values.

The LMS Update Equation

The expression below represents the LMS update equation for both complex and real valued inputs.

$$C^T[n+1] = C^T[n] + 2\mu \cdot e[n] \cdot Y[n]^T$$

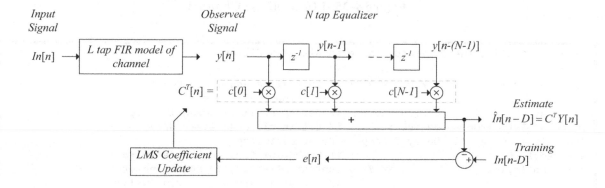

Figure 4-24: Equalizer with Adaptive Coefficient Updates

Example 4.4: *Least Mean Squares Algorithm*

Now it's time to take our LMS algorithm for a quick spin around the block. We will go straight to the most demanding case of the complex input stream combined with a complex distortion channel of tap length 6. The equalizer will feature $N = 13$ complex coefficients.

```
%% 1. Defining the complex input stream and Channel Coefficients

InputLength = 1500;
In          = sign(randn(1,InputLength)) + sign(randn(1,InputLength))*j;
Channel     = [-.19j  .14  1+.1j  -.16  +.11j + .03];
y           = filter(Channel,1, In);    %% The observation
```

The figure below shows just the real part of the input sequence and how it was distorted at the point of observation.

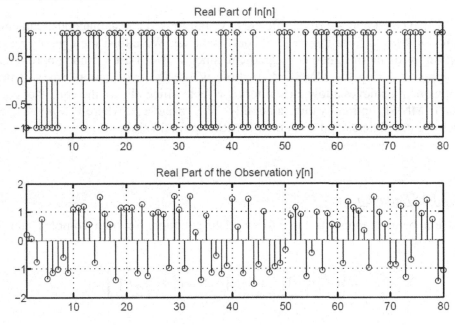

Figure 4-25: Effects of the Channel

```
%% 2. Initialize the Estimator to be an All-pass Filter

EC  = [0 0 0 0 0 0 1 0 0 0 0 0 0 ]; %% Initialize equalizer EC=C'
                                    %% as simple all pass
[M Delay_Channel]    = max(Channel);   %% Get approximate channel delay
[M Delay_Estimator]  = max(EC);        %% Get initial equalizer delay
Approximate_Delay    = Delay_Channel + Delay_Estimator;
N = length(EC);                        %% Number of taps in the estimator
Y = zeros(N,1);                %% Y -> content of estimator shift register
                                       %% Y = y[n], y[n-1] ... y[n-(N-1)]
u        = .005;                       %% LMS gain
Estimate = zeros(1,length(y));         %% Storage for the estimates
```

In section two of the code above, we initialize the equalizer coefficients to simply pass the observation without any modification whatsoever. This is called an all pass filter setup. We also find the approximate delay through both channel and estimator, which indicates the time it takes for the largest copy of any one input symbol *In*[*n*] to show up at the equalizer output. This allows us to line up the training sequence (which is of course the same as *In*[*n*]) with the estimator output for the error calculation. Note, however, that we don't necessarily have to be lined up correctly at the start. The equalizer will adjust itself by moving its main tap, or average group delay, such that the estimation and training symbols are properly lined up. This is only possible because time delay itself is just another form of linear distortion which is removed by the LMS algorithm.

```
%% 3. Start the Adaptation Loop
Error    = zeros(1,length(y));
for n=1:length(y)
    Y(2:N,1)     = Y(1:N-1,1);      %% Here we shift the observation y[n]
    Y(1,1)       = y(1,n);          %% into the estimator's shift register
    Estimate(1,n)= EC*Y;           %% Here we form the estimate of x[n]
    if(n>Approximate_Delay+3)
        TrainingSymbol    = In(1, n - Approximate_Delay + 2);
        Error(1,n)        = TrainingSymbol - Estimate(1,n);
        EC = EC + 2*u*Error(1,n)*Y';      %% LMS coefficient update
    end
end
```

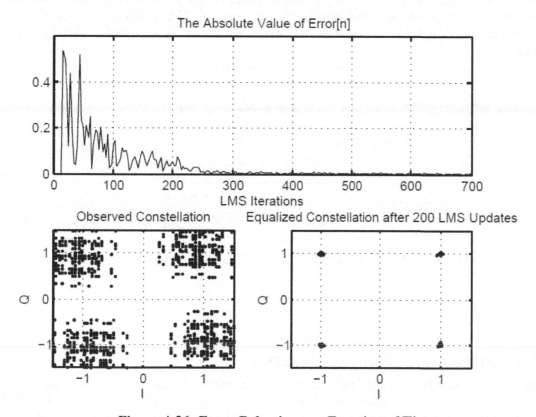

Figure 4-26: Error Behavior as a Function of Time

The absolute error reaches its minimum just past 200 iterations and the corresponding equalized constellation is a testament to the high performance of our 13 tap complex FIR equalizer. The algorithm could potentially lock more quickly and still remain stable if we were to choose a LMS gain of up to $\mu = 0.03$. However, the steady state error after lock is significantly higher than our scenario, which fixes the gain at $\mu = 0.005$. Practically speaking, the scaling factor μ should initially be set at a high but stable value to facilitate fast locking and then be reduced to minimize noise.

Decision Direction

Transmit signals used in communication systems invariably include a priori known training sequences at the beginning and during the data stream. Receivers use these training sequences to estimate the multipath channel that causes linear distortion and either program the equalizer directly or find the optimum coefficients iteratively via the LMS algorithm. Making training sequences too long or repeating them too often to reset or retrain the equalizer reduces the data throughput of our communication link. As soon as the training sequences are over, however, coefficient adaptation stops and any subsequent change in the channel characteristics will reduce the receiver performance. A simple solution to this dilemma is to guess at the most likely symbol indicated by the current estimate and use it in place of the training sample. For an estimate of $0.9 - j0.7$, a QPSK decision element would produce $[1 - j]$ as the most likely symbol from the set of possible symbols $[1 + j]$, $[1 - j]$, $[-1 + j]$ and $[-1 - j]$. With a little luck, the system will have had enough time during the initial training period to program the equalizer coefficients to within a sufficient accuracy for these subsequent guesses to be valid. The coefficient update of the LMS algorithm is thus directed based on the output made by the decision element rather than training information.

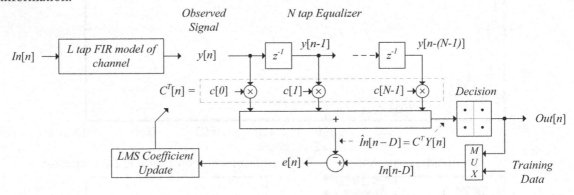

Figure 4-27: Decision Directed LMS Coefficient Adaptation with QPSK Decision Element

4.5   Decision Feedback Equalizers

In the world of mobile communication systems, the impulse response of the distorted channels follows a statistical pattern based on a Raleigh distribution. This distribution indicates that the strongest copies of the transmitted signal will likely arrive earlier at the receiver than the weaker copies. This makes intuitive sense, since, in all likelihood, the earlier copies will have had a more direct path to the transmitter and thus suffer from less attenuation. Where we have an actual line of sight between the transmit and receive antennas, the strongest copy of our TX signal will in fact arrive first. If an obstruction exists in the direct path between the antennas, then the first copy that took the direct path through the obstruction will be weaker than a later copy, which may have bounced off a smooth surface and retained much of its signal strength. Most wireless communication links don't feature a direct line of sight, which means that the strongest path arrives early but not first. The copies arriving ahead of the strongest signal are called pre-cursors, while those arriving afterwards are referred to as post-cursors. The figure below shows the absolute value of the impulse response of a typical multipath channel. Given that there are usually more post-cursors than pre-cursors, it stands to reason that they are in fact the greater contributors to the overall linear distortion.

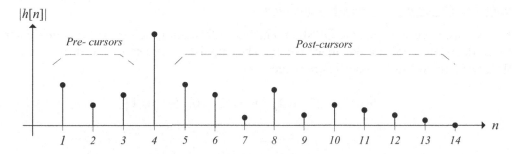

Figure 4-28: Typical Multipath Channel Impulse Response

To deal with these post-cursors better, decision feedback equalizers are used to cancel the post-cursor effects via a simple subtraction operation. Let us assume that a transmitter is sending QPSK like information with valid constellations of [1 j], [1 –j], [–1 j], and [–1 –j]. Let us further assume that a symbol value of [.9 .7j] was received and interpreted by the decision element as constellation point [1 j]. Knowing that this symbol, [1 j], will also appear in the subsequent observations and given that we know the channel impulse response, we can subtract its future contributions from those subsequent observations.

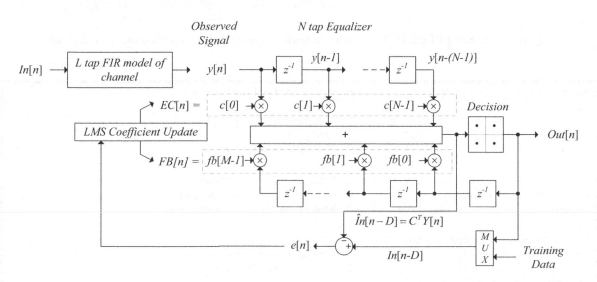

Figure 4-29: Decision Feedback Equalizer Structure

Looking at the structure, we can see the additional feedback delay path with coefficients $fb[0]$ through $fb[M\text{-}1]$, which will represent the negative of the first, second, and remaining post-cursors. In general, the FIR filter with coefficients marked $c[0]$ through $c[N\text{-}1]$ is referred to as the feed forward section, while the other is called the feedback section. To make our life easier, let us assume that our feed forward section is not a full FIR filter but a simple pass through, which may be realized by setting all coefficients $c[0]$ through $c[N\text{-}1]$ to zero except one, which takes a value of unity. In the example below, we initially examine the beneficial effects of the feedback section that cancels the post-cursors, before worrying about how to remedy the impact of the pre-cursors.

Example 4.5: *Decision Feedback Equalization*

For this example, we will assume QPSK modulated information being distorted and entering the simplified decision feedback equalizer below. The impulse response of the distorting channel is complex, with the following coefficient values.

$$h[n]=[0.05 \quad 0.1j \quad 1 \quad 0.2 \quad -0.15j \quad 0.1]$$

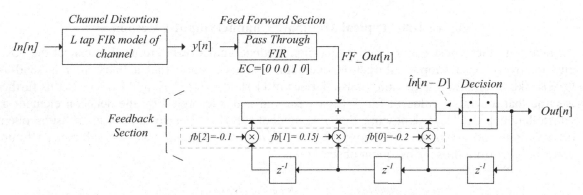

Figure 4-30: Simplified DFE Structure with Appropriate Post-Cursor Coefficients

```
%% 0. Setting up the QPSK source signal and Distortion Channel
Length   = 200;
QPSK     = sign(randn(1,Length)) + j*sign(randn(1,Length));
Channel  = [0.05  0.1*j  1.0  0.20  -0.15*j  0.1];
y        = filter(Channel ,1,QPSK);%% The observation vector


%% 1. The Feed Forward Coefficients set up as a Pass through Element
FF_Coeff = [0 0 0 1 0];            %% Pass through coefficients
FF_Out   = filter(FF_Coeff, 1, y); %% Result of feed forward FIR


%% 2. Decision Feedback Loop
FB_Pipe = zeros(1,3);              %% Feedback pipeline
FB      = [-0.2 0.15*j -0.1];      %% Feedback coefficients
Estimate= zeros(1,Length);
LastEstimate = 0;
```

```
for n = 1:Length
    Decision         = sign(real(LastEstimate))+j*sign(imag(LastEstimate));
    FB_Pipe(1,2:3)   = FB_Pipe(1,1:2);
    FB_Pipe(1,1)     = Decision;
    Estimate(1,n)    = FF_Out(1,n) + FB_Pipe*FB.';
    LastEstimate     = Estimate(1,n);
end


%% 3. EVM performance evaluation
%     At the point of observation y[n]
Ideal = sign(real(y(1,5:end))) + j*sign(imag(y(1,5:end)));
Error = y(1,5:end) - Ideal;
EVM_Observation = 10*log10(var(Error)/var(Ideal))


%     Of the estimate
Ideal = sign(real(Estimate(1,5:end))) + j*sign(imag(Estimate(1,5:end)));
Error = Estimate(1,5:end) - Ideal;
EVM_Estimate    = 10*log10(var(Error)/var(Ideal))
```

By programming the feedback coefficients $fb[0]$, $fb[1]$ and $fb[2]$ to -0.2, 0.15j and -0.1 respectively, we remove the influence of the post-cursors, which significantly cleans up the estimated constellation.

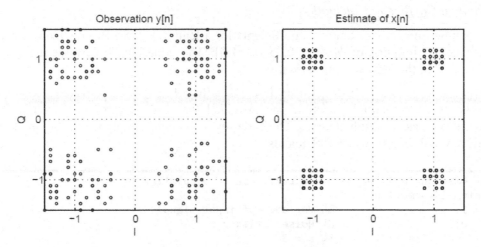

Figure 4-31: Observation (EVM = -10.7dB) versus Estimate (EVM = -19.2 dB) of the Simplified DFE Structure

Handling the Pre-Cursor Effects

The feedback portion of the DFE structure has removed the post-cursors, leaving us with the new combined channel impulse response seen on the left side of the figure below. The task now is to find the coefficient vector, *EC*, which will convolve the remaining impulse response into one single unit pulse.

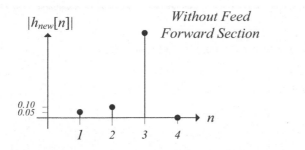
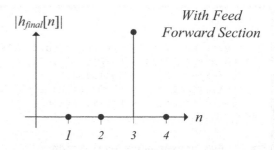

Figure 4-32: The Remaining Task for the Feed Forward FIR Filter

The task of computing the feed forward coefficients, $EC=C^T$, can be accomplished by any of the methods that we have discussed in the earlier sections of this chapter. In this example, we will compute the MMSE optimal coefficients assuming a noise free environment and evaluate the performance in terms of EVM. We will then allow the LMS algorithm to fine-tune both the feed forward and feedback coefficient vectors and reevaluate the decision feedback equalizer's performance. Baseband processors will often find far cruder coefficients that are easier to compute and then allow the LMS algorithm to fine-tune them until they reach MMSE optimal values.

MMSE Optimal Coefficient Calculation

We will compute the MMSE optimal coefficients for a length $N = 5$ FIR equalizer structure that remedy a channel impulse of $h[n] = [0.05\ 0.1\ 1.0]$. The equations below yield the equalizer coefficients, EC, in the MMSE optimal sense.

$$C^T = E(In[n - D] \cdot IN[n]^T)H^T \cdot inv(H \cdot R_{In} \cdot H^T + R_v)$$

In the previous code example, section one provided the feed forward coefficients that were implemented as a pass through element. The code segments below replace section one to produce the MMSE optimal feed forward coefficients.

```
%% 1. The MMSE Optimal Feed Forward Coefficient Calculation
%  1a. Setup information
VarIn   = 1;                    %% Variance of input signal
VarN    = 0;                    %% Noise variance
Channel = [.05 .1*j  1];        %% L = 3
N       = 5;                    %% Number of taps in feed forward FIR
L       = length(Channel);      %% Number of taps in channel
D       = N + L - 2;
```

From section 4.3, we know that R_v and R_{In} are N by N (5 by 5) and $N+L$-1 by $N+L$-1 (7 by 7) identity matrices, respectively, whereas H is the Toeplitz matrix that we have seen so many times.

$$N\ x\ L+N-1$$

$$H = \begin{bmatrix} 0.05 & 0.1j & 1 & 0 & 0 & 0 & 0 \\ 0 & 0.05 & 0.1j & 1 & 0 & 0 & 0 \\ 0 & 0 & 0.05 & 0.1j & 1 & 0 & 0 \\ 0 & 0 & 0 & 0.05 & 0.1j & 1 & 0 \\ 0 & 0 & 0 & 0 & 0.05 & 0.1j & 1 \end{bmatrix}$$

$$\overset{1 \ x \ N}{E(In[n-(D)] \cdot IN[n]^T) = \begin{bmatrix} 0 & 0 & 0 & 0 & 1 \end{bmatrix}}$$

```
%% 1b. Calculate H
row = zeros(1,N+L-1); row(1,1:L) = Channel;
col = zeros(1,N);     col(1,1)   = Channel(1,1);
H   = toeplitz(col,row);


%% 1c. Calculate RIn, Rv, Ry
RIn = VarIn*eye(N+L-1);
Rv  = VarN*eye(N);
Ry  = H*RIn*H' + Rv;


%% 1d. Calculate H E(In(n-D))IN(n)
E_InIN          = zeros(1, N+L-1);
E_InIN(1, D+1)  = VarIn;
E_InIN_H        = E_InIN*H';
EC              = E_InIN_H*inv(Ry);   %% Column Vector EC = C'

FF_Out          = filter(EC, 1, y);   %% Result of Feed Forward FIR
```

Rerunning the previous code segment with the improved MMSE optimal feed forward coefficients significantly improves the overall performance of the DFE structure, as can be seen below.

$$EC = [0.004 \ \ 0.0109j \ \ -0.06 \ \ -0.1j \ \ 1.0]$$

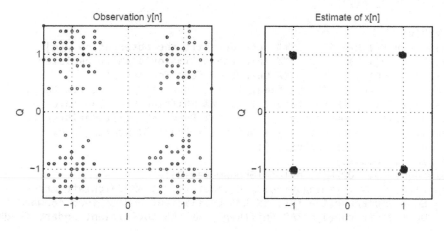

Figure 4-33: Observation (EVM = -10.7dB) versus Estimate (EVM = -31.7dB) of the Improved DFE

Example 4.6: *Combining Decision Feedback and the LMS Algorithm*

Now that we have set the feedback coefficients to perfectly cancel the post-cursors and calculated the MMSE optimal feed forward coefficients, you may ask why we would go to further trouble and continue to improve a solution that is in fact already MMSE optimal. It was the feed forward section's responsibility to eliminate the pre-cursors to yield a combined impulse response that would be perfect = [... 0 0 1 0 0 ...]. Given its limited degrees of freedom – the number of coefficients available ($N = 5$) – the feed forward section did the best it could. The new combined impulse response will have new small post-cursors that can be removed by further optimizing the feedback coefficients. It is for this reason that we decide to use the LMS algorithm to improve both the feed forward and feedback section further, and, as we will see below, there was quite a bit of performance to be had by going through this additional effort.

```
%% 0. Setting up the QPSK source signal and Distortion Channel
Length   = 1000;
QPSK     = sign(randn(1,Length)) + j*sign(randn(1,Length));
h        = [0.05  0.1*j  1.0  0.20  -0.15*j  0.1];
y        = filter(h,1,QPSK);   %% The observation vector


%% 1. Initialize the 5 tap feed forward section
EC  = [.004 .0109j -0.06 -0.1j 1.0]; %% Initialize with MMSE
                                     %% Optimal coefficients
[M GroupDealy_Channel]  = max(h);
[M GroupDelay_Estimator] = max(EC);
TotalGroupDelay          = GroupDealy_Channel + GroupDelay_Estimator;
N = length(EC);              %% Number of taps feed forward section
Y = zeros(N,1);              %% Y -> content of estimator shift register
                             %% Y = y[n], y[n-1] ... y[n-(N-1)]
u        = .005;             %% LMS gain
FF_Out       = zeros(1,length(y));  %% Output of feed forward section
Estimate     = zeros(1,length(y));  %% The estimate of In[n]


%% 2. Start the Adaptive Loop
FB_ShiftReg = zeros(3,1);          %% Feedback shift reg as column vector
DC          = [-0.2 0.15*j -0.1];  %% Feedback coefficients initialized
LastEstimate = 0;
Error   = zeros(1,length(y));

for n=1:length(y)
    Y(2:N,1)    = Y(1:N-1,1);    %% Here we shift the observation y[n]
    Y(1,1)      = y(1,n);        %% into the estimator's shift register
    FF_Out(1,n) = EC*Y;          %% Output of feed forward FIR
    Decision    = sign(real(LastEstimate)) + j *sign(imag(LastEstimate));
    FB_ShiftReg(2:3,1) = FB_ShiftReg(1:2,1); %% Shifting decision into
    FB_ShiftReg(1,1)   = Decision;           %% Feedback shift register
    Estimate(1,n)      = FF_Out(1,n) + DC*FB_ShiftReg; %% Summing node
    LastEstimate       = Estimate(1,n);
    if(n>TotalGroupDelay+3)
        TrainingSymbol= QPSK(1, n-TotalGroupDelay+2);
        Error(1,n)    = TrainingSymbol - Estimate(1,n); %% Calculate error
        EC = EC + 2*u*Error(1,n)*Y'; %% LMS coefficient update feed forward
        DC = DC + 2*u*Error(1,n)*FB_ShiftReg'; %% LMS coefficient update feedback
    end
end
```

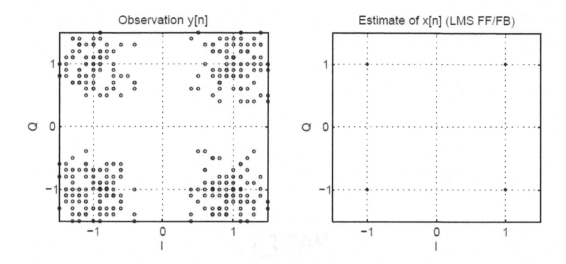

Figure 4-34: Observation (EVM = -10.7dB) versus Estimate (EVM = -60dB) of the Improved LMS based DFE Structure

References

[1] Masani, P. R. (1990), *Norbert Wiener*, Birkenhäuser Verlag, Basel, Switzerland

[2] Haykin, S., *Adaptive Filter Theory (Third Edition),* Prentice–Hall, Englewood Cliffs, NJ, 194 - 236

[3] Widrow, B., Stearns P. (1985), *Adaptive Signal Processing,* Prentice–Hall, Englewood Cliffs, NJ

[4] Haykin, S., Widrow, B. (Editors) (2003) *Least-Mean-Square Adaptive Filters,* John Wiley&Sons, Hoboken, NJ

[5] Sayed, A. H. (2003) *Fundamentals of Adaptive Filtering*, John Wiley & Sons, Hoboken, NJ, 47-51, 105-107

PART 3

Transmitter Design in Modern Wireless Communication Systems.

5   Transmitter Design in Modern Communication Systems

Modern wireless communication systems utilize a variety of different signaling formats to relay information from one transceiver to another. These formats change depending on the requirements placed on the wireless link. Do we wish to transmit high definition TV signals over the air or want our wireless speaker to receive low data rate audio waveforms? Does our signal need to reach clear across town or just to the next room? How about battery size and cost? These considerations affect the modulation format, bandwidth, and maximum output power of the transmit signal that best suits the unique application which we are tasked to develop. In the figure below, we illustrate a wireless transmitter structure that can be adapted to any signaling format and highlight the elements that are primarily involved in the modulation process.

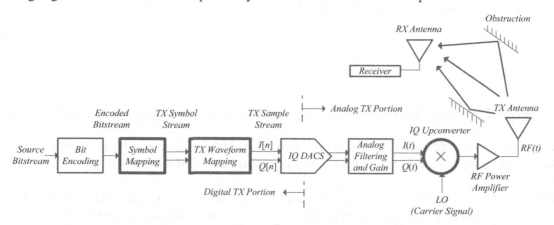

Figure 5-1: Overview of a Generalized Zero-IF Wireless Transmitter

Defining Modulation

Before we introduce the different modulation techniques at our disposal, let's first define the term modulation itself. Modulation, in our context, is the process of altering the amplitude and phase of the carrier signal in order to convey information. In the strictest sense, the modulator is the analog *IQ upconverter*, or mixer, that applies the $I(t)$ and $Q(t)$ waveforms to the carrier signal. RF engineers will actually use the terms modulator, upconverter, and mixer interchangeably. Over time the term modulation has expanded to include the processing modules which create the signals that are modulated onto the carrier. In this text, we consider the *symbol mapping* and *TX waveform mapping* blocks seen in the figure above as the digital portion of the modulator. Whereas the symbol mapping process is a relatively straight forward affair, the TX waveform mapping block is more sophisticated and features as the main topic in this chapter.

5.1   The Transmitter Architecture

The overall transmitter architecture shown in the figure above contains the main processing blocks that are required to generate virtually all modern modulation formats. As we move from one method of modulation to the next, it is primarily the internal processes in the symbol and TX waveform mapping functions that have to change. Let's take a closer look at each block in the cascaded chain of processing elements that make up the digital and analog portions of the transmitter.

5.1.1   The Digital Portion of the Transmitter

Bit Encoding

The digital portion of the transmitter starts off with a bit encoding block which usually provides the bit stream with error detection/correction capabilities as well as whitening and encryption mechanisms. In the figure below, we show how these functions could be arranged in the transmitter and their counterparts in the receiver. Let us briefly examine each submodule one by one.

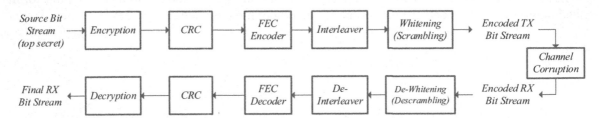

Figure 5-2: Comprehensive Bit Encoding and Decoding in the Transmitter and Receiver

→ *Encryption* scrambles the bit stream in order to avoid eavesdropping by unintended listeners. Outside of this feature, there is no benefit to the performance of the communication link.

→ *CRC* – A cyclic redundancy check is a method of detecting bit errors in data transmissions that are affected by noise. On the transmit side, a certain number of redundant bits are attached to blocks of data thereby slightly decreasing the information throughput. These redundant bits are based on the remainder of a polynomial long division of the data in the block. On the receiver side, these check bits are again calculated and compared to those that were transmitted. Should an error be detected during the comparison, the receiver must request that the data be resent. Whereas understanding the performance impact of different polynomial configurations that make up a CRC may require some effort, implementing the CRC algorithm is comparatively trivial.

→ *FEC* – Forward error correction is a bit encoding and decoding process that attempts to correct errors at the receiver and not just detect them as is the case for the CRC. Because error correction is more involved than detection, the bit encoding module that represents the FEC process will add a good deal more redundant bits than the encoding module of the CRC. Hopefully, the FEC process can correct all existing errors to which the final cyclic redundancy check must attest. If the CRC detects an error, retransmission of the data is necessary. Some typical forward error correction schemes that are used in the industry include binary convolution coding (BCC), Turbo coding, as we all popular algebraic coding schemes such as low density parity check, or LDPC, and Polar coding. Section 5.6 will cover some of these schemes.

→ *Interleaver* – In many communication channels errors appear in burst rather than being randomly distributed over time. Under these conditions, the FEC error correction performance at the receiver will suffer and retransmission of data will be required more frequently. Interleaving is the process of temporarily redistributing consecutive bits produced by the FEC encoder. In its most basic implementation, the FEC encoded bits are placed along the rows of a rectangular matrix. We then read out the bits in the matrix along the columns and pass them on to the symbol mapper. Because the deinterleaver in the receiver reverses this process, error bursts at its input are redistributed and appear randomly positioned in time thereby improving the performance of the FEC decoder.

→ *Whitening (Scrambling)* – In certain signaling schemes it is disadvantageous to transmit long sequences of successive 1's and 0's. These occurrences can temporarily introduce excessive DC bias into the transmit signal or otherwise cause signal power to concentrate excessively in certain areas of the transmit channel. Furthermore, some synchronization techniques used to acquire various signal offsets in the receiver may be hampered by this phenomenon. To counteract this problem, we randomize the transmit data using a process called whitening or scrambling. During the receiver signal processing chain, the data is eventually de-whitened or descrambled to produce the original information. In cellular communication, scrambling has an additional purpose. It helps information from one cell tower appear random compared to information from another cell tower. This is important as different cells tower at times transmit very similar information at the same time and the same frequency giving rise to correlated interference, which we very much need to avoid. When your cell phone is connected to a base station, then signals from other base stations need to look like noise. Scrambling ensures this behavior.

Symbol Mapping

The symbol mapper represents the first step in converting binary data into continuous waveforms that are eventually embedded onto the I and Q signal dimensions of the RF carrier. The symbol mapper groups incoming bits and converts them to constellation positions in the IQ plane. These positions are complex values that necessarily feature a higher resolution than the bit stream at its input.

Figure 5-3: The Symbol Mapping Block

The figure below shows how bits may be grouped to render two popular constellations called BPSK and QPSK.

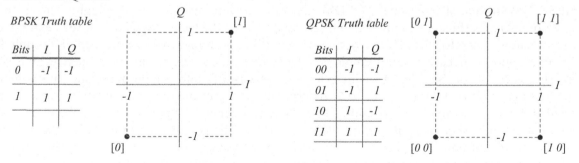

Figure 5-4: Absolute Value Symbol Mapping into BPSK and QPSK Constellations

Note that we may also map groups of bits into rotations from one constellation position to the next. For differential QPSK, or DQPSK, the input bit groups [00, 01, 10, 11] may represent phase rotations by [0, $\pi/2$, π, $3\pi/2$] radians. As we will see soon, mapping groups of input bits into rotations rather than absolute constellation positions eases certain synchronization tasks in the receiver at the cost of a slightly deteriorated bit error performance.

TX Waveform Mapping

The TX waveform mapping module embeds the stream of symbols at its input into a discrete IQ waveform that will be read out through the digital to analog converters and processed by the Analog/RF portion of the transmitter.

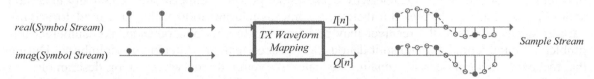

Figure 5-5: The TX Waveform Mapping Block

The majority of this chapter is dedicated to explaining four different TX waveform mapping approaches.

1. Single Tone - Pulse Shaped PSK (Phase Shift Keying) and QAM (Quadrature Amplitude Modulation)
2. Single Tone - Pulse Shaped Frequency Shift Keying (GFSK, GMSK)
3. Multi Tone - Orthogonal Frequency Division Multiplexing (OFDM)
4. Hybrid Tone - SC-FDMA (Single Carrier Frequency Division Multiple Access)

Each one of these techniques features unique attributes that make them suitable for different communication links. Single tone phase shift keying techniques are traditionally easy to implement and its variants have been successfully used in virtually all wireless communication applications including satellite links, Bluetooth, WLAN 802.11b, cellular telephony (IS-54), and cable modems. Single tone frequency shift keying schemes such as GFSK and GMSK are primarily used in hand held wireless devices where a battery supplies power to the device and therefore current consumption has to be minimized (Bluetooth, DECT, GSM). The waveforms generated by frequency shift keying modulation allow the use of very efficient RF power amplifiers. OFDM, a modern multi tone technique, has ushered in a revolution in high bandwidth, high data rate wireless links such as WLAN communication, digital HD television broadcasting and most recently the LTE/5G downlink (base station to mobile handset). However, unlike the GFSK/GMSK modulated signals, OFDM generates waveforms that require highly linear and therefore very inefficient RF power amplifiers, and as such, the technology has not seen extensive use in battery operated, handheld wireless devices. To get around this problem, the LTE uplink (mobile handset to base station) employs a technique called single carrier frequency division multiple access, or SC-FDMA, which feature aspects of both single tone and multi tone modulation. From OFDM, SC-FDMA inherits the straight forward equalization scheme to combat linear distortion due to multipath channels, whereas it maintains an amount of AM modulation that is more akin to single tone implementations and therefore avoids the use of overly inefficient power amplifiers. In this and later chapters, we will detail the implementation of these different modulation schemes and show how they benefit each communication scenario.

5.1.2 The Analog/RF Portion of the Transmitter

The DAC and Analog Filtering Sections

The spectrum of the sample stream produced by the TX waveform mapper features unique content only within the Nyquist frequency range of $\pm F_S/2$. As we know from sampling theory, sampling images appear around $\pm m \cdot F_S$ where m is a positive integer from 1 to infinity and F_S is the sampling rate. Ideally, the digital to analog conversion and low pass filtering process would simply eliminate the sampling images and yield a perfect analog reproduction of the original discrete sample stream. This, however, is not exactly what happens. The digital to analog converter, or DAC, converts the discrete sequence of numbers at its input into a continuous waveform by generating proportional voltage/current values at its output and holding those values constant for one sample period, T_S. This process introduces sample and hold, or $sin(x)/x$, distortion into the waveform as can be seen in the magnitude spectra below.

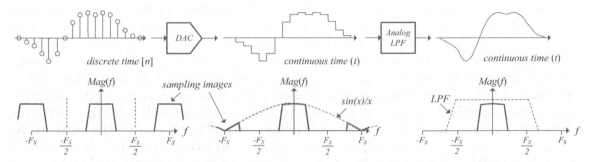

Figure 5-6: Signal Processing Through the DAC and Low Pass Reconstruction Filter

While noticeable, the $sin(x)/x$ distortion inherent in the DAC's sample and hold process diminishes quickly as we increase the sample rate of the discrete signal and digital to analog converter. It is for this reason that some transmitters will upsample the TX waveform before converting it into the analog domain. Note that the $sin(x)/x$ overlay in the magnitude spectrum of the DAC output signal reaches nulls at $\pm m \cdot F_S$ thus significantly simplifying the lowpass filter's task to eliminate the sampling images.

The IQ Upconverter (Modulator) and RF Power Amplifier

The *I and Q modulators* multiply their incoming baseband waveforms with the two RF carrier signals $cos(2\pi f_0 t)$ and $sin(2\pi f_0 t)$, which, when summed, form the final RF transmit signal as the expression below details.

$$I(t) \cdot \cos\left(2\pi f_o t\right) + Q(t) \cdot \sin\left(2\pi f_o t\right) = A(t) \cdot \cos\left(2\pi f_o t - \theta(t)\right)$$

where

$$A(t) = \sqrt{I(t)^2 + Q(t)^2}$$

$$\theta(t) = atan2(Q(t), I(t))$$

As an example, if $I(t)$ and $Q(t)$ were constants equaling 0.707 each, then $A(t) = 1$ and $\theta(t) = 45° = \pi/4$ radians.

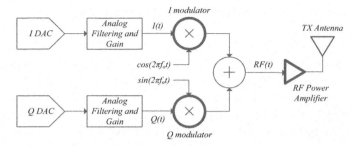

Figure 5-7: Simplified Diagram of a Transmitter's Analog Portion

The *RF power amplifier* is the topic of a great deal of research in the wireless industry due to its tendency to gain up signals in a nonlinear fashion as the required output power increases. Devices that make the extra effort to stay linear under all circumstances consume a great deal of overhead power which renders them very inefficient. This is a big problem in battery operated handheld equipment, because of the resulting shortened usage time and the excessive heat generated by the amplifier. As mentioned earlier and discussed in detail in Chapter 6, amplitude modulation creates intermodulation distortion that appears in the immediate vicinity of the TX signal frequency and spills distortion content into neighboring RF channels. Your local cellular carrier really won't like that. Therefore, a great deal of effort has been undertaken to craft modulation schemes that minimize the amount of amplitude modulation such as GFSK, GMSK, π/4-DQSPK, and modified versions of 8PSK as used in EDGE (enhanced data rates for GSM evolution). We will get to know all these techniques in this chapter. Whereas quantifying the amount of amplitude modulation is not a straight forward undertaking, some simple statistical measures are frequently used to predict the relative amount of intermodulation distortion created in the RF power amplifier for various signaling configuration. The peak to average power (PAP) ratio is a popular metric on which engineers rely. Assuming that all other condition stay unchanged, the higher the peak to average power ratio, the greater the resulting intermodulation distortion in the RF power amplifier tends to be.

$$PAP = 10\log_{10}(\frac{PeakPower}{AveragePower}) \quad dB$$

However, to get a more complete picture, the probability density function of the signal amplitude should also be examined.

5.1.3 Single Tone versus Multi Tone Modems

In general, wireless transceivers can be separated into single tone and multi tone modems. In both modem types, bits are grouped and then mapped into discrete symbols before being embedded into the discrete TX waveform (the sample stream) that is to be read out of the digital to analog converters. In the embedding process of _single tone_ modems, the symbols are often pulse shaped to form the TX sample stream. These pulses, scaled by the symbol values, appear one after another in the TX sample stream. We can actually look at the sample stream (or the analog output I/Q signals) and see the symbols represented therein. In the next figure, we recognize the first four complex values in the symbol stream, $[s_1, s_2, s_3, s_4 \ldots]$, as $1-j$, $-1+j$, $1+j$, and $-1+j$. In literature, the term single tone modem is used interchangeably with the terms _single carrier_ or _time domain_ modem.

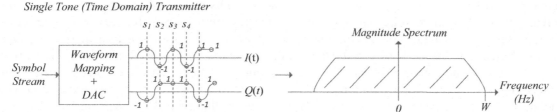

Figure 5-8: Single Tone QPSK Transmission and Related Spectrum

The pulse shape that is forced upon each symbol determines the spectrum of the final waveform. This spectrum is spread around a single tone, which is the carrier signal. Single tone systems of this type have been around for a long time and can achieve good data throughput. However, their implementation grows in complexity when the transmit signal grows in bandwidth and must operate in mobile environments that feature significant multipath distortion. Some of the more common single tone modulation techniques are pulse shaped BPSK, QPSK, pi/4-QPSK, GFSK, and GMSK. These modulation schemes have been used in many communication systems including various satellite links, 802.11b, IS-54, GSM, and Bluetooth.

OFDM, or orthogonal frequency division multiplexing, is the most commonly implemented _multi tone_ architecture to date and represents the answer to the consumer's desire for ever higher data throughput. Compared to single tone pulse shaped modulation, OFDM simplifies the equalization process in the receiver and enables antenna diversity techniques, such as MIMO, to significantly increase data reliability and throughput. OFDM's main disadvantage is the very large amount of amplitude modulation that appears on the transmit signal thus necessitating the need for highly linear and therefore very inefficient RF power amplifiers. For this reason, OFDM is primarily used in wireless applications that are not battery operated such as WLAN, digital TV broadcasting and the LTE/5G downlink (base station to mobile handset) path. Take a look at symbols $1-j$, $-1+j$, $1+j$, and $-1+j$, which are visible not in the IQ transmit waveform but in the individual tones of the signal spectrum. The multipath channel introduces nulls in select frequency locations of the received signal spectrum and in single tone modems necessitates sophisticated equalizers to correct them. However, in OFDM each individual tone of the OFDM signal may be through of as experiencing flat fading, which is trivially corrected by a applying a complex scalar to each received tone. The ease of equalization and straight forward integration into MIMO systems has made the OFDM modem the clear choice for wideband, high data rate wireless communication links. In literature, multi tone modems may also be referred to as _multi carrier_ or _frequency domain_ modems.

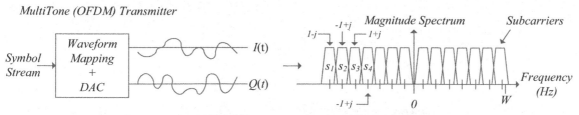

Figure 5-9: OFDM Transmission and Related Spectrum

SC-FDMA, is a _hybrid tone_ modulation technique that maintains some of the advantages of OFDM with respect to signal equalization and multiple antenna implementation while bringing the amount of amplitude modulation into the range typical of single tone, or single carrier, signaling formats. As in single tone signaling, the symbols can be observed in the time domain waveform. In the single tone pulse shaped QAM transmitter case, the successive IQ symbol values were approximated into a continuous waveform via the superposition of time staggered pulses of a certain shape. The zero-stuffing and pulse shaping steps established sample values in between the symbols to render the final continuous waveform. However, in-between sample values may also be defined by finding a set of complex tones of different frequencies, that when added, will pass through each symbol. This is akin to the FFT/IFFT interpolation technique discussed in Chapter 3. If the time domain signal is thus composed of individual subcarriers, which each experience flat fading, then the same equalization method used in OFDM applies.

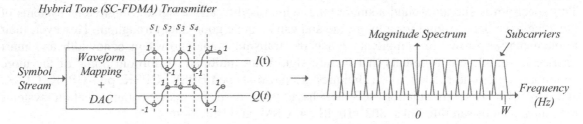

Figure 5-10: SC-FMDA Waveforms and Signal Spectra

5.2   Single Tone PSK and QAM Modulation

Single tone type modulation has been a staple of wireless communication systems for a long time and includes familiar formats such as BPSK, QPSK, π/4-DQPSK, 8PSK, 16QAM. Note that these formats refer to constellation positions which determine where the symbol mapper places his outputs in the IQ plane. To be clear, BPSK, QPSK, 16QAM, and 64QAM constellations can be used in *multi tone* modulation such as OFDM as well. The difference is how the symbols are encoded into the sample stream that finally modulates the carrier signal. In single tone modulation, the symbols that are fetched from these constellation positions are converted into time staggered pulses that are shaped to yield a signal spectrum of our liking. The figure below gives us some insight into the internal structure of the TX waveform mapping block which forms the core of the single tone phase shift keying (PSK) and quadrature amplitude modulator (QAM).

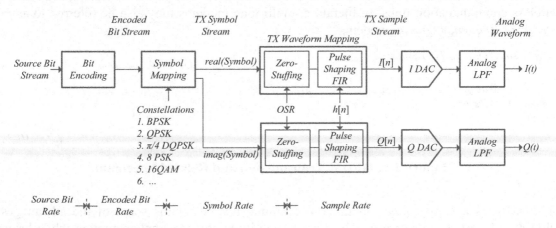

Figure 5-11: Digital Potion of a Typical Single Tone IQ Pulse Modulator

Rates in the PSK/QAM Transmitter

Before we delve into the inner working of the symbol and TX waveform mappers, note the different rates that are indicated as we move along the processing blocks above. The very first module, the bit encoding block, generally features error detection and forward error correction processes that add redundant bits to yield a bit rate that is larger than that of the source bit stream. The symbol mapper then groups the incoming encoded bits and produces constellation values at the symbol rate that, in all but the BPSK case, is less than the encoded bit rate. The zero-stuffing element in the TX waveform mapping block takes the I and Q portions of the symbol and creates discrete impulses by inserting zero sample in between the symbols. The relationship between the sample and symbol rates is called oversampling ratio, or OSR, and is described below.

$$OSR = \frac{SampleRate}{SymbolRate} = \frac{T_{Symbol}}{T_S}$$

The following figure illustrates the zero-stuffing process as well as the rate change that occurs as we create a sample stream out of a symbols stream. (The variable T_S refers to the sample period).

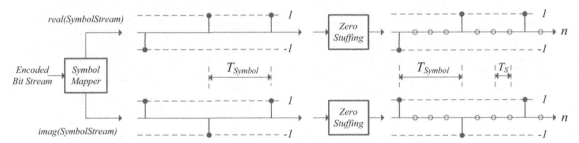

Figure 5-12: The Zero-Stuffing Process with an Oversampling Ratio of 4

The rate management in the digital portion of the transceiver can be tricky especially if modem must support many different symbol constellations.

5.2.1 Symbol Mapping (BPSK, QPSK, 8PSK and 16QAM)

Symbol mapping is one of the simplest functions to understand and implement in baseband or DSP processors. The next two figures illustrate just how groups of input bits map to each point in the constellation. Note that there is a pattern inherent in the assignment of bit groups to constellation positions. The bit group of any one position in a constellation differs in only one digit from the bit group of its neighbor. Examine the 8PSK constellation and follow the bit groups around the circle to convince yourself that this is indeed the case. This method of bit assignment, called gray coding, can be seen in all four constellations and ensures that bit errors induced by incorrect symbol detections are minimized. The constellations of each modulation format are scaled such that the variance (reference power) of their respective symbol streams remains equal. This is done so that the baseband processor can switch between modulation formats without an accompanying change in the RF output power. For a constant variance of 1, the QPSK constellation positions would be located at $\pm 0.707 \pm j0.707$, while those for 8PSK would include the QPSK positions as well as $1, j, -1$ and j.

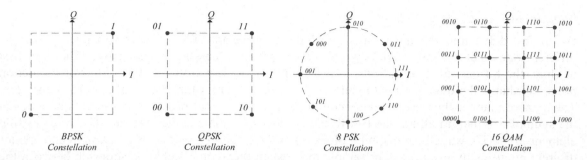

Figure 5-13: BPSK, QPSK, 8PSK, and 16QAM Mapping Diagrams

The tables below details the symbol mapping combinations for all four modulation formats such that the variance is always equal to unity.

BPSK Constellation		
Input Bits	I Value	Q Value
0	-0.707	-0.707
1	0.707	0.707

QPSK Constellation		
Input Bits	I Value	Q Value
00	-0.707	-0.707
01	-0.707	0.707
10	0.707	-0.707
11	0.707	0.707

8PSK Constellation		
Input Bits	I Value	Q Value
000	-0.707	0.707
001	-1	0
010	0	1
011	0.707	0.707
100	0	-1
101	-0.707	-0.707
110	0.707	-0.707
111	1	0

16QAM Constellation (C= 0.3063)		
Input Bits	I Value	Q Value
0000	$-3 \cdot C$	$-3 \cdot C$
0001	$-3 \cdot C$	$-C$
0010	$-3 \cdot C$	$3 \cdot C$
0011	$-3 \cdot C$	C
0100	$-C$	$-3 \cdot C$
0101	$-C$	$-C$
0110	$-C$	$3 \cdot C$
0111	$-C$	C
1000	$3 \cdot C$	$-3 \cdot C$
1001	$3 \cdot C$	$-C$
1010	$3 \cdot C$	$3 \cdot C$
1011	$3 \cdot C$	C
1100	C	$-3 \cdot C$
1101	C	$-C$
1110	C	$3 \cdot C$
1111	C	C

Figure 5-14: Symbol Mapping Tables for BPSK, QPSK, 8PSK, and 16QAM

While we won't go beyond 16 QAM in this chapter, more crowded constellations with up to 1024 positions (1024 QAM) are not uncommon in cable modems and microwave backhaul links. As the number of positions in the constellation increases, so does the information throughput and, unfortunately, the required signal to noise ratio at the receiver. In mobile communication systems, we don't have the luxury of high signal to noise ratios as we need to reliably detect distant transmit signals that have been compromised by noise and distortion.

```
function [OutputSymbols] = Symbol_Mapper(InputBits, ModulationFormat)
% Modulation format: 0, 1, 2, 3 --> BPSK, QPSK, 8PSK, 16QAM
persistent BPSK_LUT QPSK_LUT PSK8_LUT QAM16_LUT
if(isempty(BPSK_LUT))
    C         = 1/sqrt(2);
    BPSK_LUT  = 1/sqrt(2)* [-1 - 1j; 1 + 1j];
    QPSK_LUT  = 1/sqrt(2)* [-1 - 1j; -1 + 1j;  1 - 1j;  1 + 1j];
    PSK8_LUT  =            [-C + 1j*C; -1    ;    1j; C + 1j*C; ...
                           -1j        ; -C - 1j*C; C - 1j*C;      1];
    QAM16_LUT = 1/sqrt(10.66)*[-3 - 3j; -3 - 1j; -3 + 1j; -3 + 3j; ...
                              -1 - 3j; -1 - 1j; -1 + 1j; -1 + 3j; ...
                               1 - 3j;  1 - 1j;  1 + 1j;  1 + 3j; ...
                               3 - 3j;  3 - 1j;  3 + 1j;  3 + 3j];
end

NumberOfSymbols = floor(length(InputBits)/(ModulationFormat + 1));
OutputSymbols   = zeros(1,NumberOfSymbols);

for i = 1:NumberOfSymbols
    Start = 1 + (i - 1)*(ModulationFormat + 1);
    Stop  = Start + ModulationFormat;
    Bit_Group = InputBits(1,Start:Stop);
    switch(ModulationFormat)
        case 0
            Code  = Bit_Group(1,1) + 1; Symbol = BPSK_LUT(Code, 1);
        case 1
            Code  = Bit_Group(1,2)*2 + Bit_Group(1,1) + 1;
            Symbol = QPSK_LUT(Code, 1);
        case 2
            Code  = Bit_Group(1,3)*4 + Bit_Group(1,2)*2 ...
                  + Bit_Group(1,1) + 1;
            Symbol = PSK8_LUT(Code, 1);
        case 3
            Code  = Bit_Group(1,4)*8 + Bit_Group(1,3)*4 ...
                  + Bit_Group(1,2)*2 + Bit_Group(1,1) + 1;
            Symbol = QAM16_LUT(Code, 1);
    end
    OutputSymbols(1,i) = Symbol;
end
```

5.2.2   Symbol Mapping (π/4-DQPSK, D8PSK, π/8-DQPSK)

Symbol mapping using absolute value PSK requires that we demodulate the received signal coherently.

$$RF(t) = I(t)\cdot cos(2\pi f_0 t + \theta) + Q(t)\cdot sin(2\pi f_0 t + \theta)$$

Coherent demodulation entails that at the receiver we can generate local oscillator signals that are exactly equal to $cos(2\pi f_o t + \theta)$ and $sin(2\pi f_o t + \theta)$, which, when multiplied by the RF signal and after low pass filtering step, will reveal the original $I(t)$ and $Q(t)$ waveforms. Review section 1.1.8 for the mathematics behind coherent demodulation. Because the synthesizers that generate the LO signals at the transmitter and receiver use different reference oscillators, their frequencies and phases can't possibly be equal. We thus require signal processing algorithms that detect what the remaining frequency and phase offsets are and then remove them. The process isn't always straight forward and engineers looked for techniques that don't require coherency at the receiver. The result is differential encoding, in which the symbol mapper converts groups of bits into phase shifts rather than absolute constellation positions. In this section we will take a closer look at $\pi/4$-DQPSK and D8PSK as they are used in the enhanced rate modes of Bluetooth and $\pi/8$-D8PSK as used in terrestrial trunked radio, or TETRA.

$$S_k = S_{k-1} \cdot e^{j\theta_k} \qquad k = 1, 2, 3, \ldots$$

For differential encoding to work properly, the symbol mapper produces an initial point, S_0, which we arbitrarily place at one of the constellation positions (for example $S_0 = 1+j0$). Each successive position S_k ($k = 1, 2, 3, \ldots$) is reached via a rotation from the previous position S_{k-1}. In $\pi/4$-DQPSK, D8PSK, and $\pi/8$-D8PSK groups of 2 and 3 bits are mapped into phase shifts as indicated by the tables below.

$\pi/4$-DQPSK	
Input Bits	Phase Shift θ_k
00	$\pi/4$
01	$3\pi/4$
10	$-\pi/4$
11	$-3\pi/4$

D8PSK	
Input Bits	Phase Shift θ_k
000	0
001	$\pi/4$
010	$3\pi/4$
011	$\pi/2$
100	$-\pi/4$
101	$-\pi/2$
110	π
111	$-3\pi/4$

$\pi/8$-D8PSK	
Input Bits	Phase Shift θ_k
000	$\pi/8$
001	$3\pi/8$
010	$-\pi/8$
011	$-3\pi/8$
100	$7\pi/8$
101	$5\pi/8$
110	$-7\pi/8$
111	$-5\pi/8$

Figure 5-15: Symbol Mapping Tables for $\pi/4$-DQPSK, D8PSK, $\pi/8$-D8PSK

This manner of linking transmit symbols to phase shifts completely eliminates the need to acquire the phase offset of the received signal. However, unresolved frequency offset can still affect the performance of the demodulator given that it introduces an addition shift of $2\pi \cdot f_{offset} \cdot T_{Symbol}$ over a single symbol period. In the overall design, we must therefore guarantee that the transmit and receive LO frequencies are close enough such that this latent frequency offset causes a phase shift that is small compared to that of any symbol, lest we risk incurring unnecessary errors in the final symbol decision process of the receiver.

The following figure shows the constellations positions for π/4-DQPSK, D8PSK, and π/8-D8PSK as well as their idealized signal trajectories that connect them. We use the word 'idealized' because in real situations the pulse shape employed will give the signal trajectories a much smoother look as we will illustrate in the next section discussing the TX waveform mapper. For now, let's examine the π/4-DQPSK constellation consisting of 8 possible positions which are divided up into two sets of 4 points shown as black and white circles. The set of constellation positions indicated by the white circles is phase shifted by π/4 radians compared to the one represented by the black circles. Successive symbols will force the mapper to bounce back and forth between these sets of constellation positions causing the trajectories to avoid the center of the IQ plane. This scheme is specifically used to decrease the amount of amplitude variation in the signal and therefore mitigate intermodulation distortion occurring in the RF power amplifier. The D8PSK scheme shown next does not undertake that same effort and resulting signal trajectories pass through the IQ original yielding a larger amount of amplitude modulation. This issue is once again addressed in π/8-D8PSK which creates two sets of constellation points that differ by π/8 radians and results in trajectories that avoid the IQ original. The π/8-D8PSK symbol mapping technique is used in TETRA, as mentioned earlier, and a variant of it is employed in EDGE, or Enhanced Data rates for GSM Evolution.

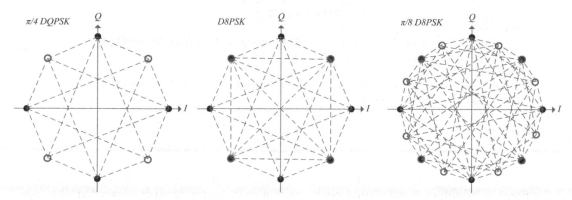

Figure 5-16: Three Popular Differential Phase Shift Keying Schemes with Ideal Trajectories

In the mapper function below, we start by defining the rotations that must be undertaken for each group of bits that is processed. The values are defined in phase tables one, two and three.

```
function OutputPositions = Differential_Symbol_Mapper(InputBits, Format)
% (modulation) Format: 0, 1, 2 --> pi_4_DQPSK, D8PSK, pi_8_D8PSK

persistent pi_4_DQPSK_LUT D8PSK_LUT pi_8_D8PSK_LUT
if(isempty(pi_4_DQPSK_LUT))
    Phase_Table1    = [pi/4; 3*pi/4; -pi/4; -3*pi/4];
    Phase_Table2    = [0;   pi;  3*pi; 2*pi; -pi; -2*pi;  4*pi; -3*pi]/4;
    Phase_Table3    = [pi; 3*pi; -pi; -3*pi; 7*pi; 5*pi; -7*pi; -5*pi]/8;

    pi_4_DQPSK_LUT  = cos(Phase_Table1) + 1j*sin(Phase_Table1);
    D8PSK_LUT       = cos(Phase_Table2) + 1j*sin(Phase_Table2);
    pi_8_D8PSK_LUT  = cos(Phase_Table3) + 1j*sin(Phase_Table3);
end % Definition of Lookup Tables (LUT)
```

Based on the phase tables, we then compute three lookup tables (LUT) that contain the complex scalars that, when multiplied by the previous position, rotate us the current constellation position.

```
BitsPerSymbol = 0;
switch(Format)
    case 0; BitsPerSymbol = 2;
    case 1; BitsPerSymbol = 3;
    case 2; BitsPerSymbol = 3;
end

NumberOfSymbols = fix( length(InputBits)/BitsPerSymbol );
OutputPositions = [1 zeros(1,NumberOfSymbols)];

for i = 2:NumberOfSymbols+1
    Start = 1 + (i - 2)*(BitsPerSymbol);
    Stop  = Start + BitsPerSymbol - 1;
    Bit_Group = InputBits(1,Start:Stop);
    switch(Format)
        case 0
            Code     = Bit_Group(1,1)*2 + Bit_Group(1,2);
            Rotation = pi_4_DQPSK_LUT(Code+1, 1);
        case 1
            Code     = Bit_Group(1,1)*4 + Bit_Group(1,2)*2+Bit_Group(1,3);
            Rotation = D8PSK_LUT(Code+1, 1);
        case 2
            Code     = Bit_Group(1,1)*4 + Bit_Group(1,2)*2+Bit_Group(1,3);
            Rotation = pi_8_D8PSK_LUT(Code+1, 1);
    end
    OutputPositions(1,i) = OutputPositions(1,i-1)*Rotation;
end
```

After the initialization phase, the bits per symbol (determined based on the *Format* parameter) and the number of output constellation positions that must be rendered is computed. Because we must start at a known constellation position in order to render the first symbol as a rotation, the number of output positions will be one greater than the number of incoming symbols. In the loop itself, the incoming bit stream is chopped into groups of bits and converted into a code which serves as an address into the lookup tables. The correct complex scalar is extracted and multiplied by the previous output constellation position to render the new one.

The Disadvantage of Differential PSK

The main disadvantage of differentially detected PSK is that mistakes often occur in pairs as compared to coherently detected PSK, which under the same circumstances will err only once. Take a look at the following figure. In this scenario, a transmit 8PSK signal was synthesized to traverse the IQ plane from *P0* to *P1*, *P2*, and *P3*. Because of noise in the system, the received signal was actually sampled by the demodulator at *P1'*, *P2'*, and *P3'*. At the transmitter we could have used 8PSK to transmit positions P1 ($= 0.707 + 0.707j$), P2 ($= -0.707 + 0.707j$), and finally P3 ($= -1$). Likewise, differential 8PSK could have been used with phase shifts of $\pi/4$, $\pi/2$, and $\pi/4$. Given the decision boundaries for coherently detected 8PSK (the dashed lines), it is clear that P2' will be interpreted incorrectly as $0+j$ rather than P2 ($= -0.707 + 0.707j$). A single mistake has been made.

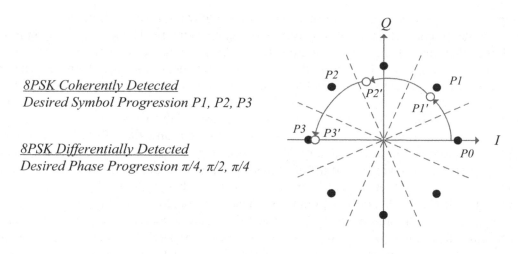

Figure 5-17: Error Progression in D8PSK and 8PSK

For the differentially encoded waveform to be received without error, the detected phase shifts have to be between π/4±π/8, π/2±π/8, and π/4±π/8. Note that an allowed phase shift can be off by as much as π/8 and still be detected properly. If you look closely, the first rotation is dead on at π/4, but the next rotation is less than the required π/2-π/8 and the final phase shift more than π/4+π/8. Two mistakes are made, rather than one. The increase in bit error rate of D8PSK is in many cases well worth the accompanying simplification at the receiver, which no longer must ascertain the frequency and phase offset in the down converted receiver signal.

5.2.3   TX Waveform Mapping in Single Tone PSK and QAM Transmitter

The constellation positions produced by the symbol mapper must somehow be mapped into a continuous waveform that can hopefully be transmitted over the communication channel in a power and spectrally efficient manner.

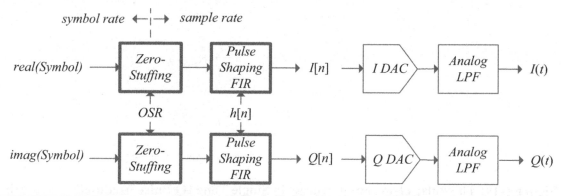

Figure 5-18: The TX Waveform Mapper with IQ Digital to Analog Converters and Analog Filters

Because in communication systems we must tightly control the spectrum of our transmit signal, we can't just take the symbol mapper outputs and directly feed them to the DAC modules. We must shape the symbols properly which requires an upsampling step that is realized by the zero-stuffing module. Review figure 12 of this chapter to see how the zero-stuffing process inserts zero valued samples to convert the symbol stream into a sample stream. The rate of the sample stream, F_S, is larger than that of the symbol stream, F_{Symbol}, by a factor we refer to as the oversampling ratio, OSR. The sample stream exiting the zero-stuffing element is a time staggered sequence of scaled impulses, whose values are equal to those of the symbols. In the figure below, we illustrate how a single impulse and a sample stream of staggered scaled impulses is processed through the pulse shaping FIR filter. Note how the impulse response of the filter passes through zero at a distance of $\pm T_{Symbol}$ from its peak value. This method of crafting the filter's impulse response, $h[n]$, allows us to ping the filter with a scaled impulse every T_{Symbol} seconds and produce an output signal, $Output[n]$, that retains the peaks of each shaped pulse. In the figure below these peaks that are preserved in the output sequence are indicated using black stems. In the analog transmit hardware, the discrete $I[n]$ and $Q[n]$ output sequences become continuous time waveforms that are transmitter over the air to finally arrive at the receive antenna. The digital portion of the receiver must determine just where these nicely preserved peaks appear, sample the receive waveform at those instances and then pass those values on to the decision element (symbol demapper). PSK and QAM transmitters use raised cosine pulse shapes to guarantee that neighboring pulses don't interfere with the peak of the current pulse. For this scenario, the transmit waveform is said to have zero ISI, or zero intersymbol interference. We will take a detailed look at the raised cosine family of pulse shapes shortly.

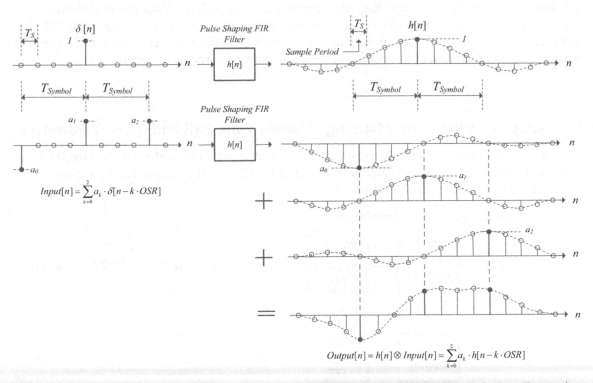

Figure 5-19: The Pulse Generation Process in Single Tone IQ Pulse Modulation Featuring Zero Intersymbol Interference

Meanwhile, it is highly informative to show how the transmit information progresses through the transmitter in both the time and frequency domains. In the figure below, we begin tracking the I component of the information at the point where it initially appears: at the output of the symbol mapper. If the produced symbols move randomly throughout their allowed constellation positions, then the magnitude spectrum of the symbol stream is a constant within the Nyquist bandwidth which covers the range of $\pm F_{Symbol}/2$. Inserting zero samples in-between changes nothing about the spectrum's shape. Zero-stuffing has simply increased the Nyquist bandwidth (now $\pm F_S/2$) by the oversampling ratio and given us an opportunity to shape the spectrum using a pulse shaping FIR filter. This expansion of the Nyquist bandwidth, while not to scale, can be seen in the two spectra at the top right of the figure. We then move on to see the effects of the pulse shaping filter, the sample and hold mechanism of the DAC module as well as the rendering of the final analog waveform by the low pass (analog reconstruction) filter.

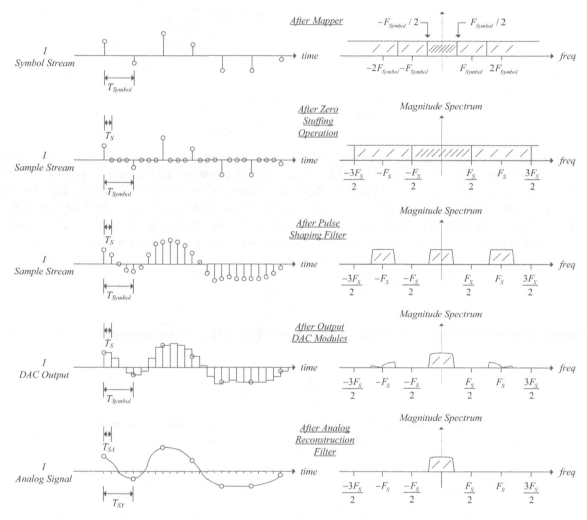

Figure 5-20: Progression of Information through the Transmitter in the Time and Frequency Domains

5.2.4   The Family of Raised Cosine (RC) Pulse Shapes

The family of raised cosine pulse shapes has enjoyed great popularity in RF communication systems that use phase shift keying (PSK) and quadrature amplitude modulation (QAM). This popularity stems from its very desirable property of passing through zero at a distance of $\pm T_{Symbol}$ from the peak of the pulse. Therefore, if the receiver properly samples the received signal at the center of each pulse, the samples will not include any interference from neighboring symbols. These *RC* pulses feature what we call zero *ISI*, or zero inter-symbol interference. The formula for the raised cosine pulse is shown below.

$$P_{RC}(t) = sinc(t/T_{Symbol}) \cdot \frac{\cos(\alpha \pi t / T_{Symbol})}{1 - (2\alpha t / T_{Symbol})^2}$$

Where

$$sinc(t/T_{Symbol}) = \frac{\sin(\pi t / T_{Symbol})}{\pi t / T_{Symbol}}$$

Notice that the pulse shape may be modified by changing the symbol period, T_{Symbol}, as well as a factor alpha, or simply α ($0 \leq \alpha \leq 1$), which determines the amount of frequency spectrum taken up by the pulse. An alpha of 0.0 takes up the least amount of bandwidth but causes the pulse to ring the longest, whereas an alpha of 1.0 takes up the most amount of bandwidth and causes the pulse to decay quickly. The figure below shows a time domain plot of three raised cosine pulses, using alpha values of 0, 0.5, and 1.0, truncated to plus and minus 3 symbol periods.

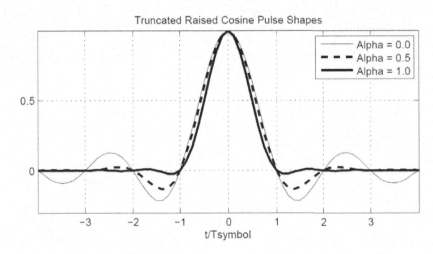

Figure 5-21: Raised Cosine Pulses for α of 0, 0.5 and 1.0 Shown over 8 Symbol Periods

To find the coefficients of the FIR filter that will generate the raised cosine pulses, we must know the oversampling ratio, OSR, as well as the available number of taps, *N*. Once established, we may calculate a vector, *t*, representing the instances in time, at which to evaluate the raised cosine pulse equation. Note how the variable *t* in the RC equation is always divided by T_{Symbol}, and in fact, the independent variable in the figure above is t/T_{Symbol}. It is for this reason that independently of what the value of T_{Symbol} actually is, we may set it to 1.0 and use the following expressions to compute the vector *t*. (*Step* = 1/*OSR*).

$$t = -floor(\frac{N}{2}) \cdot Step : Step : floor(\frac{N}{2}) \cdot Step \qquad \textit{for odd N}$$

$$t = -(\frac{N}{2}+0.5) \cdot Step : Step : (\frac{N}{2}-0.5) \cdot Step \qquad \textit{for even N}$$

```
function [h, t] = GetRaisedCosine(alpha, N, Oversampling, Root)

Tsymbol    = 1;                   %% Default = 1 second
Step       = 1/Oversampling;

if(mod(N,2) == 1)
    t = -floor(N/2)*Step : Step : floor(N/2)*Step; %% for odd N
else
    t = (-N/2 +.5)*Step : Step : (N/2 -.5)*Step;   %% for even N
end
t = t + 1e-8;   % we add this small offset to avoid divide by zero errors

%% Evaluate Raised Cosine Pulse Shape at t
h_RC = sinc(t/Tsymbol) .* ( cos(alpha*pi*t/Tsymbol)./(1 - (2*alpha*t/Tsymbol).^2));

%% Evaluate Root Raised Cosine Pulse Shape at t
RRC  = (4*alpha ./ (pi*sqrt(Tsymbol)*(1-(4*alpha*t/Tsymbol).^2)) ) .*  ...
       ( cos((1+alpha)*pi*t/Tsymbol) + (Tsymbol./(4*alpha*t)).*sin((1- ...
       alpha)*pi*t/Tsymbol) );
h_RRC  =  RRC/max(RRC);


if(Root == 1); h = h_RRC;    %% We want the Root Raised Cosine (RRC) shape
else           h = h_RC;     %% We want the Raised Cosine (RC) shape
end
```

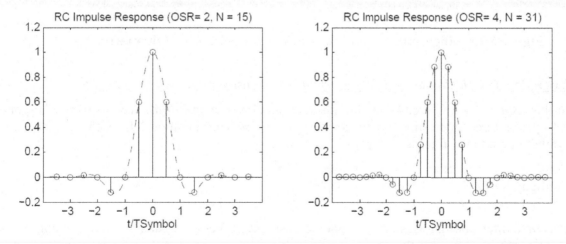

Figure 5-22: Alpha = 0.5 Raised Cosine Impulse Response for (OSR=2, N=15) and (OSR=4, N=31)

Understanding the Magnitude Response of the Raised Cosine (RC) Filter

From the figure below, which shows the magnitude responses for various filter configurations, we can immediately see that the double-sided bandwidth of the resulting pulse shaped transmit waveform will vary between F_{Symbol} and $2 \cdot F_{Symbol}$ Hz depending on the value of alpha. Note that the magnitude response for the alpha = 0 case exhibits quite a bit of ripple in both pass and stop bands. This ripple is due to the fact that the displayed magnitude responses were calculated for filters that are limited to 31 taps, and since the impulse response for the alpha = 0 case takes much more time to decay towards zero than the other cases (alpha=0.5 and 1.0), the impact of the truncation operation becomes noticeable. Given access to a longer filter, the magnitude response would look more like a brick wall. To remedy this problem, we overlay a Hanning window on the impulse response and thus gradually force it to zero at the beginning and end. The effect of the Hanning window on the magnitude response of the raised cosine filter using alpha = 0 is quite beneficial to which the right most plot attests. The ripple has disappeared and the overall shape of the magnitude response has not changed appreciably. Note further, that the Hanning overlay has not changed the zero ISI property of the raised cosine shape.

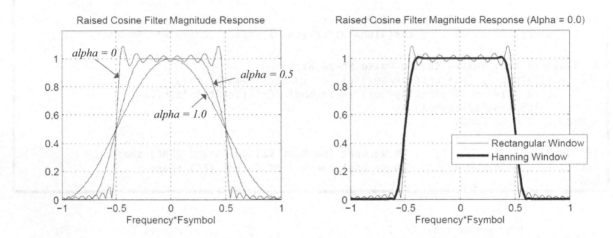

Figure 5-23: Magnitude Spectra for Various Raised Cosine Filters with N = 31 Taps

Example 5.1: *Calculating the Discrete Impulse Response for a Raised Cosine Filter*

In this example we will calculate the discrete impulse response, $h[n]$, and magnitude response, $|H[m]|$, for three interesting raised cosine pulse shape configurations. In each case, the input symbol rate into the filter is 1MSymbol per second.

Configuration 1 → N = 9, OSR = 2, Alpha = 0.5
Configuration 2 → N = 19, OSR = 4, Alpha = 0.5
Configuration 2 → N = 19, OSR = 4, Alpha = 0.5, Hanning Overlay

The first filter configuration inserts one zero-sample in between each symbol yielding a final sample rate of F_S = 2MSPS. From sampling theory, we know that the aliased images reside at $n \cdot F_S$ Hz where n is an integer value. This fact is clearly illustrated in the next figure by the sampling images at 2 and 4 MHz. Whereas we only show the positive frequencies in the magnitude response, the fundamental lobe does feature useful content between approximately -600KHz to 600KHz giving the resulting output signal a double-sided bandwidth of around 1.2MHz. In a real transmitter, the output waveform would now pass through the digital to analog converters followed by low pass filters that remove the sampling images. An analog low pass filter that passes frequencies up to 700KHz and thoroughly rejects them at 1300KHz, as would be required for the configuration 1 signal, is very complex. By raising the oversampling ratio to 4, we increase the sample rate and the center frequency of the first alias image to 4MHz as can be seen in configuration 2 of the figure below. The analog low pass filter requirements are alleviated at the expense of increased complexity ($N = 19$, and $F_S = 4$MHz) in the digital portion. Given today's deep submicron CMOS technologies, it is far better to absorb complex features into the baseband processors rather than tasking the analog hardware with unrealistic challenges. The power consumption of modern digital to analog converters barely rises with increasing sample rate, and the additional digital hardware due to the longer FIR pulse shaping filters is negligible.

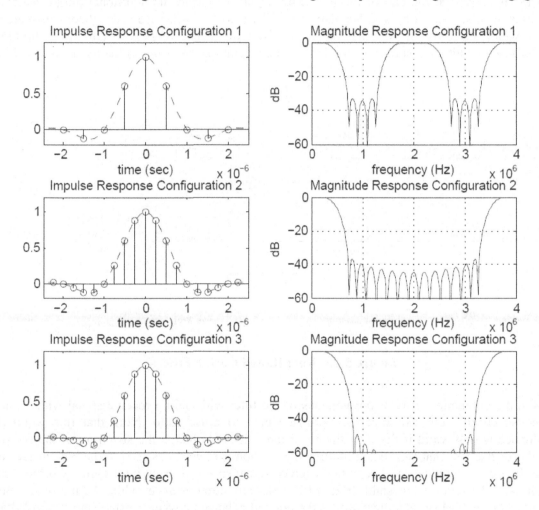

Figure 5-24: Impulse and Magnitude Responses of Raised Cosine Filter Configurations 1, 2, and 3

While the increased filter taps and sampling frequency have allowed us to increase the distance to the first alias image while maintaining the same passband behavior, the stop band performance of the configuration 2 is a little underwhelming. The ripple in the stop band is once again due to the truncation of the impulse response, which we improve in configuration 3 by overlaying a Hanning window.

The Root Raised Cosine (RRC) Pulse Shape

There is no doubt that a signal with zero intersymbol interference whose pulse peaks are properly sampled at the receiver will be optimally processed and decoded. However, there is no need to create the zero ISI condition in the transmit signal, as long as it is met at the receiver. As a matter of fact, we may create a new pulse shape, whose magnitude spectrum is the square root of the raised cosine pulse's magnitude spectrum. We can then place a filter that features said magnitude response at both the transmitter and receiver. In the figure below, we first show the raised cosine pulse being implemented at the transmitter only, while the second portion shows a root raised cosine filter at both ends of the link. Given that, in either case, a raised cosine pulse shape with zero ISI is created at the end of the link, why would we spend the additional digital hardware resources to place two filters rather than one? The answer comes from the theory of matched filters, which states that for optimal signal to noise ratio, a receive signal should pass through a filter whose magnitude response is the same as the magnitude spectrum of the received signal.

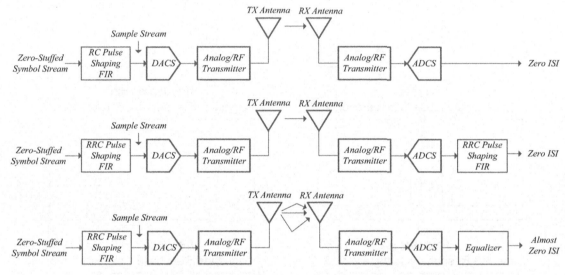

Figure 5-25: Root Raised Cosine Filters

Thus, the combination of two root versions of the filter will yield a receive signal with a raised cosine spectrum, zero ISI, and the optimal signal to noise ratio. Note that this wonderful confluence is only valid if the over the air channel doesn't change the spectrum of the transmit signal. For the case that significant multipath distortion is at work, as seen in the bottom scenario of the figure, an equalizer will bend the received signal spectrum back into shape. A zero-forcing equalizer will attempt to reestablish zero ISI under any circumstance, while the minimum mean square error, or MMSE, equalizer strikes the optimal balance between reestablishing zero ISI and best signal to noise ratio. The formula for the root raised cosine pulse shape is shown below.

$$P_{RRC}(t) = \frac{2\alpha}{\pi\sqrt{T_{Symbol}}} \cdot \frac{\cos(\frac{(1+\alpha)\pi t}{T_{Symbol}}) + \frac{T_{Symbol}}{4\alpha t} \cdot \sin(\frac{(1-\alpha)\pi t}{T_{Symbol}})}{1-(4\alpha t / T_{Symbol})^2}$$

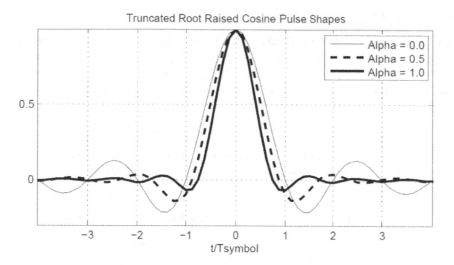

Figure 5-26: Root Raised Cosine Pulses for α of 0, 0.5 and 1.0 Shown over 8 Symbol Periods

5.2.5 TX Waveform Mapping Implemented in MatLab

The following function call implements the TX Waveform Mapping process including the zero-stuffing operation, the creation of the RC coefficients, as well as the pulse shaping operation implemented by FIR filtering the zero-stuffed symbol stream.

```matlab
function [SampleStream, Zero_Stuffed, h] = TX_Waveform_Mapping( ...
        SymbolsStream, N, OSR, Alpha, Root, Hann_Flag, Hann_Index)
% N          - Number of taps in pulse shaping filter
% OSR        - Oversampling ratio --> Best if OSR >= 4
% Alpha      - Roll-off factor --> 0 (smallest BW)  to 1 (largest BW)
% Root       - Boolean value 0/1 - RC / RRC
% Hann_Flag  - Boolean value 0/1 - Skip / Use Hanning Window
% Hann_Index - The index modifies the Hanning Window
NumberOfSymbols                 =     length(SymbolsStream);

%% 1. The Zero Stuffing Process (there is nothing to it)
Zero_Stuffed                    = zeros(1, NumberOfSymbols*OSR);
for i = 1:NumberOfSymbols
    Zero_Stuffed(1, 1 + (i*OSR)) = SymbolsStream(1,i);
end


%% 2. Build Hanning Window
n        = 0:N-1;
HannWindow = (0.50-0.50*cos(2*pi*(n+1)/(N+1)) ).^Hann_Index ;
%% 3. Synthesize the proper RC or RRC Pulse Shape
[h, t] = GetRaisedCosine(Alpha, N, OSR, Root);
if(Hann_Flag); h = h.*HannWindow; end;


%% 4. Pulse Shaping via RC filtering of Zero-stuffed Symbols
SampleStream = filter(h, 1, Zero_Stuffed);
```

Example 5.2: *Create QPSK Sample Streams for RC Alpha = 0.8, N=25, and OSR = 4*

In this example, we write a simple MatLab script that creates the sample stream for all seven methods of modulation that we have covered in this section. The different modulation methods are accessed by selecting the *Differential_Flag* and *ModulationFormat* variable appropriately.

```
%% Configuring the Symbol Mapping Process
InputBits   = round(rand(1,2000));
Differential_Flag = 0;  % 0/1 - Absolute Value/Differential Mode
ModulationFormat  = 1;
    % Differential Mode:  0/1/2   - pi/4-QPSK / D8PSK / pi/8 - D8PSK
    % Absolute Value Mode: 0/1/2/3 - BPSK / QPSK / 8PSK / 16QAM


if(Differential_Flag == 1)
 [SymbolStream] = Differential_Symbol_Mapper(InputBits, ModulationFormat);
else
 [SymbolStream] = Symbol_Mapper(InputBits, ModulationFormat);
end


%% Configuring the TX Waveform Mapping Process
N          = 25;
OSR        = 4;
Alpha      = 0.8;     %% Roll-off Factor
Root       = 0;       %% Boolean value 0/1 - RC / RRC
Hann_Flag  = 0;       %% Boolean value 0/1 - Skip / Use Hanning Window
Hann_Index = 1;       %% The index modifies the Hanning Window


[SampleStream, Zero_Stuffed, h] = TX_Waveform_Mapping(SymbolStream, N,...
                   OSR, Alpha, Root, Hann_Flag, Hann_Index);
```

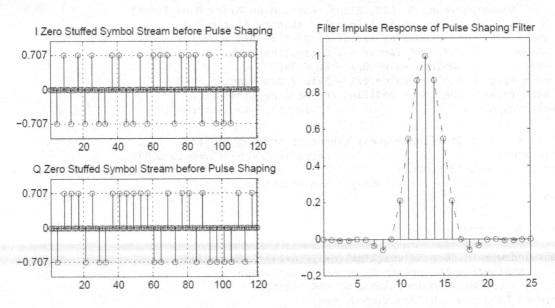

Figure 5-27: Zero-Stuffing Symbol Stream (QPSK) and RC Filter Impulse Response for N=25, OSR =4, and Alpha = 0.8

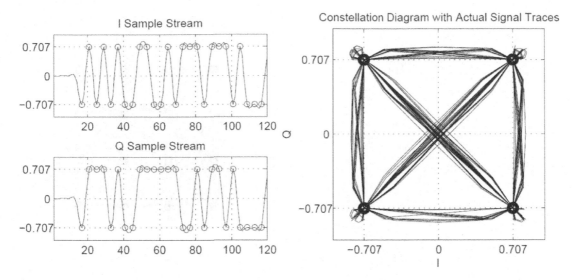

Figure 5-28: Sample Stream (QPSK) and Constellation Diagram for N=25, OSR=4, and RC Alpha=0.8

These two figures nicely illustrate the difference between the impact of the raised cosine (RC) and root raised cosine (RRC) pulse shapes on the final sample stream. The zero-stuffed symbol stream contains scaled impulses with value of ±0.707, which are accurately represented in the sample stream of the figure above using the raised cosine shape. Intersymbol interference, or ISI, is plainly visible in both the sample stream and constellation diagram of the signal using the root raised cosine to shape the transmitted pulses (see figure below).

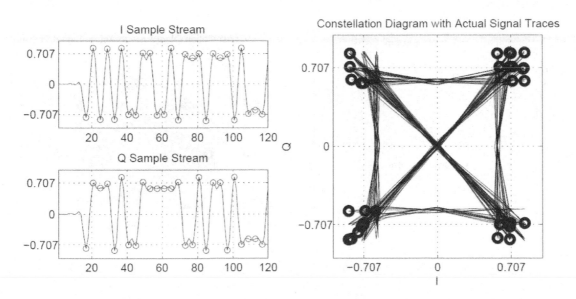

Figure 5-29: Sample Stream (QPSK) and Constellation Diagram for N=25, OSR=4, and RRC Alpha=0.8

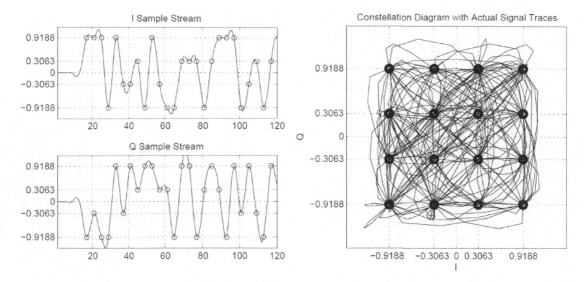

Figure 5-30: Sample Stream (16QAM) and Constellation Diagram for N=25, OSR=4, and Alpha=0.5

Notice that the constellation diagram of the QPSK signal (RC alpha = 0.8) features very little overshoot after the trajectories pass through the valid constellation positions. As we decrease the roll-off factor alpha, as is the case in these last two figures, the trajectories between constellation positions are less tightly grouped. Notice the manner in which the π/4-DQSK mapping method exposes an opening at the origin. As we have mentioned earlier, this is done in order to minimize the amount of AM modulation and subsequent intermodulation distortion produced by the RF power amplifier.

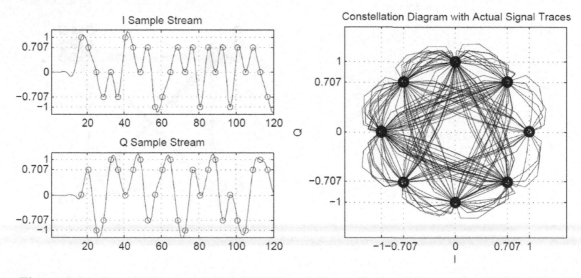

Figure 5-31: Sample Stream (π/4-DQPSK) and Constellation Diagram for N=25, OSR=4, and Alpha=0.5

5.3 Single Tone Continuous Phase Modulation

Up to now we have represented data symbols as shaped pulses which we then embedded in the I and Q transmit sample streams. These symbols are formed by passing impulses through FIR filters that reside in the I and Q paths, and the resulting complex signal features a spectrum defined by the frequency response of those FIR filters. The trajectories of these complex signals move freely throughout the IQ plane and both their amplitude and phase change with time. In wireless communication systems, where power consumption rather than data throughput is the main concern, amplitude modulation can be a real problem. An RF signal whose amplitude changes with time will cause harmonic and intermodulation distortion as it passes through non-linear processing blocks like the RF power amplifier. The harmonic distortion components are trivially removed using a passive filter ahead of the antenna. The intermodulation distortion, however, appears in the immediate vicinity of our transmit signal and will interfere with transmissions on adjacent channels. To avoid this, designers use linear RF power amplifiers, which are very power inefficient and increase the required battery size for a mobile station (handset). In many cost or weight sensitive applications, this penalty is a show stopper. Using constant envelope (amplitude) signaling avoids the problem of intermodulation distortion thereby permitting the use of highly efficient non-linear RF power amplifiers. Of course, prohibiting the amplitude from carrying information restricts the comparative data throughput of the system. This, however, is a trade-off that many communication systems such as Bluetooth, DECT, and GSM are more than happy to make. Let's take a look at a general expression representing a phase modulated signal with constant magnitude, A.

$$RF(t) = A \cdot cos(2\pi f_0 t + \theta(f_d, t))$$

The modulation formats that we will be discussing in this section exercise direct control over the frequency deviation, f_d, away from the carrier frequency, f_o, to affect the phase. Remember that in continuous time, the instantaneous phase, $\theta(f_d, t)$, is equal to the integral of the frequency.

$$\theta(f_d, t) = 2\pi \int_0^t f_d(x)dx$$

In the simplified case where f_d is a constant, the expression reduces as follows.

$$\theta(f_d, t) = 2\pi f_d t$$

5.3.1 Frequency Shift Keying (FSK)

Of course, maintaining the frequency deviation, f_d, constant over a long period of time defeats the purpose of transmitting information across the communication. Therefore, in our first modulation scheme called frequency shift keying, or FSK, we apply different frequencies over each successive symbol period, T_{Symbol}. FSK modulation is shown below.

$$RF(t) = A \cdot cos(2\pi f_o t + 2\pi \int_0^t f_d(x)dx)$$

$$f_d(t) = \sum_{n=0}^{N-1} a_n \cdot FreqPulse(t - nT_{Symbol})$$

The figure below sheds some light as to how the frequency pulse helps form the frequency deviation function, $f_d(t)$, which in this case results in the following expression.

$$f_d(t) = 3.75e6 \cdot Rect(t) - 3.75e6 \cdot Rect(t - T_{Symbol}) - 1.25e6 \cdot Rect(t - 2T_{Symbol}) + 1.25e6 \cdot Rect(t - 3T_{Symbol})$$

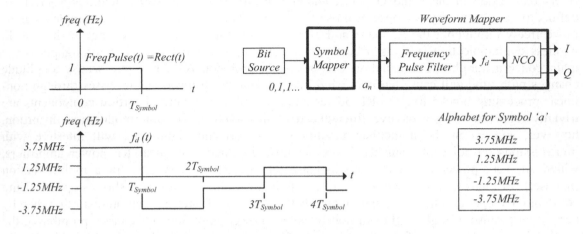

Figure 5-32: Generating the Frequency Deviation Waveform, $f_d(t)$

In the diagram above, we have chosen to employ FSK featuring a symbol set that includes 3.75MHz, 1.25MHz, -1.25MHz, and -3.75MHz. The bit source produces a stream of random ones and zeros which are then organized into groups of two bits and converted by the symbol mapper into one of these four symbols. The symbols emerge from the mapper as scaled impulses every T_{Symbol} seconds and travel through the frequency pulse shaping filter to produce $f_d(t)$. The pulse shaping filter is a rectangular function that equals 1 between [0, T_{Symbol}] and 0 otherwise.

```
Fs    = 2e8;          %% Sample Rate
Ts    = 1/Fs;         %% Sample Period
Fo    = 6.25e6;       %% Carrier Frequency = 6.25MHz
OSR   = 100;          %% Oversampling Ratio = number of samples per symbol
                      %% Tsymbol = 100*Ts


fd    = [3.75e6*(ones(1,OSR)) -3.75e6 *(ones(1,OSR)) ...
         -1.25e6*(ones(1,OSR))  1.25e6*(ones(1,OSR))];


Theta          = zeros(1,4*OSR);              %% Theta[fd,n]
for n = 2:(4*OSR)
    Theta(1,n)  = Theta(1,n-1) + 2*pi*fd(n)*Ts;  %% Accumulation
end
n          = 0:4*OSR-1;
RF_Signal = cos(2*pi*Fo*n*Ts + Theta);
```

The first part of the code starts out by defining the sample rate of our simulation followed by the carrier frequency of 6.25MHz. The oversampling ratio, OSR, of 100 will set the symbol period, T_{Symbol}, to $100 \cdot T_S = 0.5$ microseconds. Rather than writing out the code of the actual convolution, which we will do in the section on GFSK, we simply cascade row vectors that are scaled by the symbols to form $f_d(t)$.

Before we delve any further into the discrete implementation represented by the MatLab code, take a look at the figure below and notice the phase continuity at the symbol boundaries which are spaced ½ microseconds from one another. As a matter of fact, the way we define FSK via our set of equations and the implemented architecture seen on the next page, guarantees continuous phase throughout the signal progression. The form of FSK we discuss in this chapter is therefore called continuous phase frequency shift keying, or CP-FSK. Older implementations of FSK, which did not guarantee phase continuity, are hardly in use today and hence nowadays the terms FSK and CP-FSK are often used interchangeably. Whereas the phase may indeed by continuous, the frequency deviation waveform, $f_d(t)$, is not. Any type of discontinuity, or rapid change, whether in phase or frequency will cause the bandwidth of our transmit signal to unnecessarily expand, which is undesirable in today crowded radio spectrum.

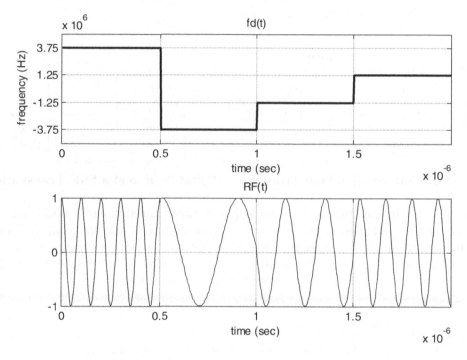

Figure 5-33: The Frequency Deviation Waveform and the Corresponding RF Output Signal

The final lines of MatLab code compute the discrete equivalent (see below) of the continuous time expressions of the last two pages. Generally speaking, the variable t is replaced by $n \cdot T_S$ ($T_S = 1/F_S$) and the summation operation used in place of the integration.

$$RF[n] = A \cdot cos(2\pi f_o n T_s + \theta[f_d, n])$$

$$\theta[f_d, n] = 2\pi \sum_{x=0}^{n} f_d[x] \cdot T_s$$

The FSK Transmitter in Detail

We have already indicated in figure 32 that the frequency deviation function is the result of the convolution of the symbol stream and the frequency pulse shaping filter.

$$f_d[n] = \sum_{k=0}^{K-1} SampleStream[k] \cdot FreqPulse[n-k]$$

However, the symbol stream must first be converted into a sample stream using zero-stuffing before they may be filtered. The figures below provide a more detailed information flow through the digital portion of a FSK IQ based transmitter.

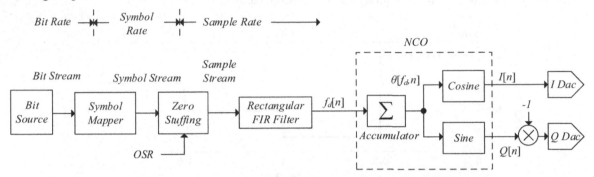

Figure 5-34: Information Flow through the Digital Portion of a FSK Transmitter

Note the changes the information is experiencing as it travels through each processing block. In the previous code we used an alphabet of 4 symbols which requires that we group 2 bits at a time to address the alphabet. The symbols thus toggle at half the bit rate and are converted into the sample stream by inserting a certain number, OSR-1, of zero samples in between. It is the oversampling ratio, OSR, which determine the final sample rate of the system.

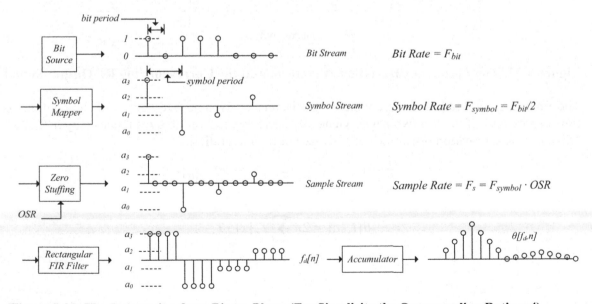

Figure 5-35: The Progression from Bits to Phase (For Simplicity the Oversampling Ratio = 4)

The Numerically Controlled Oscillator

The frequency deviation waveform, $f_d[n]$, must now be translated into a complex sinusoid using a numerically controlled oscillator, NCO, as shown in figure 34. The equation and figure below illustrate the process that computes the phase and final sinusoidal outputs. The cosine and sine functions are generally implemented using lookup tables or small ROM modules while the phase is translated into the appropriate table address. (Note that the modulo step restricts the applied phase to a dynamic range of 0 to 2π.)

$$\theta[f_d, n] = 2\pi T_{Sample} \sum_{x=0}^{n} f_d[x]$$

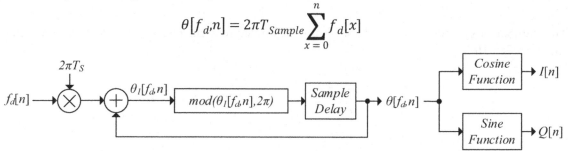

Figure 5-36: Block Diagram of Numerically Controlled Oscillator

Upconversion to Radio Frequencies

The RF signal, $RF(t) = A(t)\cdot\cos(2\pi f_o t + \theta(t))$, is said to have a complex envelope expressed by the amplitude, $A(t)$, and phase, $\theta(t)$. The baseband representation of that complex envelope is the familiar I and Q components produced by the NCO. The way $I(t)$ and $Q(t)$ map into the magnitude, $A(t)$, and phase, $\theta(t)$, of the carrier signal is governed by the angle sums trigonometric identity which we met in chapter one. The simplified block diagram below shows the implementation of the upconversion process.

$$RF(t) = A(t)\cdot\cos(2\pi f_o t + \theta(t)) = I(t)\cdot\cos(2\pi f_o t) - Q(t)\cdot\sin(2\pi f_o t)$$

$$I(t) = A(t)\cdot\cos(\theta(t))$$
$$Q(t) = A(t)\cdot\sin(\theta(t))$$

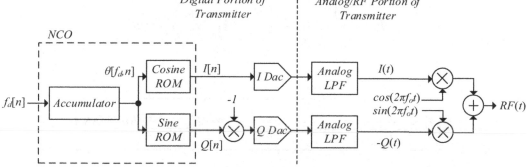

Figure 5-37: Interconnection between the Digital and Analog/RF Portions of the FSK Transmitter

5.3.2 Minimum Shift Keying (MSK)

Minimum shift keying, or MSK, is a variant of FSK in which the phase shift experienced during one symbol period, T_{Symbol}, is exactly $+\pi/2$ or $-\pi/2$. The minimum frequency deviation, f_d, away from the carrier at f_o that will achieve this phase shift is equal to $+F_{Symbol}/4$ or $-F_{Symbol}/4$.

$$f_d = \pm \frac{F_{Symbol}}{4}$$

The number of symbols in the alphabet is thus restricted to 2 entries, ± 1, represented by the phase shifts $\pm\pi/2 = \pm 90°$.

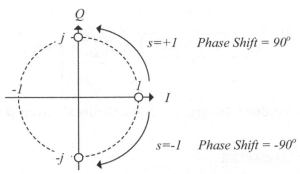

Figure 5-38: Possible Phase Shifts Allowed in Minimum Shift Keying (Start Phase is Arbitrary)

The Modulation Index h

Most communication standards will specify how large the allowed symbol phase shift is for their special implementation of FSK. They have come to use the factor h (modulation index) to indicate how many half cycles ($\frac{1}{2}$cycle $= 180° = \pi$ radians) are traversed during a single symbol period, T_{Symbol}. Note that the integral of $f_d(t)dt$ over the time period of 0 to T_{Symbol} results in the number of cycles covered. Therefore, the requirement of MSK that 90 degrees are traversed during a single symbol period results in a modulation index of h equal to $\frac{1}{2}$. In the case that $f_d(t)$ remains constant (which is the case for FSK/MSK) over a symbol period, the equation for h simplifies as follows.

$$h = \frac{2\pi \int_0^{Tsymbol} f_d(t)dt}{\pi} = 2\int_0^{Tsymbol} f_d(t)dt$$

$$h = 2f_d T_{Symbol}$$

$$\boxed{f_d = \frac{h}{2T_{Symbol}} = \frac{0.5}{2T_{Symbol}} = \frac{1}{4T_{Symbol}} = \frac{1}{4}F_{Symbol} \quad for\ MSK}$$

You may ask why this value of $h = \frac{1}{2}$ and the associated 90-degree phase shift is so special. Taking another look at the IQ plane in the above figure, it is obvious that the positions (at j and $-j$) that result from the two available symbol rotations are as far from one another as absolutely possible. Clearly, this is the optimal scenario for a receiver that processes baseband I and Q and therefore has access to the constellation positions when making a decision as to which symbol was actually sent. Notice that a modulation index of 3/2 and the associated phase shifts of ±270° would also result in constellation positions with maximum distance. However, the deviation frequency would be three times larger thus bloating the required signal bandwidth unnecessarily. The modulation index of ½ is the *minimum shift* that reaches the optimal constellation positions.

5.3.3   Gaussian FSK (GFSK) and Gaussian MSK (GMSK)

While FSK and MSK are hardly utilized in today's wireless technologies, their derivatives, namely GFSK and GMSK, have enjoyed ready adoption in Bluetooth, DECT, and GSM. The phase in FSK/MSK may be continuous, but the frequency deviation, $f_d(t)$, instantaneously jumps back and forth between the frequencies in its symbol set causing the signal spectrum to be unnecessarily wide. A way needed to be found to allow smooth frequency transitions as we move from one symbol boundary to the next. GFSK achieves this task by adding a Gaussian filter into the FSK modulator as seen below. Note that the total pulse shaping function is the convolution of the rectangular and Gaussian pulse shapes. As you can deduce from our discussion of MSK, GMSK is a special case of GFSK in which the phase shift of the carrier signal is equal to exactly ±π/2 over one symbol period.

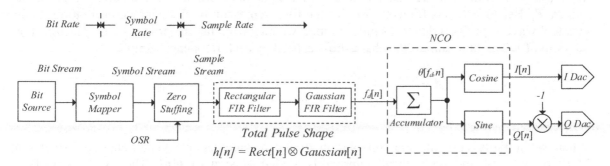

Figure 5-39: GMSK/GFSK Modulator Structure

The Gaussian filter impulse response (its pulse shape) depends on the symbol period, T_{Symbol}, as well the factor BT_{Symbol}, representing the bandwidth to bit period product. Note that the symbol and bit periods are the same in our case since the symbol mapper converts one bit (0 or 1) into one symbol (-1 or 1). The following table presents the three impulse responses we will work with.

Gaussian Pulse Shape	Rectangular Pulse Shape	Total Pulse Shape
$Gaussian(t) = C \cdot e^{-\frac{1}{2}\left(\frac{t}{\sigma}\right)^2}$ where $\sigma = \dfrac{\sqrt{\log(2)} \cdot T_{Symbol}}{2\pi \cdot BT_{Symbol}}$	$Rect(t) = 1/T_{Symbol}$ for $-\dfrac{T_{Symbol}}{2} \le t \le \dfrac{T_{Symbol}}{2}$ and zero otherwise	$h(t) = Gaussian(t) \otimes Rect(t)$

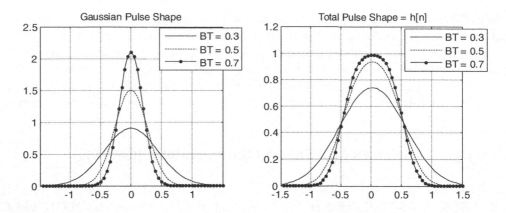

Figure 5-40: Gaussian and Total Pulse Shapes for BT$_{Symbol}$ = 0.3, 0.5, and 0.7

Pulse Shape Scaling

In the discrete MatLab implementation presented as code below, we will use a single pulse shaping filter, $h[n]$, representing the convolution of the Gaussian and Rectangular pulse shapes. The scaling of this discrete total pulse shape, $h[n]$, must be such that the sum of all coefficients multiplied by the sample period, T_s, is equal to a phase shift of ¼ cycle or $\pi/2$ radians. Note that the phase shift realized at each sample instant, n, is equal to the product of the current filter output and $2\pi \cdot T_S$. Because the coefficients, $h[n]$, of the filter represent the frequency contribution (to the eventual waveform $f_d[n]$) due to a single symbol, we may compute the proper scaling factor, A, as the ratio of $\pi/2$ and the sum of the phase shifts caused by each frequency sample.

$$A = \frac{h\pi}{2\pi \sum_{n=0}^{N-1} h_{unscaled}[n] \cdot T_S} = \frac{\pi/2}{2\pi \sum_{n=0}^{N-1} h_{unscaled}[n] \cdot T_S}$$

When we initially compute the convolution of the rectangular and Gaussian pulse shapes in MatLab, we will end up with an improperly scaled version of $h[n]$. The expression above properly scales the impulse response for the desired modulation index, h.

$$h[n] = A \cdot h_{unscaled}[n]$$

In the code example below, we have selected a modulation index, h, of 0.5 and a symbol rate of 1MSPS, which necessitates a maximum frequency deviation of ¼·F_{symbol} = 250KHz. We chose to oversample the signal by a ratio, OSR, of 16, giving rise to the sample rate of 16MHz. The code now computes the rectangular, Gaussian and total filter coefficients, $h[n]$, thus preparing to properly shape the incoming sample stream generated by the zero-stuffing operation.

```
rand('seed', 2);

%% GFSK Paramters
BT  = 0.5;  % DECT BT  = 0.5          / GSM BT  = 0.3  / Bluetooth BT = 0.5
h   = 0.5;  % Dect h = 0.5 +/- 0.05 / GSM h   = 0.5  / Bluetooth h = 0.28 to 0.35

Fsample = 16e6;           % Sample Rate
Tsample = 1/Fsample;      % Sample Period
```

```
Fsymbol   = 1e6;              % Symbol Rate
Tsymbol   = 1/Fsymbol;        % Symbol Period
OSR = Fsample/Fsymbol;        % Must be an integer


%% Rectangular Filter Definition
RectFilter       = ones(1, OSR);

%% Gaussian Filter Definition
Sigma            = sqrt(log(2))*Tsymbol/(BT*2*pi);
n                = -2*OSR:2*OSR;
t                = n*Tsample;     %% The sample instances for the filter
GaussianFilter   = exp(-0.5*(t/Sigma).^2);


%% Computing the impulse response, h[n], of the Total Pulse Shaping Filter
hn_unscaled      = conv(RectFilter, GaussianFilter);
A                = (h*pi)/(2*pi*sum(hn_unscaled)*Tsample);
hn               = A*hn_unscaled;


%% Bit Source, Symbol mapping, and zero-stuffing
BitSource             = round(rand(1,500));  % Random bit source
SymbolSource          = 2*BitSource - 1;     % Non-Return to Zero Mapping
Zero_Stuffed          = zeros(1, OSR * length(BitSource));
Zero_Stuffed(1,1:OSR:end) = SymbolSource;

%% The filtering operation (a simple convolution)
Fd  = conv(Zero_Stuffed, hn);

%% Accumulator and IQ Generation
I     = zeros(1,length(Fd));
Q     = zeros(1,length(Fd));
Phase = 0;
for i = 1:length(Fd)
    Phase       = mod(Phase + 2*pi*Fd(1,i)*Tsample , 2*pi);  % Accumulation
    I(1,i)      = cos(Phase);
    Q(1,i)      = sin(Phase);
end
```

The next figure illustrates the sample stream and its corresponding $f_d[n]$ waveforms for both the MSK and GMSK scenarios. The MSK scenario clearly shows that the frequency deviation stays constant at ±250KHz during each symbol period of 1 microsecond. Since a quarter of a cycle is traversed by a 250KHz sinusoid during a single microsecond, the MSK requirement is met.

Taking a look at the frequency deviation, $f_d[n]$, of the GMSK scenario, we can easily tell that the phase traversed is equal to $\pi/2$ only for the symbol located between the 8 and 9 microsecond time instances. We can visually integrate the frequency deviation function over successive symbol periods to verify that phase traversed is less than $\pi/2$ during all other symbol periods.

$$Phase\_Traversed = 2\pi \int_0^{Tsymbol} f_d(t)dt$$

On the following pages we will present figures that nicely illustrate the impact on the constellation positions as we apparently rotate less than 90 degrees.

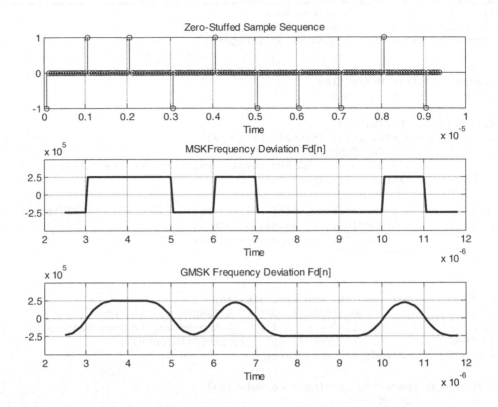

Figure 5-41: Frequency Deviations for MSK and GMSK Scenarios Based on Random Sample Stream

GFSK versus GMSK

We have mentioned earlier that GMSK is really GFSK with a modulation index, *h*, equal to ½. At the beginning of the MatLab code, we specified that the cordless phones DECT standard uses GFSK with a modulation index of $h = 0.5 \pm 0.05$. To be called GMSK, the index must be exactly maintained at 0.5 such that the constellation points that are reached as time progresses lie at their true, intended positions. If we plan to transmit training sequences over the RF link, such that the receiver may properly equalize out any multipath distortion, then we must guarantee a transmit signal that features high modulation accuracy. We can't have the modulation index, and thus the maximum frequency deviation, just hover in the neighborhood of the correct value.

Therefore, GFSK receivers won't attempt to extract the data based on IQ constellation positions observed at baseband, but simply observe the frequency deviation waveform just as the one seen in the figure above and decide whether over a particular symbol period its value is positive or negative. This method of decoding the data is simpler, requires less precision regarding the frequency deviation and thus leads to a less demanding hardware implementation. Unfortunately, its performance will be somewhat lower. Using GMSK, as is done in the GSM standard, allows us to employ sophisticated equalization and synchronization techniques which lead to higher performance.

For the GFSK case, we want to ensure that the frequency deviation waveform, $f_d[n]$, is well behaved and features a wide open eye diagram thus maximizing the distance between the possible symbol values. The waveform values at the ideal sample instances are indicated using circular markers and nicely illustrate the wide open eye. Intersymbol interference in frequency deviation waveform exists but is kept at a minimum.

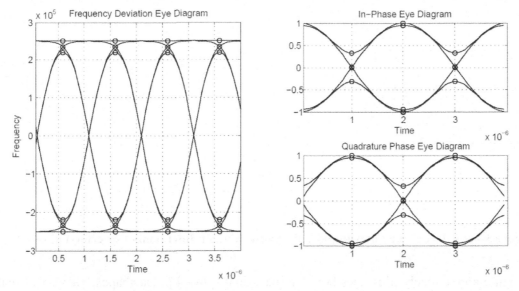

Figure 5-42: Eye Diagrams for Frequency Deviation as well as I/Q Outputs

Intersymbol Interference in the IQ domain

Our MatLab code maintains a perfectly constant modulation index of h = 0.5 during the entire simulation which results in GMSK. The IQ constellation positions therefore become meaningful which is why the I and Q eye diagrams are included in the figure above. Preferably, the I and Q values sampled at the boundary of each symbol period should be at positions [1, 0], [0, 1], [-1, 0], and [0, -1] as seen in the MSK constellation diagram below. However, due to intersymbol interference, there are additional constellation positions in the GMSK scenario.

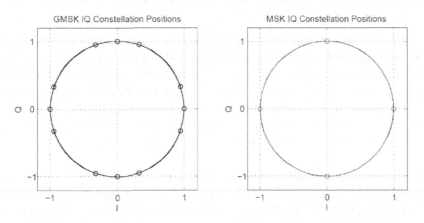

Figure 5-43: GMSK and MSK Constellations Side by Side

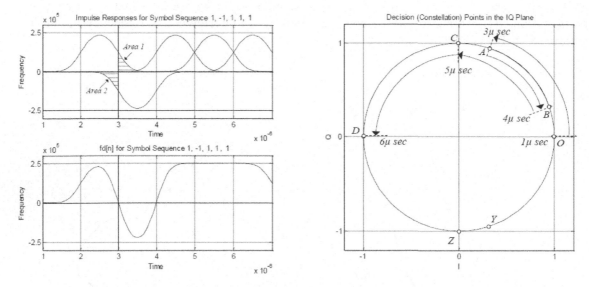

Figure 5-44: Effect of the Filter Impulse Response, h[n], on Intersymbol Interference in IQ Plane

The figure above nicely illustrates how the individual symbol pulse shapes, $\pm h[n]$, that make up the frequency deviation function, $f_d[n]$, cause intersymbol interference in the IQ plane. Remember that we had chosen a symbol rate of 1MSPS, and we subsequently sample the IQ output waveform at the end of their official symbol periods (3μsec, 4μsec, 5μsec …) even if the symbols themselves extend beyond those limits. Whereas the first pulse will eventually cause its proper phase rotation of $\pi/2$ radians, the early sampling instant at 3μsec leaves us short by $2\pi \cdot Area1$ radians. Together with the next symbol, which has already started and introduced a rotation equal to $2\pi \cdot Area2$ radians, the desired rotation from point O to C is cut short and we arrive at position A. Constellation point B at 4 microseconds is short of zero degrees as well since the second pulse hasn't completed its assigned 90 degree rotation and pulse three is already busy pushing the angle counterclockwise toward point C at 4 microseconds. Only symbol 4 traverses the full 90-degrees to location D since both the previous and following symbol force counterclockwise rotations as well.

To judge the impact of the intersymbol interference on the performance let's backtrack to the beginning at point O and look at all the possible positions that can be reached at 3 microseconds when we decide the polarity of the first symbol. Ideally, we would want to reach locations C (symbol = $1 \cdot h[n]$) and Z (symbol = $-1 \cdot h[n]$) which feature the maximum distance in the IQ plane. However, even in the worst case where we only reach either locations A or Y, the distance is only minimally smaller compared to the ideal case. Therefore, when we add noise to the constellation positions, the detection performance will not be significantly compromised.

The final figure in this chapter illustrates just what we gained by introducing the Gaussian filter into our FSK modem. The spectrum skirt of GMSK (BT = 0.5) falls off far more quickly than that of MSK, which is impractical for bandlimited wireless communication. Note that as BT increases, the GMSK spectrum will approach that of MSK and intersymbol interference is decreased. As BT decreases, the spectrum skirt is better contained at the expense of additional intersymbol interference.

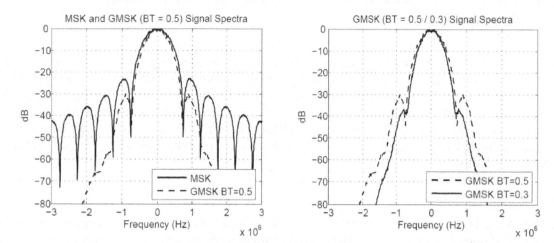

Figure 5-45: MSK, GMSK (BT = 0.5), and GMSK (BT = 0.3) Spectra

5.4 Multi Tone – OFDM (Orthogonal Frequency Division Multiplexing)

OFDM, or orthogonal frequency division multiplexing, is a technology that has recently made serious inroads in the area of communication engineering. Within the last 20 years, the technique has taken over a whole host of wireless communication standards, including WLAN (802.11a/g/n)[1][2][3], WiMAX [4], 4G LTE[5][6], 5G New Radio, as well as a variety of terrestrial television broadcasting services. OFDM facilitates certain tasks such as channel equalization and easily mates to MIMO techniques, resulting in faster, more reliable communication links.

What exactly is OFDM?

OFDM is a method of encoding complex symbols on to a time domain waveform that is to be transmitted. Remember that transmit data bits in virtually all modems are delivered serially by some type of general purpose processor. These data bits are first error encoded and then grouped and mapped into *IQ* constellation positions residing in the complex plane. OFDM differs from traditional single-tone modems in the way in which it takes these discrete time symbols to create continuous waveforms. In the figure below, we take a look at a QPSK symbol stream, $C[n]$, which is to be mapped into a continuous waveform via the traditional time domain method of pulse-shaping and the newer OFDM technique.

$$C[n] = [1\text{-}j \quad \text{-}1\text{+}j \quad 1\text{+}j \quad \text{-}1\text{+}j \quad 1\text{-}j \quad 1\text{+}j \quad \text{-}1\text{-}j \quad \text{-}1\text{+}j]$$

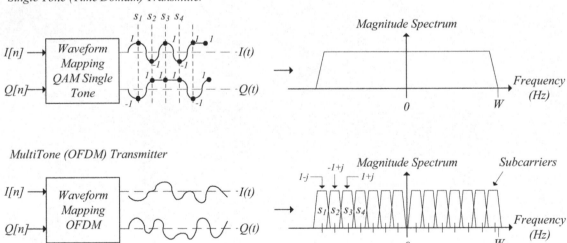

Figure 5-46: OFDM versus Pulse-Shape Mapping

If we consider the time domain waveform produced by the pulse-shaping filters, we clearly see the impact of the symbols, $C[n]$, on the continuous waveform. Each pulse-shaped symbol features frequency content that extends over the entire double-sided bandwidth of 2W. In OFDM, each symbol modulates a complex sinusoid, or subcarrier, of some particular frequency, thus confining the information to a certain region in the band. In our QPSK example, the symbols, $C[n]$, change the phase of each subcarrier, while higher order QAM constellations would modify their magnitude as well. The transmit symbols can be recognized not in the time domain waveform, as was the case for the pulse-shaping technique, but in the spectrum that is revealed via an FFT calculation in the receiver.

Why is it Better?

One of the major obstacles standing in the way of optimal performance in wireless communication links is the multipath effect and the associated selective fading phenomena that can wreak havoc on proper signal detection in the receiver. In time domain modems, selective fading at just one frequency will affect every received symbol that must be detected. Equalization is achieved by programming a digital filter featuring a response that is the approximate inverse of the multipath channel. Since OFDM transmit symbols are confined to individual subcarriers, selective fading will suppress some, leaving others largely unaffected. The selective fading model is reduced to a flat fading model when applied on a subcarrier basis. Therefore, equalizing multipath effects in OFDM is a simple matter of scaling each detected subcarrier phase and magnitude (FFT output) by a complex value that counteracts the frequency response of the channel. The overall OFDM equalization process is much simpler than that of the single-tone modem case. An additional advantage is the ability to dynamically assign more data – via higher constellation symbols – to subcarriers that are less affected by fading and therefore feature better signal to noise ratios. This flexibility provided by OFDM is not easily realized in single-tone (time domain) modems. OFDM would have been adopted earlier, had it not been for the need to incorporate hardware implementations of the Fourier and inverse Fourier transforms. These functions are notoriously hardware intensive and have only been practical in terms of implementation into digital ASIC devices for the last decade or so. With the advent of deep submicron CMOS processes, I/FFT applications have become commonplace.

OFDM and MIMO: A Match Made in Heaven

MIMO, or multiple input multiple output, is a collection of techniques that utilize multiple antennas to either remedy the effects of the multipath channel or increase the data throughput over a communication link. The fact that we can assign a single correction scalar to each OFDM subcarrier to mitigate fading due to multipath effects also facilitates the process of interpreting information sent and received in multiple antenna systems. Only with the advent of OFDM has MIMO been used in commercial systems to its maximum potential.

5.4.1   OFDM versus Time Domain Modems

In the last chapter, we introduced the digital transmit portion of a single-tone (time domain) QAM modem and examined the operation of the symbol mapping and waveform mapping processes. Remember that it was the symbol mapping operation that gathered incoming bits into groups, which would then be mapped into one of several possible constellation positions. In the QPSK case, we would group two bits and map them into one of four reference positions. The difference between the time domain and OFDM modulator is the way in which these symbols are mapped into a waveform, which is then passed out the DAC modules to the analog transmitter.

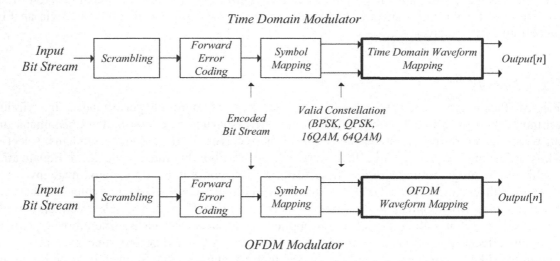

Figure 5-47: Time Domain versus OFDM Modulator

Waveform Mapping for QAM Single-Tone Modems

Section 2 of chapter 5 provides an introduction to the digital transmitter structure inside a single-tone QAM modem. The basic mechanism of translating complex symbols into analog waveforms centers on a zero-stuffing element and a subsequent pulse-shaping filter, which both shape and upsample the signal prior to being processed by the DAC or digital to analog converter. The figure below reviews the structure already presented in chapter 5 for the in-phase path.

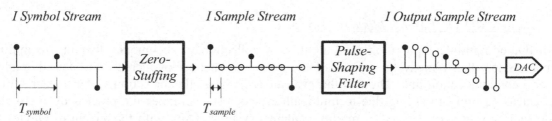

Figure 5-48: Basic Time Domain Waveform Mapping Process for QAM Single Tone Transmitter

Waveform Mapping for QAM (OFDM) Modems

The waveform mapping process for OFDM differs significantly from that of the QAM single-tone transmitter. In the OFDM case, a certain number of symbols are streamed into a shift register whose size corresponds to the IFFT length N. Once the shift register is full, the IFFT calculation starts and the input shift register begins to fill anew. This feature of processing N input symbols at a time forces a frame-like structure into the OFDM waveform. The N point IFFT output is augmented with a guard interval, which prevents interference from previous or later frames, as they are delayed through the multipath channel. The IFFT output combined with the guard interval forms an OFDM symbol. The final upsampling step once again serves to alleviate the task of the reconstruction filter following the DAC module. A more detailed description of the IFFT process, as well as the guard interval and pilot tone insertion process, is provided in the next few sections.

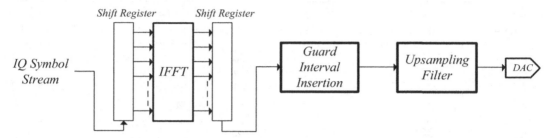

Figure 5-49: Basic Time Domain Waveform Mapping Process for OFDM Transmitter

Comparing the Two Waveform Mapping Processes

Comparing the waveform mapping processes graphically is the ideal way of appreciating the differences between the two transmitter architectures.

$$C[k] = 0.7071 \cdot [1\text{-}j \quad \text{-}1\text{+}j \quad 1\text{+}j \quad \text{-}1\text{+}j \quad 1\text{-}j \quad 1\text{+}j \quad \text{-}1\text{-}j \quad \text{-}1\text{+}j]$$

In Figure 6-6, we illustrate how the first three symbols of the symbol set above are pulse-shaped and super-positioned to form part of the final sample stream. Each symbol, indicated by the circular marker, is clearly visible in the summed output waveform.

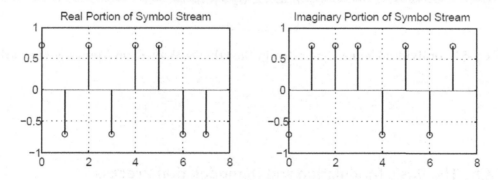

Figure 5-50: Symbol Stream and Final Analog Waveform for the Time Domain Modem

In the OFDM waveform mapping operation, each symbol $C[k]$ determines the phase and magnitude of a complex sinusoid of the form $exp(j2\pi nk/N)$. In the figure below, we show the first three complex sinusoids, as well as the summation of all eight. The impact of each symbol is noticed across the entire IFFT output sequence, and their individual values are not easily recognizable, as was the case for the single-tone QAM transmit signal. The original eight symbols, $C[k]$, will not become visible again until we take the FFT in the receiver.

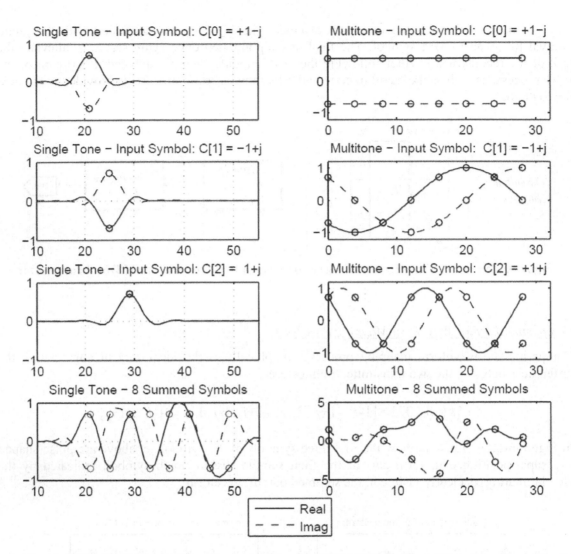

Figure 5-51: Individual Tone Generated in the OFDM Waveform Mapping Operation

5.4.2 The Basic Modulation and Demodulation Process

The basic modulation/demodulation process utilized in OFDM employs a pair of IDFT/DFT algorithms to encode and recover the transmit symbols. The IDFT operation executes every time a fresh set of N symbols has been streamed into the input shift register. A guard interval, or *GI*, is appended to the IDFT output to form an OFDM symbol. At the receiver, the guard interval is stripped and the information recovered using the DFT operation. The following figure shows the modem blocks as an IFFT/FFT pair, which is the numerically efficient implementation of the IDFT/DFT combination. You will recall from chapter 2 that the IFFT/FFT algorithm is restricted to lengths $N = 2^x$, where x is any positive non-zero integer.

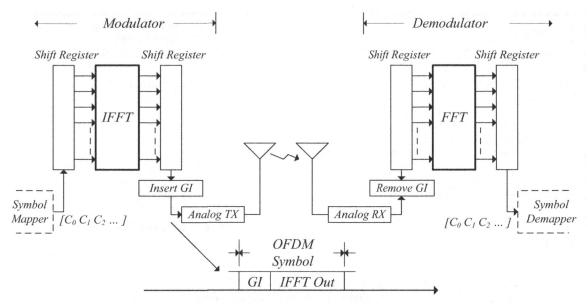

Figure 5-52: Basic Modulation and Demodulation Process in OFDM

Example 5.3: IDFT and DFT Calculation

Let us finish the example started in the last section by computing the *IDFT* output for the first symbol group of length $N = 8$.

$$C[m] = .707 \cdot [1-j,\ -1+j,\ 1+j,\ -1+j,\ 1-j,\ 1+j,\ -1-j,\ -1+j\ ]$$

The MatLab code provided next computes the inverse discrete Fourier transform and appends the guard interval to form an OFDM symbol. The guard interval generally comprises the last few samples of the IDFT output sequence. In this case we append the last two samples.

$$IDFT\_Out[n] = \sum_{m=0}^{7} C[m] \cdot e^{j2\pi nm/8} \quad for\ n = 0,1,...7$$

$$GI = IDFT\_Out[6:7]$$

$$OfdmSymbol = [GI\ IDFT\_Out]$$

Conversely, the transmit symbols are recovered by stripping away the guard interval of the OFDM symbol and taking the DFT of the IDFT output sequence.

$$C[m] = \frac{1}{8}\sum_{n=0}^{7} IDFT\_Out[n] \cdot e^{-\frac{j2\pi nm}{8}} \quad for\ m = 0,1,...7$$

```
N   = 8;                                    %% Sequence length
n   = 0:1:N-1;                              %% Time index
C   =.707*[1-j; -1+j; 1+j; -1+j; 1-j; 1+j; -1-j; -1+j ];  %% Input symbols

IDFT_Out = zeros(1,N);                      %% IDFT calculation
for i = 1:N
    m            = i - 1;
    f            = m/N;
    AnalysisTone = exp(j*2*pi*n*f);
    IDFT_Out     = IDFT_Out + C(i,1)*AnalysisTone;
end
GI          = IDFT_Out(1,7:8);              %% The guard interval
OFDM_Symbol = [GI IDFT_Out];                %% Final OFDM symbol

DFT_Out  = zeros(1,N);                      %% DFT calculation
for i = 1:N
    m            = i - 1;
    f            = m/N;
    AnalysisTone = exp(-j*2*pi*n*f);
    DFT_Out(1,i) = (1/N)*OFDM_Symbol(1,3:10)*AnalysisTone.';
end
```

The graph below shows the guard interval followed by the IFFT output. Appending the last few IFFT output samples at the front creates an OFDM symbol that is continuous at the interface between guard interval and IFFT output portion. The process of appending the guard interval is referred to as cyclic extension.

$$DFT\ Output = C[m] = .707 \cdot [1-j, -1+j, 1+j, -1+j, 1-j, 1+j, -1-j, -1+j]$$

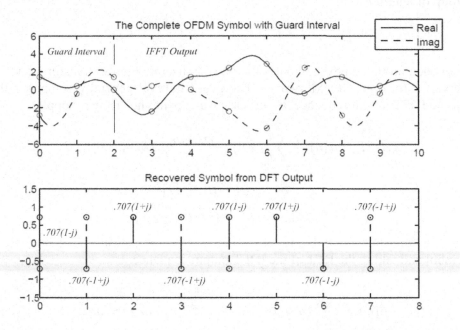

Figure 5-53: The OFDM Symbol and the Recovered Input Symbols Provided by the DFT

5.4.3 The Role of the Guard Interval

In wireless communication systems, the multipath channel causes linear distortion that is imposed on the complex envelope of the RF transmitted signal. This linear distortion, which we can conveniently model as FIR filtering of the original TX baseband signal, is described in detail in section 6.3.1. It causes a frequency response with selective fading, which in single carrier modems changes the original pulse shape and thus produces intersymbol interference. At the receiver, the equalizer simply bends the frequency response back into its preferred flat shape, thereby eliminating the intersymbol interference and providing good performance.

In OFDM systems linear distortion and intersymbol interference are separate mechanism that are solved by separate techniques. The selective fading imposed upon the frequency response is solved using very simple equalization techniques, which we discuss in section 7.1.3. However, intersymbol interference is caused by the overlapping of different copies of the signal at the receiver. As they arrive with different delays, it is possible that a later OFDM symbol from one path overlays with an earlier symbols of another path. This is very nicely shown in section 7.1.3.1. The guard interval, or cyclic prefix, eliminates allows us to eliminate the intersymbol interference. The following sections provide a large amount of detail regarding the behavior of the channel and the techniques used to remedy its impact.

Table 5-1: Preview of Section Discussing Channel Behavior and the Role of the Guard Band

Section	Content
6.3.1	Discusses the modeling of the wireless communication channel
6.3.2	Add Doppler effects to the model of section 6.3.1
7.3.1.1	Detailed explanation of the Guard Band (Cyclic Prefix)
7.3.1.4 / 5	Frequency Response Estimation and Equalization in OFDM

5.4.4 Discontinuities between OFDM Symbols

The OFDM modulation process maps a certain number of complex QAM data values into a unique OFDM symbol. As this set of QAM data values change from one OFDM symbol to the next, the waveform representing each OFDM symbol is different. Therefore, once we start appending these symbols to form the final TX output waveform, discontinuities appear at the boundary between symbols, and the transmitted spectrum will therefore suffer from spectral splatter. At the expense of some intersymbol interference rejection, the boundary between one end of one symbol and the start of the next can be smoothened, thus limiting the spectral expansion. This process is shown in detail in section 5.5.3, where we discuss an OFDM derivative called SC-FDMA.

5.5   Hybrid Tone SC-FDMA

SC-FDMA, or single carrier frequency division multiple access, is a new modulation scheme that combines the beneficial properties of both single tone and OFDM signaling formats. SC-FDMA was introduced into the mainstream communication arena, when the 3rd Generation Partnership Project, or 3GPP, included it as the choice of modulation for the LTE and 5G uplink (mobile handset to base station). SC-FDMA was selected for the uplink to address the biggest shortcoming of the incredibly popular OFDM format, which is employed in the LTE and 5G downlink. This shortcoming is the transmit signal's excessive peak to average power, or PAP, ratio, which requires the use of very linear and therefore very inefficient RF power amplifiers. The subsequent heat generated and shortened battery life are the principle hurdles that have marginalized OFDM as a modulation format for transmit waveforms of handheld communication devices.

SC-FDMA is a hybrid tone modulation format that inherits the lower PAP characteristics of single tone TX waveforms, and takes advantage of the ease with which OFDM equalizes linear distortion introduced by the wireless multipath channel. To introduce SC-FDMA, let's first compare it to the single tone PSK/QAM signals that we covered at the beginning of this chapter.

5.5.1   Similarities to Single Tone Pulse Shaped Transmit Signals

One of the hallmarks of pulse shaped single tone modulated signals is that you can see the transmit symbols represented in the final transmit waveform that is passed to the digital to analog converters. As a matter of fact, if we choose a raised cosine pulse shape, the symbols issued by the symbol mapper will end up as one of the samples produced by the waveform mapping block as we have seen in section 5.2.3. Therefore, as we change the alpha value of the raised cosine shape, only the sample values in-between the symbols are changing.

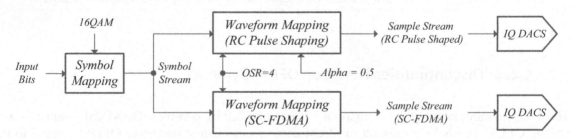

Figure 5-54: Test Setup for SC-FDMA and Pulse Shaped Single Carrier 16 QAM Comparison

The waveform mapping process for SC-FDMA is similar in that the symbol values also shows up in the output sample stream. The difference, once again, is the manner in which the sample values in-between are determined. The next figure illustrates the I portions of two 16QAM single tone (single carrier) transmit signals, one whose symbols are waveform mapped using SC-FDMA whereas the other features these same symbols mapped using raised cosine pulses. The waveforms show the original symbols issued by the symbol mapper as large circles, the output samples as small circles, and the eventual analog waveforms (continuous lines) as they should appear in the analog portion of the transmitter after the analog low pass filters. As the previous figure suggests, the symbol mappers for SC-FDMA and pulse shaped PSK/QAM modulation are the same and the techniques covered in section 5.2.1 apply to SC-FDMA.

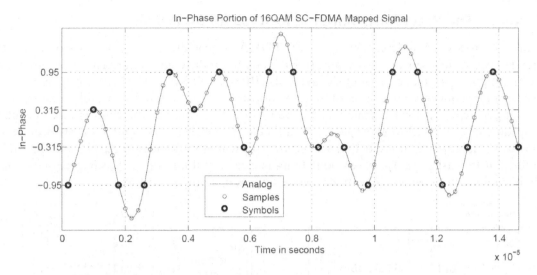

Figure 5-55: Analog Waveform, Sample/Symbol Stream for 16QAM SC-FDMA

Whereas the waveforms do look alike, the SC-FDMA modulated sample stream tends to peak a little more than the RC pulse shaped sample stream. As a matter of fact, the peak to average power (PAP) ratio for the two complex waveforms is 7.2dB for the SC-FDMA case compared to 5.5dB for RC pulse shaped case with alpha = 0.5. This compares to a similarly configured OFDM waveform whose PAP ratio would weigh in somewhere between 10 to 14dB. Note that as we decrease the alpha value of the RC pulse shaped waveform, its peak to average power ratio increases and eventually approaches that of the SC-FDMA modulated signal.

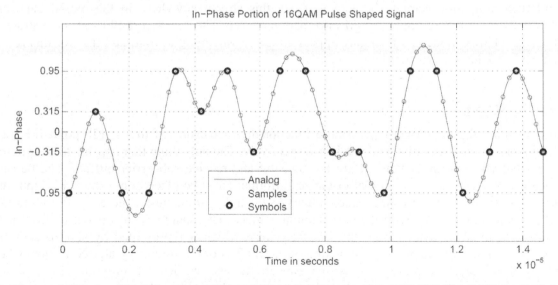

Figure 5-56: Analog Waveform, Sample/Symbol Stream for 16QAM Single Tone Pulse Shaped (RC Alpha = 0.5)

5.5.2 The Waveform Mapping Process for SC-FDMA

To start the process of understanding how SC-FDMA is generated, let us take a look at the two figures below, which pair up a DFT and IDFT block to produce an output. The figure to the left takes in a input, $x[n]$, made up of QAM symbols (BPSK, QPSK, 16QAM, … etc.) and produces the very same sequence, $x[n]$, at its output. Obviously, it is not a terribly useful structure as we could simply bypass it for the same result. The modified structure to the right is more insightful as it upsamples the QAM input sequence by a factor of four. Had the input symbols been part of the 16QAM alphabet, then the output waveform $y[k]$ would very much look like the SC-FDMA waveform of the last page. However, something is clearly still missing as upsampling could have easily been done in the single tone modem with far less computation effort.

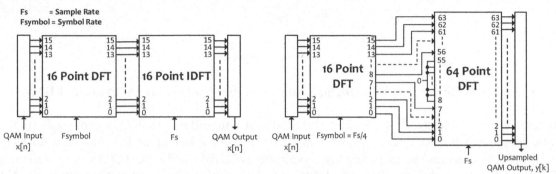

Figure 5-57: Primitive Building Blocks for SC-FDMA

SC-FDMA Modulation and Demodulation

Transmitters using SC-FDMA are not simply interested in upsampling and transmitting a QAM symbol stream, they want to be able to place this stream anywhere in the output band and potentially include traditional OFDM information as well. The next figure illustrates the structure of an SC-FDMA modem, which may appear a bit overwhelming at first, and it is for this reason that we break up the structure into two parts.

The OFDM Core

First examine the traditional OFDM core functionality consisting of the N=128-point IFFT and FFT blocks in the transmitter and receiver respectively. The only thing that appears to be different from what you saw in section 5.4 are the transmit and receive subcarrier buffers. The transmit subcarrier buffer is a construct that organizes information to be placed onto the IFFT input in a visually simple and easy to understand format. As the index of the buffer increases from 0 to 127, the frequency associated with the corresponding subcarrier monotonically increases from most negative to most positive. Notice the DC frequency, which is represented by subcarrier buffer entry 64. Remember that the input indexing of the IFFT is of the frequency aliased format, where entries 0 through 63 represent positive subcarriers and entries 64 through 127 represent subcarriers whose frequencies are aliased to negative values. The subcarrier mapping, indicated by the arrows, provides the translation from monotonically increasing to aliased frequency indexing. The LTE and 5G standards use a slightly modified version of these buffers, called resource grids, which are used both in the OFDM process of the downlink as well as in the SC-FDMA process of the uplink.

DFT Precoding and Decoding

As we saw in the illustration on the last page, to produce an SC-FDMA waveform, we require a DFT precoding step. Interestingly enough, due to this requirement, SC-FDMA is often referred to as DFT precoded OFDM in technical literature. Once again, the DFT operation will produces an output vector with frequency aliased positioning. Thus, DFT output indices 0 through 15 indicate the orientation and magnitudes of the positive frequency subcarriers, whereas indices 16 through 31 are associated with the negative frequency subcarriers. As the TX subcarrier buffer is defined for monotonically increasing subcarrier, we must remap the DFT output values just as was done with the IFFT input buffer.

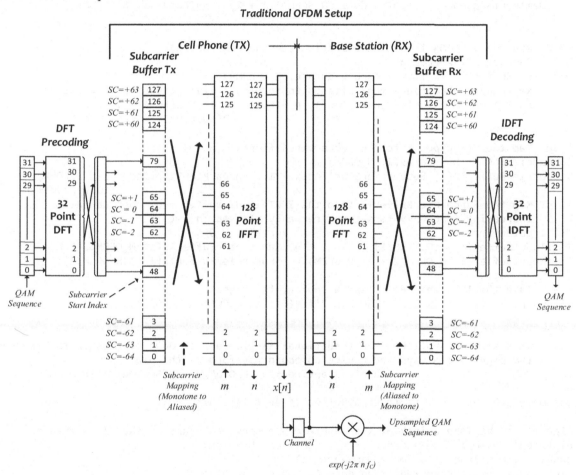

Figure 5-58: SC-FDMA Modulator and Demodulator

There are two ways to recover the QAM input sequence from the SC-FDMA modulated signal. Given the signal carrier nature of the signal, we may simply down convert the signal to baseband which is illustrated by the complex multiplication step at the bottom of the figure. Alternately, you may completely reverse the SC-FDMA process in the transmitter. The MatLab code that follows simulates the entire structure above producing both the upsampled and final QAM sequence.

```matlab
%% 1. The main FFT/IFFT parameters of the OFDM portion of the SC-FDMA processes
N_OFDM                  = 128;
N_SCFDMA                = 32;
SubcarrierStartIndex    = 48;
OSR                     = N_OFDM / N_SCFDMA; % The Oversampling ratio

%% 2. Generate BPSK (Mod Order 1) or QSPK (Mod Order 2) Input Symbols
ModOrder = 1;
if(ModOrder == 1)
    QamInputSequence = sign(randn(1, N_SCFDMA)) * (1 + 1j);
else
    QamInputSequence = sign(randn(1, N_SCFDMA)) + 1j*sign(randn(1, N_SCFDMA));
end

%% 3. Run the SC-FDMA Transmitter Section (In the cell phone)
% 3a. Compute DFT Precoding
DftOutputBuffer     = fft(QamInputSequence, N_SCFDMA);
%       Rearrange subcarriers from aliased to monotonically increasing frequencies.
DftOutputBuffer     = [DftOutputBuffer(1, N_SCFDMA/2 + 1: N_SCFDMA), ...
                                            DftOutputBuffer(1, 1:N_SCFDMA/2)];

% 3b. Map the DftOutput into the SubcarrierBufferTx
SubcarrierBufferTx = zeros(1, N_OFDM);
SubcarrierBufferTx(1, SubcarrierStartIndex + 1 : SubcarrierStartIndex + N_SCFDMA) = ...
                                            DftOutputBuffer;
% 3c. Map the SubcarrierBufferTx into the IFFT input buffer from monotonically
%       increasing frequency format to frequency aliased format
IfftInputBuffer                     = zeros(1, N_OFDM);
IfftInputBuffer(1, 1:N_OFDM/2)      = SubcarrierBufferTx(1, N_OFDM/2+1:N_OFDM);
IfftInputBuffer(1, N_OFDM/2 + 1: N_OFDM) = SubcarrierBufferTx(1, 1:N_OFDM/2);

% 3d. Take the IFFT and let's take a look at the signal.
IfftOutputBuffer  = (N_OFDM/N_SCFDMA)*ifft(IfftInputBuffer, N_OFDM);


%% 4. This step accomplishes the traditional single carrier down conversion to baseband
%       the signal to always see a nice baseband waveform in the plot.
fc                  = (SubcarrierStartIndex - (N_OFDM/2 - N_SCFDMA/2))/N_OFDM;
n                   = 0:N_OFDM - 1;
UpsampledQamSequence = IfftOutputBuffer .* exp(-1j*2*pi*n*fc);

figure(1);  %% Take a look at the waveform and convince yourself that it worked
plot(real(UpsampledQamSequence), 'r'); grid on; hold on;
Range = 1:OSR:length(UpsampledQamSequence);
plot(Range, real(UpsampledQamSequence(1, Range)), 'ro');
plot(imag(UpsampledQamSequence), 'b');
plot(Range, imag(UpsampledQamSequence(1, Range)), 'bo');


%% 5.  Pass the information through the Receiver
% 5a. Take the FFT
FftOutputBuffer  = (N_SCFDMA/N_OFDM)*fft(IfftOutputBuffer, N_OFDM);

% 5b. Map the FFT Output in to the SubcarrierBufferRx
%       Mapping from frequency aliased to monotonically increasing format.
SubcarrierBufferRx(1, 1:N_OFDM/2)       = FftOutputBuffer(1, N_OFDM/2+1:N_OFDM);
SubcarrierBufferRx(1, N_OFDM/2 + 1: N_OFDM) = FftOutputBuffer(1, 1:N_OFDM/2);
```

```
%  5c. Extract the desired subcarrier values SubcarrierBufferRx
BufferTemp       = SubcarrierBufferRx(1, SubcarrierStartIndex + 1 :
SubcarrierStartIndex + N_SCFDMA);

%  5d. Rearrange subcarriers back into frequency aliased arrangements
IdftInputBuffer  = zeros(1, N_SCFDMA);
IdftInputBuffer(1, 1:N_SCFDMA/2)             = BufferTemp(1, N_SCFDMA/2 + 1: N_SCFDMA);
IdftInputBuffer(1, N_SCFDMA/2 + 1: N_SCFDMA) = BufferTemp(1, 1:N_SCFDMA/2);
QamOutputSequence = ifft(IdftInputBuffer, N_SCFDMA);

Error         = QamInputSequence - QamOutputSequence;
MeanSquareError = mean( Error .* conj(Error));   %% This must be tiny
```

The question still beckons: Why not just use single carrier modulation and demodulation? What is the purpose of the fancy DFT precoding followed by OFDM? The reason for this is that virtually all SC-FDMA transmit scheme will also include some amount of traditional OFDM to which the figure below attests.

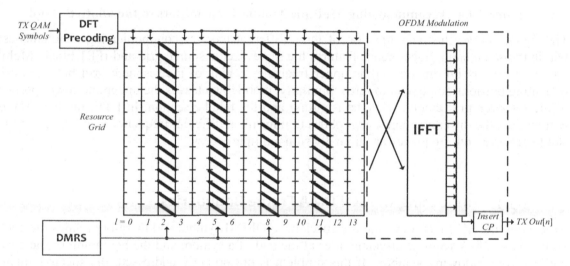

Figure 5-59: Typical SC-FDMA Setup seen in LTE C-V2X

The figure above shows fourteen subcarrier transmit buffers organized into a matrix we call a resource grid. Each buffer will be mapped to the IFFT to produce a total fourteen OFDM symbols, which are index using the variable *l*. Of these fourteen, only ten carry information that has been DFT precoded to produce low peak to average power waveforms at the output. The other OFDM symbols carry fixed demodulation reference information that is used by the receiver to estimate the channel equalize the information. The DMRS information is inserted directly into the transmit subcarrier buffers, meaning that at the receiver, we can see the original DMRS values and compute the frequency response of the channel by simply dividing the received by the transmitted values at each subcarrier location. If you recall, this feature is one of the big advantages of OFDM modulation. So how about the peak to average power ration that we have made such an effort to maintain with our SC-FDMA scheme? Luckily, the DMRS information can be crafted to also provide a low peak to average power ratio. A popular type of DMRS signal is the Zadoff-Chu sequence, which is used in many different communication systems for just this purpose.

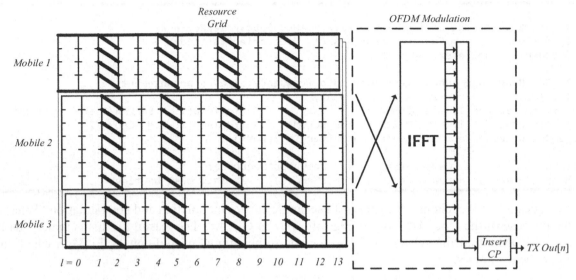

Figure 5-60: Accommodating Multiple Mobile Transmitters in the allotted Band

The figure above illustrates the case of three different mobiles communicating simultaneously within the same band. Each mobile features the same sized resource grid and IFFT block. Mobile one transmits its information in the upper frequency region of the resource grid but leaves the remaining portions empty (set to zero). Similarly, mobile two leaves the upper and lower portions of the resource grid empty, while transmitting and the middle section. In LTE, the base station will instruct the different connected mobiles to transmit at different frequencies and different time slot to achieve both frequency and time division multiple access.

5.5.3 Smoothing the SC-FDMA Frame Boundaries

At its core, the SC-FDMA symbol is a specialized OFDM symbol and therefore retains the cyclic prefix. SC-FDMA features discontinuities at the end of a symbol and the beginning of the cyclic prefix of the following symbol . If this problem is not properly addressed, the spectral splatter produced in neighboring channels may be unacceptable in certain situations. To illustrate the solution to this problem we modify the cyclic prefix region slightly which we illustrate using the 128 point IFFT output sequence. We will include 16 samples in the cyclic prefix.

→ Extract the last 16 samples of the 128-point IFFT output of the current symbol (call it A).

→ Extract the first 16 samples of the 128-point IFFT output of the preceding symbol (call it B).

Normally, our cyclic prefix would simply consist of A[112:127], but we change the expression as follows.

$$CP = Mask1 \cdot B[0{:}15] + Mask2 \cdot A[112{:}127]$$

Note that $B[0{:}15]$ connects to the end of the IFFT output of symbol B in a continuous fashion, whereas $A[112{:}127]$ connects to the start of the IFFT output of symbol A in a continuous fashion . The masks are constructed such that the information provided from the earlier frame, $B[0{:}15]$, dominates the first few samples of the cyclic prefix, whereas the information from the current frame $A[112{:}127]$ dominates the remaining samples.

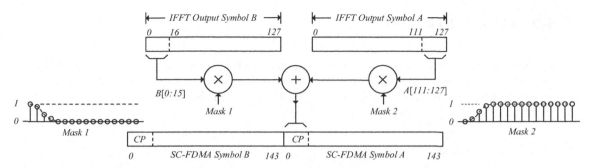

Figure 5-61: Boundary Smoothing Technique for SC-FDMA

In the next figure, we illustrate the spectra of a single tone raised cosine pulse shaped 16QAM signal, as well as two similar SC-FDMA waveforms of which one of them features smoothing by modifying the cyclic prefix. The figure below clearly illustrates that the SC-FDMA signal without smoothing features significant spectral splatter beyond its intended bandwidth. Constructing the cyclic prefix such that information from the past and current symbols transition within the first four samples of the CP improves the splatter to the point where the overall transmit spectrum is good enough for most applications. The transition region of four symbols reduces spectrum splatter at the expense of intersymbol interference robustness.

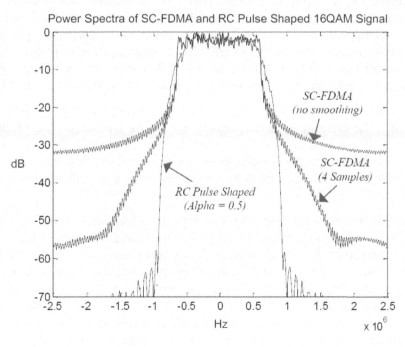

Figure 5-62: Power Spectra for SC-FDMA with and without Smoothing

5.6   Error Correction

Attempting to prevent errors in a wireless communication system is a sophisticated task, to which the multiple processing blocks below attest. Even a single error can render the entire received message meaningless and a retransmission must be requested. In the transmitter, error handling is handled in the processing chain below.

The Transmit Bit Chain

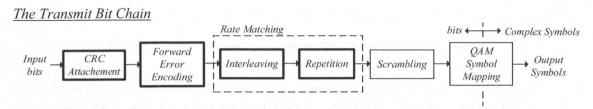

Figure 5-63: Typical Transmit Bit Chain from a 4G/5G Cellular Link

→ Error detection: The receiver must know whether an error occurred in order to decide whether a retransmission is necessary or not. Cyclic redundancy checks are commonly used for this task.

→ Error correction: Even in good signal to noise ration situations, errors will occasionally happen and we need the ability to correct them. Without error correction, the number of errors will gradually increase as the SNR decreases. This type of gradual deterioration is undesirable. Ideally, error correction would prevent all errors until a particular signal to noise ratio is reached, below which communication becomes utterly impossible. Error correction provides this type of non-linear behavior where we move from error free communication to massive numbers of error within a very small range of two or three dB of SNR. Convolutional, Turbo, Polar and LDPC (low density parity check) coding techniques are typically employed to correct bit errors.

→ Interleaving: Interleaving is a very simple process that scrambles the position of bits within the message. Usually, the message is subdivided into bit blocks, whose bits are interleaved according to the same rules. This is beneficial as the channel has the tendency to disturb bits that are close in proximity, and the error correction mechanisms can easily fail in those scenarios. Error correction mechanisms work best when they need to correct bit errors that a nicely spread out rather than appearing in bursts.

→ Repetition: Repetition of bits provides two separate advantages. When bits are processed in the receiver, they are always disturbed by noise. Having multiple copies of the same bit allows us to average/sum the copies thereby increasing the signal to noise ratio of the original bit ahead of the error correction module. An additional advantage of repetition is one of opportunity. In modern OFDM communication systems, messages are meant to be placed over a convenient grid which spans a certain amount of time and frequency. This grid will be able to support a certain number of final coded bits, and interleaved bits are thus repeated until they fill out the entire grid area that is assigned to them. The total rate of the FEC scheme is matched to the designated grid area.

→ Scrambling can serve different purposes. It can be used to whiten (avoiding long runs of 0s or 1s) the bits stream, and make bit sequences appear random in nature. This randomization is important in cellular system and makes interfering signal appear as noise at your receiver.

The Receive Bit Chain

The receiver inverts the process undertaken in the transmitter with one important exception. The received bits first appear after the QAM symbol demapping steps as noisy floating point values (equal to ±1 plus a noise component) called softbits as compared to hard bits, which take values of 0 and 1. Maintaining these softbit values provides additional information regarding just how confident we are regarding the orientation of the bits. Let's assume that a singe bit value 0/1 was BPSK mapped to constellation positions -1 + 0j / +1 + 0j. BPSK symbol demapping is then the simple step of extracting the real value from the input symbols yielding floating point values equal to ±1 plus noise. Ideally, the receiver would process a received value of 0.1 differently than a value of 1.5 even though they both indicate that they represent an input bit of 1. Clearly, the receiver has far more confidence that a received value of 1.5 is really a bit value of 1, than that a received value of 0.1 is really a bit value of 1. Taking advantage of this knowledge is a key piece of information in the error correction process and provides up to 2dB of processing gain.

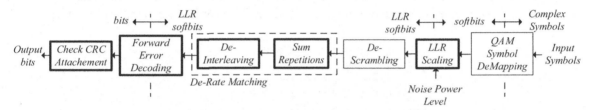

Figure 5-64: The Receive Bit Chain (from right to left)

In communication systems of the past, the QAM symbol demapping step would have provided hard bits immediately, thus discarding these confidence values.

→ LLR (log likelihood scaling) – is a scaling technique that modifies the noisy softbit values produced by the QAM demapper to more accurately express confidence/belief and improve overall decoding error correction performance. (See section 5.6.2 for a detailed discussion)

→ The original bit repetition step is now reversed by summing the softbits back into a single value thereby achieving a processing gain of $10 \cdot log_{10}(NumberOfRepetitions)$. Some of the more important bit messages in LTE are repeated up to 16 times providing an improvement in signal to noise ratio of up to $10 \cdot log_{10}(16) = 12$dB.

→ Deinterleaving reverses the interleaving process and places bit back into their nominal position.

→ The forward error correction decoding technique will take up the bulk of this section. They have the ability to both detect and correct error.

→ Finally, the receiver takes the bit stream and separates the actual message bits from the CRC bits. It recomputes the CRC from the message bits and compares it to the received CRC bits. If they match then it is assumed all errors in the bit stream have been corrected successfully.

5.6.1 Cyclic Redundancy Checking

A cyclic redundancy check (CRC) is an error-detection method commonly used in digital networks and storage devices to detect accidental changes in content. We take a block of message bits and attach a check value, which is based on the remainder of a polynomial division of the message content. The receiver recomputes the CRC given the received message bits and then compares them to the received CRC bits. If the computed and received CRC bits match, then the transmission is deemed error free. The CRC calculation requires a generator polynomial and a bit message. Generator polynomials are changed to bit blocks according to the following example.

→ Generator Polynomial: $x^3 + x^2 + 1$ yields [1 1 0 1]

→ Generator Polynomial: $x^5 + x^4 + x^3 + 1$ yields [1 1 1 0 0 1]

The degree of the generator polynomial is equal to the size of the CRC. Thus, a generator polynomial equal to $x^3 + x^2 + 1$ has a degree 3 and produces 3 CRC bits. The CRC is the remainder of a polynomial division, which we illustrate here for the message [1 1 1 0 1 1 0].

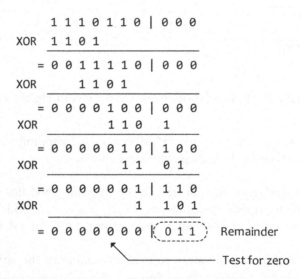

Figure 5-65: CRC Generation via Polynomial Division

The figure above shows how the polynomial sequence of [1 1 0 1] is used to XOR certain sections of the message bit sequence. Note the addition of 3 zeros at the right of the message to make room for the remainder at the end of the procedure. After each XOR operation, the polynomial sequence is moved to the right with the objective to turn all message bits into zeros at the end, which terminates the algorithm. Note that the divisor (our polynomial sequence) may move to the right by more than one position in order to align with the next 1 in the dividend as can clearly be seen in the first two steps. The LTE specification in TS36.212 Section 5.1.1 provides 4 polynomial generators with 24, 16 and 8 bits.

→ Generator 24A = [1, 1, 0, 0, 0, 0, 1, 1, 0, 0, 1, 0, 0, 1, 1, 0, 0, 1, 1, 1, 1, 1, 0, 1, 1]
→ Generator 24B = [1, 1, 0, 0, 0, 0, 0, 0, 0, 0, 0, 0, 0, 0, 0, 0, 0, 0, 0, 0, 1, 1, 0, 0, 0, 1, 1]
→ Generator 16 = [1, 0, 0, 0, 1, 0, 0, 0, 0, 0, 0, 1, 0, 0, 0, 0, 1]
→ Generator 8   = [1, 1, 0, 0, 1, 1, 0, 1, 1]

```matlab
function Remainder = GenerateCRC(Message, ModeString, InputPolynomial)

assert(nargin == 2 || nargin == 3, 'Not enough arguments to execute this function');

%% 0. The CRC generator can use the default LTE CRC polynomials or the one supplied
%      See TS36.212 Section 5.1.1 for details on the LTE CRC polynomials
switch lower(ModeString)
    case 'ltecrc24a'
      Polynomial = [1,1,0,0,0,0,1,1,0,0,1,0,0,1,1,0,0,1,1,1,1,1,0,1,1];
    case 'ltecrc24b'
      Polynomial = [1,1,0,0,0,0,0,0,0,0,0,0,0,0,0,0,0,0,0,1,1,0,0,0,1,1];
    case 'ltecrc16'
      Polynomial =  [1, 0, 0, 0, 1, 0, 0, 0, 0, 0, 0, 1, 0, 0, 0, 0, 1];
    case 'ltecrc8'
      Polynomial = [1, 1, 0, 0, 1, 1, 0, 1, 1];
end

%% 1. Error checking
[MessageRows, MessageColumns] = size(Message);

% 1a. Ensure the both the message and the polynomial are vectors.
assert(MessageRows == 1 || MessageColumns == 1, 'The input message must be a vector');

% 1b. Ensure that the MSB of the Polynomial is a one. It wouldn't make sense otherwise
assert(Polynomial(1,1) == 1, 'The first bit of the polyomial must be equal to 1.');

% 1c. Convert to row vector and ensure that both feature only bits
Message     = Message(:).';

for i = 1:length(Message)
    assert(Message(1,i)==0 || Message(1,i)==1,'The Message must be composed of bits');
    if(i <= length(Polynomial))
        assert(Polynomial(1,i)==0 || Polynomial(1,i)==1, 'Polynomial must be bits');
    end
end

%% 2. Compute the CRC process
%      Create a temporary message by appending length(Polynomial) - 1 zeros to the
original message.
TempMessage = [Message, zeros(1, length(Polynomial) - 1)];
for i = 1:length(Message)
    Range = i:(i+length(Polynomial)-1);
    if(TempMessage(1,i) ~= 0) % The modulo 2 addition
        TempMessage(1, Range) = mod(TempMessage(1, Range) + Polynomial, 2);
    end
    if(sum(TempMessage) == 0); break; end
end

% CRC remainder as a row vector
Remainder = TempMessage(1, (end - length(Polynomial) + 2):end);

% Change the remainder to a column vector if the original message was also a column
vector.
if(MessageColumns == 1); Remainder = Remainder.'; end
```

5.6.2   Forward Error Correction and the Log Likelihood Scaling Process

Forward error correction refers to a set of techniques that can both detect and correct bit errors that occur during the transmission of data over a communication channel, or the storage of data on a CD, hard drive, or other media. These methods augment the source bit stream at the transmitter by adding redundant bits, which are used in the receiver to properly recover the original TX information. The example below illustrates a simple FEC scheme to illustrate the process of error detection and correction.

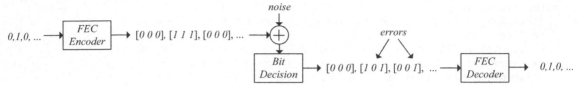

Figure 5-66: Bluetooth Rate 1/3 Forward Error Correction Feature

This method of simply repeating each input bit three times seems overly simple, but surprisingly this is the technique Bluetooth uses to implement its rate 1/3 forward error correction feature. The rate of the encoder is defined as the ratio of the number of input bits to output bits produced by the encoder. At the receiver, the error decoder looks at each group of three bits and decides on what was actually sent based on a majority rule. Thus, the received bit groups [0 0 0], [1 0 1], and [0 0 1] are interpreted as bits 0, 1, and 0, and we come to the conclusion that bit 2 of group [1 0 1] and bit 3 of group [0 0 1] must be in error. This example illustrates a common rule regarding forward error correction in general. It is easier to correct errors that are spread out over time rather than those that occur in successive bits, and certainly in this case, encountering two errors in a single group of three received bits will cause the FEC decoder to fail. More sophisticated methods of forward error correction don't use this simple technique of bit repetition, but employ methods that use algebraic or convolutional coding to add redundant bits to the input stream.

Maximum Likelihood Decoding

The diagram below illustrates a concept called maximum likelihood decoding, which represent the perfect technique for recovering the original input bits.

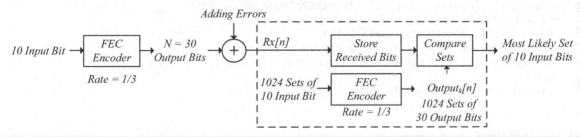

Figure 5-67: Maximum Likelihood Decoding

The figure illustrates a rate 1/3 encoder that transforms a set of 10 input bits into a set of $N = 30$ output bits, which, once passed through the channel, experience additive noise causing occasional errors. The decoder features an identical copy of the encoder and excites it with all $2^{10} = 1024$ possible 10 bit input sequences to produce 1024 30 bit output sequences. The decoder now compares the received bit sequence, $Rx[n]$, with the 1024 possible output bit sequences,

$Output_k[n]$. The index k extends from 0 to 1023, whereas the index n extends from 0 to 29. The output of the decoder is the 10-bit input sequence that produced the output sequence that was closest the match to the received bit stream $Rx[n]$.

Even a novice will quickly notice that this technique is doomed as the number of potential output sequences that must be compared to the received bit stream grows geometrically with the number of input bits. However doomed, the techniques that we will look at attempt to achieve similar performance to the maximum likelihood technique without the geometrically growing processing effort involved. But before we look at these techniques, let's understand how we compare these sequences to the received bit stream. To understand this, we need to realize that the 0/1 valued bits are transmitted over an analog communication channel. Therefore, bits are mapped to convenient floating point values, as shown below where a binary value of 1 is mapped to a floating point value of +1.0, whereas the binary value 0 is mapped to -1.0. Once noise is added to these bits by the channel, the received softbit values can be expressed statistically by their probability density functions.

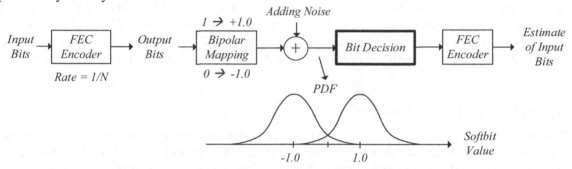

Figure 5-68: Hard of Soft Bits Decisions

Hard Bit Decision

Hard bit decision involves mapping all received softbits with values above 0.0 to a binary value of 1, and values less than 0.0 to a binary value of 0. The comparison between the many possible output sequences, $Output_k[n]$, and the received hard bits stream, $Rx[n]$, can be done by computing the XOR operation of the bits of each potential output sequence and the received bits and then computing the sum. Remember that the XOR operation of like bits produces a 0, whereas the XOR operation of unlike bits produces a 1. The potential output sequence that produced the smallest sum is the one that is most similar to the received bits.

$$Sum_k = \sum_{n=0}^{N-1} XOR(Output_k[n], Rx[n])$$

Soft Bit Decision

In soft bit decision decoding, the noisy received bipolar bit values are passed on directly to the FEC decoder. Note that the softbit values carry far more information than the hard bit decision value as their size reveals confidence/belief that the decoder might be able to use. We can be very confident that a received softbit value of 1.3 was originally equal to 1.0, whereas a received value of 0.05 doesn't allow us this type of certainty. From the PDFs we can conclude that it is only minimally more likely that the received value of 0.05 was originally a 1.0 than a -1.0.

How would we compare this sequence of received softbits, $Rx[n]$, to the 1024 possible output bit vectors that the decoder has computed for us? Once we map the potential output bit sequences, $Output_k[n]$, from binary bit values 0/1 to floating point values of -1.0/+1.0, it is a simple matter of just correlating them against the received sequence $Rx[n]$. The highest correlation result indicates the winning output sequence and the corresponding input sequence is the output of the decoder. Note that the confidence value represented by the magnitude of the received softbit is automatically included in the equation below, as softbits with small magnitude contribute less to the overall correlation than softbits with large magnitudes and therefore large confidence.

$$Corr_k = \sum_{n=0}^{N-1} Output_k[n] \cdot Rx[n]$$

→ Note that some communication systems map the binary values 0/1 to floating point values of +1.0/-1/0 (4G/5G) whereas others map them to floating point values of -1.0/+1.0 (IEEE 802.11). This simple mapping works particularly well for the BPSK and QPSK constellations. Higher level QAM constellation require more sophisticated mapping as shown in a few pages from now.

LLR Scaled Soft Decision Decoding

Soft bit decision decoding, as discussed above, provides superior error correction performance, which results in extended transmit range or conversely the option to reduce output power at the transmit antenna. However simple, the softbit decision scheme we have introduced above does not express confidence optimally in cases when different received bits suffer from different amounts of noise. If we knew the noise power value (the variance of the Gaussian probability density functions) associated with each softbit value, then we could compute a more reliable confidence metric. The next figure nicely illustrates how this knowledge can help us to better ascertain confidence. We are presented with two scenarios in which the noise variance, σ_n^2, varies from 0.25 to 1.0. Observe the markers, which represent a situation in which the softbit value is equal to 0.5. After examining the 'x' marker, it is clear that the likelihood that the softbit was originally equal to -1 is far smaller for the plot on the left (featuring less noise power) than for the plot on the right. Without knowledge regarding noise power, this quite obvious fact is not available to us. It turns out that the optimum metric for expressing confidence/belief in this situation is the log likelihood ratio expression below, where y represents the received softbit value and S the bipolar transmit bit value.

$$LLR\ Confidence = ln(\frac{Prob(S=+1|y)}{Prob(S=-1|y)})$$

Before we start deriving the expression for the LLR scaling process, we should mention that the LLR confidence is a linear function with respect to the softbit value, y. This was of course also the case for our previous confidence metric where *Confidence* = y. Whereas for the old metric, the slope of this linear function was equal to 1.0, the slope of the LLR confidence changes with respect to noise power.

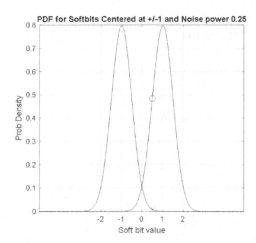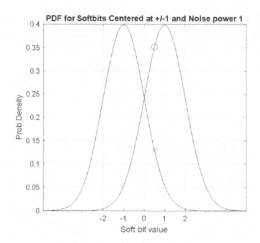

Figure 5-69: Softbits with Varying Noise Powers of 0.25 and 1.0

The first hurdle in our derivation of the LLR confidence metric is the fact that the probabilities mentioned in the LLR confidence equations can't be read off the PDF directly. The metric is the natural logarithm of the probability that the observation, y, was originally a 1.0 divided by the probability that the observation was originally -1.0. What we can apparently read of the PDF plot is the ratio of the probability of y given that the original noise free value was equal to +1.0 and the probability of y given that the original noise free value was equal to -1.0. Notice that the PDFs don't provide probabilities directly as they are probability density function, which need to be integrated over a range of y value, to yield a probability. However, the ratio of probability densities is equal to the ratio of probabilities in this case and the ratio below is therefore valid.

$$\frac{Prob(y|S=+1)}{Prob(y|S=-1)} = \frac{ProbDensity(y|S=+1)}{ProbDensity(y|S=-1)}$$

The second hurdle is the fact that the probabilities of y given S is not necessarily the same as the probabilities of S given y, which is the formulation in the LLR confidence equation. This problem can be solved by recalling Baye's rule, which defines the probability that the observation has value y and the original bipolar bit value was equal to S as follow.

$$Prob(y,S) = Prob(y) \cdot Prob(y \mid S) = Prob(S) \cdot Prob(S \mid y)$$

Therefore

$$Prob(y \mid S) = \frac{Prob(S) \cdot Prob(S \mid y)}{Prob(y)}$$

And

$$\frac{Prob(y|S=+1)}{Prob(y|S=-1)} = \frac{Prob(S=+1) \cdot Prob(S=+1 \mid y)}{Prob(S=-1) \cdot Prob(S=-1 \mid y)}$$

Now, assuming that the transmitted bits assume values of +1 and -1 with equal likelihood ($Prob(S = +1) = Prob(S = -1)$), the expression reduces to our desired result.

$$\frac{Prob(y|S=+1)}{Prob(y|S=-1)} = \frac{Prob(S=+1 \mid y)}{Prob(S=-1 \mid y)}$$

Note that binary bits values of 0/1 are not always mapped to ±1 as is the case for BPSK, but may be mapped to other values such as ±0.7071 as is the case for QPSK. We will therefore use the variable 'a' to indicate the size of the transmitted softbit. Armed with this fact, we can now recall the expression of the Gaussian PDF which states the following.

$$Prob(y \mid S) = \frac{1}{\sqrt{2\pi\sigma^2}} exp\left(-\frac{(y-S)^2}{2\sigma^2}\right)$$

Thus

$$LLR\ Confidence = ln\left(\frac{Prob(y|S = +a)}{Prob(y|S = -a)}\right) = ln\left(\frac{exp\left(-\frac{(y-a)^2}{2\sigma^2}\right)}{exp\left(-\frac{(y+a)^2}{2\sigma^2}\right)}\right)$$

$$= -\frac{(y-a)^2}{2\sigma^2} + \frac{(y+a)^2}{2\sigma^2} = -\frac{y^2 - 2ya + a^2}{2\sigma^2} + \frac{y^2 + 2ya + a^2}{2\sigma^2} = \frac{4ya}{2\sigma^2}$$

$$LLR\ Confidence = \frac{2a}{\sigma^2}y$$

The BPSK Scenario

Input Bits (b0)	I Out	Q Out
0	-1	0
1	1	0

Figure 5-70: BPSK Constellation Mapping (IEEE Definition)

Let's assume that the BPSK symbol, x, was OFDM modulated at the transmitter and demodulated at the receiver to produce the complex observation Y.

$$Y = hx + n$$

The BPSK transmit symbol, x, was altered by the channel frequency response h, which is also a complex scalar, and a complex noise quantity n, with variance σ^2_n. We now compute the estimate of the transmit symbol as follows.

$$\hat{x} = \frac{1}{h}Y = \frac{1}{h}(hx + n) = x + \frac{n}{h}$$

The final bit observation, y, is extracted by taking the real value of the complex observation of x.

$$y = real\left(x + \frac{n}{h}\right)$$

The bit estimate, y, has the original mean magnitude of $a = 1.0$ and the following variance σ^2.

$$\sigma^2 = \frac{1}{2}\frac{1}{|h|^2}\sigma^2_n$$

Note, the complex noise term n/h has a variance of $\sigma^2_n/|h|^2$, which is reduced by ½ as we are only considering the real part of y. LLR scaling for BPSK modulated signals is as follows:

$$LLR\ Confidence = \frac{2a}{\sigma^2}y = \frac{4|h|^2}{\sigma_n^2}real(\hat{x})$$

→ y is the real portion of the estimate of x, which is the equalized symbol observation Y.

→ h is the frequency response estimate at the frequency and time location where the original BPSK symol x was transmitted. All receivers have to undertake this channel estimate and communication system designers take great pains to ensure that there are enough reference/pilot symbols available such that this estimate can be computed with good accuracy.

→ a is the magnitude of the bipolar bit mapped onto the real axis and equals 1.0.

→ σ_n^2 is the complex noise and interference power of the observation Y. Determining this quantity is an interesting proposition. Ideally, we would estimate this power, which can be done using the same reference/pilot symbols used to determine the frequency response h. If we are reasonably certain that term σ_n^2 is mostly due to thermal antenna noise (interference from other signals is small) then we can determine the noise power directly from the gain setting of the analog receiver. The gain setting in the receiver sets the receiver's noise figure, and noise power at its ADC outputs can be simulated and is often available in tabular form. If the options above are too involved, then a simply trick may be employed. We can safely assume that the sequence of bit estimates, y, that we will pass off to the FEC decoder stem from complex received symbols that are located within a short time span. If the gain of the receiver doesn't change over this small time interval, then the term σ_n^2 will be constant for all observations, y. Arbitrarily setting the term σ_n^2 to 1.0 simply scales all LLR softbit values by the same amount, which in turn scales the correlations that are undertaken in the FEC decoder when determining the best reference output sequence $Output_k[n]$. As long as all correlation results are scaled equally, the determination of the most likely sequence $Output_k[n]$ is unaffected.

The QPSK Scenario

Input Bits (b0)	I Out	Input Bits (b1)	Q Out
0	-0.7071	0	-0.7071
1	0.7071	1	0.7071

Figure 5-71: QPSK Constellation Mapping (IEEE Definition)

The QPSK case is identical to the scenario above with the exception of the bit mapping magnitude $a = 1/sqrt(2)$ rather than 1.0. The LLR confidence metric reduces to the following.

$$LLR\ Confidence(Bit0) = \frac{2a}{\sigma^2}y = \frac{4|h|^2}{\sqrt{2}\sigma_n^2}real(\hat{x})$$

$$LLR\ Confidence(Bit1) = \frac{2a}{\sigma^2}y = \frac{4|h|^2}{\sqrt{2}\sigma_n^2}imag(\hat{x})$$

→ Depending on which bit we wish to recover, y is either the real portion (for bit 0) or the imaginary portion (for bit 1) of the estimate of x, which is the equalized symbol observation Y.

→ The quantities h and σ_n^2 are identical to the BPSK scenario.

The 16QAM Scenario

The 16QAM scenario is naturally a bit more involved as the constellation features 4 positions along the real and imaginary axis. The mapping shown below ensures that the overall power of the 16QAM constellation is equal to that of the BSPK and QPSK variants.

Input Bits (b0 b1)	I Out	Input Bits (b2 b3)	Q Out
00	-3/sqrt(10)=-0.949	00	-3/sqrt(10)=-0.949
01	-1/sqrt(10)=-0.316	01	-1/sqrt(10)=-0.316
11	1/sqrt(10) = 0.316	11	1/sqrt(10) = 0.316
10	3/sqrt(10) = 0.949	10	3/sqrt(10) = 0.949

Figure 5-72: 16 QAM Constellation Mapping (IEEE Definition)

From the table above and the figure below, it is clear that bit 0 should be interpreted as a 1 if y (the real portion of the estimate of x) originally started out as value $S = 0.316$ or $S = 0.949$. Conversely, bit 0 should be interpreted as a 0 if y originally started out as $S = -0.316$ or $S = -0.949$. Therefore,

$$LLR(bit0) = ln\left(\frac{Prob(y\,|\,S=+0.949) + Prob(y\,|\,S=+0.316)}{Prob(y\,|\,S=-0.949) + Prob(y\,|\,S=-0.316)}\right)$$

Similarly, the LLR scaling factor for bit1 would be as follows.

$$LLR(bit1) = ln\left(\frac{Prob(y\,|\,S=+0.316) + Prob(y\,|\,S=-0.316)}{Prob(y\,|\,S=-0.949) + Prob(y\,|\,S=+0.949)}\right)$$

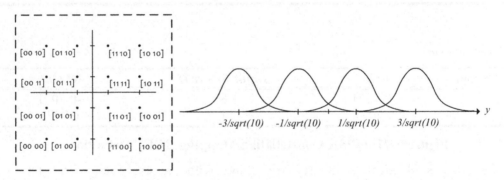

Figure 5-73: 16QAM Constellation with Softbit PDFs

Bits 2 and 3 have a similar form except that y represents the imaginary portion of estimate of x.

$$LLR(bit2) = ln\left(\frac{Prob(y\,|\,S=+0.949) + Prob(y\,|\,S=+0.316)}{Prob(y\,|\,S=-0.949) + Prob(y\,|\,S=-0.316)}\right)$$

$$LLR(bit3) = ln\left(\frac{Prob(y\,|\,S=+0.316) + Prob(y\,|\,S=-0.316)}{Prob(y\,|\,S=-0.949) + Prob(y\,|\,S=+0.949)}\right)$$

The probabilities have to be computed using the Gaussian probability density function rather than some convenient simplifications that we were able to apply for BPSK and QPSK.

The 64QAM Scenario

The 64QAM scenario is a natural extension of the 16QAM case and features 8 positions along the real and imaginary axis. Once again, the mapping shown below ensures that the overall power of the 64QAM constellation is equal to that of the BSPK, QPSK and 16QAM variants.

Input Bits (b0 b1 b2)	I Out	Input Bits (b3 b4 b5)	Q Out
000	-7/$sqrt$(42)= 1.080	000	-7/$sqrt$(42)= 1.080
001	-5/$sqrt$(42)= 0.771	001	-5/$sqrt$(42)= 0.771
011	-3/$sqrt$(42) = 0.463	011	-3/$sqrt$(42) = 0.463
010	-1/$sqrt$(42) = 0.154	010	-1/$sqrt$(42) = 0.154
110	1/$sqrt$(42)= -0.154	110	1/$sqrt$(42)= -0.154
111	3/$sqrt$(42)= -0. 463	111	3/$sqrt$(42)= -0. 463
101	5/$sqrt$(42) = -0.771	101	5/$sqrt$(42) = -0.771
100	7/$sqrt$(42) = -1.080	100	7/$sqrt$(42) = -1.080

Figure 5-74: 64QAM Constellation Mapping (IEEE Definition)

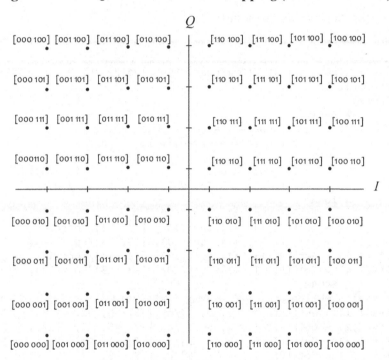

Figure 5-75: 64QAM Constellation Diagram (IEEE Definition)

Similarly to the 16QAM case, we can read off bits 0 through 2 from the constellation diagram above and determine the log likelihood ratios assuming that y is the real part of the estimate of x. The constant c in the expressions below represents the scalar 1/$sqrt$(42).

$$LLR0 = ln\left(\frac{Prob(y \mid S = 1c) + Prob(y \mid S = 3c) + Prob(y \mid S = 5c) + Prob(y \mid S = 7c)}{Prob(y \mid S =- 1c) + Prob(y \mid S =- 3c) + Prob(y \mid S =- 5c) + Prob(y \mid S =- 7c)}\right)$$

$$LLR1 = ln\left(\frac{Prob(y \mid S =- 3c) + Prob(y \mid S =- 1c) + Prob(y \mid S = 1c) + Prob(y \mid S = 3c)}{Prob(y \mid S =- 7c) + Prob(y \mid S =- 5c) + Prob(y \mid S = 5c) + Prob(y \mid S = 7c)}\right)$$

$$LLR2 = ln\left(\frac{Prob(y \mid S = -5c) + Prob(y \mid S = -3c) + Prob(y \mid S = 3c) + Prob(y \mid S = 5c)}{Prob(y \mid S = -7c) + Prob(y \mid S = -1c) + Prob(y \mid S = 1c) + Prob(y \mid S = 7c)}\right)$$

The expressions for bits 3, 4, and 5 are identical to those for bits 0, 1, 2 except for the fact that y represents the imaginary portion of the estimate of x. The function GetLLR() implements the general LLR expressions using the Gaussian PDF, which we need to do for 16 and 64QAM. The function LlrDemapperIEEE() computes the LLR confidence for all IEEE 802.11a constellations.

```
function LLR_Value = GetLLR(Positions1, Positions0, y, NoiseVariance, h)
    Numerator   = 0; % The denominator can remain at zero for very low noise variance.
    Denominator = 1e-6;  % That's bad, so initialize it to a small value.
    NoiseVarEstimate = NoiseVariance / (h * conj(h)); % Noise variance of the estimate.
    C1          = 1/sqrt(2*pi*NoiseVarEstimate);
    C2          = 2*NoiseVarEstimate;
    for i = 1:length(Positions1)  % Positions1 and Positions2 are row vectors.
        a           = Positions1(1,i);
        Numerator   = Numerator   + C1*exp(-((y-a)^2)/C2);
        a           = Positions0(1,i);
        Denominator = Denominator + C1*exp(-((y-a)^2)/C2);
    end
    LLR_Value = log(Numerator/Denominator);
end % GetLLR()
```

```
function LlrBits = LlrDemapperIEEE(QamIndex, SymbolValue, h, NoiseVar)
    LlrBits = zeros(1, QamIndex);
    switch(QamIndex)
        case 1        % BPSK
            y_I           = real(SymbolValue);
            LlrBits       = (4*h*conj(h)/NoiseVar)*y_I;
        case 2        % QPSK
            y_I           = real(SymbolValue);    y_Q           = imag(SymbolValue);
            LlrBits       = (4*h*conj(h)/(0.707*NoiseVar))*[y_I, y_Q];
        case 4        % 16QAM
            y_I           = real(SymbolValue);    y_Q           = imag(SymbolValue);
            LlrBits(1,1) = GetLLR([+3, +1]/sqrt(10), [-3, -1]/sqrt(10), y_I, NoiseVar, h);
            LlrBits(1,2) = GetLLR([-1, +1]/sqrt(10), [-3,  3]/sqrt(10), y_I, NoiseVar, h);
            LlrBits(1,3) = GetLLR([+3, +1]/sqrt(10), [-3, -1]/sqrt(10), y_Q, NoiseVar, h);
            LlrBits(1,4) = GetLLR([-1, +1]/sqrt(10), [-3,  3]/sqrt(10), y_Q, NoiseVar, h);
        case 6        % 64QAM
            y_I = real(SymbolValue); y_Q = imag(SymbolValue); c = sqrt(42);
            LlrBits(1,1) = GetLLR([ 7,  5, 3, 1]/c, [-7, -5, -3, -1]/c, y_I, NoiseVar, h);
            LlrBits(1,2) = GetLLR([-3, -1, 3, 1]/c, [-7, -5,  5,  7]/c, y_I, NoiseVar, h);
            LlrBits(1,3) = GetLLR([-5, -3, 3, 5]/c, [-7, -1,  1,  7]/c, y_I, NoiseVar, h);
            LlrBits(1,4) = GetLLR([ 7,  5, 3, 1]/c, [-7, -5, -3, -1]/c, y_Q, NoiseVar, h);
            LlrBits(1,5) = GetLLR([-3, -1, 3, 1]/c, [-7, -5,  5,  7]/c, y_Q, NoiseVar, h);
            LlrBits(1,6) = GetLLR([-5, -3, 3, 5]/c, [-7, -1,  1,  7]/c, y_Q, NoiseVar, h);
    end
end % LlrDemapperIEEE()
```

5.6.3 Binary Convolutional Coding

Binary convolutional codes, BCC, have been the hallmark of forward error correction techniques since the 1960. Whereas newer technique using Turbo codes and low density parity check, LDPC, codes are slightly better, binary convolutional coding has advantages even in today's communication systems. Currently, BCC is used in all WLAN and in 4G LTE. BCC is important to use for the following reasons.

→ The decoding process is simpler than that of Turbo coding or LDPC coding. The result is less hardware, power consumption and design time. Admittedly, with ever decreasing CMOS device sizes, some of these advantages have diminished somewhat.

→ Understanding BCC is essential as it features prominently in Turbo code implementations.

→ Binary convolutional decoding can finish the majority of its task before the last bit has arrived. Therefore, the decoding process finishes shortly after the arrival of the last bit, which allows the receiver to respond to messages very quickly thus achieving very low RX-TX turn-around time. This is a critical feature in WLAN systems. Turbo and LDPC coding implementations need to wait until the last bit has arrived to start the bulk of their very large processing task.

→ Another reason to prefer BCC over Turbo coding or LDPC is the fact that the later methods provide no performance advantage when the message is relatively short. It is for this reason that the PDCCH, physical downlink control channel, and the PBCH, or physical broadcast channel, in LTE are protected by a BCC code. The PBCH only contains 24 data bits and is protected by a 16-bit CRC to form a 40 bit message to be encoded. The figure below, illustrates the different available FEC technologies currently used in the industry.

Table 5-2: Popular Forward Error Correction Technologies

FEC Technology	Type	Application	Description
Convolutional (state machine based)	BCC (1960)	802.11 a/g and beyond) 4G LTE	The standard FEC in all IEEE 802.11 Technology. Protect PDCCH, and PBCH channels in LTE
	Turbo (1993)	4G LTE	Protects PDSCH Channel in LTE
Block Codes (Algebraic)	Hamming(1950)	Computer Memory	Detects two errors and corrects one in a 7 bit block.
	Reed-Solomon (1960)	CD/MP3/ Satellite, DVB	Was used for a long time in digital storage and satellite communications.
	LDPC (1960 (Gallagar) / 1996 (Rediscovered))	WLAN (11n and beyond) 5G NR	Was introduced into WLAN as of 802.11n as backup FEC. Used to protect the PDSCH in 5G.
	Polar Codes (2009)	5G NR	Protects the PBCH in 5G

Convolutional encoders are constructed using shift registers and modulo 2 adders as can be seen in the next figure. Convolutional encoders are defined by the number of input and output bits, the number of memory elements in the shift register as well as the manner in which the memory elements are connected to the modulo 2 adders.

c → *number of output bits*

x → *number of input bits*

m → *number of memory elements*

From these quantities we may declare the rate and constraint length specifications of the encoder.

$$EncoderRate = \frac{x}{c} \qquad ConstraintLength = L = m + 1$$

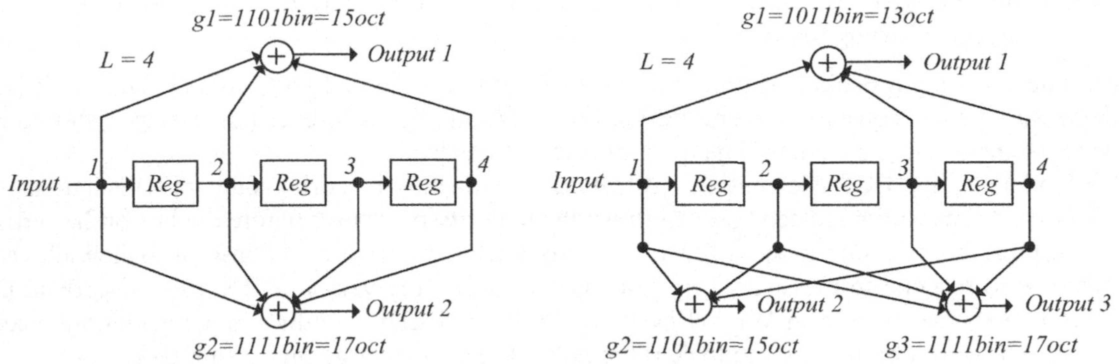

Figure 5-76: Rate ½ and Rate 1/3 Convolutional Encoders with Constraint Length = 4

The manner in which the shift register positions connect to the modulo-2 adders is defined by the generator polynomials g1, g2, g3, … and so on. In the rate ½ (constraint length = 4) encoder example of the figure above, g1 = [1101] indicates that nodes 1, 2, and 4 are connected to the top adder, whereas g2 = [1111] specifies that all nodes should connected to the bottom modulo-2 adder. In literature, these generator polynomials are expressed either in binary or octal form. If you are a bit rusty on octal notation, remember that you convert groups of three bits into digits as for example 15oct=1'101bin and 145oct = 1'100'101bin. Not all generator polynomials result in great encoding properties that protect the communication channel from error events. The table below lists optimal binary convolutional encoder configurations that have been used in various communication systems. (See Creonic, Viterbi Decoder User Guide, V1.0.0, Jan 16, 2012)

Table 5-3: Popular Binary Convolutional Code Configurations

Standard	Constraint Length	Rate	g0 (oct)	g1 (oct)	g2 (oct)	g3 (oct)
GSM / EDGE	5	½	33	23		
	5	1/3	33	25	37	
IEEE 802.15.3c, WiMax, LTE, DVB	7	1/3	133	171	165	
DAB	7	1/4	133	171	145	133
IEEE 802.11a/b/g/n, GSM	7	½	133	171		
	7	1/3	133	171	145	
CDMA 2000, UMTS	9	½	753	561		
	9	1/3	557	663	561	
	9	1/4	765	671	513	473
Other Optimal Polynomials to Consider	4	½	15	17		
	4	1/3	13	15	17	
	5	1/3	25	33	37	
	9	½	561	753		
	9	1/3	557	663	711	
	10	1/3	1117	1365	1633	

The Encoder States

The state of the encoder is the content of the memory elements of the embedded shift register. For the ½ and 1/3 rate encoders shown earlier, the state is defined by the values at nodes 2, 3, and 4. Given that the shift register contains 3 memory elements, the total number of possible states is equal to $2^3 = 8$.

Example 5.4: *Progression of Encoder States*

In the following example, we will illustrate how the encoder state and its outputs progress given an input sequence of [1 0 1 1 0 1 0 0]. We will use the ½ rate encoder using generator polynomials $g1 = 1101$ and $g2 = 1111$ for this example. This is the ½ rate encoder shown in the last figure on the left.

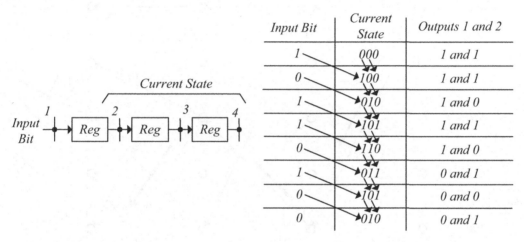

Input Bit	Current State	Outputs 1 and 2
1	000	1 and 1
0	100	1 and 1
1	010	1 and 0
1	101	1 and 1
0	110	1 and 0
1	011	0 and 1
0	101	0 and 0
0	010	0 and 1

Figure 5-77: State Transitions in a Convolution Encoder of Constraint Length = 4

Given that the constraint length is equal to 4, there are a total of 16 different bit combinations that produce outputs. In the next figure, we illustrate how the encoder transitions from the current to the next state and realize that from any single state only two others may be reached. Assuming that the encoder is currently in state [0 0 0], the only possible outputs that can be produced are [0 0] and [1 1]. The fact that neither [0 1] nor [1 0] are possible output combinations will be used by the decoder to better detect and correct errors.

Input Bit	Current State	Next State	Current Encoder Out1 Out2		Input Bit	Current State	Next State	Current Encoder Out1 Out2	
0	000	000	0	0	0	100	010	1	1
1		100	1	1	1		110	0	0
0	001	000	1	1	0	101	010	0	0
1		100	0	0	1		110	1	1
0	010	001	0	1	0	110	011	1	0
1		101	1	0	1		111	0	1
0	011	001	1	0	0	111	011	0	1
1		101	0	1	1		111	1	0

Figure 5-78: All Possible Encoder States and Input Bit Combinations along with their Output Values

The Trellis Diagram

The trellis diagram is an intuitive method of showing how the encoder progresses through different states and what the associated limitations are. Assume for a moment, that at discrete time equal to 0 the encoder shift register is reset to zero, and we find ourselves in state 000. From the diagram below, we can see that the next possible states that can be reached are 000 and 100 depending on what the current input bit value is. The progression from one state to the next is indicated by solid arrows if the input bit was a 0, and dashed arrows if the input bit was a 1. Situated along the arrows are bit pairs indicating the encoder output.

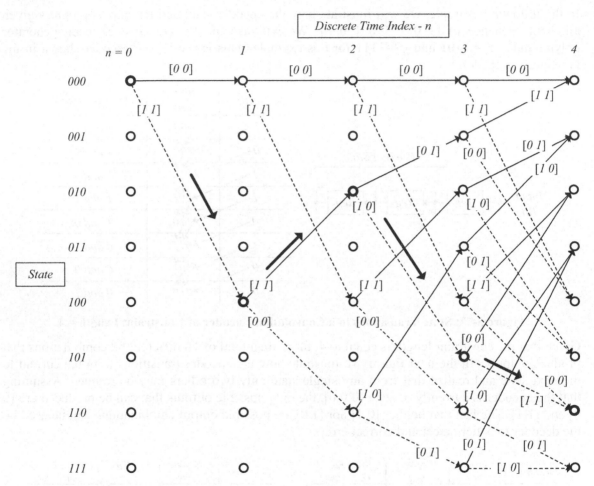

Figure 5-79: Trellis Diagram for Rate ½ Convolutional Encoder with g1 = 1101 and g2 = 1111

Note that whichever state the encoder currently occupies, its two possible outputs are as different as they can be. Meaning that if you are in states 000, 001, 100, and 101, the possible outputs are either [1 1] or [0 0]. Likewise, states 010, 011, 110, and 111 will produce either [0 1] or [1 0]. On the previous page, we presented a tabulated example of how the encoder changes and produces outputs based on an input sequence equal to [1 0 1 1 0 1 0 0]. This example is partially illustrated via the thick arrows that twist through the trellis to produce output values of [1 1], [1 1], [1 0], [1 1], … [1 0], [0 1], [0 0], [0 1] in the figure above.

MatLab Implementation of Convolutional Encoder

```
%% Defining and Setting up the Rate 1/2 Convolutional Encoder
g1          = [1 1 0 1];      %% Generator polynomial 1
g2          = [1 1 1 1];      %% Generator polynomial 2
L           = length(g1);     %% Constraint Length = 4
m           = L - 1;          %% Number of Memory Elements = 3

ShiftReg    = zeros(1, m);            %% Memory of Encoder, holds the State
InputBits   = [1 0 1 1 0 1 0 0];      %% The Input Bits
In_Length   = length(InputBits);
En_States   = zeros(1, In_Length);    %% Storage for Traversed States
OutputBits  = zeros(1, 2*In_Length);  %% Storage for Output Bits

%% Running the Encoder
for i = 1:In_Length
    InputBit                = InputBits(1,i);
    En_States(1,i)          = sum(ShiftReg.*[4 2 1]);
                                    % Save state as Decimal Number
    Array                   = [InputBit ShiftReg]; % Nodes 1,2,3,4
    Out1                    = mod(sum(bitand(Array, g1)), 2);
    Out2                    = mod(sum(bitand(Array, g2)), 2);
    OutputBits(1,(i-1)*2 + 1) = Out1;  % Transferring Outputs to
    OutputBits(1,(i-1)*2 + 2) = Out2;  % Storage
    ShiftReg                = [InputBit ShiftReg(1,1:m-1)];  % Update State
end

%% Result and Debug Information
display(['InputBits:      ' num2str(InputBits)]);  % InputBits:  1  0  1  1  0  1  0  0
display(['OutputBits:     ' num2str(OutputBits)]); % OutputBits:11 11 10 11 10 01 00 01
display(['Encoder States: ' num2str(En_States)]);  % Enc States: 0  4  2  5  6  3  5  2
```

The encoder shown above is quite specialized and will only work for the ½ rate constraint length = 3 case. At the end of this section, after we have gained some additional understanding regarding convolutional encoding and decoding, we present MatLab code for a full featured convolutional encoder that will elegantly work for all possible configuration.

5.6.4 Convolutional Decoder (The Viterbi Algorithm)

The output bit stream created by the convolutional encoder travels across the noisy communication channel and arrives at the decoder input in the receiver. Given the noise in the system, the encoded bit stream may arrive at the decoder with embedded errors. It is the function of the convolutional decoder to correctly estimate the information bit stream that entered the encoder on the transmit side, given the noise corrupted received bit stream. The decoder is fully aware of the encoder structure and initial state and can therefore recreate conditions under which the encoder operated. In fact, the decoder will attempt to determine which path was taken through the encoder's trellis structure and thus estimate the unknown input bits that caused said path. Take a look at the trellis diagram of the last page once more and notice that the encoder, whose memory elements are initialized to zero, starts in state zero and can reach all 8 possible states after 3 iterations (discrete time index = 3). Note further, that at a discrete time index of 3, there are a total of 8 unique paths through the trellis, each one leading to one of 8 states.

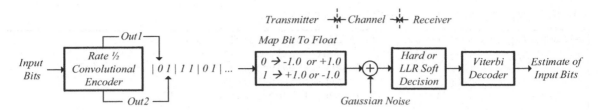

Figure 5-80: The Combination of Convolutional Encoding and Decoding

There are two possible exit directions out of each state, and the number of unique paths through the trellis increases to 16 as we arrive at discrete time index $n = 4$. As a matter of fact, each subsequent time step will increase the unique number of paths through the trellis by factor of two. Each path is unique in the way it moves from state to state and also in the sequence of output bits that it generates. The convolutional decoder in the receiver knows about all these paths and decides to compare the incoming bit stream with the output sequences of each possible path to find the best match. Once found, the path whose output sequence best matches the incoming bit stream is selected as the true, corrected transmit information. If the match isn't perfect, then the decoder assumes an error has occurred during transmission. This method of comparing the actual bit stream entering the decoder with all possible encoder output bit streams and picking the one that is the closest match is called maximum likelihood decoding. As mentioned earlier, this method makes perfect sense but is doomed to failure given that the number of paths, and thus the potential encoder output bit streams, increase geometrically as discrete time progresses, and we don't have infinite amounts of memory or time to store and compare all the possible path information.

Before we delve into the Viterbi algorithm, which solves the above-mentioned dilemma, take a look at how bits 0/1 are mapped to floating point numbers. Depending on the specification, the bits 0/1 may be mapped to -1.0/+1.0 (WLAN) or +1.0/-1.0 (LTE). You may now assume that Gaussian noise is added to the floating point or soft bits by the channel. At the receiver, we either do a hard demapping operation taking us from the floating point numbers back to bits, or we may retain the softbit values for which we compute the LLR ratio as shown in the last section. Using the LLR bit beliefs yields a 2dB advantage in decoding performance.

The Viterbi Algorithm

Take a look at the trellis diagram and note that at a discrete time index of 4, every state can be reached from two other states. All in all, there are 16 paths that arrive at 8 states at time index 4. The figure below illustrates that the only way to reach state 0 at a discrete time index of 4 is from states 0 or 1 and then only if the input bit is equal to zero.

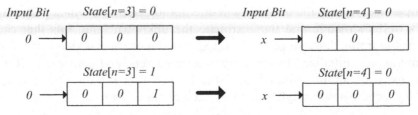

Figure 5-81: Reaching State 0 from State 0 and 1

Viterbi suggested that as two paths arrive at each state, the one with the greater difference or distance to the incoming bit stream should be discarded, whereas the other path survives. Therefore, with every iteration, 16 paths are trimmed back to 8 making the overall memory requirements manageable. Before we show how the survival process works in the trellis diagram, let's define a few terms that will help us better understand the process.

→ Hamming distance: The metric is the number of bits that differ between the received sequence and a particular path through the trellis. Observe the receive sequence [1 1], [1 1], [1 0] and compare it to the top path that starts in state 000 and ends in state 000 at $n = 3$. That path features encoder output bits [0 0], [0 0], [0 0] producing a Hamming distance of 5 bits.

→ Branch metric: This metric is the Hamming distance between the received bits and the encoder output bits of any path as we progress from one discrete time instance to the next. Take a look at the transition from $n=3$ to $n=4$. There are two paths that enter state 000 at $n = 4$. The first comes from the lower valued state 000 and would produce an encoder output of [0 0], whereas the second arrives from upper valued state 001 and would produce an encoder output of [1 1]. Comparing these encoder bits to bits [1 1] received during the transition from $n=3$ to $n=4$, we end up with two branch metrics $BML[0]=2$ and $BMU[0]=0$ respectively, where L/U means lower/upper.

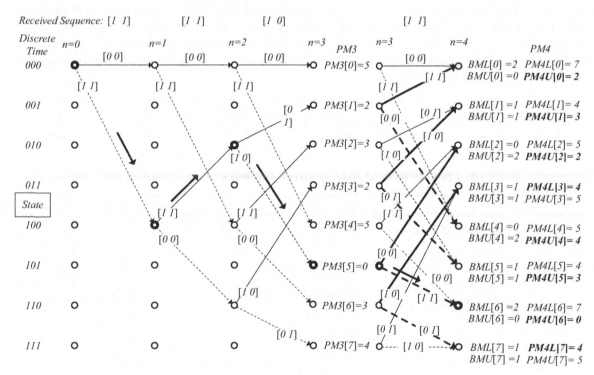

Figure 5-82: Trellis Diagram with Path and Branch Metrics (Survivors in Bold Type)

→ Path Metric: This metric is the Hamming distance between an entire path and the received input bit sequence. Take another look at the explanation of Hamming distance, which found that the path metric of the path starting in state 000 (at $n = 0$) and ending in state 000 (at $n = 3$) was equal to 5. The figure above shows the different path metrics at discrete time $n = 3$ labeled PM3[0], PM3[1], PM3[2], ..., PM3[7].

The Key Shortcut of the Viterbi Algorithm

Take a look at the eight nodes (states) at time $n = 4$. Two paths enter each state producing two path metrics BML[state] and BMU[state]. The states from which these two paths originated at $n = 3$ each feature a path metric of PM3[0] and PM3[1]. We now add those path metrics valid at $n = 3$ to the branch metrics to produce two new path metrics valid at $n = 4$. Take a look at state 000 at $n = 4$ and take note how the path metrics are updated.

$$PM3[0] + BML[0] = PM4L\ [0] = 5 + 2 = 7$$

$$PM3[1] + BMU[0] = PM4U[0] = 2 + 0 = 2$$

The Viterbi algorithm decides that the smaller path metric, $PM4U[0]$, indicates the better path and retains the path originating in state 1 at $n = 3$ and discards the one originating in state 0, yielding PM4[0] = 2. The decoder will thus never have to track more than 2^m paths, were m is the number of memory elements in the encoder. The decoder will examine every state at every discrete time instance n, and eliminate half the paths. Sounds like a risky proposition, but it works very well.

The Mechanics of Viterbi Decoding

In this section, we enumerate the different structures and mechanisms that we need to implement the Viterbi decoder in software. The following features are applicable for any convolutional encoder configuration. As a matter of fact, we will soon show a MatLab implementation that works for any BCC setup. Remember that we always have $N = 2^m$ states where m is the number of memory elements in the encoder.

→ We need two vectors to hold the path metrics for each state. The first will hold the path metrics for the state we are about to reach (right most column in table below), whereas the second holds the path metrics of the state we are just leaving (the center column in table below). As we move from $n = 3$ to $n = 4$, these two vectors carry the following values.

Table 5-4: Path Metric of Surviving Path at n = 3 and n = 4

State	Path Metric at $n = 3$	Path Metric at $n = 4$
000 = 0	PM3[0] = 5	PM4[0] = 2
001 = 1	PM3[1] = 2	PM4[1] = 3
010 = 2	PM3[2] = 3	PM4[2] = 2
011 = 3	PM3[3] = 2	PM4[3] = 4
100 = 4	PM3[4] = 5	PM4[4] = 4
101 = 5	PM3[5] = 0	PM4[5] = 3
110 = 6	PM3[6] = 3	PM4[6] = 0
111 = 7	PM3[7] = 4	PM4[7] = 4

As we move from $n = 4$ to $n = 5$, we simply copy the values in the third column to the second column before calculating the branch metrics and then updating the path metrics at $n = 5$.

→ We need a traceback matrix with dimensions $N \times NumDecodedBits$. This traceback matrix will hold the previous state value of the surviving path. From the earlier trellis diagram, the traceback matrix at column 4 (for $n = 4$) should hold the states from which the survivors originated at $n = 3$. Compare the trellis diagram and the trace back unit in the next figure.

Table 5-5: Portion of Traceback Unit

State	n =1	n=2	n=3	n=4
000	? (0)	? (0)	? (0)	1
001	? (2)	? (2)	? (2)	3
010	? (4)	? (4)	? (4)	5
011	? (6)	? (6)	? (6)	6
100	? (0)	? (0)	? (0)	1
101	? (2)	? (2)	? (2)	3
110	? (4)	? (4)	? (4)	5
111	? (6)	? (6)	? (6)	6

You need to compute the two path metrics before you can determine the survivor and find out from which state it originated. Notice that it is not until $n = 4$ that we are even in a position to figure out survivors. So what happens to columns one through three? Do we need to execute some special procedure to figure out these values? A little further down we will show you a very elegant trick that smoothly and correctly fills in these state values, which for now are shown in parenthesis. Given that the best path metric was $PM4[6] = 0$ (see table 5-4), the arrows in the table above indicate the trace back operation starting in state six.

The Viterbi Decoder Loops

For our software implementation of the Viterbi decoder, we set up two nested loops. The outer loop steps us from discrete time $n = 0$ to $n = NumBitsToDecode$. The inner loop, will compute the branch metric, the path metric and the entry into the trace back matrix for each one of the N states. However, before we enter the outer loop, it is advisable to compute all possible encoder outputs, so that they don't need to be constantly recomputed during operation. We store the encoder outputs as a function of their state and their input bit value in the following two matrices. (See section 1c in the MatLab code.)

→ *EncoderOutputInput0[State]* where *State* = 0, 1, … *N*-1   (Encoder output for input bit = 0)

→ *EncoderOutputInput1[State]* where *State* = 0, 1, … *N*-1   (Encoder output for input bit = 1)

We now begin the first iteration of the outer loop, which takes us from time $n = 0$ to $n = 1$. In the inner loop, we iterate from *StateIndex* = 0 to *StateIndex* = N/2-1 and identify *StateA* and *StateB* as follows. (See the two loop at the start of section 2 in the MatLab code)

→ *StateA = StateIndex*   (See section 2a in the code)

→ *StateB = StateA + N/2*

If you take another look at the trellis diagram, you will notice that both *StateA* and *StateB* are reachable from two other states and these states are identical. (Check section 2b in the code).

→ *PreviousLowerState* = 2·*StateA*   (the previous state with the lower integer value)

→ *PreviousUpperState* = 2·*StateA*+1 (the previous state with the larger integer value)

We now compute the branch metrics going into *StateA* as follows (See Section 2c in the code):

➔ *BML[StateA] = sum(RxBits* **XOR** *EncoderOutputInput0[PreviousLowerState])*

➔ *BMU[StateA] = sum(RxBits* **XOR** *EncoderOutputInput0[PreviousUpperState])*

We now update the path metrics for *StateA* as follows.

➔ *PML [StateA] = BML[StateA] + PM[PreviousLowerState]*

➔ *PMU [StateA] = BMU[StateA] + PM[PreviousUpperState]*

We then determine which path metric is better, select the survivor accordingly and thereby determine which state, the *PreviousLowerState* or *PreviousUpperState*, the survivor path came from. We now enter that state into the trace back matrix for discrete time *n* and *StateA*. Section 2c in the MatLab code shows these computations for both *StateA*, as shown here, as well as *StateB*. The diagram below illustrates the relationships between states (and some of the quantities required for branch and path metric calculations) as discrete time increments by one unit.

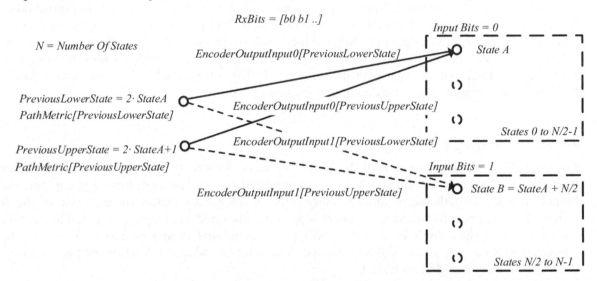

Figure 5-83: Viterbi Iteration for the Inner Loop

Once we repeat the steps indicated above for *StateB*, we have executed the Viterbi decoder loop for a single advance in time. We need to repeat the procedure for each time step. This doesn't seem to be a big problem as we continue with *n* = 5, *n* = 6, and so on. But what about the columns 1 through 3 which we filled with question marks in the last table. Given that we know that the encoder starts in state 0 (as we initialize it as such), we can preprogram the path metric to favor state zero at time *n* = 0. We thus bias state zero by assigning it a very small path metric and assign large ones to the others. We now start the Viterbi loop to compute path metrics and survivors for the transitions to *n* = 1, *n* = 2, and so on. The poor initial path metric of state 1 through 7 will keep the algorithm from generating survivor paths that lead back to those states.

Table 5-6: Progression of Path Metrics from n = 0 to n = 3

Path Metric at $n = 0$	Path Metric at $n = 1$	Path Metric at $n = 2$	Path Metric at $n = 3$
0	2	4	5
100	100	100	2
100	100	0	3
100	100	100	2
100	0	2	5
100	100	100	0
100	100	2	3
100	100	100	4

What do to when we are done?

Once we have applied the Viterbi loop, we will end up with a path metric vector of length N for the last decoding time instance. We now need to trace back the final survivor path through the trace back matrix to find out the correct bits that were encoded at the transmitter. Should we start the trace back procedure at the state with the best path metric? No. In most cases we append tail bits (all zeros) to the message thus clearing the encoder's memory elements. Thus, the encoder is guaranteed to end in state 0 and it is at this state that we being our trace back procedure. As you work through the traceback unit, it is a simple matter to determine which bit must have been present at the encoder input. States 0 through $N/2 - 1$ must have been reached by an encoder input bit = 0, whereas states $N/2$ to N -1 must have been reached by an encoder input bit = 1. This makes sense as the encoder input bit will be the most significant bit of the state that follows.

Hard Decoding vs Softbit Decoding

In our previous discussion, a smaller path metric is better as it represents the Hamming distance (the number of positions in which the received bits differ from the encoded bits of the selected path). For hard bits values 0/1, the XOR operation is used during the branch metric calculation produces the Hamming distance. (c is the number of output bits for every input bit.)

$$BM = \sum_{n=0}^{c-1} ReceivedBit[n] \; XOR \; EncodedBit[n]$$

In real communication systems, bit values 0/1 are translated to either +1.0/-1.0 or -1.0/+1.0 depending on the standard (IEEE, 3GPP, 3GPP2 … etc.). These bits are transmitted, then noise corrupted during transmission and finally arrive as softbits at the receiver. As mentioned earlier in this chapter, log likelihood ratio processing is usually applied to these bits, which scales them somewhat. However, they are still floating point values where a positive value indicates the belief that a 1 has arrived rather than a 0. The branch metric needs to be calculated via correlation rather than the XOR operation (modulo 2 addition).

$$BM = \sum_{n=0}^{c-1} ReceivedBit[n] \cdot EncodedBit[n]$$

In this case, a large positive branch metric indicates that the received and encoded bits are similar. Thus, survivors are selected based on how large the path metric is rather than how small.

If done correctly, LLR soft decoding will provide a 2dB performance advantage over hard decoding.

The Final Implementation

The following Viterbi Decoder implementation will work for hard and soft decoding as well as for both bit to float mapping schemes and any rate and constraint length.

```
function DecodedBits = ViterbiDecoder(EncodedBits,   ...
                                ConstraintLength,   ... % Num memory elements + 1
                                PolynomialsOct,     ... % Lte -> [133, 171, 164]
                                TailBitingBool,     ... % True-> Indicates TailBiting
                                ModeString) % Hard, SoftNonInverting, SoftInverting
%% 0. Change input bit format to SoftNonInverting.
if   (strcmp(ModeString, 'Hard'))
    EncodedBits = 2*EncodedBits - 1; % Convert to non inverting softbits (0/1 -> -1/+1)
elseif(strcmp(ModeString, 'SoftInverting'))
    EncodedBits = -EncodedBits;  % Convert to non inverting softbits  (+1/-1 -> -1/+1)
elseif(~strcmp(ModeString, 'SoftNonInverting'))
    error('Unsupported ModeString');
end


% ------------------------------------------------------------------
%% 1. Initializations before we run the Viterbi Decoder loop
% 1a. Initialize Constants
NumberOfStates     = 2 ^ (ConstraintLength - 1);
NumOutBitsPerInBit = length(PolynomialsOct);

% 1b. Translate the polynomials from octal form to binary vector form. In Lte, the
% PolynomialsOct are [133, 171, 164]oct thus [1'011'011, 1'111'001, 1'110'101]bin
PolynomialsBin = zeros(NumOutBitsPerInBit, ConstraintLength);
PolynomialsOct = PolynomialsOct(:); % Guarantee a column vector
for i = 1:length(PolynomialsOct)
    PolynomialsBin(i, :) = oct2poly(PolynomialsOct(i, 1));
end

% 1c. We will now build two matrices each featuring dimensions NumberOfStates by
%     NumOutBitsPerInBit
% The first  matrix holds the encoder output for state i with input bit = 0.
% The second matrix holds the encoder output for state i with input bit = 1.
% We pre-build these matrices for speed so that during the Viterbi loop we don't
% have to keep recalculating the same values.
EncoderOutputForInput0 = zeros(NumberOfStates, NumOutBitsPerInBit);
EncoderOutputForInput1 = zeros(NumberOfStates, NumOutBitsPerInBit);
for i = 0:NumberOfStates-1
  StateBin = de2bi(i, ConstraintLength - 1, 'left-msb');
  for b = 1:NumOutBitsPerInBit  % The *2-1 factor changes hard to NonInverting Softbits
    EncoderOutputForInput0(i+1,b)=mod(sum([0, StateBin].*PolynomialsBin(b,:)),  2)*2-1;
    EncoderOutputForInput1(i+1,b)=mod(sum([1, StateBin].*PolynomialsBin(b,:)),  2)*2-1;
  end
end
```

```
% 1d. Build the Traceback matrix. The first column of the traceback matrix indicates
%      the state we are in at the very start.
NumDecoderOutputBits = length(EncodedBits)/NumOutBitsPerInBit;
DecodedBits          = zeros(1, NumDecoderOutputBits);
TraceBackUnit        = zeros(NumberOfStates, NumDecoderOutputBits);

% 1e. Define the PathMetric Array. The path with the largest path metric is the winner.
%      The Viterbi decoder always assumes that the encoder started in state 0. For Tail
%      biting mode, this is not true. We will find a work around for this mode.
PathMetricArray      = -1000*ones(NumberOfStates, 1);
PathMetricArray(1,1) = 0;         % We favor state 0 (the first element in the vector)
PathMetricCopy       = PathMetricArray; % A copy of the array for update purposes
%      PathMetricMatix exist for debugging purposes and shows path metric progression.
PathMetricMatrix     = zeros(NumberOfStates, NumDecoderOutputBits + 1);
PathMetricMatrix(:,1) = PathMetricArray;

%% 2. Let's start the Viterbi Decoder Loop (the outer loop)
for OutputBitIndex = 1:NumDecoderOutputBits
    Range        = (OutputBitIndex-1)*NumOutBitsPerInBit + (1:NumOutBitsPerInBit);
    ReceivedBits = EncodedBits(1, Range);  % Grab NumOutBitsPerInBit encoded bits

    % Run through each state (the inner loop)
    for StateAsInt = 0:NumberOfStates/2 - 1
        % 2a. We will process two states per loop iteration. StateA and StateB
        % No matter what the previous state before StateA was, it's input bit was 0.
        StateAAsInt = StateAsInt;
        % No matter what the previous state before StateB was, it's input bit was 1.
        StateBAsInt = StateAAsInt + NumberOfStates/2;

        % 2b. We could only have gotten to StateA via two other States.
        % Interestingly, the two states that lead to StateA, also lead to StateB.
        % One of the states has a lower integer value, and one a higher integer value
        PreviousLowerStateAsInt = 2*StateAsInt;
        PreviousUpperStateAsInt = 2*StateAsInt + 1;

        % -------------------------------------------------------------------------
        % 2c. Compute branch and path metrics for StateA.
        % Let's first find the encoder outputs generated during the transition from
        % these two previous states to StateA. Remember, because StateAAsInt is always
        % NumberOfStates/2, the input bit to get to StateA must have been a 0.
        EncoderOutputLower = EncoderOutputForInput0(PreviousLowerStateAsInt + 1, :);
        EncoderOutputUpper = EncoderOutputForInput0(PreviousUpperStateAsInt + 1, :);

        % New path metric = old path metric + branch metric
        BranchMetricLower  = sum(ReceivedBits .* EncoderOutputLower);
        BranchMetricUpper  = sum(ReceivedBits .* EncoderOutputUpper);
        NewPathMetricLower = PathMetricArray(PreviousLowerStateAsInt+1,1) + ...
                                                        BranchMetricLower;
        NewPathMetricUpper = PathMetricArray(PreviousUpperStateAsInt+1,1) + ...
                                                        BranchMetricUpper;
        % And the survivor for StateA is????
        if(NewPathMetricLower >= NewPathMetricUpper)
            SurvivorPathMetric        = NewPathMetricLower;
            SurvivorPreviousStateAsInt = PreviousLowerStateAsInt;
        else
            SurvivorPathMetric        = NewPathMetricUpper;
```

```
                SurvivorPreviousStateAsInt = PreviousUpperStateAsInt;
        end
        TraceBackUnit(StateAAsInt + 1, OutputBitIndex) = SurvivorPreviousStateAsInt;
        PathMetricCopy(StateAAsInt + 1, 1)             = SurvivorPathMetric;


        % ------------------------------------------------------------------
        % Compute branch and path metrics for StateB.
        EncoderOutputLower = EncoderOutputForInput1(PreviousLowerStateAsInt + 1, :);
        EncoderOutputUpper = EncoderOutputForInput1(PreviousUpperStateAsInt + 1, :);

        % New path metric = old path metric + branch metric
        BranchMetricLower  = sum(ReceivedBits .* EncoderOutputLower);
        BranchMetricUpper  = sum(ReceivedBits .* EncoderOutputUpper);
        NewPathMetricLower = PathMetricArray(PreviousLowerStateAsInt+1,1) + ...
                                                        BranchMetricLower;
        NewPathMetricUpper = PathMetricArray(PreviousUpperStateAsInt+1,1) + ...
                                                        BranchMetricUpper;

        % And the survivor for StateB is????
        if(NewPathMetricLower >= NewPathMetricUpper)
            SurvivorPathMetric       = NewPathMetricLower;
            SurvivorPreviousStateAsInt = PreviousLowerStateAsInt;
        else
            SurvivorPathMetric       = NewPathMetricUpper;
            SurvivorPreviousStateAsInt = PreviousUpperStateAsInt;
        end
        TraceBackUnit(StateBAsInt + 1, OutputBitIndex) = SurvivorPreviousStateAsInt;
        PathMetricCopy(StateBAsInt + 1, 1)             = SurvivorPathMetric;
    end
    % Copy the updated path metrics into the original array
    PathMetricArray = PathMetricCopy;
    PathMetricMatrix(:,OutputBitIndex + 1) = PathMetricArray;
end

%% 3. Work your way backwards through the trace back unit. If the transmitted bit
%     stream has padding bits that forced the encoder into the zero state then we know
%     to start the trace back from state 0. If we don't know what the finat state was,
%     then begin the traceback from the state with the largest (best) path metric.
FinalStateAsInt = 0;
if(TailBitingBool == true)
    [~, Temp] = max(PathMetricArray);
    FinalStateAsInt = Temp - 1;
end


% Start the traceback
CurrentStateAsInt    = FinalStateAsInt;
for CurrentOutputBitIndex = NumDecoderOutputBits:-1:1
    if(CurrentStateAsInt < NumberOfStates/2); LastBitEnteringEncoder = 0;
    else;                                     LastBitEnteringEncoder = 1; end
    DecodedBits(1, CurrentOutputBitIndex) = LastBitEnteringEncoder;

    % The CurrentStateAsInt is now the previous state as indicated by trace back unit
    CurrentStateAsInt = TraceBackUnit(CurrentStateAsInt + 1, CurrentOutputBitIndex);
end
```

Survivor Path Behavior

In the diagram below, we follow a survivor path for the case were the receiver knows that the encoder both started and ended in state 000. Take a look at the final result at discrete time $n = NumDecodedBits$. For each state, we may now look into the trace back matrix and follow its survivor path back in time. We make the following observations:

→ The survivor paths, ending in each state at $n = NumDecoderBits$, merge into one path a certain number of time steps in the past.

→ Whereas the final state was indeed equal to 000, the final path metric indicated the smallest (best for hard decoding) value at state 100. Had we not known what the final state was, we would have started our trace back from the wrong position.

→ Because the survivors originated from a single path a few time steps in the past, the final path metric differences are likely due to differences in branch metric calculations during the last couple of discrete time steps. Errors in the received bit sequence during the last few time steps can easily cause this different in final path metric.

→ The number of time steps between the point in time that the common path slit into many and the final time is called trace back depth.

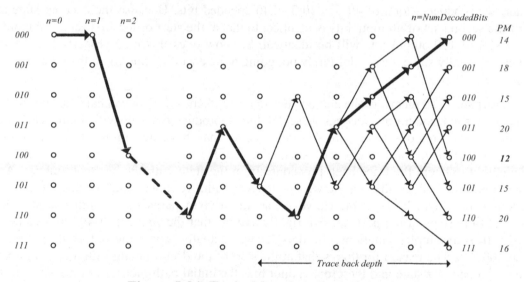

Figure 5-84: Typical Survivor Path Behavior

For a particular encoder configuration, we can easily simulate the behavior of the Viterbi algorithm to find out what this trace back length typically is. For a software implementation, we typically don't need to worry about this length as we can easily store the entire trace back matrix even when the number of transmitted bits is large. However, in a hardware implementation, this is not the case and we take care only to remember trace back information for a number of past time instances equal to the trace back length. We don't show the hardware implementation of the Viterbi algorithm in this text, but with some knowledge of digital logic, it is not hard to figure out from the software implementation that we presented.

Understanding Tail-Biting Operation

In order for us to begin the trace back operation from the proper state, the Viterbi decoder must know the final state of the convolutional encoder at the transmit side. You don't necessarily want to pick the state with the best path metric to start the trace back as we saw in the last figure. For this reason, we append a certain number of tail bits (all zeros) at the end of the transmit message to force the convolutional encoder into the zero state when finished. Clearly, these tail bits must be encoded and transmitted, which introduces unnecessary overhead. For example, the master information block, or MIB, in LTE consists of 40 bits (24 message bits and 16 CRC bits), which need to be convolutionally encoded. LTE uses a convolutional encoder with constraint length of 7 thus requiring 6 tail bits to be inserted at the end of the 40 MIB bits. This overhead was deemed too much by the architects of the 3GPP specification. To avoid transmitting these bits, we use a scheme called tail biting. Tail biting consists of a small modification to the encoder initialization, and a modified way of using the normal Viterbi decoder we developed.

At the transmitter, we first determine the final state of the encoder after having pushed in all 40 input bits. Whereas the convolutional encoder is usually initialized to all zeros (the zero state), we now initialize the memory elements in the encoder to the projected final state. This guarantees that the initial and final states of the encoder are equal, yielding an interesting property. Imagine repeating the 40 MIB bits two times to produce an 80-bit input bit sequence. Given a rate $1/3^{rd}$ encoder, we produce a total of $40 \cdot 3 + 40 \cdot 3 = 240$ encoded bits. Because the encoder state at the start of the second set of 40 input bits is identical to that at the start of the first set of 40 input bits, the two sets of 120 output bits will be identical. So, how does this nice property help us at the receiver, especially since we are definitely not going to be encoding any more than the original 40 bits?

The receiver sees 120 received bits and decides to repeat them for a concatenated sequence of 240 bits. The Viterbi decoder now decodes these 240 bits to produce two sets with 40 bits each. As we have seen in the previous figure, the Viterbi decoder may have accumulated path metrics that have become erroneous at the very end. As the decoder doesn't know the correct final state, it has no choice but to begin the trace back procedure from the state with the best path metric. As this can potentially lead to identifying the wrong survivor path, we must conclude that the second set of decoded bits might be in error. However, as all survivor paths merge into one at a certain number of time steps in the past, we can rightly assume that the first set of 40 bits was decoded properly. In some implementations, the decoder will actually repeat the output three times and decode 360 bits. The reason for this is that at the start of the decoding algorithm, we are unaware of the initial encoder state and therefore cannot bias the initial path metrics at time $n = 0$ to favor any initial state as we would for the non tail-biting version. Therefore, an error in the first 40 decoded bits is not unthinkable. Try it out yourself by changing the MatLab code provided. The tail-biting process trades improved throughput at the expense of computational effort.

Updated Convolutional Encoder

The following code segment demonstrates an updated convolutional encoder that works for all configurations including different rates, polynomials, bit formats and operational mode (tail-biting or normal).

```matlab
function EncodedBits = ConvolutionalEncoder(InputBits,          ... % 0s and 1s
                                   ConstraintLength,   ... % Number of memory elements + 1
                                   PolynomialsOct,     ... % Lte -> [133, 171, 164]
                                   TailBitingFlag,     ... % True or False
                                   ModeString)     % Hard, SoftNonInverting, SoftInverting
% The number of output bits per input bits = 1/EncoderRate
NumOutBitsPerInBit = length(PolynomialsOct);

% Translate the polynomials from octal form to binary vector form. In Lte, the
% PolynomialsOct are [133, 171, 164]oct and [1'011'011, 1'111'001, 1'110'101]bin
PolynomialsBin = zeros(length(PolynomialsOct), ConstraintLength);
PolynomialsOct = PolynomialsOct(:); % Guarantee a column vector
for i = 1:length(PolynomialsOct)
    PolynomialsBin(i, :) = oct2poly(PolynomialsOct(i, 1));
end

% Is the convolutional encoder/decoder of the Tail-biting type?
if(TailBitingFlag == true)
    % The initial register state must be equal to the final projected state
    Reg = fliplr(InputBits(1, end - ConstraintLength + 2:end));
else
    % The initial register state of the encoder must be equal to all zeros.
    Reg = zeros(1, ConstraintLength - 1);
end

% Run the convoluational encoder
EncodedBits = zeros(1, length(InputBits) * NumOutBitsPerInBit);
for i = 1:length(InputBits)
    InputBit   = InputBits(1,i);
    OutputBits = zeros(1, NumOutBitsPerInBit);
    for b = 1:NumOutBitsPerInBit        % Compute each output bit
        OutputBits(1,b) = mod( sum([InputBit, Reg] .* PolynomialsBin(b, :)), 2);
    end
    Range               = NumOutBitsPerInBit*(i-1) + (1:NumOutBitsPerInBit);
    EncodedBits(1, Range) = OutputBits;
    Reg                 = [InputBit, Reg(1,1:ConstraintLength-2)];
end

% Format the output bits
if( strcmp(ModeString, 'SoftNonInverting') == 1)
    EncodedBits = 2* EncodedBits - 1;        % Maps from 0/1 to +1/-1
elseif( strcmp(ModeString, 'SoftInverting') == 1)
    EncodedBits = -(2* EncodedBits - 1);     % Maps from 0/1 to -1/+1
elseif (~strcmp(ModeString, 'Hard') == 1)    % No reformatting if ModeString == 'Hard'
    error('Unsupported Mode String');
end
```

5.6.5   Low Density Parity-Check (LDPC) Codes

Low density parity-check (LDPC) codes are a class of linear block codes that have the ability to correct bit errors in a received binary message. These codes were first described by Robert Gallager in his Doctor of Science thesis in 1960. Due to the restricted amount of computing power back then, these codes were left unused in favor or Reed-Solomon codes, binary convolutional codes and later Turbo Codes. More recently, however, LDPC codes have gained significantly in popularity and are now found in different WLAN implementations, WiMax as well as the cellular 5G specifications. Note that both Turbo Codes and LDPC codes allow operation close to Shannon's predicted capacity limit. We may even claim that these codes are as good as it gets in terms of providing error free communication. The performance of these codes is especially good when the messages to be transmitted have many bits. This compares to binary convolution codes which perform equally well for short bit messages but can no longer compete with Turbo or LDPC codes as these messages become longer. The LTE specification uses binary convolutional codes only for shorter messages. Before we dive into the background and implementation of LDPC codes, we will first review parity check concepts and algebraic codes, on which LDPC codes are based.

Single Parity Check Codes

Parity checks have been used for a very long time and in its simplest implementation, the single parity check counts the number of 1s in a bit message b_k, where $k = 0, 1, \ldots K\text{-}1$, and appends a single parity bit, p_0, according to the following expression.

$$mod(\sum_{k=0}^{K-1} b_k, 2) = p_0 \qquad (0)$$

The parity bit is equal to zero for an even number of 1s in the message and equal to one for an odd number of 1s in the message. The following bit message features three ones yielding a single parity bit equal to 1.

Bit Message = 010011 → Parity Bit = 1

As an obvious but nevertheless useful consequence, the modulo-two sum of both the message bits and the parity bit must be equal to zero, as an even number of 1s is required.

$$mod((\sum_{k=0}^{K-1} b_k) + p_0, 2) = 0 \qquad (1)$$

The single parity check construct is rather simplistic, and if at the receiver, the sum of the parity and received message bits is 1, then there is in fact no way to correct the error as we don't know in which bit position it occurred. We are simply able to detect an error but nothing more.

5.6.5.1 The Hamming Code

A way needed to be found to increase the amount of bit error protection, which is done by applying multiple single parity checks to the message bit sequence. In the late 1940's Richard Hamming developed a family of algebraic error correcting codes that were based on the application of multiple single parity checks. The following is the famous 4, 7 Hamming code featuring $K = 4$ message bits and $L = 3$ parity bits to yield 7 transmitted bits. The code features three single parity check sets applied to four message bits. The three check sets include the following bits.

→ Parity check set 0: [b0, b1, b2 → p0]
→ Parity check set 1: [b0, b2, b3 → p1]
→ Parity check set 2: [b0, b1, b3 → p2]

The 4,7 Hamming code is able to correct the occurrence of a single error in the set of seven transmitted bits. Assume for a moment that b_1 was in error causing equation (1) to fail for the parity check sets 0 and 2. As parity check set 1 passes, we know that b_0, b_2, b_3 are correct, which indicates that b_1 must have been in error for parity checks 1 and 3 to fail. Certainly, an error in both p_0 and p_2 could have led to the same situation, but this represents two errors in the transmitted sequence, which is a situation beyond the correction abilities of the 4,7 Hamming code. The figure below illustrates the parity bit generation process via equation (0) and the error detection process of each parity check set as defined in equation (1).

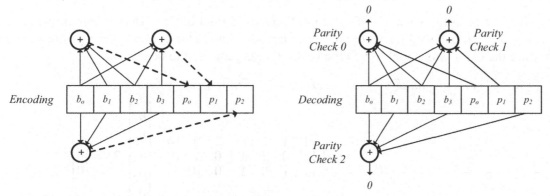

Figure 5-85: Modulo Two Addition Configuration for K = 4, $N=K+L$ = 7 Hamming Code

Note that we define the rate of an error correction code as the ratio of the number of message bits and the total number of encoded bits. The encoding rate for the 4,7 Hamming code is equal to 4/7.

$$EncodingRate = \frac{NumMessageBits}{NumEncodedBits} = \frac{4}{7}$$

5.6.5.1 Parity Check and Generator Matrices

The Generator Matrix, G

Whereas the two illustrations in the figures above describe the same configuration, let us first take a look at the one to the left, which describes the computation of the parity bits via modulo two arithmetic.

$$b_0 \oplus b_1 \oplus b_2 = p_0$$
$$b_0 \oplus b_2 \oplus b_3 = p_1$$
$$b_0 \oplus b_1 \oplus b_3 = p_2$$

This formulation leads us to what we refer to as the generator matrix, G, which maps the message bit row vector, B, from a 4 dimensional space into the encoded bit vector, C, existing in 7 dimensional space. Clearly, the matrix G is handy in the encoding process.

$$mod(\ B \qquad\qquad \cdot \qquad\qquad \boldsymbol{G} \qquad\qquad ,2)\ = \qquad\qquad C$$

$$mod([b_0 \quad b_1 \quad b_2 \quad b_3] \cdot \begin{bmatrix} 1 & 0 & 0 & 0 & 1 & 1 & 1 \\ 0 & 1 & 0 & 0 & 1 & 0 & 1 \\ 0 & 0 & 1 & 0 & 1 & 1 & 0 \\ 0 & 0 & 0 & 1 & 0 & 1 & 1 \end{bmatrix}, 2) = [b_0 \quad b_1 \quad b_2 \quad b_3 \quad p_0 \quad p_1 \quad p_2]$$

The Parity Check Matrix, H

The rightmost illustration of the last figure presents an equivalent set of modulo two expressions leading us to the parity check matrix, H. The situation below illustrates the same set of equations after a modulo two subtraction of p_0, p_1 and p_2 from both sides of the equations.

$$b_0 \oplus b_1 \oplus b_2 \oplus p_0 = 0$$
$$b_0 \oplus b_2 \oplus b_3 \oplus p_1 = 0$$
$$b_0 \oplus b_1 \oplus b_3 \oplus p_2 = 0$$

Given that our message vector is of length 4, there are a total of $2^4 = 16$ message bit vectors B producing $2^4 = 16$ encoded bit vectors C. Each encoded bit vector C obeys the equations above, as they are the expressions with which the vector C was generated to begin with.

$$mod(\qquad\qquad \boldsymbol{H} \qquad\qquad \cdot \quad C^T, \ 2) = Zero\ Vector$$

$$mod(\begin{bmatrix} 1 & 1 & 1 & 0 & 1 & 0 & 0 \\ 1 & 0 & 1 & 1 & 0 & 1 & 0 \\ 1 & 1 & 0 & 1 & 0 & 0 & 1 \end{bmatrix} \cdot \begin{bmatrix} b_0 \\ b_1 \\ b_2 \\ b_3 \\ p_0 \\ p_1 \\ p_2 \end{bmatrix}, 2) = \begin{bmatrix} 0 \\ 0 \\ 0 \end{bmatrix}$$

Clearly, the parity check matrix H is handy during the decoding process at the receiver, as every encoded message vector C must satisfy the equation above. Let's once again assume that bit b_1 was received in error. Not just do we know that an error occurred, given that our matrix arithmetic did not yield the zero vector, we also know which bit was in error as the resulting column vector is equal to column two of the matrix H.

$$mod\left(\begin{bmatrix} 1 & 1 & 1 & 0 & 1 & 0 & 0 \\ 1 & 0 & 1 & 1 & 0 & 1 & 0 \\ 1 & 1 & 0 & 1 & 0 & 0 & 1 \end{bmatrix} \cdot \begin{bmatrix} b_0 \\ b_1 \oplus 1 \\ b_2 \\ b_3 \\ p_0 \\ p_1 \\ p_2 \end{bmatrix}, 2 \right) = mod\left(\begin{bmatrix} 0 \\ 0 \\ 0 \end{bmatrix} \oplus \begin{bmatrix} 1 \\ 0 \\ 1 \end{bmatrix}, 2 \right) = \begin{bmatrix} 1 \\ 0 \\ 1 \end{bmatrix}$$

Deriving the Generator Matrix from the Parity Check Matrix

For our simple Hamming code example, it is quite easy to write out both the generator and parity check matrices directly from the encoder structure. However, as we will find out soon, it is helpful to have a more formal conversion procedure. Note that the construction of the H matrix, seen above, is a concatenation of an LxK parity matrix, P, and an LxL identity matrix, I.

$$H = \begin{bmatrix} 1 & 1 & 1 & 0 & 1 & 0 & 0 \\ 1 & 0 & 1 & 1 & 0 & 1 & 0 \\ 1 & 1 & 0 & 1 & 0 & 0 & 1 \end{bmatrix} = \begin{bmatrix} P_{LxK} & | & I_{LxL} \end{bmatrix}$$

It is clear from our corresponding generator matrix that the conversion is as follows

$$G = \begin{bmatrix} 1 & 0 & 0 & 0 & 1 & 1 & 1 \\ 0 & 1 & 0 & 0 & 1 & 0 & 1 \\ 0 & 0 & 1 & 0 & 1 & 1 & 0 \\ 0 & 0 & 0 & 1 & 0 & 1 & 1 \end{bmatrix} = \begin{bmatrix} I_{KxK} & | & P^{T} \end{bmatrix}$$

Even a college freshman knows how to take the transpose of a matrix, thus finding G appears to be trivial. Later on, however, we will see parity check matrices, H, that don't have this nice form, which features the identity matrix on the right, and thus allows for an easy transformation to G. How do we handle these cases? The answer lies in the parity check equation $H \cdot C^{T} = 0$, which we saw earlier. We apply Gauss-Jordan elimination, in which we add or subtracts rows from one another until the identity matrix appears at the right of the matrix H. Of course, the zero vector on the right of the equal sign never changes while we are adding and subtracting rows (we are adding and subtracting zeros from one another after all). We end up with a new matrix, H', which has the right form and still satisfies $H' \cdot C^{T} = 0$. We now transform H' into the generator matrix G, as shown via the simply procedure above. We will see the Gauss-Jordan elimination soon.

Binary Symmetric Channels vs Binary Erasure Channels

When engineers work with techniques that enable us to detect and correct binary errors, they have to make some assumptions about how the channel causes these errors and what type of information is available at the decoder. Engineers and scientists have done much of their work using a simple channel model called the binary symmetric channel, which takes in bits and produces bits. It successfully conveys the bits' values given a probability of p, whereas it produces errors by flipping the polarity of these bits given a probability of $1 - p$. The likelihood of committing an error is the same whether the input but was a 0 or 1. However, the important aspect of this channel is that it presents the decoder with 0's and 1's, but nothing else. All our discussions up to this point, have assumed this binary symmetric channel.

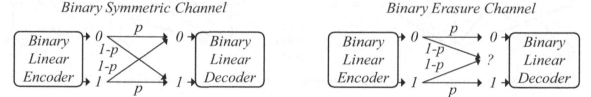

Figure 5-86: The Binary Symmetric, BSC, and Binary Erasure (BEC) Channels

The binary symmetric channel, or BSC, is a nice generalization that was useful for engineers some decades ago, but it is in fact not what the binary linear decoder will actually see in a real communication system. In real life, bits 0 and 1 are mapped to complex values such as QAM symbols. Additive noise will superimpose on these symbols and their demapping operation will yield noisy bits. The binary erasure channel, or BEC, is a better generalization of the channel model and takes into account that the receiver isn't always sure whether the bit presented to the decoder is in fact a 1 or 0. These occurrences are called erasures, and they lead us to new and interesting methods of working with linear codes providing superior performance given that the binary erasure channel provides additional information.

Single Parity Check Codes (Binary Erasure Channels)

Binary erasure channels assume that there are three rather than two outputs states. We either have a 0 bit, a 1 bit or an erasure bit, which is an indication by the receiver that it is unsure of the bits polarity. Let us take a look at the bit message below, which features an erasure in bit position 2. Given that the receiver is sure about the polarity of message bits 0, 1, 3, 4, 5 and the parity bit, we can easily determine the polarity of the erased bit to be 0, as the number of ones must be even.

$$\text{Bit Message} = 01?011 \rightarrow \text{Parity Bit} = 1$$

Compared to this same example at the start of the section using the binary symmetric channel, the use of the binary erasure channel allows us to not only detect but also correct the error.

Decoding and the Tanner Graph

Before proceding with LDPC codes, let us introduce the Tanner graph, which is the preferred method of illustrating a linear code. The Tanner graph features bit nodes, which represent the bits of the received code word, and check nodes which compute the modulo-two addition of all bit node values to which that they are connected. Remember that the modulo-two sums computed by the check nodes must always be zero. The two tanner graphs below, illustrate the Hamming code that we have seen earlier. In the graph on the left, notice that bit, b_2, has been erased (we are not sure what it is). To fix the problem, we look at check node c_1, which requires that the modulo-two sum of bits b_0, b_2, b_3 and p_1 is 0. Since the bit values of b_0, b_3 and p_1 are 1, then b_1 must also feature a value of 1 and we replace the erased bit accordingly. Check node c_0 could have also been used to fix the erasure.

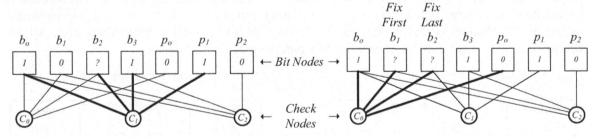

Figure 5-87: Illustration of Error Correction in Binary Erasure Channels Using the Tanner Graph

The figure on the right is even more interesting as it features two erasures. Given the simple binary symmetric channel, the Hamming code was only able to correct a single error. This is no longer the case. Notice that check node c_0 features two erasures, which can't be simultaneously fixed. However, we can use check node c_1 to first fix bit b_2, and then use check node c_0 to fix bit b_1 as b_2 is no longer erased. This leads us to the following important insights.

→ A single check node can only fix a single erasure present in the bit nodes to which it is connected.
→ We can use an iterative decoding strategy, by looking at each check node and first repairing those that have only a single erasure associated with them. Once all check nodes have been used

to fix single erasures, you can repeat the process as check nodes that initially featured multiple erasures may no longer have that problem.

Protecting the Parity Bits

Connecting a bit node to several check nodes provides superior protection when one or more of the message bits are erased. Looking at the Tanner graph of the Hamming code above, we realize that each parity bit is only assigned to one check node. This feature makes erasures among the parity bits harder to deal with than erasures in the message bits. To fix this problem, we assign multiple connections not just to the message bits but the parity bits as well. The rule that the modulo-two additions at the check nodes must produce a 0 still stands of course. Consider the following parity check matrix representing a code rate of ½ producing three parity check bits for three message bits. Notice that the parity bit p_0 is connected to multiple check nodes.

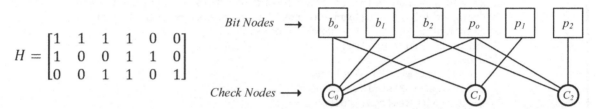

$$H = \begin{bmatrix} 1 & 1 & 1 & 1 & 0 & 0 \\ 1 & 0 & 0 & 1 & 1 & 0 \\ 0 & 0 & 1 & 1 & 0 & 1 \end{bmatrix}$$

Figure 5-88: Parity Check Matrix and Tanner Graph with Multiple Parity Connections

As was mentioned earlier in this section, to find the encoder we must use the Gauss-Jordan elimination to determine an equivalent matrix that has the proper form allowing us to transform to the generator matrix G. In this case, we simply add row 1 to row 3 and then add row 1 to row 2.

$$H = \begin{bmatrix} 1 & 1 & 1 & 1 & 0 & 0 \\ 1 & 0 & 0 & 1 & 1 & 0 \\ 0 & 0 & 1 & 1 & 0 & 1 \end{bmatrix} \rightarrow \begin{bmatrix} 1 & 1 & 1 & 1 & 0 & 0 \\ 1 & 0 & 0 & 1 & 1 & 0 \\ 1 & 1 & 0 & 0 & 0 & 1 \end{bmatrix} \rightarrow \begin{bmatrix} 1 & 1 & 1 & | & 1 & 0 & 0 \\ 0 & 1 & 1 & | & 0 & 1 & 0 \\ 1 & 1 & 0 & | & 0 & 0 & 1 \end{bmatrix} = [P_{3x3} \mid I_{3x3}]$$

According to our formulas mentioned earlier, the generator matrix G reduces to the following.

$$G = \begin{bmatrix} 1 & 0 & 0 & 1 & 0 & 1 \\ 0 & 1 & 0 & 1 & 1 & 1 \\ 0 & 0 & 1 & 1 & 1 & 0 \end{bmatrix} = [I_{3x3} \mid P_{3x3}{}^T]$$

As we proceed in this chapter, we will primarily encounter parity check matrices H that require Gauss-Jordan elimination. In anticipation of this fact, we provide the MatLab code below that executes the Gauss-Jordan elimination. The code will not produce a valid generator matrix for just any H. Some of the rows may be linearly dependent on one another resulting in a row with all zero bits during the elimination. When crafting a parity check matrix, H, the designer should ensure that a generator matrix can be computed properly. The parity check matrices for the 802.11n standard, which we will see soon, can easily be transformed in this way.

```
function H_New = GaussJordanElimination(H)
    H_New = H;
    [rows, columns] = size(H_New);
    N              = columns;  % The number of encoded bits
    L              = rows;     % The number of parity bits
    K              = N - L;    % The number of message bits

    % --------------------------------------------------------------------
```

```
    % Step 1: The forward elimination step
    for column = K+1 : N
        % [RowOfInterest, column] are the coordinates of the diagonal of the LxL
        % square matrix all the way to the right of H_New. The point of the exercise
        % is to place 1's in these positions. If we don't find a 1 there, we need
        % to look to see whether one of the rows below has a one in this column
        % and then pivot.
        RowOfInterest = column - K;  % Increments as follows: 1, 2, 3 .... L

        % 1a. The pivoting step (if we need to pivot)
        % Determine whether we need to pivot or not.
        if(H_New(RowOfInterest, column) ~= 1)
            % We will now look at the rows below until we find a 1 in this column.
            % Rather than switching the current row with the one containing the 1,
            % we add the row below to the current on. This is fine as we are
            % doing modulo 2 addition. We could have switched them as well.
            RemainingRowsBelow = L - RowOfInterest;
            PivotSuccessful = false;
            if(RemainingRowsBelow > 0)
                % We attempt to pivot rows
                for row = RowOfInterest+1:L
                    if(H_New(row, column) == 1)
                        H_New = AddRows(H_New, row, RowOfInterest);
                        PivotSuccessful = true;
                        break;
                    end
                end
            end
            assert(PivotSuccessful, 'The Pivot operation failed. Go make a better H.');
        end

        % At this point, we have ensured that the diagonal of the square matrix on
        % the right of H_New features ones everywhere.
        % 1b. Forward elimination
        for row = RowOfInterest+1:L
          if(H_New(row, column) == 1); H_New = AddRows(H_New, RowOfInterest, row); end
        end
    end


    % ----------------------------------------------------------------
    % Step 2: The backward elimination step
    for column = N : - 1: K + 2
        RowOfInterest = column - K;   % Decrements as follows: L, L-1, L-2, ... 2
        % If any of the rows above feature a 1 in this column, get rid of it by adding.
        for row = 1:RowOfInterest - 1
          if(H_New(row, column) == 1); H_New = AddRows(H_New, RowOfInterest, row); end
        end
    end
end

% The following functions adds the source row of InputM to the target row of InputM.
function OutputM = AddRows(InputM, SourceRow, TargetRow)
    OutputM              = InputM;
    OutputM(TargetRow, :) = mod(OutputM(TargetRow, :) + OutputM(SourceRow, :), 2);
end
```

5.6.5.2 LDPC Codes

To extend Hamming's work with linear error correction codes, Robert Gallager focused on two important questions regarding how connections should be made between bit and check nodes when the number of encoded bits becomes large.

→ Should each parity check set include many encoded bits or few? In our last example, parity check node c_0 was connected to four of the encoded bits, whereas node c_1 and c_2 were connected to 3 of the bits. Thus each check node was connected to half or more bit nodes. Should this trend continue as we encode messages with hundreds or thousands of bits? Using many bits in a check set increases the probability that we encounter more than one erasure, thus potentially invalidating our iterative scheme that fixed one error at a time. It is thus advantageous to use fewer message bits in our parity check sets. The 'low density' portion of the LDPC acronym stems from this approach that attempts to keep the ratio of the number of encoded bits used in each parity check set to the total number of encoded bits relatively low. This causes the parity matrix to feature a low density of ones.

→ The second question that needed to be answered was how we should select the bits nodes that pertain to each parity check set. As it turns out, the bit nodes for each parity check set can be selected more or less randomly and still provide good performance.

The iterative error correction scheme, called message passing, used in most LDPC decoders, does not work with erasures as real channels provide us with yet more information. Therefore, before looking at LDPC decoding, we introduce the binary belief channel.

Binary Belief Channels

The binary erasure channel introduced earlier is a more realistic approximation of a true communication channel, but it still falls short of the information that we have in real situations, where a received bit is a floating point number, which indicates the belief or confidence that the bit is equal to 1. In modern communication receivers this belief is computed using log likelihood ratio metrics. Take a look at the channel model shown below, which starts with a message bit vector $b[k]$, where $k = 0, 1, \ldots K\text{-}1$. The linear encoder transforms this input array into an $N = K+L$ length encoded bit vector $c[n]$, where $n = 0, 1, \ldots N\text{-}1$. These bits are now mapped into QAM symbols and the channel adds white Gaussian noise. We now apply the LLR scaling techniques introduced in section 5.6.2 yielding a vector of received bit beliefs $r[n]$. The LDPC decoder discussed later provides the estimate, $b'[k]$, of the initial message bit vector $b[k]$.

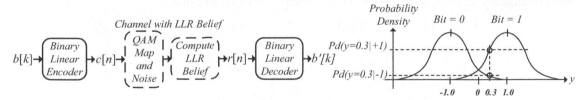

Figure 5-89: The Binary Belief Channel

The receiver can estimate the noise power at its input and thus the variance of the two Gaussian probability density functions, which we show above for the BPSK scenarios. As we have shown in section 5.6.2, the log likelihood belief for a soft bit value, y, can be computed according to the

---

following formula (valid for BPSK and QPSK constellations). For 16 and 64 QAM constellation, see section 5.62 for the LLR belief expressions.

$$LLR\ Belief = \ln\left(\frac{P(Bit = +1 \mid y)}{P(Bit = -1 \mid y)}\right) = ln(\frac{Pd(y \mid Bit = +1)}{Pd(y \mid Bit = -1)})$$

The binary belief channel provides us with more information than either the binary symmetric or binary erasure channel and will therefore provide even better decoding performance.

The Single Parity Check in the Binary Belief Channel

For the binary symmetric channel (BSC) and binary erasure channel (BER), we showed that the modulo-two sum of the bits, r, connected to any single parity check note must equal to zero, thus guaranteeing an even number of 1s. The received bits produced by the binary belief channel (*BBC*) are floating point values, where a positive number indicates a belief that the received bit was a 1, and a negative number indicates that the received bit was a 0. In the case of floating point beliefs, the single parity check node must also indicate the existence of an even number of positive beliefs. Compare the single parity check equations for the binary symmetric and binary belief channels below. Convince yourself that expression $SPC_{BBC}(r)$ yield +1 by applying two valid vectors $r = [1, 1, -1]$ and $r = [1, 1, -1, -1]$ (both have an even number of positive beliefs) to the expression

$$SPC_{BSC}(r) = mod\left(\sum_{k=0}^{K-1} r, 2\right) = 0 \qquad SPC_{BBC}(r) = \prod_{k=0}^{K-1} sign(-r) = +1$$

Computing Extrinsic Belief using a Soft-Input Soft-Output Single Parity Check Decoder

The extrinsic belief of any single belief in the receive vector r is computed using the remaining believes in the vector. Therefore, the computation of the extrinsic belief of $r[0]$, does not include the belief $r[0]$ itself. Take a look at beliefs $r[1] = -1.0$, $r[2] = -0.9$, and $r[3] = 1.2$ in the receive vector below. Because we know that the total number of positive beliefs in the received vector must be even, we can conclude that the sign of $r[0]$ must be positive. Therefore, we can further conclude that the sign of the received belief $r[0] = -0.1$ is likely in error. Furthermore, compared to the absolute value of $r[0]$, the absolute values of $r[1]$, $r[2]$, and $r[3]$ are rather healthy. Therefore, we can conclude that the size of the extrinsic belief, $l[0]$, of the received belief $r[0]$ must be healthy as well, no matter what the actual size of $r[0]$ indicates.

$$r = [\ r[0] \quad r[1] \quad r[2] \quad r[3]\ ] = [-0.1 \quad -1.0 \quad -0.9 \quad 1.2]$$

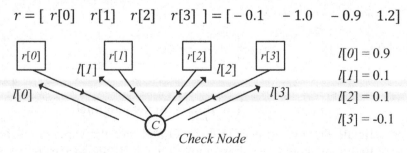

Check Node

Figure 5-90: Single Parity Check Node Connected to Received Belief Vector r_k

Let's now consider the extrinsic belief of $r[1]$, based on the values of $r[0] = -0.1$, $r_2 = -0.9$, and $r_3 = 1.2$. Given the fact that the number of positive beliefs must be even, we conclude that the sign of the extrinsic belief, $l[1]$. of $r[1]$ is positive as well. However, the size of the extrinsic belief of $r[1]$ is not so obvious as the absolute value of $r[0]$ is far less than that of r_2 and r_3. Clearly, the small absolute value of $r[0]$ indicates uncertainty regarding the belief of $r[1]$. As a matter of fact, the size of the extrinsic belief of $r[1]$ is only as large as the smallest belief of the remaining values $r[0]$, $r[2]$, and $r[3]$. The following rules summarize the concepts above regarding extrinsic beliefs, which we will express using the variable l. The extrinsic belief, $l[x]$, for any received bit belief $r[x]$ will be computed as follows:

$$sign(l[x]) = -\prod_{x \neq y} sign(-r[y])$$

$$abs(l[x]) = min(abs(r \mid without \; r[x]))$$

When we combine these rules, we get the soft-input soft-output (SISO) single parity check decoder which obeys the following expressing.

$$l[x] = -\prod_{x \neq y} sign(-r[y]) \cdot min(abs(r \mid without \; r[x]))$$

Note that the soft-input soft-output, or SISO, nomenclature simply indicates that the input and output values are of floating point format. From the discussion above, it is clear that some of the extrinsic beliefs $l[x]$ are different from the actual received beliefs $r[x]$. It is our goal to use the extrinsic beliefs $l[x]$ to update the intrinsic beliefs $r[x]$. This is done using a soft-input soft-output repetition decoder, which we will discuss next.

```
% The received belief vector rn
rn = [-0.1, -1, -0.9, 1.2];
disp(SISO_SPC_Decoder(rn));

% r - Is the vector of intrinsic beliefs at the bit nodes
% l - Is the vector of extrinsic belief for the single parity check node at index x
function l = SISO_SPC_Decoder(r)    % The SISO single parity check decoder
    l   = zeros(size(r));
    for x = 1:length(r)
        r_other     = r;
        r_other(x) = [];                    % Remove rx from vector r and store in r_other
        Sign_Other = -sign(prod(-r_other));% Compute sign of other beliefs
        Mag_Other  = min(abs(r_other));   % Compute minimum abs value of other beliefs
        l(x)        = Sign_Other * Mag_Other; % Build the extrinsic belief vector
    end
end
% -> The output is = [l0, l1, l2, l3] = [0.9, 0.1, 0.1, -0.1]
```

Updating Intrinsic Information using a Soft-Input Soft-Output Repetition Decoder

The intrinsic information of any received belief $r[x]$ in the received belief vector is itself $r[x]$. In this section, we wish to update the intrinsic belief using the calculated extrinsic beliefs from each check node to which $r[x]$ is connected. Because the extrinsic beliefs are separately computed without $r[x]$ as an input, we can view the original belief $r[x]$ as well as the set of extrinsic beliefs as independent observations of the same quantity. It is for this reason that we can add them up to get an update intrinsic belief. The vector k in the expression below provides the check node indices of those check nodes connected to $r[x]$. The vector k for each check node can be read off directly from the rows of the parity check matrix, as we have shown multiple times in the various Tanner graphs. The sequence $k = 0$ through N in the figure below is meant as an example only.

$$r[x]_{new} = r[x] + \sum_k l_k[x]$$

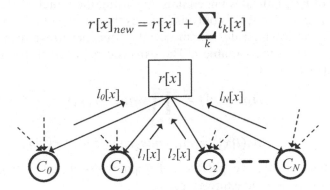

Figure 5-91: Updating Intrinsic Belief

```
% rx  -  The current instrinsic belief regarding the received bit
% lx  -  A vector of extrinsic beliefs provided the check nodes to which rx connects.
function UpdatedIntrinsicBelief = SISO_Repetition_Decoder(rx, lx)
    UpdatedIntrinsicBelief = sum([rx, lx]);
end
```

5.6.5.3 The LDPC Message Passing Decoder

Now that we have covered the SISO SPC and Repetition decoder concepts, the iterative mechanism featured in the message passing decoder is straight forward. The iteration involves these steps.

Step 1: Pass intrinsic beliefs (the received beliefs) from the bit nodes to the check nodes.
Step 2: Calculate the all extrinsic beliefs for all check node using the SISO-SPC decoder.
Step 3: Pass the extrinsic beliefs back up to the bit nodes where they are summed with the intrinsic beliefs (the received LLR bit values) to form updated intrinsic beliefs. This is the SISO repetition decoder.
Step 4: At this point, we pass the updated intrinsic beliefs from the bit nodes to the check nodes (as done in step 1) with one important exception. Take a look at the last figure and picture the updated intrinsic belief at the bit node as $r[x]_{new} = r[x] + l_0[x] + l_1[x] + l_2[x] + \cdots + l_N[x]$. When we pass $r[x]_{new}$ back to check node C_0, we must first subtract $l_0[x]$ from $r[x]_{new}$. Similarly, before passing $r[x]_{new}$ to C_1, we must first subtract $l_1[x]$. This process of subtraction, which must be undertaken for all check nodes, prevents an improper build-up of beliefs in the message passing process.
Step 5: Repeat step 2.
Step 6: Repeat step 3.

We now return to step 4. Note that the new extrinsic values that have been calculated are once again summed with the **original** received LLR beliefs, $r[x]$, not the previously updated intrinsic beliefs $r[x]_{new}$.

Example 5.5: LDPC Example

Consider the following 7 by 14 parity check matrix implementing a ½ rate LDPC encoder. The parity check matrix is only meant to demonstrate the message passing decoder process and was not designed for high performance. In this example, the variables N, L, and K represent the number of encoded bits, the number of parity bits and the number of message bits respectively.

$$H = \begin{bmatrix} 1 & 0 & 0 & 1 & 0 & 0 & 1 & | & 1 & 0 & 0 & 0 & 0 & 0 & 0 \\ 0 & 1 & 0 & 0 & 1 & 0 & 0 & | & 1 & 1 & 1 & 0 & 0 & 0 & 0 \\ 0 & 0 & 1 & 0 & 0 & 1 & 0 & | & 0 & 0 & 1 & 1 & 0 & 0 & 0 \\ 0 & 1 & 0 & 1 & 0 & 0 & 1 & | & 0 & 0 & 0 & 1 & 1 & 0 & 0 \\ 1 & 0 & 0 & 0 & 1 & 0 & 0 & | & 0 & 0 & 0 & 0 & 1 & 1 & 0 \\ 0 & 0 & 1 & 0 & 0 & 1 & 0 & | & 0 & 0 & 0 & 0 & 0 & 1 & 1 \\ 1 & 0 & 0 & 0 & 0 & 0 & 1 & | & 0 & 0 & 0 & 0 & 0 & 0 & 1 \end{bmatrix}$$

```
%% 1. Define the Parity Check Matrix, H
H = [1 0 0 1 0 0 1 1 0 0 0 0 0 0; ...
     0 1 0 0 1 0 0 1 1 1 0 0 0 0; ...
     0 0 1 0 0 1 0 0 0 1 1 0 0 0; ...
     0 1 0 1 0 0 1 0 0 0 1 1 0 0; ...
     1 0 0 0 1 0 0 0 0 0 0 1 1 0; ...
     0 0 1 0 0 1 0 0 0 0 0 0 1 1; ...
     1 0 0 0 0 0 1 0 0 0 0 0 0 1];
N = size(H,2);   % Number of encoded bits = number columns in matrix
L = size(H,1);   % Number of parity bits (check nodes) = number of rows in matrix
K = N - L;       % Number of message bits
```

Notice the diagonal pattern of 1's in the right most portion of the parity check matrix. The pattern makes it easy to compute the Gauss-Jordan elimination to get to the generator matrix G. We can simply add row 1 to row 2 to eliminate the 1 in row 2 column 8, and then use backward elimination to create the L by L identity matrix in the right portion of H.

```
%% 2. Compute the Gaussian elimination to force H into desired format.
H_Modified = UseGaussianElimination(H);
P          = H_Modified(1:L, 1:K);
G          = [eye(K), P'];          % The generator matrix G

% To ensure that the generator matrix is valid perform the following check.
Error= sum(sum(mod(G*H', 2), 1), 2);
assert(Error == 0, 'An error occurred while generating G.');
```

$$G = \begin{bmatrix} 1 & 0 & 0 & 0 & 0 & 0 & 0 & | & 1 & 1 & 0 & 0 & 0 & 1 & 1 \\ 0 & 1 & 0 & 0 & 0 & 0 & 0 & | & 0 & 0 & 1 & 1 & 0 & 0 & 0 \\ 0 & 0 & 1 & 0 & 0 & 0 & 0 & | & 0 & 0 & 0 & 1 & 1 & 1 & 0 \\ 0 & 0 & 0 & 1 & 0 & 0 & 0 & | & 1 & 0 & 1 & 1 & 0 & 0 & 0 \\ 0 & 0 & 0 & 0 & 1 & 0 & 0 & | & 0 & 0 & 1 & 1 & 1 & 0 & 0 \\ 0 & 0 & 0 & 0 & 0 & 1 & 0 & | & 0 & 0 & 0 & 1 & 1 & 1 & 0 \\ 0 & 0 & 0 & 0 & 0 & 0 & 1 & | & 1 & 1 & 0 & 0 & 1 & 1 & 1 \end{bmatrix}$$

At this point, let's generate the encoded bits for an input bit message = [1, 0, 1, 0, 1, 0, 1]. We then map the encoded bits into floating point numbers as follows: $0 \rightarrow -1$ and $1 \rightarrow +1$.

```
%% 3. Generate the encoded bits based on the following Input bit vector
InputBits              = [1, 0, 1, 0, 1, 0, 1];
OutputBits             = mod(InputBits*G, 2);
TxBipolarBits          = (2*OutputBits - 1);        % Convert from 0/1 to -1/+1
```

To emulate the belief of the received bits, we simply add Gaussian noise to the transmitted bipolar bits. Unless the various transmit bits experience different amounts of attenuation in the channel indicated by the value h in the expression below, there is need to use the LLR expression as the fraction preceding the received bit value is a constant. Using the equation evenly scaled all noisy received bits values providing no gain in performance. The LLR equation (shown here for the BPSK scenario) need only be used when different bits experience different amounts of attenuation in the channel, or we are using 16QAM or higher constellations.

$$LLR\ Confidence = \frac{4|h|^2}{\sigma_n^2} \cdot ReceivedBitValue$$

In our example here, we simply assign the received bit beliefs directly, thereby forcing errors in certain positions, and test whether the algorithm can correct them. Feel free to play with the number of errors to check at what point the decoder begins to fail.

```
% TxBipolarBits = [ 1  -1   1   -1 1  -1   1    -1   -1   1  -1   1  1  -1
% Define Rx bit beliefs forcing errors are in positions 2, 7, 12
RxBeliefs     = [ 2  .1  1.5  -1  1  -1  -0.2  -0.8  -0.8  1  -1  -0.3  1  -1];
```

In order to compute the message passing algorithm described in the seven steps shown earlier, we will use two matrices called M and E. We first copy the received beliefs into the non-zero positions of the parity check matrices as follows. The entries in the rows of the matrix below indicate the beliefs provided to the check nodes thus constituting step 1. E is an all zero matrix.

$$M = \begin{bmatrix} 2.0 & -- & -- & -1 & -- & -- & -.2 & | & -.8 & -- & -- & -- & -- & -- & -- \\ -- & 0.1 & -- & -- & 1.0 & -- & -- & | & -.8 & -.8 & 1.0 & -- & -- & -- & -- \\ -- & -- & 1.5 & -- & -- & -1 & -- & | & -- & -- & 1.0 & -1 & -- & -- & -- \\ -- & 0.1 & -- & -1 & -- & -- & -.2 & | & -- & -- & -- & -1 & -.3 & -- & -- \\ 2.0 & -- & -- & -- & 1.0 & -- & -- & | & -- & -- & -- & -- & -.3 & 1.0 & -- \\ -- & -- & 1.5 & -- & -- & -1 & -- & | & -- & -- & -- & -- & -- & 1.0 & -1 \\ 2.0 & -- & -- & -- & -- & -- & -.2 & | & -- & -- & -- & -- & -- & -- & -1 \end{bmatrix} - L$$

In step 2 we now compute the extrinsic beliefs for each row, which would look as follows.

$$SisoSpcDecoder([2 \quad -1 \quad -.2 -0.8]) \quad = [-0.2 \quad 0.2 \quad 0.8 \quad 0.2]$$
$$SisoSpcDecoder([0.1 \quad 1 \quad -.8 -0.8 \quad 1]) \quad = [-.8 \quad -.1 \quad -.1 \quad .1 \quad .1 -.1]$$
$$SisoSpcDecoder([1.5 -1 \quad 1 \quad -1]) \quad = [1 \quad -1 \quad 1 -1]$$
$$SisoSpcDecoder([0.1 -1 \quad -.2 \quad -1 -.3]) = [-.2 \quad 0.1 \quad 0.1 \quad 0.1 \quad 0.1]$$
$$SisoSpcDecoder([2.0 \quad 1 -0.3 \quad 1]) \quad = [-.3 \quad -.3 \quad 1 \quad -.3]$$
$$SisoSpcDecoder([1.5 -1 \quad 1 \quad -1]) \quad = [1 \quad -1 \quad 1 \quad -1]$$
$$SisoSpcDecoder([2 \quad -.2 \quad -1]) \quad = [-.2 \quad 1 \quad 0.2]$$

These extrinsic values are populated into the E matrix as seen below.

$$E = \begin{bmatrix} -0.2 & -- & -- & 0.2 & -- & -- & 0.8 & | & 0.2 & -- & -- & -- & -- & -- & -- \\ -- & -.8 & -- & -- & -.1 & -- & -- & | & 0.1 & 0.1 & -0.1 & -- & -- & -- & -- \\ -- & -- & 1.0 & -- & -- & -1.0 & -- & | & -- & -- & 1.0 & -1.0 & -- & -- & -- \\ -- & -0.2 & -- & 0.1 & -- & -- & 0.1 & | & -- & -- & -- & 0.1 & 0.1 & -- & -- \\ -0.3 & -- & -- & -- & -.3 & -- & -- & | & -- & -- & -- & -- & 1.0 & -.3 & -- \\ -- & -- & 1.0 & -- & -- & -1.0 & -- & | & -- & -- & -- & -- & -- & 1.0 & -1 \\ -0.2 & -- & -- & -- & -- & -- & 1.0 & | & -- & -- & -- & -- & -- & -- & 0.2 \end{bmatrix}$$

At this point, we sum all extrinsic values at the bit nodes (along the columns of E) and add them to the received bits beliefs. This represents steps 3 and 7

$$sum(E) = \begin{bmatrix} -0.7 & -1.0 & 2.0 & 0.3 & -0.4 & -2.0 & 1.9 & | & 0.3 & 0.1 & 0.9 & -0.9 & 1.1 & 0.7 & -0.8 \end{bmatrix}$$

$+$

$$RxBeliefs = r[x] = \begin{bmatrix} 2.0 & 0.1 & 1.5 & -1 & 1 & -1 & -0.2 & | & -.8 & -.8 & 1 & -1 & -.3 & 1 & -1 \end{bmatrix}$$

$=$

$$r_{new}[x] = \begin{bmatrix} 1.3 & -.9 & 3.5 & -.7 & 0.6 & -3 & 1.7 & | & -.5 & -.7 & 1.9 & -1.9 & 0.8 & 1.7 & -1.8 \end{bmatrix}$$

Note that in the MatLab code, we refer to $r_{new}[x]$ simply as intrinsic beliefs. These updated intrinsic beliefs are now used to repopulate the matrix M. However, this time when we subtract the matrix E, it is no longer filled with zero. This subtraction process represents step 4 as explained at the start of this section. We repeat this iteration several times (usually no more than 10 times) for optimum performance.

```
%% 4. Let's do the Min Sum LDPC decoding
NumberOfIterations    = 10;
IntrinsicBeliefs      = RxBeliefs;
E                     = zeros(size(H));
for Iteration = 1:NumberOfIterations
    % Step 1 and Step 4
    M = IntrinsicBeliefs .* ones(size(H));    %% Populate the M Matrix
    %  Zero out positions that are not interesting to us. This also helps
    %  when we add values column wise to find the total extrinsic belief
    M = M .* H;
    M = M - E;          % During the first pass, E  = 0 (Step 1)
                        % During later passes,   E ~= 0 (Step 4)

    % Step 2 and Step 5:
    for CheckNodeIndex = 1:L
        % Grab the current row in the parity check matrix i.e. [0 1 0 0 1 1 0 0 ...]
        CheckNodeConnectionsVector = H(CheckNodeIndex, :);
        % Find the indices of those connections           i.e. [2, 5, 6, ...]
        CheckNodeConnectionIndices = find(CheckNodeConnectionsVector);
        % Fetch the extrinsic beliefs at that row and send to SISO_SPC_Decoder
        r                  = M(CheckNodeIndex, CheckNodeConnectionIndices);
        l                  = SISO_SPC_Decoder(r);
        % This matrix remembers the intrinsic beliefs that we need to subtract
        E(CheckNodeIndex, CheckNodeConnectionIndices) = l;
    end
    % Step 4:
    % Find the sum of the extrinsic beliefs for a particular bit node.
    TotalExtrinsicBeliefs = sum(E, 1);
    % Add that sum to the intrinsic belief (original received bit beliefs) to get
```

```
      % the new updated intrinsic belief = r[x]new in the text.
    IntrinsicBeliefs      = SISO_Repetition_Decoder(RxBeliefs, TotalExtrinsicBeliefs);
end

%% 5. Map received beliefs back to bits.   Minus sign --> 0    and Plus sign --> 1
RxBitEstimates     = 0.5*(sign(IntrinsicBeliefs(1, 1:K))+1);
FinalMessageBitErrors  = sum(mod(RxBitEstimates + InputBits , 2));
```

The final intrinsic belief after 10 iteration is as follows:

$$IntrinsicBeliefs = [6.1 \ -2.2 \quad 4.6 \ -3.3 \ \ 2.2 \ \ -4.1 \ \ 4.5 \ | -3.3 \ -2.2 \ \ 3.5 \ -3.5 \ \ 2.2 \ \ 4.1 \ -4.5]$$

The signs of the final intrinsic beliefs above properly match the polarities of the original encoded transmit bits [1 0 1 0 1 0 1 | 0 0 1 0 1 0 0], of which the first seven are our message.

5.6.5.1 Parity Check Matrices in 802.11n

The following is the prototype matrix that is used to configure the 802.11n ½ rate LDPC encoder [1] featuring 648 encoded bits based on 324 message bits. As can clearly be seen, the matrix below features only 12 rows and 24 columns. In fact, each entry in the prototype represents a specific variation of a 27 by 27 identity matrix or zero matrix as follows:

→ A 0 represents a 27 by 27 identity matrix.

→ A positive number, x, represents an identity matrix, whose entries have been cyclically shifted by x position to the right.

→ A -1 indicates a 27 by 27 zero matrix.

$$H_{648\_½} =$$
$$
\begin{bmatrix}
0 & -1 & -1 & -1 & 0 & 0 & -1 & -1 & 0 & -1 & -1 & 0 & 1 & 0 & -1 & -1 & -1 & -1 & -1 & -1 & -1 & -1 & -1 & -1 \\
22 & 0 & -1 & -1 & 17 & -1 & 0 & 0 & 12 & -1 & -1 & -1 & -1 & 0 & 0 & -1 & -1 & -1 & -1 & -1 & -1 & -1 & -1 & -1 \\
6 & -1 & 0 & -1 & 10 & -1 & -1 & -1 & 24 & -1 & 0 & -1 & -1 & -1 & 0 & 0 & -1 & -1 & -1 & -1 & -1 & -1 & -1 & -1 \\
2 & -1 & -1 & 0 & 20 & -1 & -1 & -1 & 25 & 0 & -1 & -1 & -1 & -1 & 0 & 0 & -1 & -1 & -1 & -1 & -1 & -1 & -1 & -1 \\
23 & -1 & -1 & -1 & 3 & -1 & -1 & -1 & 0 & -1 & 9 & 11 & -1 & -1 & -1 & -1 & 0 & 0 & -1 & -1 & -1 & -1 & -1 & -1 \\
24 & -1 & 23 & 1 & 17 & -1 & 3 & -1 & 10 & -1 & -1 & -1 & -1 & -1 & -1 & -1 & -1 & 0 & 0 & -1 & -1 & -1 & -1 & -1 \\
25 & -1 & -1 & -1 & 8 & -1 & -1 & -1 & 7 & 18 & -1 & -1 & 0 & -1 & -1 & -1 & -1 & -1 & 0 & 0 & -1 & -1 & -1 & -1 \\
13 & 24 & -1 & -1 & 0 & -1 & 8 & -1 & 6 & -1 & -1 & -1 & -1 & -1 & -1 & -1 & -1 & -1 & -1 & 0 & 0 & -1 & -1 & -1 \\
7 & 20 & -1 & 16 & 22 & 10 & -1 & -1 & 23 & -1 & -1 & -1 & -1 & -1 & -1 & -1 & -1 & -1 & -1 & -1 & 0 & 0 & -1 & -1 \\
11 & -1 & -1 & -1 & 19 & -1 & -1 & -1 & 13 & -1 & 3 & 17 & -1 & -1 & -1 & -1 & -1 & -1 & -1 & -1 & -1 & 0 & 0 & -1 \\
25 & -1 & 8 & -1 & 23 & 18 & -1 & 14 & 9 & -1 & -1 & -1 & -1 & -1 & -1 & -1 & -1 & -1 & -1 & -1 & -1 & -1 & 0 & 0 \\
3 & -1 & -1 & -1 & 16 & -1 & -1 & 2 & 25 & 5 & -1 & -1 & 1 & -1 & -1 & -1 & -1 & -1 & -1 & -1 & -1 & -1 & -1 & 0 \\
\end{bmatrix}
$$

As an example, the following matrix is a 5 by 5 identity matrix that has been cyclically shifted to the right by 2 positions.

$$
I_2 = \begin{bmatrix}
0 & 0 & 1 & 0 & 0 \\
0 & 0 & 0 & 1 & 0 \\
0 & 0 & 0 & 0 & 1 \\
1 & 0 & 0 & 0 & 0 \\
0 & 1 & 0 & 0 & 0 \\
\end{bmatrix}
$$

The H648_½ prototype matrix therefore expands to a 12·27 by 24·27 or 324 by 648 parity check matrix.

The following MatLab code expands any prototype matrix with the above format into a parity check matrix.

```matlab
% Produce the full parity matrix based on a prototype matrix
% M         - The prototype matrix
% Dimension - The dimensions of each submatrix
function H = CreateParityCheckMatrix(M, Dimension)
    SizeH = size(M)*Dimension;   % The size in rows/columns of the parity check matrix
    H     = zeros(SizeH);        % The paritiy check matrix initialized to all zeros
    for row = 1:size(M, 1)
        for col = 1:size(M, 2)
            % Extract the cyclic shift from the prototype matrix
            CyclicShift   = M(row, col);
            % The cyclically shifted identity matrix
            I_Shifted     = circshift( eye(Dimension), CyclicShift, 2);
            % Figure out the row/column ranges inside H where we would insert I_Shifted
            RowRange      = (1:Dimension) + (row - 1)*Dimension;
            ColRange      = (1:Dimension) + (col - 1)*Dimension;
            % Actually insert it only if the cyclic shift is positive.  A negative
            % shift in the prototype matrix indicates that this submatrix is all zeros.
            if(CyclicShift >= 0)
                H(RowRange, ColRange) = I_Shifted;
            end
        end
    end
end
```

Example 5.6: LDPC Example using the 802.11n 648 Bit Rate ½ Encoder

Note that in section three of the code below, the user may set the signal to noise ratio of the bipolar received bits. The expectation is that error free communication should begin at approximately 2dB to 2.5dB of signal to noise ratio. An interesting feature of LDPC codes is that once we reach a certain signal to noise ratio, communication is suddenly error free even for long bit messages. This is quite different compared to binary convolutional coding, where the error performance continues to improve more modestly with increasing SNR as shown in the figure on the next page.

```matlab
%% 1. Define the Prototype Matrix (see 802.11n specification - reference [1])
ProtoType = [  0 -1 -1 -1  0  0 -1 -1  0 -1 -1  0  1  0 ...
                                 -1 -1 -1 -1 -1 -1 -1 -1 -1 -1; ...
              22  0 -1 -1 17 -1  0  0 12 -1 -1 -1 -1  0 ...
                                  0 -1 -1 -1 -1 -1 -1 -1 -1 -1; ...
               6 -1  0 -1 10 -1 -1 -1 24 -1  0 -1 -1 -1 ...
                                  0  0 -1 -1 -1 -1 -1 -1 -1 -1; ...
               2 -1 -1  0 20 -1 -1 -1 25  0 -1 -1 -1 -1 ...
                                 -1  0  0 -1 -1 -1 -1 -1 -1 -1; ...
              23 -1 -1 -1  3 -1 -1 -1  0 -1  9 11 -1 -1 ...
                                 -1 -1  0  0 -1 -1 -1 -1 -1 -1; ...
              24 -1 23  1 17 -1  3 -1 10 -1 -1 -1 -1 -1 ...
                                 -1 -1 -1  0  0 -1 -1 -1 -1 -1; ...
              25 -1 -1 -1  8 -1 -1 -1  7 18 -1 -1  0 -1 ...
                                 -1 -1 -1 -1  0  0 -1 -1 -1 -1; ...
```

```
                  13  24  -1  -1   0  -1   8  -1   6  -1  -1  -1  -1  -1 ...
                                       -1  -1  -1  -1  -1   0   0  -1  -1  -1; ...
                   7  20  -1  16  22  10  -1  -1  23  -1  -1  -1  -1  -1 ...
                                       -1  -1  -1  -1  -1  -1   0   0  -1  -1; ...
                  11  -1  -1  -1  19  -1  -1  -1  13  -1   3  17  -1  -1 ...
                                       -1  -1  -1  -1  -1  -1  -1   0   0  -1; ...
                  25  -1   8  -1  23  18  -1  14   9  -1  -1  -1  -1  -1 ...
                                       -1  -1  -1  -1  -1  -1  -1  -1   0   0; ...
                   3  -1  -1  -1  16  -1  -1   2  25   5  -1  -1   1  -1 ...
                                       -1  -1  -1  -1  -1  -1  -1  -1  -1   0];

SubmatrixDimension = 27;
H = CreateParityCheckMatrix(ProtoType, SubmatrixDimension);
N = size(H,2);   % Number of encoded bits = number columns in matrix
L = size(H,1);   % Number of parity bits (check nodes) = number of rows in matrix
K = N - L;       % Number of message bits

%% 2. Compute the Gaussian elimination to force H into desired format.
H_Modified = UseGaussianElimination(H);
P          = H_Modified(1:L, 1:K);
G          = [eye(K), P'];           % The generator matrix G

%     To ensure that the generator matrix is valid perform the following check.
Error= sum(sum(mod(G*H', 2), 1), 2);
assert(Error == 0, 'An error occurred while generating G.');


%% 3. Generate the encoded bits based on the following Input bit vector
InputBits           = zeros(1, K);
InputBits(1, 2:2:end) = ones(1, K/2);  % InputBits = [1,  0, 1,  0, 1,  0, ...., 1,  0]
OutputBits          = mod(InputBits*G, 2);
TxBeliefs           = (2*OutputBits - 1);   % Convert from 0/1 to -1/+1

% Generate and add noise to the output bits
rng(0);
SnrDb         = 2;
SnrLinear     = 10^(SnrDb/10);
NoiseVariance = 1/SnrLinear;
NoiseVector   = sqrt(NoiseVariance)*randn(size(TxBeliefs));
RxBeliefs     = (TxBeliefs + NoiseVector);


%% 4. Let's do the Min Sum LDPC decoding
NumberOfIterations  = 10;
IntrinsicBeliefs    = RxBeliefs;
E                   = zeros(size(H));
for Iteration = 1:NumberOfIterations
    % Step 1 and Step 4
    M = IntrinsicBeliefs .* ones(size(H));   %% Populate the M Matrix
    % Zero out positions that are not interesting to us. This also helps
    % when we add values column wise to find the total extrinsic belief
    M = M .* H;
    M = M - E;         % During the first pass, E  = 0 (Step 1)
                       % During later passes,   E ~= 0 (Step 4)

    % Step 2 and Step 5:
    for CheckNodeIndex = 1:L
        % Grab the current row in the parity check matrix i.e. [0 1 0 0 1 1 0 0 ...]
```

```
        CheckNodeConnectionsVector = H(CheckNodeIndex, :);
        % Find the indices of those connections         i.e. [2, 5, 6, ...]
        CheckNodeConnectionIndices = find(CheckNodeConnectionsVector);
        % Fetch the extrinsic beliefs at that row and send to SISO_SPC_Decoder
        r                         = M(CheckNodeIndex, CheckNodeConnectionIndices);
        l                         = SISO_SPC_Decoder(r);
        % This matrix remembers the intrinsic beliefs that we need to subtract
        E(CheckNodeIndex, CheckNodeConnectionIndices) = l;
    end
    % Step 4:
    % Find the sum of the extrinsic beliefs for a particular bit node.
    TotalExtrinsicBeliefs = sum(E, 1);
    % Add that sum to the intrinsic belief (original received bit beliefs) to get
    % the new updated intrinsic belief.
    IntrinsicBeliefs      = SISO_Repetition_Decoder(RxBeliefs, TotalExtrinsicBeliefs);
end


%% 5. Map received beliefs back to bits.  Minus sign --> 0   and Plus sign --> 1
RxBitEstimates       = 0.5*(sign(IntrinsicBeliefs(1, 1:K))+1);
FinalMessageBitErrors = sum(mod(RxBitEstimates + InputBits , 2));
```

Notice the very desirable bit error rate performance of the Polar, LDPC and Turbo codes, which improve from poor to phenomenal error correction abilities within a single dB of signal to noise ratio.

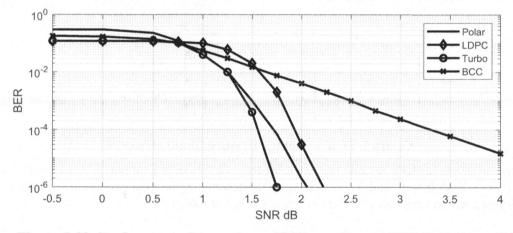

Figure 5-92: Performance Comparison of Different Rate ½ FEC Techniques [2]

5.6.1   Polar Codes

Polar codes represent a channel coding technology that was first introduced by Erdal Arikan in 2009 and is now used for purposes of error protection of the control channels in 5G cellular networks. Like low density parity check (5G and WLAN) and Turbo codes (4G LTE), polar codes operate close to Shannon's channel capacity limit indicating that these three coding techniques are likely as well as we can do in terms of forward error correction performance over a communication channel. Whereas polar codes have an elegant information theory based proof, our goal in this section is to illustrate their implementation.

The Binary Erasure Channel

In wireless communication systems, bits are generally mapped into QAM (BPSK, QPSK, 16QAM, ... etc.) symbols and are transmitted over channel, W, that is disturbed by additive white Gaussian noise. If we ignore multipath distortion for a moment, then each QAM symbol, and thus each bit, will be equally affected by the noise and would experience similar error behavior at the receiver. We will use the binary erasure channel, which we have met in the last section on LDPC codes, to model each instance of W. Remember that the binary erasure channel takes in a bit and then indicates at the receiver whether it believes the input bit to be a 0, a 1, or whether it is unsure of the input bit's polarity resulting in an erasure indicated by the question mark in the figure below. The probability of a correct detection at the receiver is equal to 1-p, whereas the probability of an erase will be equal to p.

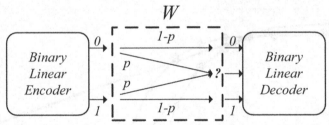

Figure 5-93: The Binary Erasure Channel Model

As a mental exercise, let's assume that each bit traverses its own channel, W, as indicated in the figure below. Both bits X_1 and X_2 will be correctly decoded with equal probability 1- p and result in an erasure with probability p as defined by our channel model.

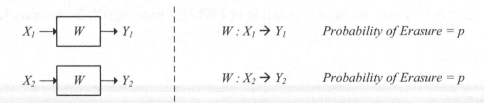

Figure 5-94: Decoding Bits via the Binary Erasure Channel W

The expressions $W : X_1 \rightarrow Y_1$ and $W : X_2 \rightarrow Y_2$ indicate that we will decode X_1 based on the available result Y_1 via channel W. Similarly, we will decode X_2 based on Y_2 via the second channel W.

5.6.1.1 Polarization

Polarization is the process of biasing the parallel channels W into new channels that are both better and worse that W in terms of error behavior. In the figure below, we introduce the polar transform for the two-bit construction, which linearly remaps input bits U_1 and U_2 into the familiar bits X_1 and X_2, which once again traverse the two channels marked W.

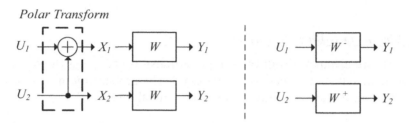

Figure 5-95: The Basic Two Input Polar Transform

The intention is that the polar transform in combination with the identical channels W will create two new channels W^- and W^+, which are inferior (W^-) and superior (W^+) to W in terms of error probability. To understand this better, let's write the probabilities of the various results that may occur at the output. The first line below indicates the probability that the result Y_1 successfully decodes as $X_1 = U_1 + U_2$ and Y_2 successfully decodes as $X_2 = U_2$ is equal to $(1-p)^2$, whereas the last line indicates the probability that both Y_1 and Y_2 resulted in erasures is equal to p^2.

$$\begin{bmatrix} Y_1 = U_1 \oplus U_2 \: and \; \; Y_2 = U_2 \; \; with \, probability \; \; (1-p)(1-p) \\ Y_1 = \quad ? \quad and \quad Y_2 = U_2 \; \; with \, probability \qquad p(1-p) \\ Y_1 = U_1 \oplus U_2 \: and \; \; Y_2 = \; \; ? \; \; with \, probability \; \; (1-p)p \\ Y_1 = \quad ? \quad and \quad Y_2 = \; \; ? \; \; with \, probability \; \; p^2 \end{bmatrix}$$

Note that in order to recover U_1, we must have properly decoded both outputs Y_1 and Y_2 as without knowledge of U_2, which appears at Y_2, we won't be able to reverse $Y_1 = U_1 + U_2$. Thus, the probability of recovering U_1 is dictated by the first line above.

$$Prob(U_1 \mid Y_1, Y_2) = (1-p)^2$$

In order to recover U_2 only Y_2 must be decoded properly (the sum of the first two lines above).

$$Prob(U_2 \mid Y_1, Y_2) = (1-p)^2 + p(1-p) = 1-p$$

At this point, we should point out that the probability to recover U_1 is poor, whereas as the probability to decode U_2 is no better than that of the normal channel W. Clearly, something is missing. Thus, at this point we make the daring assumption that in order to recover U_2, actually know what U_1 was. Now, the probability is the sum of the first three lines above.

$$Prob(U_2 \mid Y_1, Y_2, U_1) = (1-p)^2 + 2p(1-p) = 1-p^2$$

The new probability to recover U_2 is significantly better than the old one, but in order to realize this potential, we must either have already successfully recovered U_1, or have known what it is beforehand. Successfully recovering U_1 seems counterintuitive as the probability of doing so is not great, and knowing U_1 beforehand reduces the amount of information conveyed by half.

Once we increase the number of transmit bits and thus move to higher order polar transform constructs, we will actually use both of the possibilities listed above. Knowing U_1 beforehand does reduce the throughput, but as is the case with all other forward error correction schemes, redundant bits are used to enable superior error behavior for those bits that remain. Furthermore, decoding U_1 successfully seems unreliable in our simple example above, but improves during the successive decoding scheme that we will use later.

Example 5.7: Simple Two-Bit Polarization

For our binary erasure channel W, our two-bit polar transform achieves the following polarization indicating performance that is pushed to the extremes. The probability of recovering the input bits of channels W, $W-$ and $W+$ are as follows for erasure probability $p = 0.3$.

$$W \rightarrow \quad Prob(X_1 \mid Y_1) = (1 - p) = 0.7$$
$$W^- \rightarrow \quad Prob(U_1 \mid Y_1, Y_2) = (1 - p)^2 = 0.49$$
$$W^+ \rightarrow \quad Prob(U_2 \mid Y_1, Y_2, U_1) = 1 - p^2 = 0.91$$

The channels have been polarized as W^- and W^+ are worse and better than W respectively. Note further that the new probabilities of erasure for W- and W+ are as follows;

$$W^- \rightarrow \quad p^- = 1 - (1 - p)^2 = 2p - p^2 = 0.51$$
$$W^+ \rightarrow \quad p^+ = 1 - 1 + p^2 = p^2 = 0.09$$

Moving to Larger Number of Input Bits

The manner in which we increase the number of input bits is rather trivial as can clearly be seen in the figure below. The two diagrams on the left-hand side are especially helpful as they indicate how the basic 2-bit structure is repeated twice, once for inputs U_1 and U_2 and again for inputs U_3 and U_4. Notice the rearranged indexing of the input bit U_1 through U_4. Due to the existence of the two 2-bit structures that are enclosed by the thick dashed lines, U_1 must be decoded prior to U_3 and U_2 must be decoded prior to U_4. Further, due to the remaining two modulo-two adders, the input U_1 must be decoded prior to U_2 and U_3 must be decoded prior to U_4.

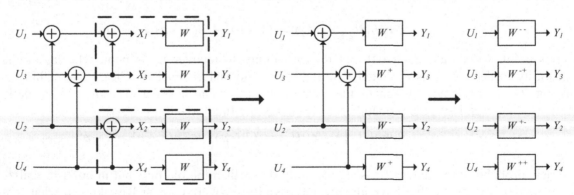

Figure 5-96: Size Four Polar Transform Construction

Rewiring into Standard Construction

In order to get around the strange order, in which we need to decode the input bits, we reorder the structure into the standard form and decode the input bits sequentially from U_1 to U_4. While confusing to look at initially, the only thing that was done to get to the standard construction was the positional exchange of U_3 and U_2. Take a moment to convince yourself that modulo-adder structures do indeed take on the new form after this simple input position switch.

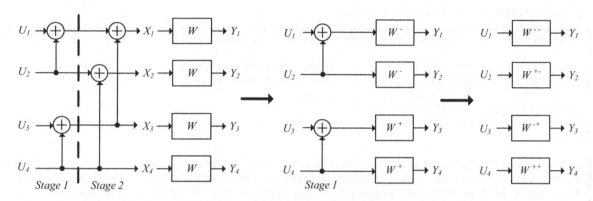

Figure 5-97: Standard Form of Size Four Polar Transform Construction

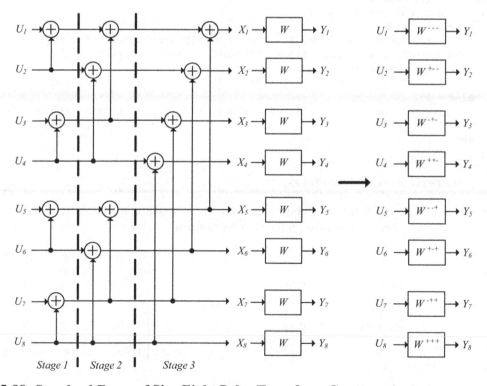

Figure 5-98: Standard Form of Size Eight Polar Transform Construction

Computing the Erasure Probabilities of the Polarized Channels

On the previous pages we illustrated that the poor channel alters the erasure probability, p, of the previous channel via the expression $2p - 2p^2$, whereas the good channel's erasure probability improves as p_2. Take another look at figure 5-97 and observe how stage two of the structure is absorbed into channels W to form instances of W^- and W^+. Afterwards, the modulo-two structures of stage one are absorbed into the instance W- and W+ to form instances of W^{--}, W^{+-}, W^{-+}, W^{++}. The following MatLab code will produce a matrix of erasure probabilities as we move from right to left starting with the probabilities of the original channels W in the last column to the final polarized probabilities in the first column.

```
function Output = FindChannelProbabilities(N, OriginalErasureProbability)

% N - is the number of bits that are being processed in the polar transform
% p - is the erasure probability of the original channel W
p            = OriginalErasureProbability;
NumberOfStages = log10(N)/log10(2);
if(rem(NumberOfStages, 1) ~= 0); error('N must be 2^i, where i is an integer > 0.');
end

% The ChannelErasureProbabilities Matrix will hold the erasure probabilities, where the
% last column holds the erasure probabilities of the original 2^N channels W, whereas
% column 1 will contain the erasure probabilities of the final polarized channels.
ChannelErasureProbabilities = zeros(N, NumberOfStages + 1);
ChannelErasureProbabilities(:, NumberOfStages + 1) = p*ones(N, 1);

% Iterate through each stage
for stage = NumberOfStages   :-1:1
    for channel = 1:N
        pPreviousChannel = ChannelErasureProbabilities(channel, stage+1);
        MakeChannelBetter = mod( floor( (channel - 1)/(2^(stage-1)) ), 2);
        if(MakeChannelBetter == 1)
            ChannelErasureProbabilities(channel, stage) =  pPreviousChannel^2;
        else
            ChannelErasureProbabilities(channel, stage) =  2*pPreviousChannel -
                                                 pPreviousChannel^2;
        end
    end
end
end
Output = ChannelErasureProbabilities;
```

For the size four and eight input structures and an initial erasure probability of 0.3 the channel erasure probability matrices will take on the following form.

$$
ErasureMatrix4 = \begin{bmatrix} 0.760 & 0.51 & 0.3 \\ 0.260 & 0.51 & 0.3 \\ 0.172 & 0.09 & 0.3 \\ 0.0081 & 0.09 & 0.3 \end{bmatrix}
\qquad
ErasureMatrix8 = \begin{bmatrix} 0.9424 & 0.760 & 0.51 & 0.3 \\ 0.5774 & 0.760 & 0.51 & 0.3 \\ 0.4525 & 0.260 & 0.51 & 0.3 \\ 0.0677 & 0.260 & 0.51 & 0.3 \\ 0.3143 & 0.172 & 0.09 & 0.3 \\ 0.0295 & 0.172 & 0.09 & 0.3 \\ 0.0161 & 0.008 & 0.09 & 0.3 \\ 0.0001 & 0.008 & 0.09 & 0.3 \end{bmatrix}
$$

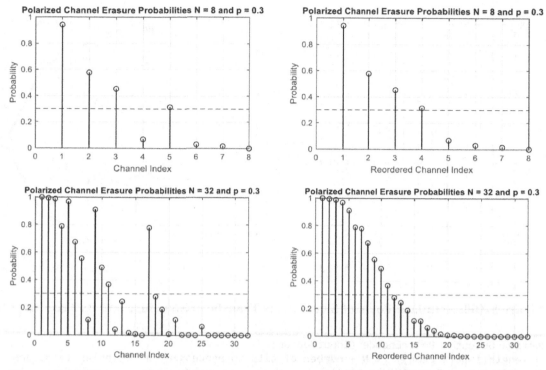

Figure 5-99: Erasure Probabilities for the Size 8 and 32 Polar Transform Structures

Looking at the figure above, the plots to the left display the erasure probabilities versus the proper channel indices. Note that the erasure probability performance does not improve (decrease) monotonically as the channel index increases. In the plots to the right, we have sorted the channels to monotonically improve in erasure probability performance. Note that as the number of bits to be transformed increases, the polarization also increases. Thus, with increasing bit size, the number of very good channels and very bad channels increases at the expense of the mediocre channel in between making the transition between the good and bad channels sharper.

In order to make the polar coding work properly, we will insert input bits equal to 0 into those channels with poor and mediocre erasure probabilities and transmit information on those with good performance. In the top left diagram in the figure above, we would force U_1, U_2, U_3 and U_5 to zero. These zero bits are also called **frozen** bits.

5.6.1.2 Running the Polar Encoder in MatLab

To run the polar encoder in MatLab we introduce the concept of the tree diagram, which features different stages each containing a certain number of nodes indicates by the circles. The concept of the tree diagram will come in handy when we discuss the successive cancellation decoder. Each node encompasses a certain number of size-2 polar transform structures. The example size-8 polar encoding structure on the left of the next figure nicely illustrates the four nodes of stage 1, with each node containing just one size-2 polar transform structure. Stage 2 features two nodes each holding two size-2 polar transform structures. Finally, stage 3 features just one node holding four size-2 polar transform structures.

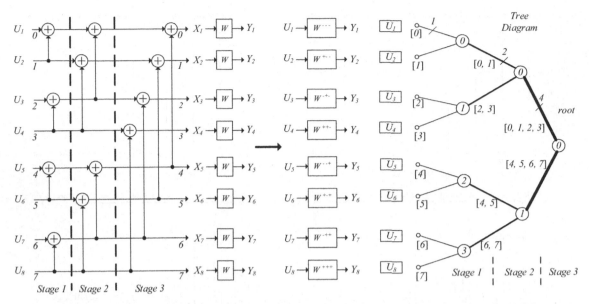

Figure 5-100: Standard Form of Size-8 Polar Transform Structure and its Tree Diagram

```
function Output = PolarEncoder(InputVector)
N = length(InputVector);  % N - number of bits to processed in the polar transform
NumberOfStages = log10(N)/log10(2);
if(rem(NumberOfStages, 1) ~= 0);
    error('N must be of the form 2^i, where i is an integer > 0.');
end
% The matrix, M, shows the progression of the input bits through the Polar structure.
M        = zeros(N, NumberOfStages + 1);
M(:, 1) = InputVector(:);              % Load input vector into first column
% The polar encoding process
for stage = 1:NumberOfStages
    NumberInputsPerNode = 2^stage;                 %  2,   4,   8,   16, ....
    NumberOfNodes       = N / NumberInputsPerNode; % N/2, N/4, N/8, N/16, ....
    for node = 1:NumberOfNodes
        U1_StartIndex           = 2^(stage) * (node - 1) + 1;
        U1_StopIndex            = U1_StartIndex + NumberInputsPerNode/2 - 1;
        U2_StartIndex           = U1_StopIndex + 1;
        U2_StopIndex            = U2_StartIndex + NumberInputsPerNode/2 - 1;
        U1_Indices              = U1_StartIndex : U1_StopIndex;
        U2_Indices              = U2_StartIndex : U2_StopIndex;
        U1                      = M(U1_Indices, stage);
        U2                      = M(U2_Indices, stage);
        X1                      = mod(U1 + U2, 2); % This is the set of size 2 polar
        X2                      = U2;                    % transform structures
        M(U1_Indices, stage + 1) = X1;
        M(U2_Indices, stage + 1) = X2;
    end
end
Output = M(:, NumberOfStages + 1);
```

Assume that the input bits U1 through U8 are located at the 0-based indices 0 through 7. Each stage that follows will have the same output indices 0 through 7. The U1 indices in the code above reference those outputs that will be connected to the upper inputs of the size-2 polar transform structures, whereas the U2 indices refer to those output that connect to their lower inputs.

→ Stage 1 – 4 Nodes (U1_Indices): [0], [2], [4], [6]   -   (U2_Indices): [1], [3], [5], [7]
→ Stage 2 – 2 Nodes (U1_Indices): [0,1], [4,5]   -   (U2_Indices): [2,3], [6,7]
→ Stage 3 – 1 Node (U1_Indices): [0, 1, 2, 3]   -   (U2_Indices): [4, 5, 6, 7]

Encoder Rate

As we have indicated before, the structure size of the polar encoder will always be $N = 2^n$, where n must be an integer larger than 0. Of the N input bits, K positions will actually feature transmit bits, whereas the N-K positions with the worst erasure probability will be frozen to 0. The resulting encoder is said to be of type (N, K) and feature the following rate.

$$EncoderRate = \frac{K}{N}$$

5.6.1.3 Simplified Successive Cancellation Decoding of Polar Codes

The successive cancellation decoder begins with the knowledge of the received softbit values, and begins the process of decoding U1. With this new knowledge it then decodes U2 through U8 assuming the size-8 polar transform structure of our example. With each decoded value of U, the decoding process becomes more robust.

The Binary Belief Channel

The binary erasure channel was a wonderful model that we used to find the erasure probabilities for each channel. However, as we discussed at length in section 5.6.2, the quantity the receiver actually produces are log likelihood ratio beliefs. Note that a positive LLR belief indicates the that a 1 was likely transmitted, whereas a negative LLR belief represents the presence of a 0 was likely transmitted. The magnitude of the LLR believe indicates the amount of confidence that we have in our belief.

Polar Decoder for the Basic Size N=2 Structure

Since even the larger polar encoding structures are subdivided into individual size-2 transform structures, we look at the decoding process of this simple example first.

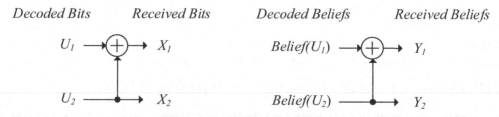

Figure 5-101: Basic Building Block of Successive Cancellation Decoder for N = 2

Note that our basic $N=2$ sized polar transform on the left side is invertible and given X_1 and X_2 we can figure out U_1 and U_2 as follows.

$$X_1 = U_1 \oplus U_2 \quad and \quad X_2 = U_2$$

We start by taking the equation for X_1 and solve for U_1. Given that $X_2 = U_2$, we substitute for U_2 in the expression below to yield the new inverted transform.

$$U_1 = X_1 \oplus U_2$$

$$U_1 = X_1 \oplus X_2 \quad and \quad U_2 = X_2$$

Unfortunately, we do not have access to X_1 and X_2 directly but only their estimates Y_1 and Y_2, which are floating point values expressing the log likelihood belief that the received bit was a 1. In order to emulate the binary modulo two operation using floating point belief, where a positive value would indicate that the transmitted bit was a 1 and a negative value would indicate that the transmitted bit was a 0, we provide the following truth tables expressing the XOR operation for both binary bits and LLR beliefs.

Table 5-7: Truth Table $U_1 = \text{mod}(X_1 + X_2, 2)$ Using Bits and Belief Orientations

X_1	X_2	U_1		Y_1	Y_2	$Belief(U_1)$
0	0	0		-	-	-
0	1	1		-	+	+
1	0	1		+	-	+
1	1	0		+	+	-

 The orientation, or sign, of the belief of U_1 is governed by the following expression which implements the truth table to the right.

$$sign(Belief(U_1)) = - \, sign(Y_1) \cdot sign(Y_2)$$

The magnitude of the belief of U_1 can nicely be approximated as the smaller of the magnitudes of Y_1 and Y_2. If for example we had two received beliefs $Y_1 = -1.1$ and $Y_2 = -1.2$, then we could with good confidence conclude that the modulo two addition will also yield good confidence as $U_1 \approx -1$. However, two received beliefs $Y_1 = -1.1$ and $Y_2 = -0.1$ cannot possible produce a good belief in the result, U_1, as the small magnitude of Y_2 indicates a significant amount of added noise and thus large uncertainty in its sign. The magnitude of the belief of U_1 is computed below.

$$|Belief(U_1)| = min(|Y_1|, |Y_2|)$$

The Minsum Decoder

Combining these two concepts produces the Minsum decoder, which we have already encountered in the chapter on LDPC codes. The Minsum decoder has the following form.

$$Belief(U_1) = MinSum(Y_1, Y_2) = - \, sign(Y_1) \cdot sign(Y_2) \cdot min(|Y_1|, |Y_2|)$$

```
function OutputBelief = MinSumDecoder(Y1, Y2)
OutputBelief        = -sign(Y1).*sign(Y2).*min(abs(Y1), abs(Y2));
```

The Repetition Decoder

As this is a successive cancellation decoder, for this small size-2 example we make a hard decision regarding the belief of U_1 at this point to produce a bit (0 or 1). As the size of the polar transform increases, intermediate floating-point beliefs such as these won't be prematurely truncated but are passed on to lower stages and help us in the decoding process. From the truth tables you can see that if the sign of the belief of U_1 is negative or positive then Y_1 and Y_2 will have the same signs or opposite signs respectively. Therefore, if the belief of U_1 is negative and thus the bit $U_1 = 0$, then we know that the signs of beliefs Y_1 and Y_2 should be the same and we may add Y_2 and Y_1 to produce the belief of U_2, otherwise we subtract them.

$$Belief(U_2) = Rep(Y_1, Y_2, U_1) = \begin{bmatrix} Y_2 + Y_1 & if & U_1 = 0 \\ Y_2 - Y_1 & if & U_1 = 1 \end{bmatrix}$$

Knowing U_1 and the belief of X_1, which is Y_1, provides us with additional or repeated information that allows us to better compute the belief of U_2. It is this repeated information that gives the decoder its name and allows us to sum the two beliefs to yield the belief of U_2, for which we may now also apply a hard decision to produce $U_2 = \text{sign}(Belief(U_2))$.

```
function OutputBelief = RepetitionDecoder(Y1, Y2, U)
OutputBelief = zeros(length(U), 1);
for Index = 1:length(U)
    if(U(Index, 1) <= 0); OutputBelief(Index, 1) = Y2(Index,1) + Y1(Index, 1);
    else             ; OutputBelief(Index, 1) = Y2(Index,1) - Y1(Index, 1); end
end
```

Example 5.8: A Simple Size-2 Successive Cancellation Decoder

As we have seen in countless diagrams, even large sized polar encoding structures are in fact composed of collections of size-2 structures, which we group into nodes. Therefore, the decoding steps involved in the size-2 structure is the corner stone of the entire decoding process. The structure below illustrates a simple size-2 polar transform structure, whose received belief $Y_1 = 1.1$ and $Y_2 = -0.5$ we intend to decode to produce bits U_1 and U_2. The first four steps represent the decoding process, whereas the last two steps represent the re-encoding process, whose output will guide the repetition decoding of the stage above.

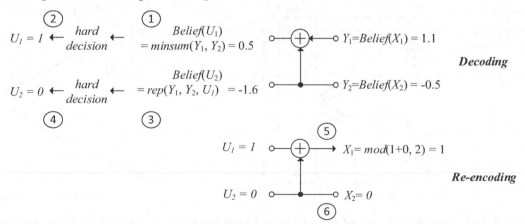

Figure 5-102: Successive Cancellation Decoding for Simple Size-2 Polar Transform

We start the process with the quantities Y_1 and Y_2, which represent the received beliefs of X_1 and X_2 (the originally transmitted bits).

1: Compute *Belief*(U_1) as the *MinSum*(Y_1, Y_2) = -sign(Y_1)·sign(Y_2)·min($|Y_1|$, $|Y_2|$).
2: Compute U_1 as a hard decision of *Belief*(U_1). $U_1 = 0$ if *Belief*(U_1) < 0 and $U_1 =1$ otherwise.
3: Given U_1, compute *Belief*(U_2) = *Rep*(Y_1, Y_2, U_1).
4: Compute U_2 as a hard decision of *Belief*(U_2). $U_2 = 0$ if *Belief*(U_2) < 0 and $U_2 =1$ otherwise.

At this point, the decoding process is done as we have the values for U_1 and U_2. We only reencode U_1 and U_2 into X_1 and X_2 as these bits are needed by the repetition decoders of the size-2 structures in the stage above. The necessity of this step will become clear in the next examples.

5: Compute $X_1 = \text{mod}(U_1 + U_2, 2)$
6: $X_2 = U_2$.

Example 5.9: A Size-4 Simplified Successive Cancellation Decoder

The diagram below illustrates the standard form of the size-4 polar transform structure along with its tree diagram. The tree diagram features a root node, where we start the decoding process, as well as leaves, which represent the computed beliefs $B(U_x)$. The processing flow in this larger structure is a generalized version of what we saw in the last example. Note the following rules.

→ Each node at a particular stage will have 2^{stage} input and output values.

→ Each node at a particular stage will contain $2^{\text{stage-1}}$ size-2 polar transform structures.

→ When a node first begins processing, it computes the $2^{\text{stage-1}}$ MinSum decoding results and passes them upward in the tree structure. If the node is located in stage 2 or higher, it will have to wait for processing in the lower stages before continuing with the repetition decoding.

→ Once the node receives hard decisions back from the node in the lower stage, it will compute beliefs using the repetition decoder and pass the results downwards in the tree.

→ Hard decision bits computed in a node, will always be passed back to the node in the stage above where it had originally received the MinSum decoded results.

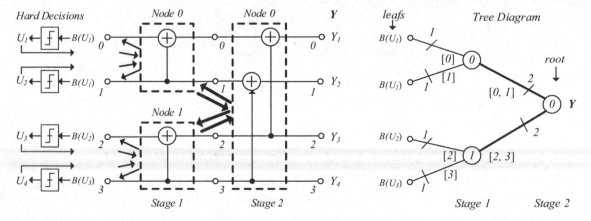

Figure 5-103: Standard Form of Size-4 Polar Transform Structure and its Tree Diagram

→ Once we reach a leaf and the bit U_x is frozen to a value of 0, we naturally return 0 back to the node, from which the calculated MinSum belief originated no matter what that belief was.

Note that the decoding and production of beliefs moves from right to left, starting at the root node and moving towards the leaves, whereas the re-encoding of bits moves from left to right. The following MatLab functions determines the next stage and node that we reach given the current node and stage as well as the direction of movement and whether we are computing a Minsum (only interesting if we move left).

```matlab
function [NextStage, NextNode] = GetNextNode(CurrentNode    ...
                                           , CurrentStage   ...
                                           , MovingLeft     ...
                                           , NodeState)
% If we are moving left, which is the decoding direction producing beliefs
if(MovingLeft == true)
    NextStage   = CurrentStage - 1;    % As we are moving left
    NextNode    = CurrentNode * 2;     % If MinSum    decoder (Upper node)
    if(NodeState == 1)                 % If repetition decoder (Lower node)
        NextNode = NextNode + 1;
    end
% Else we are moving right, which is the re-encoding direction producing 0/1 bits.
else
    NextStage   = CurrentStage + 1;        % As we are moving right
    NextNode    = floor(CurrentNode / 2);
end
```

Each node in a stage has a certain number of input and output ports. The following MatLab function computes the port indices.

```matlab
function [AllIndices, UpperIndices, LowerIndices] = ...
                                    GetPortIndices(CurrentNode, CurrentStage)
NumberOfPorts = 2^CurrentStage;
StartPort     = CurrentNode * NumberOfPorts;
AllIndices    = StartPort : StartPort + NumberOfPorts - 1;    % -1 because this is a
UpperIndices  = AllIndices(1, 1:NumberOfPorts/2);             % 0 based index
LowerIndices  = AllIndices(1, NumberOfPorts/2 + 1 :NumberOfPorts);
end
```

In this example, let's assume that input bits U_1 and U_2 are frozen to zero, whereas input bits U_3 and U_4 carry information as $U_3 = 1$ and $U_4 = 0$. We thus encode the vector $[0, 0, 1, 0]$ via *PolarEncoder*(*InputVector*) to produce an output of $[1, 0, 1, 0]$. Let's map this into beliefs $[1, -1, 1, -1]$ and add some small amount of noise yielding the received beliefs $Y = [0.9, -1.1, 0.8, -1.0]$. In this exercise, we attempt to decode Y to produce the originally transmitted sequence $[0, 0, 1, 0]$.

To make our life easier, we provide two different matrices that track the port values within the decoding structure. The first matrix, L, holds the beliefs, which are filled in as we move from right to left, whereas the second matrix, B, hold bits, which are partially known (frozen bits) or generated via hard decisions as we move from left to right. We will also provide a state value for each one of the nodes in each stage to indicate what the node still needs to do. The state values will be 0 (nothing has been done), 1 (the MinSum decoding is done), 2 (The repetition decoding is done), and 3 (the re-encoding is done and we are finished with this node). The value -10 in matrix B indicates that no bit re-encoding or hard decision has been done.

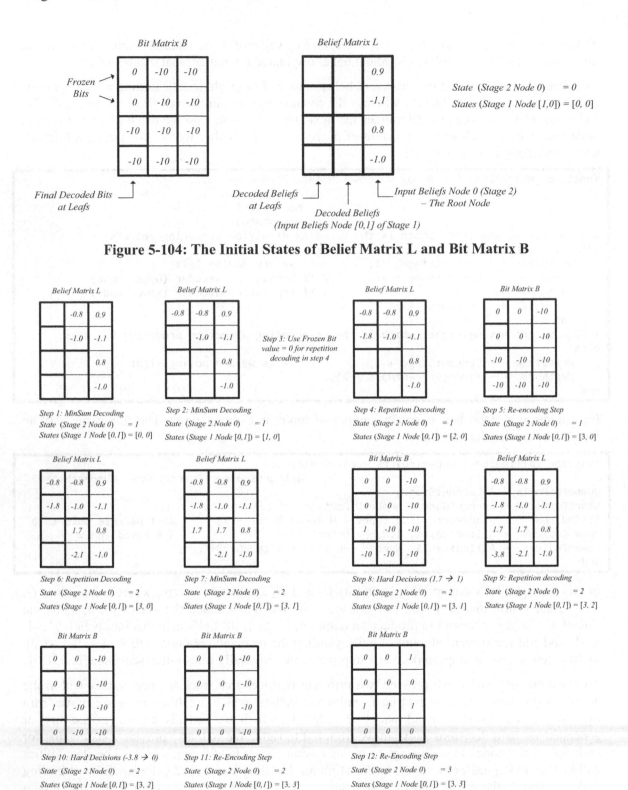

Figure 5-104: The Initial States of Belief Matrix L and Bit Matrix B

Figure 5-105: Progression of Matrices L and B Throughout the Decoding Process

Note that at step 10 we are actually done as the received beliefs have all been decoded, and we only proceed to step 12 for the sake of completeness. The following MatLab command and scripts will produce the decoded beliefs.

```
>> PolarDecoder([0.9, -1.1, 0.8, -1.0], [0, 1])
>> ans = [0; 0; 1; 0]
```

```
function Output = PolarDecoder(Y, FrozenPositions0Based)

N              = length(Y);
NumberOfStages = log10(N)/log10(2);  % Number of stages in structure
if(rem(NumberOfStages, 1) ~= 0); error('N must be equal to 2^integer'); end

% Define state of the nodes:
% Nothing Done = 0, MinSum Decoding Done = 1, Repetition Decoding Done = 2
States = cell(NumberOfStages, 1);
for stage = 1:NumberOfStages
    States{stage, 1} = zeros(1, 2^(NumberOfStages - stage) );
end

% Define the Belief matrix and place the received beliefs into the right most column
L = zeros(N, NumberOfStages + 1);
L(:, NumberOfStages + 1)        = Y(:);

% Define the bit matrix and place the known frozen values into the left most column
B = -10*ones(N, NumberOfStages + 1);
B(FrozenPositions0Based+1, 1)   = zeros(length(FrozenPositions0Based), 1);

CurrentNode  = 0;                 % The start node
CurrentStage = NumberOfStages;    % The start state

while true
    % Check the state of ths current node and see what needs to be done
    NodeState = States{CurrentStage, 1}(1, CurrentNode + 1);
    % If 0 -> We need to run the MinSum Decoding    (Moving right to left)
    % If 1 -> We need to run the repetition decoder (Moving right to left)
    % If 2 -> Re-encode Bit information             (Moving left to right)
    % If 3 -> The node is done. Current configuration should never be reached.
    % Get the ports for the current node and stage
    [~, CurrentUpperIndices, CurrentLowerIndices] = ...
                                GetPortIndices(CurrentNode, CurrentStage);
    switch(NodeState)
        case 0    % We will do the MinSum decoding operation
            UpperBeliefs          = L(CurrentUpperIndices + 1, CurrentStage+1);
            LowerBeliefs          = L(CurrentLowerIndices + 1, CurrentStage+1);

            if(CurrentStage == 1)
                % If we are in stage 1:
                % 1. Run the MinSumDecoder and deposit result in L
                OutputBelief        = MinSumDecoder(UpperBeliefs, LowerBeliefs);
                L(CurrentUpperIndices + 1, CurrentStage) = OutputBelief;
                % 2. Make a hard decision unless we have reached a frozen bit.
                if(B(CurrentUpperIndices + 1,1)~=0)  % Hard decision unless frozen bit
                    B(CurrentUpperIndices + 1, CurrentStage) = ...
                                        0.5*(sign(OutputBelief + 1e-10) + 1);
                end
                U             = B(CurrentUpperIndices + 1, 1);
```

```
        % 3. Run the repetition decoder.
        OutputBelief = RepetitionDecoder(UpperBeliefs, LowerBeliefs, U);
        L(CurrentLowerIndices + 1, CurrentStage) = OutputBelief;
        % 4. Make a hard decision unless we have reached a frozen bit.
        if(B(CurrentLowerIndices + 1, 1)~=0)  % Hard decision unless frozen bit
            B(CurrentLowerIndices + 1, CurrentStage) = ...
                                    0.5*(sign(OutputBelief + 1e-10) + 1);
        end
        % 5. Run the re-encoding step.
        B(CurrentUpperIndices + 1, 2) = mod(B(CurrentUpperIndices + 1, 1)+ ...
                                B(CurrentLowerIndices + 1, 1), 2);
        B(CurrentLowerIndices + 1, 2) = B(CurrentLowerIndices + 1, 1);
        % MinSum and Repetition done
        States{CurrentStage, 1}(1, CurrentNode + 1) = 2;
    else
        % If we are in stage > 1
        % 1. Run the MinSumDecoder and deposit result in L
        OutputBeliefs        = MinSumDecoder(UpperBeliefs, LowerBeliefs);
        L(CurrentUpperIndices + 1, CurrentStage) = OutputBeliefs;
        States{CurrentStage, 1}(1, CurrentNode + 1) = 1; % MinSum decoding done
    end
case 1    % We will do the Repetition decoding operation for stages > 1
    UpperBeliefs    = L(CurrentUpperIndices + 1,  CurrentStage + 1);
    LowerBeliefs    = L(CurrentLowerIndices + 1,  CurrentStage + 1);
    U               = B(CurrentUpperIndices  + 1, CurrentStage);
    OutputBeliefs   = RepetitionDecoder(UpperBeliefs, LowerBeliefs, U);
    L(CurrentLowerIndices + 1, CurrentStage) = OutputBeliefs;
    States{CurrentStage, 1}(1, CurrentNode + 1) = 2; % Repetition decoding done
case 2
    % Now that we finished the repetition decoding operation we will now do the
    % re-encoding
    UpperBits       = B(CurrentUpperIndices + 1, CurrentStage);
    LowerBits       = B(CurrentLowerIndices + 1, CurrentStage);
    B(CurrentUpperIndices + 1, CurrentStage+ 1) = mod(UpperBits+LowerBits, 2);
    B(CurrentLowerIndices + 1, CurrentStage+ 1) = LowerBits;
    if(CurrentStage == NumberOfStages); break; end
otherwise
    error('The node is done with all computations.');
end


% Let's the current node and stage
% If we are already in stage 1, then we must move from right to left
MovingLeft = false;
if((NodeState == 0 || NodeState == 1) && CurrentStage ~= 1); MovingLeft = true; end
[CurrentStage, CurrentNode] = GetNextNode(CurrentNode   ...
                                , CurrentStage ...
                                , MovingLeft   ...
                                , NodeState);
% disp(L);            % Uncomment for debugging
% disp(B);
% Breakpoint = true; % And set your breakpoint here
end
Output = B(:,1);
```

Notice the three display lines at the end of the code that are commented out. Uncomment them and set your breakpoint to see the *L* and *B* matrices change after every step.

The following code implements a test bench that will allow you to run the decoder for any size and noise condition. The most important parameters that you will want to change are *N*, *K* and the signal to noise ratio in *dB*. Notice that at 3dB of signal to noise ratio and above, the decoder is basically flawless (encoder rate = 1024/2048 = 1/2), whereas at 1dB it is dead in the water. This type of abrupt change in performance is very desirable (also seen in the LDPC decoder), unlike the convolutional coding algorithm whose performance doesn't improve as quickly with increasing signal to noise ratio. Now reduce *K* to 680 (encoder rate = 680/2048 = 1/3) and the performance is solid as a rock at 1dB of signal to noise ratio.

```
% Configure the polar encoder
N                = 2048;                % Number of total encoded bits
K                = 1024;                % Number of message bits
NumberFrozenBits = N - K;               % Number of frozen positions set to 0
EncoderRate      = K/N;                 % The encoder rate
MessageBits      = randi([0 1], K, 1);

% Let's determine the erasure probability of each channel
p                        = 0.2;    % The erasure probabilities might change a little with p.
ErasureProbabilitiesMatrix = FindChannelProbabilities(N, p );
ErasureProbabilities        = ErasureProbabilitiesMatrix(:,1);
[SortedProbabilities, SortedIndices] = sort(ErasureProbabilities, 'descend');


IndexOfFrozenBits  = SortedIndices(1:NumberFrozenBits, 1);
IndexOfMessageBits = SortedIndices(NumberFrozenBits+1: end, 1);


% Produce the bits to be transmitted through the polar encoder
TxBits = zeros(1, N);
TxBits(1, IndexOfMessageBits) = MessageBits;
EncodedBits        = PolarEncoder(TxBits);
EncodedBipolarBits = 2*EncodedBits - 1;     % Map from 0/1 to -1/1

% The Encoded Bipolar Bits are now disturbed by noise to produce received beliefs
SNR_dB      = 2;
SNR_Linear  = 10^(SNR_dB/10);
SignalPower = 1;     % = MeanSquare(EncodedBipolarBits)
NoisePower  = SignalPower/SNR_Linear;
AWG_Noise   = sqrt(NoisePower) * randn(N, 1);
Y           = EncodedBipolarBits + AWG_Noise;   % Received beliefs

% Run the Successive Cancellation Decoder
IndexOfFrozenBits0Based = IndexOfFrozenBits - 1;
DecodedBits             = PolarDecoderBook(Y, IndexOfFrozenBits0Based);
DecodedMessageBits      = DecodedBits(IndexOfMessageBits, 1);

% Compute number of errors and bit error rate
NumberOfErrors = sum(mod(MessageBits + DecodedMessageBits, 2));
BER            = NumberOfErrors / K;
disp(['Number of Message Bits: ' num2str(K)]);
disp(['Number of Errors:       ' num2str(NumberOfErrors)]);
disp(['BER:                    ' num2str(BER)]);
```

5.6.1.4 Higher Performance with List Decoding

As engineers are apparently never satisfied with performance, a modification to the successive cancellation decoder was developed that is based on the decades old concept of list decoding. This addition improves the performance slightly ($\sim$ 1dB more or less), but it is worth it in commercial application. Polar coding is used to protect the broadcast channel in 5G signals, and the guiding principle in cell phone communication is to squeeze out every last bit of performance even at the expense of lots of extra resources. Given that semiconductors are getting ever smaller and more powerful, resources are certainly abundant. We will not develop any code for this here, but as the concept is quite simple, a few diagrams should suffice to explain the content in enough detail for the reader to expand the code of the last section.

Let's take a look at the top most portion of the figure below, which illustrates the MinSum decoding operation at the first non-frozen node U_x. Notice the healthy belief of 1.1, and the rather underwhelming belief of -0.2, which produce a MinSum result of $-\text{sign}(1.1)\cdot\text{sign}(-0.2)\cdot\min(|1.1|, |-0.2|) = 0.2$. If the sign of the -0.2 value is incorrect, then the MinSum operation fails yielding an incorrect hard decision and the entire decoding operation is wasted even if no other mistake is made during the subsequent decoding. Therefore, we will decide to retain both decisions, the one with our belief, which indicates that the hard decision of 0.2 should be $U_x = 1$, as well as the other decision (against our belief) as $U_x = 0$. As the decision $U_x = 0$ is more likely to be wrong than the decision $U_x = 1$, we will assign a penalty (DM = decision metric) to the result, which is equal to the absolute value of the calculated belief used for the hard decision. There is no penalty for the decision that goes along with our belief and $DM = 0$ for this case.

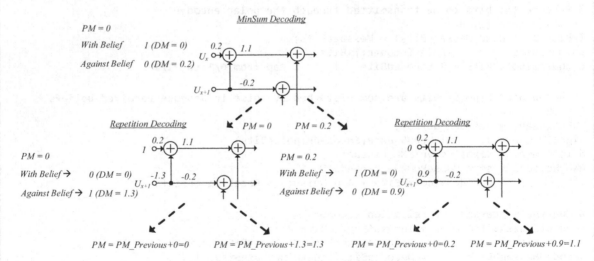

Figure 5-106: Making Both Decisions During the Hard Decision Process

If we wish to continue with both decisions, then we need to instantiate an additional decoder (mainly a new pair of L and B matrices) and copy the content of the original L and B matrices along with the decision information against the belief. In both decoders, we now compute the subsequent repetition decoding results, and once again during the hard decoding step we retain the answer according to the belief at hand as well as retaining the opposite answer. This is illustrated as U_{x+1} in the two structures at the bottom of the figure above. To continue, we now need a total of 4 decoders to hold all combinations of decisions. It quickly becomes clear that we can't continue on retaining both decisions for every non-frozen bit, as the number of L and B

matrices would grow geometrically. Therefore, we must find a way to prune the possibilities to a reasonable number of parallel decoders such as 4 or 8. To make this pruning operation possible, we maintain a path metric, *PM*, which keeps track of the accumulated decision metrics (penalties) as we make our successive decisions. Let's take the situation of the lasts figure where we arrived at four decoders with path metrics 0, 1.3, 0.2, and 1.1. Each decoder now produces the decision metric with the belief as *DM_W*, which is always 0, and against the belief as *DM_A*. We didn't use the nomenclature of *DM_W* and *DM_A* above, but here it is convenient. With a set of 8 path metrics, those paths that yield the smallest *PM* are retained and their associated *L* and *B* matrices are updated and copied over. The concept of eliminating certain paths is not necessarily difficult and we have seen a similar survivor algorithm in the Viterbi decoder. The harder task is developing techniques that minimize the amount of memory and memory related processing required for the list decoding process. Please note that the simplified successive cancellation decoder is quite good on its own.

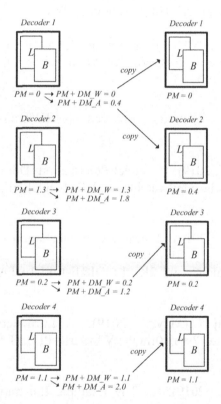

Figure 5-107: Update Mechanism in List Decoding for Simplified Successive Cancellation Decoder

If the four decoders encounter a frozen bit, then each will only compute a single decision metric. If the hard decision of the belief is equal to the frozen bit value of 0, then *PM* remains unchanged, otherwise it is increased by the absolute value of the belief (DM). The frozen bit value of 0 will always configure the next decoding operation, and the *L* and *B* matrices in the 4 decoders are updated but no copying is needed.

Take a look at the plot comparison of different ½ rate FEC coding schemes in figure 5-92. The polar code curve includes the list decoding. Turbo, LDPC, and Polar coding perform quite similarly, whereas convolutional coding lags somewhat in performance.

Please review the many outstanding youtube videos on LDPC and Polar coding published by professor Andrew Thangaraj of the Indian Institute of Technology in Madras. I have included some of them in the reference section below [5] [6] [7] [8] [9].

References

[1] IEEE Std 802.11n-2009 (Oct 2009) Amendment 5: Enhancements For Higher Throughput, New York, NY

[2] Tahir, A., Schwarz, S., Rupp, M., (2017) *BER Comparison Between Convolutional, Turbo, LDPC, and Polar Codes*, Technische Universität (TU) Wien

[3] Arikan, E. (July 2009). "Channel Polarization: A Method for Constructing Capacity-Achieving Codes for Symmetric Binary-Input Memoryless Channels". IEEE Transactions on Information Theory. 55 (7): 3051–73

[4] R. G. Gallager, "Low density parity check codes," IRE Trans. Inf. Theory, vol. IT-8, no. 1, pp. 21- 28, Jan. 1962.

[5] Andrew Thangaraj (May, 2019), "Channel Polarization, Definition of (N,K) Polar Code and Encoding",https://www.youtube.com/watch?v=1uYEq4ueOok&list=PL1mTAEwKQZ4PfR2tZ16pY1hoSwqQf7EQb

[6] Andrew Thangaraj (May, 2019), "Successive Cancellation (SC) Decoder for Polar Codes: Illustration of its Building Blocks with N=2, 4 ", https://www.youtube.com/watch?v=wK2KI2LtdQI&list=PL1mTAEwKQZ4PfR2tZ16pY1hoSwqQf7EQb&index=2&t=169s

[7] Andrew Thangaraj (May, 2019), "Introduction to Polar Codes", https://www.youtube.com/watch?v=rB0rhQKyV34&list=PL1mTAEwKQZ4PfR2tZ16pY1hoSwqQf7EQb&index=3

[8] Andrew Thangaraj (May, 2019), "A Toy Example Illustration of SISO MinSum Iterative Message Passing Decoder", https://www.youtube.com/watch?v=rB0rhQKyV34&list=PL1mTAEwKQZ4PfR2tZ16pY1hoSwqQf7EQb&index=3

[9] Andrew Thangaraj (May, 2019), "Illustration of SISO decoder for (3,2) SPC code and min-sum approximation.", https://www.youtube.com/watch?v=2xgGWEGJ7NM&list=LL&index=74

6 Distortion, Offsets, and Noise in Communication Systems

You are sitting in the lab. The baseband processor you designed is running in an FPGA right in front of you and is hooked up snugly to your company's highly proprietary RF transceiver IC. The baseband processor is sucking up the IQ stream provided by the analog receiver's output ports, processing it, and showing you a receive constellation on your laptop screen. Out of nowhere, your boss leans over your shoulder and grunts, "The EVM stinks!". "Yeah, I've noticed," you reply. "It must be something with the RF transceiver chip." (Of course, it couldn't be on your end.) The chief analog IC designer finally shows up and examines the constellation on your screen with a puzzled expression on his face. "What am I looking at?" he thinks to himself, then turns to your boss and says, "Hey, I designed this thing exactly to the specification," and escapes back to his office. "Great," you think, "I'm left holding the short end of the stick."

The truth is that as a DSP/communication engineer you are well suited to solve these types of problems. First, you'll need a crash course in what can go wrong in the analog portion of a transceiver and in the process learn how to properly diagnose and fix these types of issues. Unfortunately, most books on communication systems limit their discussion to the effects of white Gaussian noise on bit error rate by illustrating large numbers of BER versus E_b/n_o curves for different modulation formats. This oversimplified view shields us from the large majority of real life effects. In practice, your baseband or DSP processor will need to deal with multipath effects, phase noise, DC offset, and a whole host of other nightmares that will cause you plenty of restless nights. And none of these has anything to do with the design of the core detector of your baseband processor. It is your job to build synchronization modules into the processor that can handle these defects and present the detector with a quality signal that it can use to make optimal symbol decisions. To cover the various effects that afflict the transceiver, the presentation in this chapter is subdivided into four sections:

1. In the first section, we provide an overview of the defects that afflict both analog/RF transmitters and receivers and show how to quantify these problems as measurement metrics.

2. The second section introduces us to the communication/modulation model that will serve as the vehicle of our MatLab simulations. Here we introduce MatLab code of a simple QAM (quadrature amplitude modulation) based, single-tone modem that provides the constellations we will be compromising using the defect models developed in the third section.

3. The third section provides a detailed discussion of each defect and how it needs to be modeled using MatLab code. The transmitted information generated by the QAM-based, single-tone modem will be distorted by each defect (one at a time), and the behavior of the received constellation studied. Finally, we unify the MatLab defect code into a single function, which can be easily called by either the single-tone modem code platform of this chapter or the OFDM based-platform of chapter 6.

4. We end the chapter by reviewing dynamic DC offset issues, digital to analog conversion and nonlinear effects that were not integrated into the MatLab defect model previously.

Modeling Effort

One of the goals of this chapter is to create a MatLab defect model that can be bolted to modem blocks of any communication standard. The figure below shows the defect model connected to the single-tone modulation blocks developed in this chapter, as well as the OFDM modulation/demodulation models developed in the next chapter.

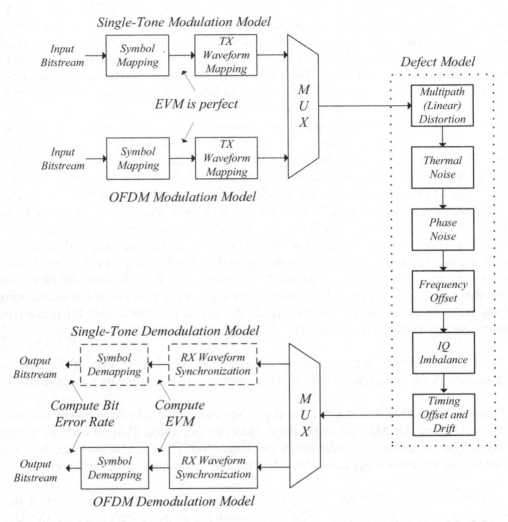

Figure 6-1: Interaction between Modulation, Defect, and Demodulation Models

6.1 Distortion, Offsets and Noise: An Overview

Having a thorough understanding of the analog portion of a transceiver is a splendid bonus for a DSP engineer working in the wireless industry. Your coworkers will start calling you a system engineer and you end up getting invited to meetings in which the staff scientists plan your company's next wireless communication architectures. And let's not forget that a nice pay increase usually accompanies your expanded abilities and knowledge.

6.1.1 Distortion, Offsets and Noise in Analog Transmitters

Let us start expanding our knowledge with an overview of how signals generated in the transmit path deviate from their idealized formulation.

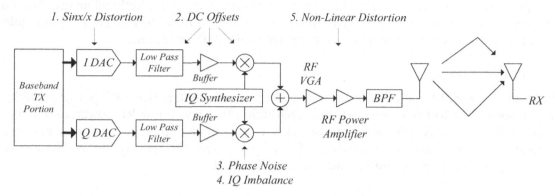

Figure 6-2: Distortion, Offset and Noise Sources in a Zero-IF Transmitter

1. Sinx/x Distortion

Inexperienced engineers may encounter their first signs of trouble by incorrectly choosing the digital to analog converter (DAC) sample rate. Because the DAC maintains its output value steady for one sample period, the generated analog IQ signals will suffer from sample and hold (*sinx/x*) distortion, which is particularly troublesome at low sampling rates. At the end of this chapter, we will provide a mathematical explanation of this phenomenon, which can easily be remedied by upsampling the signal prior to the digital to analog conversion process.

2. DC Offsets

The next unavoidable problems in our transmitter are the numerous positions at which DC offset can sneak into the system. DC offsets that make it to the modulator inputs will cause carrier leakage in the final RF output. With carrier leakage in the transmit signal, the received constellation will appear not to be centered correctly and some of the synchronization algorithms in the receiver will no longer work optimally. This problem is mitigated by careful analog design and the ability to add counteracting offsets into the digital signal stream just ahead of the DAC modules.

3. Phase Noise

Phase noise is produced by frequency drift in the local oscillators, which are used to up and down convert the signals of interest between the RF and baseband domains. Phase noise is completely independent of the received signal strength. From the stand-point of the DSP and communication systems designer, it is important to know how to simulate phase noise and see the effect it has on the demodulated constellation and subsequent error performance. Once the effect is known, proper phase noise specifications can be provided to the RF engineers in charge of designing the synthesizer, which generates the LO signal. But, even without knowing specifics about the communication standard of a particular radio system, we can make some statements regarding phase noise that are globally true. Since the flicker-like phase noise is restricted to low frequencies, it can be more easily removed from signals featuring large bandwidths.

---

4. IQ Imbalance

IQ phase imbalance occurs when the quadrature local oscillator (LO) signals are not completely orthogonal (the sine / cosine components of the LO signals are not exactly 90 apart). Amplitude imbalance occurs if the output of the *I*-modulator is different from than that of the *Q*-modulator. IQ imbalance can be simulated and even mitigated using DSP algorithms.

5. Nonlinear Distortion

By far the dominant trouble maker in modern wireless transmitters is the RF power amplifier. A large amount of effort is invested in designing high efficiency, linear RF power amplifiers that meet the demanding ACPR (adjacent channel power ratio) and emission mask requirements set by modern RF communication standards. The DSP engineer can remedy this problem to some extent by introducing pre-distortion into the transmit signal.

6.1.2 Distortion, Offsets, and Noise in Analog Receivers

The next figure illustrates the RF/analog portion of a modern Zero-IF receiver, along with the entry points of the different afflictions from which the receiver may suffer. It is evident from the figure that the signal has been linearly distorted by the transmission channel, even before arriving at the receiving antenna. To complicate things further, the antenna and low noise amplifier may add healthy amounts of white Gaussian noise to the signal, which limits the absolute sensitivity of our receiver. Next, phase noise is introduced by the local RF oscillator signal, which, instead of remaining fixed, starts to slowly drift around its intended center frequency. As mentioned earlier, the LO signal is actually split into two paths that are injected into separate In-phase and Quadrature-phase modulators. If these two LO signals do not exhibit a phase offset of 90 degrees, as they should, an IQ imbalance occurs, which causes cross-talk between the *I* and *Q* data streams. The final type of distortion at RF is of the nonlinear type. In general, the active RF components – LNA (low noise amplifier), RF VGA (variable gain amplifier) and mixer – will cause nonlinear distortion if their input signals become large. The remaining problems reside in the baseband frequency section of our radio. They include linear distortion due to overly aggressive filtering, DC offsets, and timing drift in the sample clock of the analog-to-digital converters.

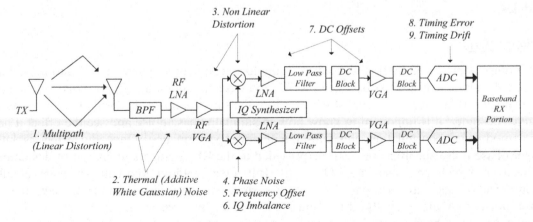

Figure 6-3: Distortion, Offset and Noise Sources in a Zero-IF Receiver

1. Multipath Channel

The transmitted RF signal radiates in many different directions, reflects off different physical structures, and finally arrives at the receiving antenna via multiple paths and associated time delays. Multipath effects cause a significant amount of linear distortion in wireless communication systems and must be mitigated in all high performance wireless communication systems. In chapter 4, we learned that linear distortion processes like multipath channels are modeled as complex finite impulse response (FIR) filters. It is the focus of this chapter to find out what the complex impulse response would be for different communication environments.

2. Thermal Gaussian Noise

In properly designed analog receivers, the amount of additive white noise present in the signal at the ADC inputs is due to the thermal noise produced by the antenna, the noise figure of the radio, and the low pass filter (LPF) bandwidth inside the analog section of the receiver. Thermal noise power originates in the resistive part of the antenna and is further exacerbated by the noise figure of the receiver. Since thermal noise is white, properly filtering the signal at baseband frequencies limits the amount of noise seen at the ADC inputs. Short of cooling the antenna/receiver assembly, there is little that can be done to avoid thermal noise in well designed analog receivers.

3. Nonlinear Distortion

Nonlinear distortion is a problem that afflicts both the transmitter and receiver portions of analog radios, causing frequency content of the signal to spread into neighboring channels. If we are attempting to receive a weak transmission, and a strong signal resides in a neighboring RF channel, then the spectral splatter (spectral regrowth) of the high power waveform can drown out the smaller signal. There is little that can be done in the digital portion of the receiver to make up for this problem, beyond correctly controlling whether the LNA is turned on or off and setting the gain in the RF VGA (variable gain amplifier) properly.

4. Phase Noise

The effect of phase noise in the receiver is identical to that in the transmitter. Since phase noise manifests itself at very low frequencies, it can be tracked out to some extent in the digital section of the modem. This tracking operation acts like a high pass filter by removing slowly varying phase drift but leaving magnitude information intact. Nevertheless, this tracking operation will remove low frequency information from the signal. If the signal bandwidth is large and the high pass corner small, little to no harm is done.

5. Frequency Offset

Frequency offset is an unavoidable problem, which must be dealt with in any communication link. Introductory communication books always gloss over the fact that the LO (local oscillator) signals in the transmitter and receiver do not truly feature identical frequencies. The LO signals are created by phase lock loops that gain their reference clock signals from crystal oscillators – one at the receiver and the other at the transmitter. These devices do not oscillate at identical frequencies and cannot possibly lead to identical LO signals.

6. IQ Imbalance

The effects of IQ imbalance in the receiver are identical to those in the transmitter that were previously discussed.

7. DC Offsets

DC offsets are particularly bothersome in newer direct conversion (zero-IF) receivers. Once the signal has been down-converted, every subsequent amplifier and filter will add DC offsets into the signal. Zero-IF receivers are particularly vulnerable to its effects because of the large amount of gain required in the analog baseband section of the radio. The older super heterodyne receivers included IF sections at intermediate frequencies, where the waveforms could be conveniently gained up and AC coupled before the second down conversion process would take the signals to baseband. The problem gets worse for modern burst modems, where the DC offset must be removed very quickly. The digital portion of the receiver must coordinate sophisticated high pass filtering operations in the analog radio, in concert with the automatic gain control operation (which sets the gains of our RF and baseband variable gain blocks). At best, unresolved DC offsets in the transmitter and receiver may cause the received constellation to be offset slightly from its center. For complex constellations, this can negatively influence the detection of the transmitted symbols. At worst, mismanaging DC offsets will cause the receiver to completely miss incoming packets. There's a detailed discussion on this later in this chapter.

8. Timing Offset

Sample timing offset in the analog-to-digital converter (ADC) modules is due to the fact that, initially, the receiver does not know the perfect sampling instant. As we will see in the next chapter, it is important to know the waveform values at positions that facilitate detection of the transmitted symbol. There is no way to align the sample clocks driving the DAC at the transmitter and the clock driving the ADC at the receiver. Understanding sample timing error is best done by looking at example waveform plots, which will be presented in this chapter. Timing error is a type of linear distortion that can be removed by an interpolator or equalizer.

9. Timing Drift

Timing drift is produced via the same mechanism that causes frequency offset. The main reference clocks driving the DAC and ADC modules at the transmitter and receiver do not feature identical frequencies, even if their specification claims that they do. These oscillators will always be slightly different causing somewhat different sample rates at the transmitter and receiver. Timing drift causes timing offset, which needs to be tracked out.

6.1.3   Modulation Accuracy (EVM) and Spectrum Emission

Modulation accuracy, measured as error vector magnitude (EVM), and spectrum emission are two common metrics used to determine the quality of a test signal. Modulation accuracy is affected by every defect in the system, whereas spectrum emission primarily measures nonlinear effects in transmitters and, to a lesser degree, in receivers.

Modulation Accuracy

Modulation accuracy is a measure of how far a received or transmitted waveform has deviated from its original reference counterpart. The setup for measuring modulation accuracy is illustrated below for the case of an RF/analog test receiver. The receiver under test is surrounded by high performance hardware and software components, such that degradation in modulation accuracy may be isolated to the test device. Note, before the modulation accuracy measurement, the received signal first passes through a synchronization step, which mitigates (via discrete mathematics) some of the analog defects, so that the EVM calculation even becomes possible.

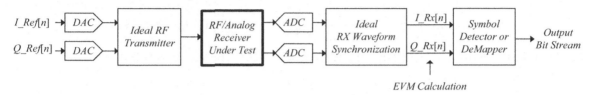

Figure 6-4: Demodulation Accuracy Measurement for a Receiver under Test

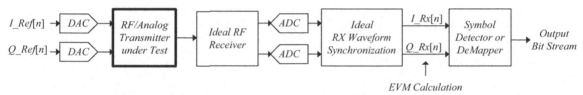

Figure 6-5: Modulation Accuracy Measurement for a Transmitter under Test

A high performance signal generator is normally utilized as the ideal transmission source, whereas a vector signal analyzer, or VSA, is used to implement the ideal RF receiver and synchronization functions. Note that most VSA instruments feature both an RF input port, for the transmitter accuracy test, and an IQ base band port, for the receiver accuracy test.

Error Vector Magnitude

Modulation and demodulation accuracy are computed as EVM by comparing the reference IQ symbols encoded into the transmit waveform with those produced by the receiver's synchronization module.

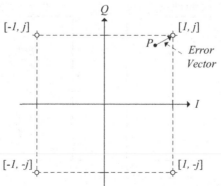

Figure 6-6: Error Vector in a QPSK Constellation

The figure above shows the error vector, which is defined as the difference between the complex received symbol $P = I\_Rx + jQ\_Rx$ and its original reference location $I\_Ref + jQ\_Ref$.

$$ErrorVector = I\_Rx - I\_Ref + j(Q\_Rx - Q\_Ref)$$

The final EVM calculation requires knowledge of the average error vector power calculated over a large number of decision samples N, and the constellation's reference power.

$$AverageErrorVectorPower = \frac{1}{N}\sum_{k=0}^{N-1} ErrorVector[k] \cdot ErrorVector[k]^*$$

$$ReferencePower = \frac{1}{NumRef}\sum_{x=1}^{NumRef} Reference[x] \cdot Reference[x]^*$$

For the QPSK case of the last figure, the reference power is computed as follows.

$$ReferencePower_{QPSK} = \frac{1}{4}\sum_{x=1}^{4} ReferencePosition[x] \cdot ReferencePosition[x]^*$$

$$= \frac{1}{4}[(1+j)\cdot(1+j)^* + (1-j)\cdot(1-j)^* + (-1-j)\cdot(-1-j)^* + (-1+j)\cdot(-1+j)^*]$$

$$= \frac{1}{4}[2+2+2+2] = 2$$

EVM is defined as the ratio of the average error vector power and the reference power of the constellation, and may be expressed in terms of dB or percent root mean square (*rms*) error.

$$EVM_{dB} = 10 \cdot \log 10(\frac{AverageErrorVectorPower}{ReferencePower})$$

$$EVM_{\%rms} = \sqrt{\frac{AverageErrorVectorPower}{ReferencePower}} \cdot 100\%$$

The figure below shows received QPSK symbols groups featuring EVM values of -34 and -14dB, respectively. Clearly, an EVM of -14dB is good enough for error free QPSK symbol detection.

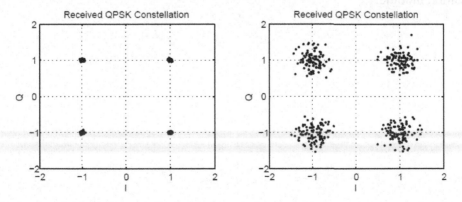

Figure 6-7: QPSK Constellations with EVM of -34 dB (2.1%) and -14 dB (20.7%)

ACPR and Spectrum Emission Mask

Spectrum emission is a defect metric that mostly concerns analog/RF transmitters. In most modern wireless communication systems, users are assigned frequency channels within which their transmit signal must reside. Nonlinear distortion, mainly due to the RF power amplifiers, causes signal power to spill into neighboring channels, thus creating interference for other users. All modern communication standards have strict specifications regarding these emissions as this type of interference limits the number of paying customers that can communicate on any given wireless network.

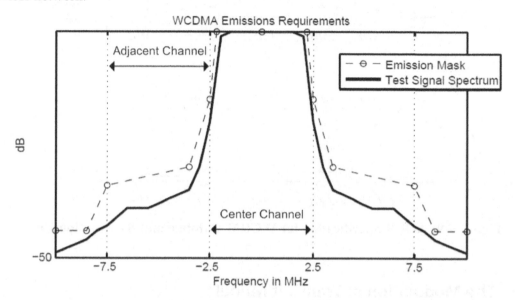

Figure 6-8: Typical Spectrum Emission Requirements for WCDMA Handset

The first such specification is called the spectrum emission mask. The emission mask, which is a function of frequency, limits the amount of frequency content residing outside the center channel. The figure above illustrates the normalized power spectrum, shown as the dark thick line, as well as the emission mask, the dotted line, which sets the limit for all frequencies in the immediate vicinity of the main transmission lobe. Testing for emission mask failures is time intensive which leads us to a faster test metric used heavily in mass production settings.

This second metric is called the ACPR (adjacent channel power ratio) or ACLR (adjacent channel leakage ratio). Different communication standards use different terminology for concepts that are basically the same. These metrics compare the total amount of power that resides in the current center channel to that in the adjacent channels. The following example explains the ACLR measurement for WCDMA mobile station (handset) and that of the base station. Referenced to the current center frequency, the total signal power in the center channel is that confined to a bandwidth of 3.84MHz, while the power in the adjacent channels is measured within that same 3.84MHz bandwidth but at frequency offsets of +/-5MHz and +/-10MHz. The specification for WCDMA cell phones indicates that the total power within the center band must be 33dB greater than that in the channels centered at +/-5MHz. Furthermore, it requires that the total power within the 3.84MHz center band must be 43dB greater than that inside the 3.84MHz wide channels centered at +/-10MHz [1]. As can be seen in the next figure, the specifications for the base station unit are even more demanding than those for the mobile unit. This should come as no surprise, since equipment cost and current consumption are nowhere near as critical in the base station transmitter, and higher quality components, coupled with an ample supply of current, may be utilized to significantly improve linearity performance.

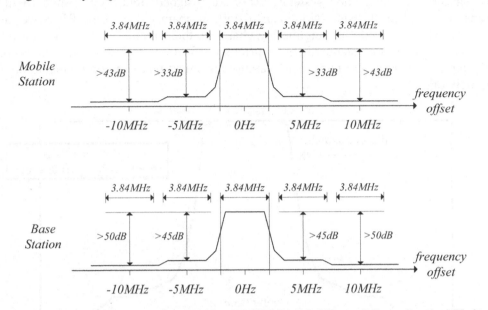

Figure 6-9: ACLR Specification for WCDMA Mobile and Base Station Unit

6.2 The Modulation or Transmit Model

To best illustrate the negative impact that the different defects have on the receive constellation, we introduce a simple, single-tone modulation model that will serve as the sample source of the transmit information stream. The MatLab code for this transmitter will be used in conjunction with the defect model code, developed in section 5.3, to show the distorted *RX* constellations for four separate modulation formats: BPSK, QPSK, 8PSK, and 16QAM. The modulation model consists of a symbol mapping step, which translates groups of input bits into complex symbols that fit on to the reference positions of the current modulation format's constellation, and a TX waveform mapping step that integrates each symbol into the TX waveform or sample stream.

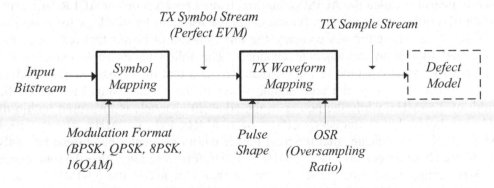

Figure 6-10: The Generic Modulation Model

The *symbol mapping* process for BPSK, QPSK, 8PSK, and 16QAM follows the simple steps shown in the figure below. Random bits are generated and mapped into the symbols that are appropriate for the selected modulation format. During the *TX waveform mapping* process, the *I* and *Q* symbol streams are passed through a zero-stuffing module, which inserts zero samples in between the generated symbols. Once the information exits the zero-stuffing element, it is no longer referred to as a symbol stream but as a sample stream, with an accompanying change in rate. Clearly, the rate at which a new sample appears is larger than the rate at which a new symbol is generated. The ratio between the sample and symbol rates is referred to as oversampling ratio.

$$OSR = \frac{Rate_{Sample}}{Rate_{Symbol}} = \frac{T_{Symbol}}{T_{Sample}}$$

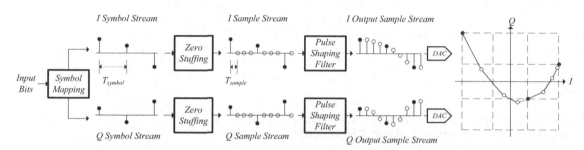

Figure 6-11: The Modulation Model with an Oversample Ratio (OSR) of 4

The sample stream is finally pulse-shaped using a raised cosine FIR filter and sent on to the digital to analog converter modules.

Symbol Mapping

Symbol mapping is one of the simplest functions to understand and implement in base band or DSP processors. The figure below illustrates just how groups of input bits map to each point in the constellation. Note that there is a pattern inherent in the assignment of bit groups to constellation positions. The bit group of any one position in a constellation differs in only one digit from the bit group of its neighbor. Examine the 8PSK constellation and follow the bit groups around the circle to convince yourself that this is indeed the case.

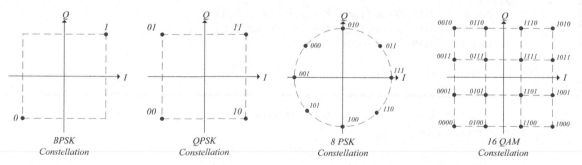

Figure 6-12: BPSK, QPSK, 8PSK, and 16QAM Mapping Diagrams

This method of bit assignment, called gray coding, can be seen in all four constellations and ensures that bit errors induced by incorrect symbol detections are minimized.

The constellations of each modulation format are scaled such that the variance (reference power) of their respective symbol streams remains equal. This is done so that the baseband processor can switch between modulation formats without an accompanying change in the RF output power. For a constant variance of 1, the QPSK constellation positions would be located at ±0.707 $\pm j0.707$, while those for 8PSK would include the QPSK positions as well as $1, j, -1$ and j.

```
function [OutputSymbols] = Mapper(InputBits, ModulationFormat)
% Modulation format: 0, 1, 2, 3 --> BPSK, QPSK, 8PSK, 16QAM
persistent BPSK_LUT QPSK_LUT PSK8_LUT QAM16_LUT
if(isempty(BPSK_LUT))
    C        = 1/sqrt(2);
    BPSK_LUT = 1/sqrt(2)* [-1 - j; 1 + j];
    QPSK_LUT = 1/sqrt(2)* [-1 - j; -1 + j;  1 - j;  1 + j];
    PSK8_LUT =                [-C + j*C; -1     ;        j; C + j*C; ...
                           -j      ; -C - j*C; C - j*C;      1];
    QAM16_LUT = 1/sqrt(10.6)*[-3 - 3*j; -3 - j; -3 + j; -3 + 3*j; ...
                              -1 - 3*j; -1 - j; -1 + j; -1 + 3*j; ...
                               1 - 3*j;  1 - j;  1 + j;  1 + 3*j; ...
                               3 - 3*j;  3 - j;  3 + j;  3 + 3*j];
end

NumberOfSymbols = length(InputBits)/(ModulationFormat + 1);
OutputSymbols   = zeros(1,NumberOfSymbols);


for i = 1:NumberOfSymbols
    Start = 1 + (i - 1)*(ModulationFormat + 1);
    Stop  = Start + ModulationFormat;
    Bit_Group = InputBits(1,Start:Stop);

    switch(ModulationFormat)
        case 0
            Code   = Bit_Group(1,1) + 1; Symbol = BPSK_LUT(Code, 1);
        case 1
            Code   = Bit_Group(1,2)*2 + Bit_Group(1,1) + 1;
            Symbol = QPSK_LUT(Code, 1);
        case 2
            Code   = Bit_Group(1,3)*4 + Bit_Group(1,2)*2 ...
                   + Bit_Group(1,1) + 1;
            Symbol = PSK8_LUT(Code, 1);
        case 3
            Code   = Bit_Group(1,4)*8 + Bit_Group(1,3)*4 ...
                   + Bit_Group(1,2)*2 + Bit_Group(1,1) + 1;
            Symbol = QAM16_LUT(Code, 1);
    end
    OutputSymbols(1,i) = Symbol;
end
```

The mapper function above takes in an array of input bits to produce an output array of symbols for the desired modulation format.

TX Waveform Mapping (Zero-Stuffing and Pulse-Shaping)

The array of output symbols is then handed to the TX_Waveform_Mapping function, which zero-stuffs the array and pulse-shapes each symbol into the final transmit IQ sample stream. Note that the function requires an oversampling ratio, OSR, and a delay, as well as a raised cosine index called alpha used in the calculation of the pulse-shaping filter's impulse response.

```
function [OutputSamples] = TX_Waveform_Mapping(InputSymbols, OSR, Alpha, Delay)

NumberOfSymbols   =   length(InputSymbols);
%% 1. The Zero Stuffing Process (there is nothing to it)
SampleStream                 = zeros(1, NumberOfSymbols*OSR);
for i = 1:NumberOfSymbols
    SampleStream(1, 1 + (i*OSR)) = InputSymbols(1,i);
end

%% 2. The Pulse Shaping Process
[num,den]    = rcosine(1,OSR,'default',Alpha, Delay);
OutputSamples = filter(num, den, SampleStream);
```

In the example plots below, 30 QPSK symbols were generated and zero-stuffed via an oversampling ratio of 4. This process causes three zeros to be inserted between neighboring QPSK symbols. The impulse response of the raised cosine FIR filter, shown to the right, provides an interesting insight into the motivation behind using the raised cosine family of pulse shapes. Note that, four samples before and four samples after the main peak, the impulse response is equal to zero. Thus as a non-zero sample, the symbol, travels through the FIR filter and produces a peak, the symbols four samples before and four samples after produce values equal to zero. The peak of each symbol entering the FIR remains intact at the output. The pulse shape is said to produce no inter-symbol interference, or ISI.

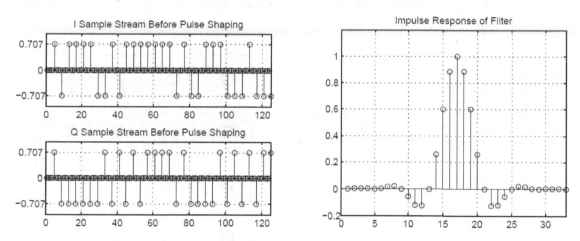

Figure 6-13: The QPSK Zero-Stuffed Sample Stream Prior to Pulse-Shaping

This aspect is clearly demonstrated in the next figure showing the pulse-shaped *I* and *Q* sample streams, which feature sample positions that are equal in size to the corresponding symbols that spawned them. The receiver needs to find these perfect positions and then determine the quadrant in which they belong, to transform the received symbols back to bits, in what we call a demapping or detection process. The figure below shows the IQ signal trace as it moves in the complex plane, emphasizing the key positions representing the original symbols as circles.

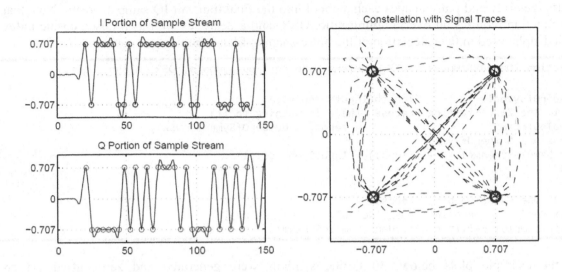

Figure 6-14: Ideal Pulse-Shaped QPSK Sample Stream and its Associated Constellation

As the frequency content of the transmitted signal is altered by multipath or other filtering processes, the raised cosine shape is lost, and inter-symbol interference is introduced. The samples that were once neatly positioned at $\pm 0.707 \pm j \, 0.707$ now appear to deviate from that reference spot, making it harder for the detection/demapping process to determine which reference constellation position each symbol belongs to. As a matter of fact, every defect discussed in this chapter will move the ideal sample value away from the reference positions, as we will soon see.

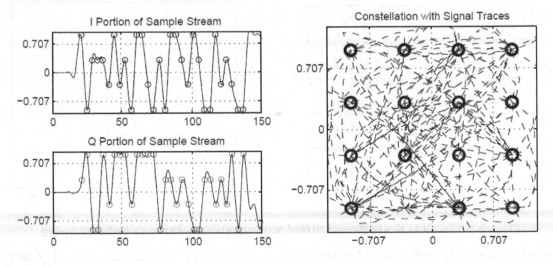

Figure 6-15: Ideal Pulse-Shaped 16-QAM Sample Stream and its Associated Constellation

6.3   The Defect Model

We now have a good grasp of the method of generating the complex transmit symbols that make up our constellation and the waveform mapping process that embeds them in a sample stream. Next, let us develop the defect models that represent the problem inherent in the analog portion of our transceiver. The figure below illustrates the different defects that we will introduce into the transmitted sample stream, along with the parameters we need to provide to model them accurately in MatLab.

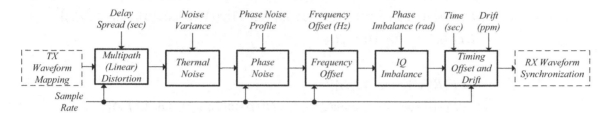

Figure 6-16: Overview of the Defect Models and Order of Insertion

Note that while dynamic DC offset phenomena and nonlinear distortions will be discussed and the former modeled using MatLab, these effects are not integrated into the cascaded model structure illustrated in the figure above. Dynamic DC offset is especially problematic in Zero-IF receivers and must be resolved before even taking a look at the receive constellation. We will present a detailed discussion of the problem and solution to dynamic DC offset in section 5.4.

The effects of nonlinear distortion are discussed but not modeled in MatLab. The primary reason for this omission is the significant difficulties in choosing a sufficiently sophisticated model and computing correct model parameters. This is a science all to itself. If the receiver is designed and operating properly, we may safely ignore nonlinear effects for now.

Pinpointing Each Defect

To more precisely locate the entry point of each defect in the analog/RF portion of the communication link, we will show modified versions of the diagram below at the beginning of each section

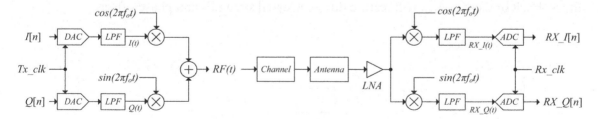

Figure 6-17: Simplified Model of the Modulation/Demodulation Process

The discrete $I[n]$ and $Q[n]$ input signals may be traced through the DAC and low pass filter modules, which reconstruct the analog waveforms, $I(t)$ and $Q(t)$, from their discrete progenitors. From there, each waveform is up-converted to the RF frequency domain via analog multipliers and added to form the following expression.

$$RF(t) = I(t) \cdot \cos(2\pi f_o t) + Q(t) \cdot \sin(2\pi f_o t)$$

Assuming unity gain for all receiver blocks and negligible contributions from distortions, noises and offsets, the analog $I(t)$ and $Q(t)$ waveforms are recovered using simple down-conversion and low pass filtering steps.

$$\begin{aligned}
Rx\_I(t) &= LPF[RF(t) \cdot \cos(2\pi f_o t)] \\
&= LPF[I(t) \cdot \cos(2\pi f_o t)\cos(2\pi f_o t) + Q(t) \cdot \sin(2\pi f_o t) \cdot \cos(2\pi f_o t)] \\
&= LPF[I(t) \cdot 0.5(\cos(0) + \cos(4\pi f_o t)] + LPF[Q(t) \cdot 0.5(\sin(0) + \sin(4\pi f_o t)] \\
&= I(t) \cdot 0.5 \cdot \cos(0) = 0.5 \cdot I(t)
\end{aligned}$$

$$\begin{aligned}
Rx\_Q(t) &= LPF[RF(t) \cdot \sin(2\pi f_o t)] \\
&= LPF[I(t) \cdot \cos(2\pi f_o t)\sin(2\pi f_o t) + Q(t) \cdot \sin(2\pi f_o t) \cdot \sin(2\pi f_o t)] \\
&= LPF[I(t) \cdot 0.5(\sin(0) + \sin(4\pi f_o t)] + LPF[Q(t) \cdot 0.5(\cos(0) - \cos(4\pi f_o t)] \\
&= Q(t) \cdot 0.5 \cdot \cos(0) = 0.5 \cdot Q(t)
\end{aligned}$$

Review the section on trigonometric identities in chapter 1 to convince yourself of the validity of the above statements.

6.3.1 Multipath Distortion (No Mobility)

In wireless communication systems, the transmitted signal will likely bounce off many different physical structures such as buildings or mountains before arriving at its destination. Because of this phenomenon, copies of the transmitted signal will arrive at the receiving antenna at different times. Each of these copies will feature different signal strengths and phase offsets.

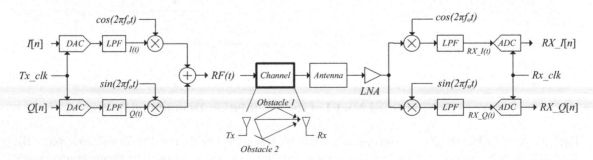

Figure 6-18: The Location of Multipath Effects in Our Communication Model

Conversely, we may say that the complex envelope of each of these distinct RF carriers was multiplied by a complex scalar featuring some unique magnitude and phase. These copies then add via superposition at the receiving antenna. The transversal filter shown below is a great model for this process. The complex envelope of the transmitted signal is depicted as $x[n]$, which is extracted at various stages of delay and multiplied by the complex factors $h[0]$ through $h[N-1]$.

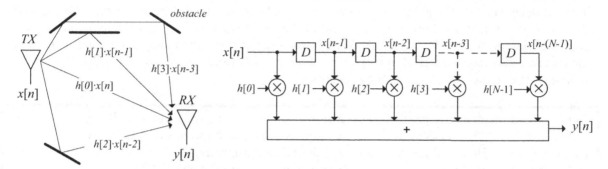

Figure 6-19: Transversal Filter Model of Linear Distortion

The scalars $h[n]$ (the impulse response of the transversal filter model) are the complex weights that change the envelope of each copy of the transmitted RF signal. Many multipath models have been developed to achieve believable propagating conditions in urban, suburban, rural and office environments. All these models differ in the way the complex weights are calculated. While the models are highly dependent on the nature of the physical environment, we may nevertheless make some generalizations regarding the complex coefficients $h[n]$.

→ It is likely that the strongest transmitted RF copy will arrive early. This is especially true if a direct line of sight exists between the transmitter and the receiver.

→ If an object exists between the transmitter and receiver, then the most direct path must penetrate the obstacle and may arrive with less power than a copy that was reflected.

→ The phase offset of the different RF copies can be thought of as a random variable that spans 0 to 2π radians and features a uniform probability density function.

Not only are the generalizations above common sense but they have also been verified in countless experiments. Unlike the relatively benign amounts of linear distortion caused by analog channel filtering in the baseband frequency sections of the radio, RF multipath distortion can cause significant amounts of selective fading in the frequency domain and conversely ISI (inter-symbol interference) in the time domain. Some of the most sophisticated DSP algorithms that reside in modern digital baseband processors are dedicated to mitigating the problem of multipath. The expressions below summarize the Naftali model [6, 7] used commonly for indoor wireless LAN applications.

$$h[n] = N(0, \frac{1}{2}\sigma_n^2) + j \cdot N(0, \frac{1}{2}\sigma_n^2)$$

Note that $N(0, 0.5\sigma_n^2)$ refers to a Gaussian distribution with a mean equal to zero and a variance equal to $0.5\sigma_n^2$. The variance, as a function of the tap index n, is shown below.

$$\sigma_n^2 = e^{\left(\dfrac{-n \cdot T_s}{T_{rms}}\right)}$$

In the equation above, T_s refers to the sample period, while T_{rms} is called the *rms* delay spread of the multipath channel. The *rms* delay spread is a measure of the severity of the multipath channel. Typical delay spreads for different environments can be seen below. The larger the delay spread, the larger the time delay between the arrival of the first path and that of other appreciably large copies of the transmitted RF signal. As we will soon see, it is not the number of copies arriving at the receiver that causes us grief; it is the size of the delay between them, which produces selective fading effects in the signal's frequency spectrum.

Delay spread < 100 nanoseconds	→ Inside office buildings & residential homes
Delay spread < 200 nanoseconds	→ Large commercial buildings & warehouses
Delay spread = 150 to 400 nanoseconds	→ Suburban environments
Delay spread = 1 to 3 microseconds	→ Dense urban environments like Manhattan
Delay spread = 2 to 25 microseconds	→ Urban environments (distant large buildings)

Now, it should be obvious that the paths arriving at the antenna will feature delays that do not line up with some arbitrary sampling period that we have chosen for our model. Our filter model, however, is a slave to that fixed delay between taps, and we therefore lose a dimension of realism in the model. Luckily, realism to that extent is not required.

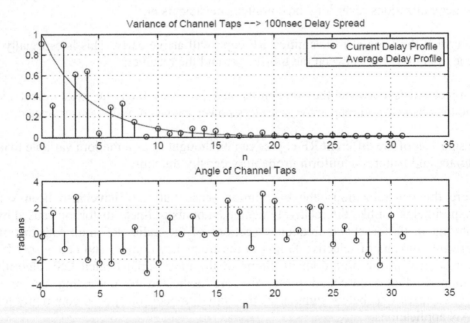

Figure 6-20: Example Delay Profile with RMS Delay Spread of 100 Nanoseconds

The figure above illustrates the magnitude and phase values of the FIR taps that make up our model. We chose to model the channel with a delay spread of 100 nanoseconds, using 32 taps at a sampling period of 1/40MHz. As expected, the strongest path occurs early on at tap 2. We can further conclude that the channel could just as well have been modeled with 20 taps rather than 32. The taps at 0 and 1 ahead of the strongest path at tap 2 are called the pre-cursors of the model while those after tap 2 are referred to as post-cursors. Clearly, there are more post-cursor taps than pre-cursor taps, which again makes sense if we assume that the strongest signal will arrive relatively early at the receive antenna.

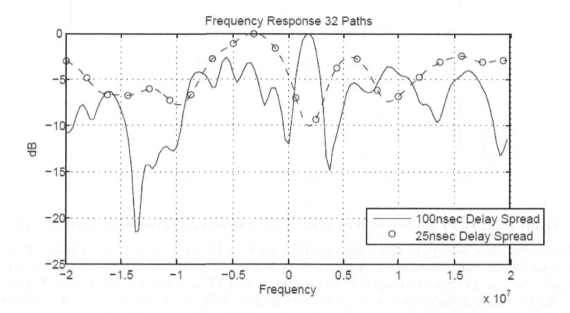

Figure 6-21: Magnitude Response of Channels with 100nsec and 25nsec *RMS* Delay Spread

To better illustrate just how ugly the multipath model's frequency response can look, we will calculate the response of channels with 100 and 25 nanosecond delay spreads, respectively. Quite obviously, the 100 nanosecond response will be more difficult to correct than its 25 nanosecond counterpart.

Correcting the detrimental effect of multipath channels requires the use of inverse filters that bend the frequency response back to their flat ideal shapes. Unless, however, the signal to noise ratio is stellar, amplifying frequencies that have been severely attenuated will significantly increase noise. Filters that blindly reshape the frequency spectrum into its original flat shape are called *zero-forcing equalizers*. Filters that take the present noise power into account and strike a balance between equalization and final signal to noise ratio are called *minimum mean square error (MMSE) equalizers*. When the signal to noise ratio is high, the two structures would be almost identical. (See chapter 4.)

Number of Paths and Frequency Response

Surprisingly, it is not the number of total paths that causes the severe selective fading phenomena, but the *rms* delay spread. To illustrate this fact, we synthesize a multipath model with 32 taps and then reduce that model to 14 taps by replacing 18 of them with nulls.

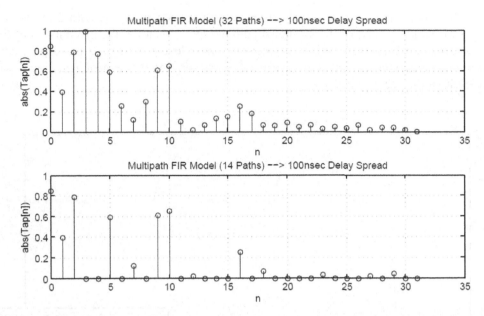

Figure 6-22: Two Channels with Identical Delay Spread but Different Numbers of Paths

Correcting these effects in single-tone modems such as the one we have established in this chapter can be quite hair-raising. A thorough overview of how this is done using time domain equalizers was presented in chapter 4. The advantage of OFDM is that the equalization happens in the frequency domain rather than in the time domain. This simplified equalization process is easily married to MIMO principles, which has made OFDM the standard of choice for most high speed wireless application. (See chapters 7 and 8.)

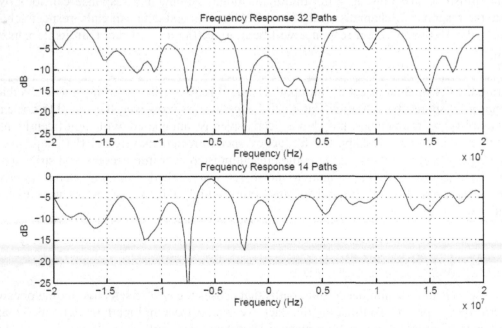

Figure 6-23: Magnitude Response for Different Numbers of Paths (Same Delay Spread)

Different communication standards use different techniques to deal with selective fading. The basic single-tone modulation formats used in this chapter require time domain equalizers, as discussed in the previous chapter. Single-tone modulation formats that rely on spread spectrum communication use a powerful technique called Rake reception to counteract multipath fading, while OFDM easily combines frequency domain equalization with antenna diversity and MIMO techniques to improve the channel.

The code below produces the FIR equivalent coefficients of our multipath filter. Depending on the delay spread and sample rate, the number of FIR taps, N, will change. We may have to run the function a few times to make sure N is large enough to capture all coefficients of meaningful magnitude. Furthermore, as we have already seen, we can set some of these taps to zero without changing the delay spread. It may not be realistic to have 50 paths arriving at the receiver, even if the delay spread is large.

```
function FIR_Taps = GetMultipathFilter(SampleRate, DelaySpread, N)
Ts          = 1/SampleRate; %% Sampling period in seconds
Trms        = DelaySpread;  %% Delay spread in seconds
n = 0:N-1;
ExpVariance = exp(-n*Ts/Trms);
FIR_Taps    = zeros(1,N);

for i = 1:N
    FIR_Taps(1,i) = sqrt(ExpVariance(1,i))*randn(1,1) + ...
                    j*sqrt(ExpVariance(1,i))*randn(1,1);
end
```

Example 5.1: *Receive Constellation for a Multipath-Affected QPSK Signal*

In this example, we consider a QPSK signal running at a symbol rate of 10MSPS and sample rate of 40MHz. Anything but a benign multipath channel would create a complete mess of the constellation. The multipath coefficients for this example are [0.25 1 -0.25].

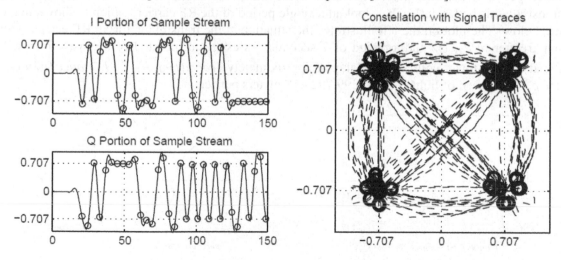

Figure 6-24: QPSK Signal with Benign Multipath

6.3.2   Multipath Distortion with Doppler

A Doppler shift is an apparent frequency offset seen by the receiver when either the TX or RX terminal are in motion. This effect is especially felt in mobile communication environments like cell phone networks, where a mobile phone is communicating with a cell tower from a moving vehicle. This Doppler shift is most noticeable when the vehicle is driving directly toward or away from the cell tower, whereas no Doppler shift is generated when the vehicle is driving in a perpendicular direction in relation to the tower. The figure below, illustrate the relative motion of three cell phones in relationship to the cell tower. The link between the tower and mobile 1 will not experience any Doppler, whereas the link with mobile 3 and mobile 2 will experience positive and negative Doppler respectively.

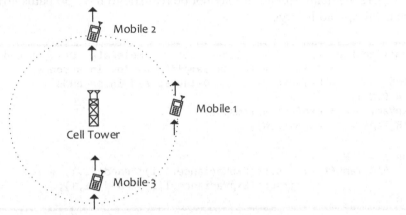

Figure 6-25: Motion of Mobile Phones in Relationship to a Cell Tower

Contraction of Time

Take a look at Mobile 3, which is moving directly toward the cell tower and note that as the distance between the two decreases so does the transit or flight time of the RF signal. To demonstrate this better, let's take a look at a single period of the RF carrier, which is shown in the figure below. The plot on the left illustrates the situation at the transmitter for an RF carrier with center frequency F_{Tx} Hz and a period of T seconds. As expected, point A of the waveform will arrive at the cell tower with a slight delay of approximately $TransitTime1 = c \cdot DistanceToTower$, where c is the speed of light equal to 299,792,458 meters per second.

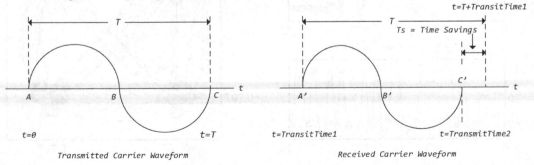

Figure 6-26: Contraction of Time

Point C of the transmit waveform leaves the antenna T seconds later, but requires less flight time than point A as the distance to the tower is now less. The time saving, Ts, is directly related to the distance in meters that the mobile traveled during T seconds, and this distance is equal to the velocity of the mobile toward the tower in meters per second multiplied by the carrier period T.

$$Dist = Vel \cdot T \; meters$$

The time savings, Ts, is equal to velocity in meters per second of the mobile moving toward the tower divided by the speed of light.

$$T_s = \frac{Dist}{C} \; seconds$$

The apparent carrier period seen at the receiver is Ts less than the carrier period, T, at the transmitter, and the corresponding carrier frequency is the reciprocal of that new period.

$$F_{RX} = \frac{1}{T_{RX}} = \frac{1}{T - Ts} = \frac{1}{T - \dfrac{Dist}{C}} = \frac{1}{T - \dfrac{Vel \cdot T}{C}}$$

$$F_{RX} = \frac{1}{T} \frac{C}{C - Vel}$$

The Doppler shift is the difference in frequency seen at the receiver and transmitter and is expressed as follows.

$$F_{Doppler} = F_{RX} - F_{TX}$$
$$= \frac{1}{T} \frac{C}{C - Vel} - \frac{1}{T}$$
$$= \frac{1}{T}(\frac{C}{C - Vel} - 1)$$
$$= F_{TX}(\frac{C}{C - Vel} - \frac{C - Vel}{C - Vel})$$
$$= F_{TX}(\frac{Vel}{C - Vel})$$
$$\cong F_{TX}(\frac{Vel}{C})$$

The phenomenon described above appears as a frequency offset, but it is actually a contraction of time that affects the signal modulated onto the carrier as well. To the receiver, it appears as if the reference clock source at the transmitter that is controlling the signal generation (baseband processor and digital to analog converters) is running slightly too fast for a positive Doppler shift or two slow for a negative Doppler shift. Thus, aside from the apparent frequency offset, the signal also suffers from timing drift, which in most cases is far less problematic than the error in center frequency. Because the two phenomena go hand in hand, the receiver can estimate the Doppler shift by determining the timing drift and visa versa. This fact is exploited in cellular telephony technologies like LTE, which continually monitors the time of arrival of synchronization signals. If these signals appear at a period that is smaller than expected, then we may estimate the Doppler shift based on how much the time is compressed.

Doppler and Error in the Time Base

The figure below illustrates a crude diagram of a transmitter and receiver with both its digital and analog/RF components. Notice how the TCXO, or temperature compensated crystal oscillator, on the transmitter side provides a 10MHz reference clock signal to the synthesizer and the base band processor / DAC combination. The synthesizer mulitplies the reference frequency by a programmable value to produce the RF local oscillator signal, which will determine to the center frequency of your final RF transmit signal. Let us assume that the reference frequency of 10MHz is multiplied inside the synthezier by 190 to produce an LO carrier of 1900MHz. Reality dictates that the reference frequency is not exactly 10MHz, and this error is specified in ppm, or parts per million. A 10MHz reference oscillator specified with an error of ±1ppm, will oscillate in a range equal to 10MHz·(1 ± 1e-6) = 10MHz ± 10Hz. Similarly, the transmit local oscillator signal will feature a frequency of 1900MHz(1± 1e-6) = 1900MHz ± 1900Hz.

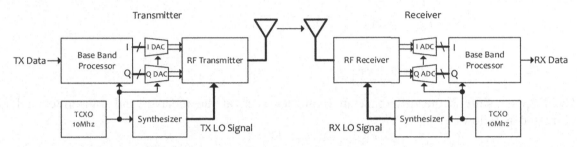

Figure 6-27: Simplified Diagram of a Wireless Transmitter and Receiver

For a positive time base error of 1ppm, not just will the TX LO signal oscillate too quickly, the entire IQ signal produced by the base band processor is read out slightly too quickly throught the digital to analog converters. If the RX TCXO was perfect at 10MHz, the receiver would see a positive 1900Hz frequency offset and read in an IQ waveform that is compressed by 100%·1e-6 = 0.0001%. This timing drift between the transmit and receive base band processors and the frequency offset occur in unison as long as a single oscillator supplies the time base as shown.

Interestingly enough, a Doppler shift of 1900Hz from a 1900MHz carrier signal due to a moving mobile phone is due to the the associated 1ppm = 0.0001% time contraction. At the receiver, the effects due to the motion of the mobile phone can not be distinguished from the effect due to the time base difference in the TXCOs at the transmitter and receiver, and both may be parameterized as a time base error in ppm.

The New Multipath Model including Doppler Effects

As mentioned at the start of this section, the Doppler shift is proportional to the speed of the mobile and the direction in relation to the cell tower. Remember, however, that the transmit signal of the mobile phone can take many paths toward the receiver in the cell tower, and not all those paths are line of sight. The following figure illustrates four paths each contracting differently as the mobile moves toward the receiver. As a matter of fact the distance along path 3 increases with the motion of the mobile causing a negative frequency offset (Doppler shift).

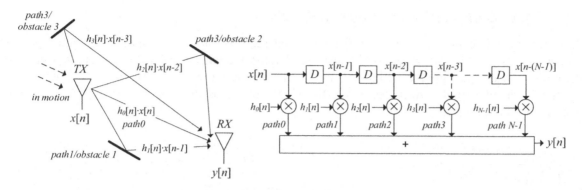

Figure 6-28: Multipath Model with Doppler Shift

In the last section, this very same figure was shown for the multipath model without mobility. The only difference is that the coefficients $h[0]$, $h[1]$, … etc now change with time as $h_0[n]$, $h_1[n]$,…etc. Note also that T_s represents the sampling period.

$$h_0[n] = h_0[0] \cdot exp(j2\pi n T_s F_{Doppler0})$$
$$h_1[n] = h_1[0] \cdot exp(j2\pi n T_s F_{Doppler1})$$
$$\vdots$$
$$h_{N-1}[n] = h_{N-1}[0] \cdot exp(j2\pi n T_s F_{Doppler\ N-1})$$

The complex coefficients $h_0[0]$, $h_1[0]$, $h_2[0]$, through $h_{N-1}[0]$ are identical in every way to the coefficients $h[0]$, $h[1]$, through $h[N-1]$ of the last section. Their magnitude tends to decrease with the path delay and their phase is random. However, this time these coefficients spin around the complex plane give the amount of Doppler shift.

As before, our model is somewhat constained by the fact that the path delays are always multiples of the fixed time delay $D = T_s$ seconds, but this construction allows us to use our simple FIR filter structure. Note that the model does not include the timing drift that we mentioned in this section. Beside the fact that this is not the right place to insert it, the timing drift is small and we can ignore it for now small stretches of time. We will use this model when designing and evaluating equalizers for mobile applications (see the LTE equalizer section). The table below gives us an idea of some typical values for the Doppler shift (frequency offset) given the velocity toward the receiver and the center frequency, at which we communicate. Note that for reasonable driving speeds, the Doppler shift for the 1900MHz center frequency stays within a range or ± 250Hz.

Equivalent Velocity	Doppler for Fc=800MHz	Doppler for Fc=1900MHz	Drift in ppm = 1e6 · Doppler / Fc
10.8 kmh	8 Hz	19 Hz	0.01 ppm
54 kmh	40 Hz	95 Hz	0.05 ppm
108 kmh	80 Hz	190 Hz	0.1 ppm
540 kmh	400 Hz	950 Hz	0.5 ppm
1080 kmh	800 Hz	1900 Hz	1 ppm

Figure 6-29: Table of Doppler Shifts and Equivalent Time Base Error

6.3.3 Gaussian White Noise in Receivers

Anyone who has ever dealt with radios knows that there is some minimum input signal power threshold below which no reception is possible. That is the case even if there is enough gain in the radio to scale the signal to within the correct dynamic range of its analog to digital converters. As the received signal power decreases, the signal to noise ratio decreases as well, causing eventual loss of reception.

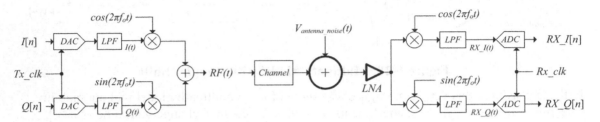

Figure 6-30: Main Insertion Point of Thermal Noise

So, where does this noise come from that makes reception of small signals so difficult? Is the antenna picking up galactic noise that radiates everywhere in space, or are the semiconductors in the radio generating the noise? If you don't know, the real answer will surprise you. The Gaussian white noise is primarily generated by the resistive part of the antenna and exacerbated by the RF modules that follow [5]. It is the thermal noise component of resistive devices that we must examine to better understand noise in receivers. The figure above illustrates the position at which the dominant contributions of thermal noise occur. The noise model of a resistor is included into the antenna model below to illustrate the generation of both the noise component and the noiseless received signal.

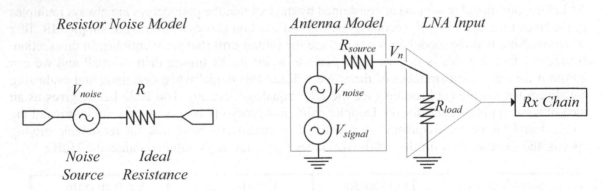

Figure 6-31: Resistor and Antenna Noise Models

The antenna, therefore, produces two superimposed signals: the thermal noise voltage of its resistive element and the received signal itself. The thermal noise voltage (*rms*) of a resistor obeys the following expression.

$$V_{noise} = \sqrt{4KTBR}$$

In the equation K is the Boltzman constant ($1.38 * 10^{-23}$ Joules/Kelvin), T is the temperature in Kelvin ($0°C = 273°K$ and $0°K = -273°C$), B is the bandwidth in Hertz, and R is the resistance in Ohms. In order to transfer maximum power from the antenna into the low noise amplifier (LNA), the input resistance R of the LNA must be the same as the source resistance of the antenna. If V_n is the voltage due to the antenna noise at the node between the antenna and the LNA, then the maximum available noise power that can be transferred to the LNA is equal to KTB.

$$Power = V_n \cdot I = \frac{V_{noise}}{2} \cdot \frac{V_{noise}}{2 \cdot R} = \frac{4KTBR}{4R} = KTB \quad watts$$

The thermal noise is flat and evenly spread across all frequencies, which suggests that the integrated noise power is almost infinite. In fact, the noise that is transferred from the antenna to the low noise amplifier is band limited from the very start. The matching or tuning elements connecting the antenna and RF LNA module are the first components that limit the noise bandwidth, followed by the various filters in the receive chain. By the time the received signal has made its way deep into the baseband processor, it has been filtered to pass only those frequencies at which the transmitted signal has useful energy. Eventually the noise is band-limited to the same frequency range as is occupied by the signal of interest. B is therefore equal to the double-sided bandwidth (bandwidth at RF frequencies) of the received signal.

The Noise Floor

The noise floor is the measure of a waveform created from the sum of all noises and unwanted signals within the measurement system. The noise floor, which can be measured in terms of dBm or Vrms, can readily be seen on spectrum analyzer displays. In our measurement system, the analog receiver, the unwanted signal is dominated by resistive thermal noise. If the LNA and all subsequent receiver blocks were perfectly noise free, then the noise floor would be the thermal noise power, KTB watts. The power is measured within a certain RF or IF bandwidth B, which in the case of a spectrum analyzer would be the resolution bandwidth and in the case of our receiver is the double-sided channel filter bandwidth. As we change the resolution bandwidth of a spectrum analyzer, the reduction of the noise floor can readily be seen. For a noise free receiver, the noise floor is dominated by the antenna resistance as follows.

$$Power_{dBm} = 10 \cdot \log_{10}(KTB \cdot 1000)$$

Since dBm is a logarithmic expression of the power in milliwatts, KTB is first multiplied by $1000 mW/W$. Note here that the maximum available power from a resistive load, KTB, is independent of the value of the resistances themselves. The load and source resistances must merely be equal. Below are some example noise floors for different bandwidths.

$$-174 dBm \leftarrow 1 Hz$$
$$-164 dBm \leftarrow 10 Hz$$
$$-114 dBm \leftarrow 1 MHz$$

Noise Figure

Unfortunately, we don't live in a perfect world, and every radio component adds to the total noise content within our signals frequency band. These noise contributions give rise to a parameter called noise figure, which is quantified in *dB*. The noise figure of an RF device or an entire receiver is defined as the signal to noise ratio of the input signal minus the signal to noise ratio of the output signal.

$$NoiseFigure = NF = SNR_{in} - SNR_{out} \quad \leftarrow dB$$

$$NoiseFactor = F = \frac{SNR_{in}}{SNR_{out}} \quad \leftarrow linear$$

$$NF = 10 \cdot \log_{10}(F)$$

Note that the noise figure of a device is measured if – and only if – the noise component of the input signal is the maximum available thermal noise power from a resistive source. Therefore, the input signal may not contain amplified noise from a previous RF block, and the RF device must be matched to the source impedance for the noise figure measurement to be valid.

We may assume that each RF device adds some small amount of noise to the maximum received noise from the source and then processes the new signal/noise combo in a noise free fashion. The extra device noise added at the input of the RF module decreases the signal to noise ratio that can be seen at the output terminal. The signal to noise ratio can be expressed using the following equations.

$$SNR_{out} = 10 \cdot \log_{10}(\frac{Gain_{RFDevice} \cdot SignalPower_{Input}}{Gain_{RFDevice} \cdot (NoisePower_{Input} + NoisePower_{RFDevice})})$$

$$= 10 \cdot \log_{10}(\frac{SignalPower_{Input}}{NoisePower_{Input} + NoisePower_{RFDevice}})$$

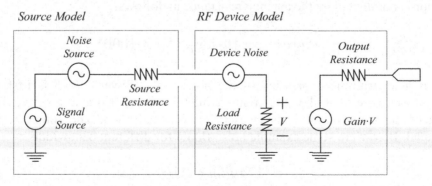

Figure 6-32: Noise Figure Test Setup

We now substitute the expression for output signal to noise ratio into our noise figure equation to yield the following formulation.

$$NF = 10 \cdot \log_{10}(\frac{SNR_{in}}{SNR_{out}})$$

$$= 10 \cdot \log_{10}(\frac{SignalPower_{Input} \Big/ NoisePower_{Input}}{SignalPower_{Input} \Big/ (NoisePower_{Input} + NoisePower_{RFDevice})})$$

$$\boxed{NF = 10 \cdot \log_{10}(\frac{NoisePower_{Input} + NoisePower_{RFDevice}}{NoisePower_{Input}})}$$

So there you have it. The noise figure of an RF device is equal to the sum of the input noise power (thermal noise due to source resistance) and the input referred self-generated noise power of the RF device divided by the input noise power. The device noise is generated in different areas of the RF device, but for convenience we think of it as being produced at the input.

Cascaded Noise Figure

So, what happens if we need to know the noise figure of the entire receiver consisting of many cascaded RF devices attached to the antenna? The first RF device sees thermal noise superimposed with the received signal from the antenna. The second device would likely see an input waveform composed of amplified signal and amplified noise. While the self-generated noise power of the second RF device might be appreciable compared to the original thermal noise power from the antenna, it may look quite small compared to the amplified noise combo at its input. We therefore can't just add all the noise figures in our RF receiver chain. The correct formulas for the overall cascaded noise factor, F, and subsequent noise figure, NF, are shown below.

$$\boxed{F = F_1 + \frac{F_2 - 1}{Gain_1} + \frac{F_3 - 1}{Gain_1 \cdot Gain_2} + \cdots + \frac{F_n - 1}{Gain_1 \cdot Gain_2 \cdots Gain_{n-1}}}$$

$$NF = 10 \cdot \log_{10}(F)$$

MatLab Model

The code snippet provided next illustrates how white Gaussian noise is generated. The variance of the input signal is first found and used, together with the desired SNR (signal to noise ratio), to calculate the appropriate noise power. Whereas the discussion above has centered mostly on the initial RF blocks of the receiver, our DSP calculations deal with the equivalent IQ waveform at baseband frequencies. Therefore, the input to the function below is assumed to be a complex signal, and the code generates complex noise via the two '*randn*' calls. We need to keep in mind that the additive white Gaussian noise returned by the function features a flat spectrum. If we need to generate noise with frequency content that is not flat (colored noise), we simply filter white noise through an appropriate FIR filter. For example, if noise must be added to a signal that has just passed through a baseband channel filter, then it would be inappropriate for that noise to feature a flat spectrum. In that case, we have to generate the additive white Gaussian noise, pass it through the same filter as the input signal, and then rescale it until the desired SNR is met.

```
function Noise = Generate_AWGN(Input, SNR)

MeanSquare = (1/length(Input))*(Input*Input');
                                    %% Total signal power
NoisePower = MeanSquare/(10^(SNR/10)); %% Required AC noise power
STDNoise   = sqrt(NoisePower);         %% Standard deviation of noise

Noise = STDNoise*(0.70711*randn(1,length(Input)) + j*0.70711*randn(1,length(Input)) );
```

In the figure below, we plot the receive constellation of our now familiar QPSK signal, as well as a 16QAM variant, both at a signal to noise ratio of 17dB. For the QPSK case, the detector determines inside which quadrant each received symbol resides and makes a decision as to which reference symbol ($\pm 0.707 \pm j0.707$) was likely sent by the transmitter. From the plot, we can tell that the detector will have no problem demapping the received symbols into the correct bit groups. The 16QAM case, however, is a different matter entirely. For reference, we outlined the 16 decision quadrants used by the detector to determine which reference symbol was sent. The reference symbols are at the center of each quadrant. The received symbols deviate from the perfect reference position at the quadrant's center to the point where, at one instant or another, a received symbol is bound to cross a quadrant boundary and be erroneously demapped by the detector. Clearly, to maintain error free reception for the 16QAM case, the SNR ought not to drop below 17dB.

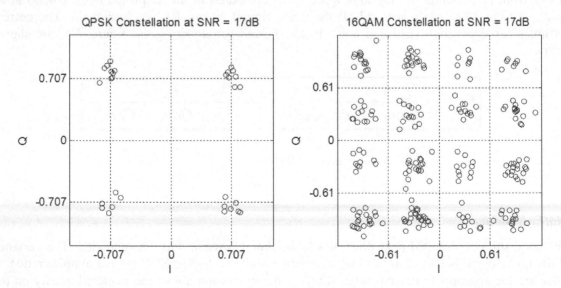

Figure 6-33: QPSK and 16QAM Signal Constellation in the Presence of White Gaussian Noise

6.3.4 Phase Noise

Phase noise is a defect that causes the angle of the LO signals to drift around its intended instantaneous phase of $2\pi f_o t$. For the purpose of IQ demodulation, the analog circuitry generating the LO signal produces two orthogonal waveforms.

$$LO_{In\text{-}phase} = cos(2\pi f_o t + \theta(t))$$
$$LO_{Quadrature} = sin(2\pi f_o t + \theta(t))$$

The phase noise component, $\theta(t)$, is seen in the figure below modifying the phases of both LO waveforms in the same manner and by the same amount.

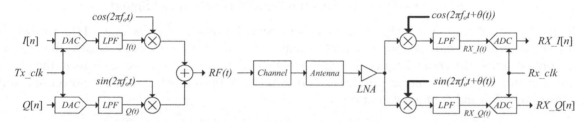

Figure 6-34: Insertion Points of Phase Noise (Shown Only at the Receiver)

Note that degradation of modulation accuracy due to phase noise is independent of the received signal strength. This is in stark contrast to the antenna and RF LNA supplied thermal noise which degrades the modulation accuracy of weak signals far more than that of its stronger counterparts.

Modeling Phase Noise at Baseband Frequencies

Modeling phase noise, $\theta(t)$, at baseband frequencies is a trivial operation and reduces to the simple expression below.

$$Output(t) = Input(t) \cdot e^{j\theta(t)}$$

The harder task, which we address in this section, is synthesizing the waveform $\theta(t)$, or its discrete counterpart $\theta[n]$, given a standard phase noise profile as seen on the display of phase noise analyzers or on the pages of textbooks and data sheets. The eventual goal for the system engineer is to generate $\theta(t)$, based on any phase noise profile, and study its effect on the demodulated constellation. This information can then be used to guide the development of the transceiver's synthesizer, which generates the LO signals.

Understanding the Phase Noise Profile

As many engineers may already know, phase noise can be read straight off a spectrum analyzer whose noise floor is low enough not to mask the phase disturbances. Special phase noise modules are often sold as add-ons to generic spectrum analyzers to automate phase noise testing and provide the low noise floor performance required by the test.

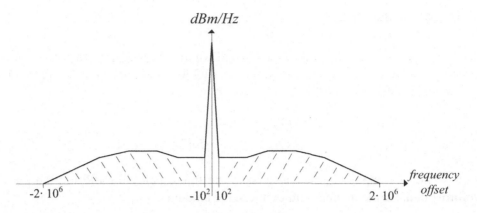

Figure 6-35: Power Spectral Density (PSD) of LO Signal

Internally, the phase noise analyzer determines the signal power in dBm within a 1Hz bandwidth for some range around the carrier's center frequency. This power spectral density function is integrated across the entire frequency range to yield total carrier power, and leads us to the phase noise profile. The phase noise profile, $L(f)$, is defined as the ratio between the power spectral density function and the total carrier power.

$$L(f) = 10 \cdot \log_{10}(\frac{Power\ in\ 1Hz\ Bandwidth}{TotalCarrierPower})\ dBc/Hz$$

$L(f)$ is therefore power spectral density, PSD, normalized to the total power in the carrier, to produce units of *dBc*/Hz rather than *dBm*/Hz. The profile is generally graphed as a single-sided frequency plot and, as shown below, indicates that at 10^4 Hz away from the carrier frequency, the power within a 1 Hz band is 85dB less than the total power of the carrier signal.

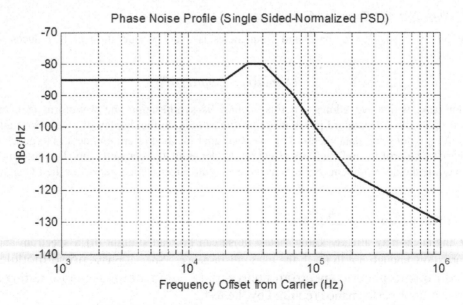

Figure 6-36: Typical Single-Sided LO Phase Noise Profile

The phase noise profile may be equivalently expressed as follows.

$$L(f) = P(f)_{1Hz} \quad (dBm/Hz) \quad - \quad P_{Total} \quad (dBm)$$

Being a spectrum analyzer at heart, however, the phase noise analyzer doesn't measure $\theta(t)$ but generic signal power within its resolution bandwidth of 1Hz. How do we get from a standard phase noise profile to a phase waveform, $\theta(t)$, that we may incorporate into our model?

From the Phase Noise Profile to $\theta(t)$

The normalized power spectral density, of which we show the positive frequency side as the phase noise profile, integrates to a total unitless power of 1.0. For our simulation at baseband, we consider the LO spectrum down-converted to 0Hz which yields the following complex unity magnitude vector.

$$S(t) = 1 \cdot e^{j(\phi + \theta(t))}$$
$$= 1 \cdot e^{j\phi} \cdot e^{j\theta(t)}$$
$$= C \cdot e^{j\theta(t)}$$

The argument ϕ refers to a random static phase offset, which, when coupled with the magnitude of 1.0, becomes a complex but static number, C, that is then phase modulated by $\theta(t)$. It is the power of the complex error induced by the $e^{j\theta(t)}$ components that is being captured by the phase noise analyzer. We define the complex error as follows.

$$IQ\_Error(t) = C \cdot e^{j\theta(t)} - C = C\cos(\theta(t)) + jC\sin(\theta(t)) - C$$

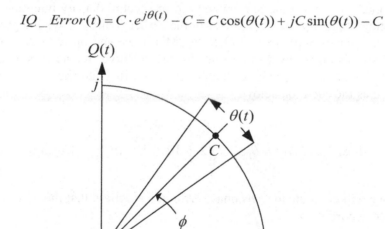

Figure 6-37: Deviation away from Nominal Position

Unless the phase noise of the LO signal is absolutely horrendous, the phase deviation $\theta(t)$ will not vary by more than a couple of degrees. For small phase deviations, the expressions $\cos(\theta(t)) \approx 1$ and $\sin(\theta(t)) \approx \theta(t)$ radians lead us to the final equations for IQ_Error and $|IQ\_Error|$.

$$IQ\_Error(t) = jC \cdot \theta(t) \quad \rightarrow \quad |IQ\_Error(t)| = |\theta(t)|$$

> The result above is meaningful because it tells us that the magnitude and consequently the power spectral density of the *IQ_Error(t)* signal measured by the analyzer is the same as that of the function $\theta(t)$ we need to use for our simulation. The key, then, is to craft a noise signal with the same magnitude spectrum as the *IQ_Error* of the analyzer, and use it as $\theta(t)$.

Generating θ(t) Step by Step

1 → Sample the phase noise profile over frequency, making sure to record the normalized power spectrum values at negative frequencies as well.

2 → Generate the magnitude spectrum by taking the square root of the power spectrum values.

3 → Use the magnitude spectrum as the magnitude response to generate a FIR filter via either the frequency sampling or least squares method of FIR design.

4 → Create white Gaussian noise and pass it through the FIR to generate colored noise that obeys the same magnitude spectrum as the measured phase noise.

5 → Normalize the resulting noise signal such that it has the proper *rms* phase error.

RMS Phase Error

Beyond the phase noise profile, the analyzer also provides a standard metric called *rms phase error*. We mentioned earlier that the normalized power spectral density integrates to unity power across the frequency range of interest ($\pm$2MHz in the earlier figure). If we integrate over only the actual noise content (for example from 100Hz to 2MHz), we end up with the single-sided noise power. We multiply this power by 2 to account for the negative frequency side and arrive at the total unitless phase noise power. Taking the square root will yield the *rms phase error metric*, which is expressed as the equation below.

$$RmsPhaseError = \theta_{rms} = \sqrt{2\int_{1e2}^{2e6} 10^{\frac{PSDdBc}{10}}\,df} \qquad (radians)$$

In step 5 above, we will compute the *rms* phase error and ascertain that the output, $\theta(t)$, of the FIR filter is normalized accordingly.

Generating Phase Noise (a Block Diagram)

The block diagram shown below illustrates how to generate $\theta(t)$ according to the five steps listed above. The only additional step shown in the diagram is the upsampling operation. To keep the number of coefficients of the FIR filter within reason, we initially choose a sample rate of 4MSPS, which enables the processing of signals with a single-sided bandwidth of no more than the needed 2MHz. The eventual IQ signal to which we need to add the phase noise will likely run at a higher rate, requiring us to upsample the noise waveform.

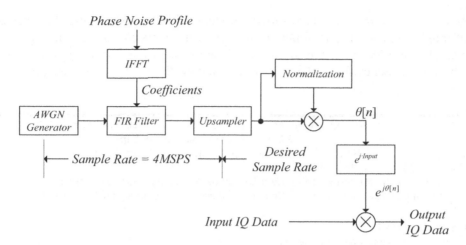

Figure 6-38: The Creation of Phase Noise

Scaling the final discrete noise signal $\theta[t]$ to the proper *rms* phase error is the final step in the noise creation chain before its introduction into the signal. Simply scaling the noise at the end is preferable to controlling the power of the original white noise and DC gain of the FIR filter.

```
function  [Out Pow_Pos Freq_Pos RMS_PE] = ...
          PhaseNoiseGenerator(SampleRate, NumberOfSamples, dBc, Freq)


%% 1. Sample the Phase Noise Profile (Don't include 0Hz
%                       or frequencies > 2e6 in the Freq input vector)
MaxFreq         = Freq(1,length(Freq));
FreqIndex       = [0 Freq MaxFreq+1 2.1e6]; %% Conditioned frequency index
dBc_Power       = [-200 dBc -200 -200];     %% Conditioned profile


%% 1a. Because the Phase Noise profile's frequency axis is shown as log
%                     format, we linearly interpolate at dB frequencies
FreqIndex_dB       = 20*log10(FreqIndex+1);
NewFrequencies_dB  = 60:126;  %% 20*log10(1e3Hz) to about 20*log10(2e6Hz)
New_dBc_Power_dB_Freq  = interp1(FreqIndex_dB, dBc_Power, ...
                                 NewFrequencies_dB, 'linear');
%% 1b. Now we interpolate linearly to get a small frequency spacing of 1KHz
NewFrequencies = 0:1e3:2e6;
New_dBc_Power  = interp1(10.^((NewFrequencies_dB)./20),  ...
                New_dBc_Power_dB_Freq, NewFrequencies,'linear','extrap');
NewLinearPower = 10.^(New_dBc_Power/10);


%% 1c. While we are at it, let's find the rms phase error
SSB_Power = 0;    %% Find rms phase error via trapezoidal integration
for i=1:length(NewFrequencies)-1
    Rise = 0.5*NewLinearPower(1,i) + 0.5*NewLinearPower(1,i+1);
    Run  = NewFrequencies(1,i+1) - NewFrequencies(1,i);
    SSB_Power = SSB_Power + Rise*Run;
end
RMS_PE = sqrt(2*SSB_Power);  %% rms phase error in radians
```

In section one, the input frequency and dBc vectors are terminated by forcing -200dBc at the boundaries of 0 and 2e6Hz. We then sample the normalized power spectral density in 66 even log frequency steps before re-interpolating to small steps of 1KHz. Not doing this causes the noise profile of our synthesized $\theta(t)$ function to curve in between the given input [*freq*, *dBc*] coordinates when plotting x-axis in log format. You can interpolate directly to the 1KHz spacing and skip the log frequency interpolation, but the final synthesized phase noise profile won't look as pleasing.

```
%% Steps 2 and 3. Converting power to magnitude in a format for the IFFT
                          %% FIR Filter has 2000 taps
Freq_Pos = (0:999)*4e6/2000;       %% Set up frequencies for IFFT
Freq_Neg = (-1000:1:-1)*4e6/2000;
Pow_Pos  = interp1(NewFrequencies, NewLinearPower, Freq_Pos,'linear');
Pow_Neg  = interp1(NewFrequencies, NewLinearPower, abs(Freq_Neg),'linear');


%% Preparing for IFFT
LinearPower     = [Pow_Pos Pow_Neg];
Magnitude       = sqrt(LinearPower);  %% Set up gains for IFFT
Temp            = ifft(Magnitude);


%% Step 4: Compute the FIR filter taps and filter White Gaussian Noise -WGN
FIR_Taps        = [Temp(1,1001:2000) Temp(1,1:1000)];
H               = hann(2002);
H1              = H(2:2001,1)';
FIR_Taps        = FIR_Taps.*H1; %% Overlay the Hanning window

SampleRatio     = SampleRate/4e6;
SamplesAt4MHz   = ceil(NumberOfSamples/SampleRatio);
WGN             = randn(1,2000 + SamplesAt4MHz);  %% Generate WGN


FilterOutput    = filter(FIR_Taps,1,WGN);
Out4MHz         = FilterOutput(1,1000:end);
```

Before the IFFT is taken to arrive at the FIR coefficients, the power spectral density is converted to linear units of power followed by square root operation to yield the required magnitude response. After the IFFT operation, the impulse response is scaled by a Hanning window.

The final operations in step 4 are the generation of white Gaussian noise and its subsequent filtering by the FIR structure. Finally, the phase noise signal sampled at 4MHz is upsampled to the target rate and normalized to have the desired root mean square error value.

```
%% 5. Upsample from 2MHz to the desired sample rate
SampleTime4MHz = 1:length(Out4MHz);
NewSampleTime  = (1:NumberOfSamples)*1/SampleRatio;
Output         = interp1(SampleTime4MHz, Out4MHz, NewSampleTime,'spline');
OutTemp        = Output(1,1:NumberOfSamples);
RMSOutput      = sqrt(var(OutTemp));
Out            = OutTemp*RMS_PE/RMSOutput;
```

Example 5.2: *Generating and Analyzing Phase Noise*

To convince ourselves that the function we developed does indeed produce phase noise with the proper shape, let's exercise it using the example profile of figure 5-31 which we quantify below.

$$Freq = [\ 1e3\ \ 20e3\ \ 30e3\ \ 40e3\ \ 70e3\ \ 100e3\ \ 200e3\ \ 1000e3]$$

$$dBc = [\ -85\ \ -85\ \ -80\ \ -80\ \ -90\ \ -100\ \ -115\ \ -130\ \ ]$$

As additional inputs, we will task the function to produce one million samples at a rate of 10MHz. Remember that, internally, the PhaseNoiseGenerator() function will first produce the noise waveform at a static 4MHz and then upsample it to the desired rate. The figure below compares the input noise profile with the normalized carrier's normalized power spectral density. Except at low frequencies, where the spectrum calculation detects the large carrier signal, the profiles match well.

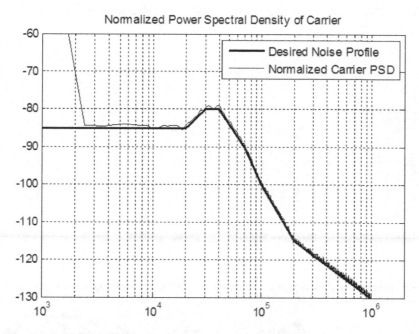

Figure 6-39: Comparing the Desired Phase Noise Profile to the Normalized PSD of the Calculated Carrier Waveform

```
%% Example 5.2: Creating and analyzing phase noise

SampleRate      = 10e6;   %% Output sample rate
NumberOfSamples = 1e6;    %% Number of samples in simulation


dBc  = [ -85 -85  -80  -80  -90 -100    -115 -130 ]; %% Desired phase
Freq = [ 1e3 20e3 30e3 40e3 70e3 100e3  200e3 1000e3]; %% Noise profile


[Out RMS_PE] = PhaseNoiseGenerator(SampleRate, NumberOfSamples, dBc, Freq);
```

```
Carrier = exp(j*Out);    %% Phasor expression of the carrier
display(['Calculated RMS phase noise (degrees): ' num2str(RMS_PE*57.29)]);
[Mag f] = Spectrum(Carrier, 2048, SampleRate);

figure(10);
semilogx(Freq, dBc, 'k', 'LineWidth', 2); hold on; grid on;
semilogx(f, Mag - 10*log10(SampleRate/Points), 'k');  title('Normalized Power Spectral
Density of Carrier'); axis([1e3 2e6 -130 -60]);
legend('Desired Noise Profile', 'Normalized Carrier PSD');
```

Example 5.3: *Receive Constellations Affected by Phase Noise*

In this example, we show the impact of two separate noise profiles on BPSK and 8PSK modulated signals. Depending on the bandwidth of the signal, some of the phase noise in the received waveform may be removed via phase tracking, resulting in better behaved constellations with lower EVM. Phase noise can be more easily tracked out in wideband than narrow band signals.

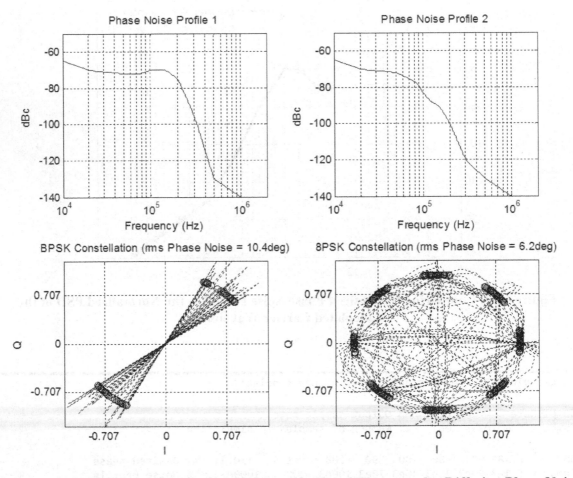

Figure 6-40: BPSK and 8PSK Constellation and Signal Traces under Differing Phase Noise Conditions

6.3.5   Frequency Offset

Frequency offset is an irrevocable reality that stems from the fact that the master clocks on the transmitter, *Tx_clk*, and receiver, *Rx_clk*, are not exactly the same. The LO signals, cos(2πf_o t) and sin(2πf_o t), of both the transmitter and receiver are derived from these slightly dissimilar master clocks which result in slightly dissimilar LO frequencies indicated by Δf in the figure below.

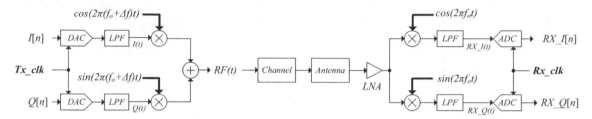

Figure 6-41: Frequency Offset Model

Modeling the effect of frequency offset on the complex signal at the output of the ADC converters is straight forward. If we assume that the received signal $Rx(n) = RX\_I(n) + jRX\_Q(n)$, and the transmit signal $Tx(n) = I(n) + jQ(n)$, then the following expression may be formulated.

$$Rx[n] = Tx[n] \cdot e^{j2\pi(\Delta f)nT_s}$$

$$Rx[n] = Tx[n] \cdot (\cos(2\pi(\Delta f)nT_s) + j\sin(2\pi(\Delta f)nT_s))$$

The effect of frequency offset is modeled by multiplying the received IQ signal by a complex sinusoid spinning in the negative direction at the offset rate Δf.

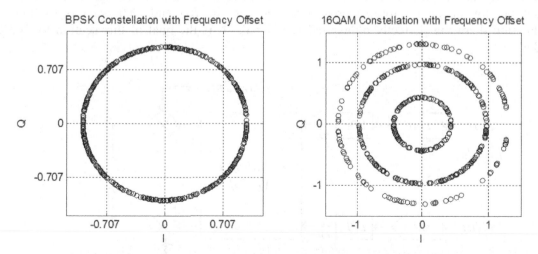

Figure 6-42: Typical BPSK and 16QAM Constellation without Signal Traces when Frequency Offsets Are Present

6.3.6 Imbalances in IQ Modulators

Phase Imbalance in IQ Modulators

Another common flaw in RF communication circuits is amplitude and phase imbalance in the IQ up-converters and down-converters. Phase imbalance occurs when the cosine and sine LO signals are not perfectly orthogonal (90 degrees apart), whereas amplitude imbalance occurs when one modulator produces a larger signal than the other. This may be due to slight mismatches in modulator gain or differences in the *rms* level of the various input signals.

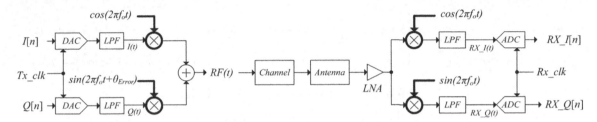

Figure 6-43: IQ Phase Imbalance Model in the Transmit Section

Because the effect of IQ imbalance is the same for both the modulator and demodulator, we will limit our discussion to the transmitter side of the link. An RF signal carrying IQ information that was up-converted using a phase imbalanced modulator is mathematically expressed in the equation below.

$$RF(t) = I(t) \cdot \cos(2\pi f_o t) + Q(t) \cdot \sin(2\pi f_o t + \theta_{Error})$$

The imbalance is modeled by a phase error, θ_{Error}, which, in this case, we insert into the quadrature portion of the LO signal. Phase imbalance in the IQ modulator creates cross-talk between the I and Q portions of the complex envelope riding on the carrier signal. It is possible to model the imbalance by forcing cross talk between the $I(t)$ and $Q(t)$ waveforms and then modulating them via perfectly orthogonal LO signals. It is the goal of this section to find this model.

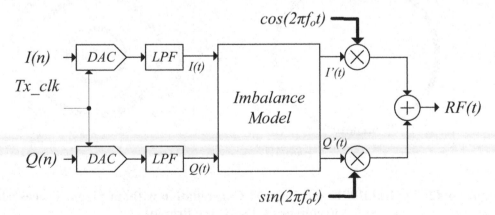

Figure 6-44: IQ Phase Imbalance Models

To understand the impact of the phase error in the quadrature portion of the LO signal, we employ the following trigonometric identity (see section 1.1.8).

$$\sin(\alpha(t) + \theta_{Error}) = \sin(\alpha(t)) \cdot \cos(\theta_{Error}) + \cos(\alpha(t)) \cdot \sin(\theta_{Error})$$

$$
\begin{aligned}
RF(t) &= I(t) \cdot \cos(2\pi f_o t) + Q(t) \cdot \sin(2\pi f_o t + \theta_{Error}) \\
&= I(t) \cdot \cos(2\pi f_o t) + Q(t) \cdot [\sin(2\pi f_o t) \cdot \cos(\theta_{Error}) + \cos(2\pi f_o t) \cdot \sin(\theta_{Error})] \\
&= (I(t) + Q(t) \cdot \sin(\theta_{Error})) \cdot \cos(2\pi f_o t) + (Q(t) \cdot \cos(\theta_{Error})) \cdot \sin(2\pi f_o t)
\end{aligned}
$$

$$RF(t) = I'(t) \cdot \cos(2\pi f_o t) + Q'(t) \cdot \sin(2\pi f_o t)$$

The cross-talk generated due to the phase imbalance can be summarized in the *imbalance model* shown here.

$$
\boxed{
\begin{aligned}
I'(t) &= I(t) + Q(t) \cdot \sin(\theta_{Error}) \\
Q'(t) &= Q(t) \cdot \cos(\theta_{Error})
\end{aligned}
}
$$

Imbalance Model

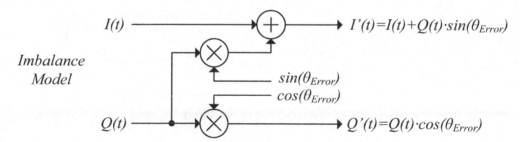

Figure 6-45: IQ Phase Imbalance Model

Mitigating Phase Imbalance in IQ Modulators

Once we have studied the imbalance model, mitigating the phase problem turns out to be straight-forward. The *rebalance function* that alters the source IQ information to eliminate the phase imbalance effect in the IQ modulators is implemented in the digital domain as follows.

$$
\boxed{
\begin{aligned}
I(n) &= I'(n) - Q'(n) \cdot \tan(\theta_{Error}) \\
Q(n) &= Q'(n) / \cos(\theta_{Error})
\end{aligned}
}
$$

Next we show the diagram indicating where the rebalancing module is placed and how it functions internally. If you insert the *I'* and *Q'* term into the rebalance model, you will see that the initial *I* and *Q* terms are correctly produced.

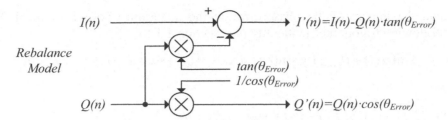

Figure 6-46: IQ Phase Rebalance Model

<u>*Mitigating Amplitude Imbalance in IQ Modulators*</u>

Fixing amplitude imbalance in IQ modulators is a trivial task, which involves inserting an in-phase and quadrature gain element after the phase rebalance module. In general, amplitude imbalance in IQ modulators is less common than phase imbalance.

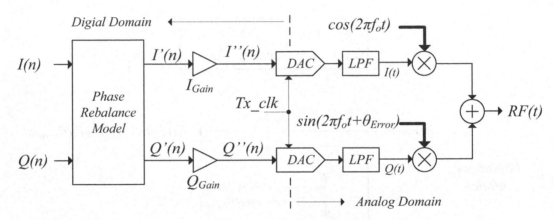

Figure 6-47: Phase and Amplitude Rebalancing for IQ Modulators

<u>*MatLab Model*</u>

```
function Output = IQImbalance(Input ,IQPhaseError, I_Gain, Q_Gain)

I       = real(Input);
Q       = imag(Input);

%% Phase imbalance model
I_Temp  = I+Q*sin(IQPhaseError);        %% Simulate IQ phase
Q_Temp  = Q*cos(IQPhaseError);          %% Imbalance

%% Amplitude imbalance model
I_Out   = I_Gain*I_Temp;                %% Simulate IQ
Q_Out   = Q_Gain*Q_Temp;                %% Amplitude imbalance

Output  = I_Out + j*Q_Out;
```

Example 5.4: *Receive Constellations Affected by Phase and Amplitude IQ Imbalances*

In this example, we will illustrate the effect of 15 degrees phase imbalance and 3dB magnitude imbalance on a QPSK and 16QAM constellation, respectively.

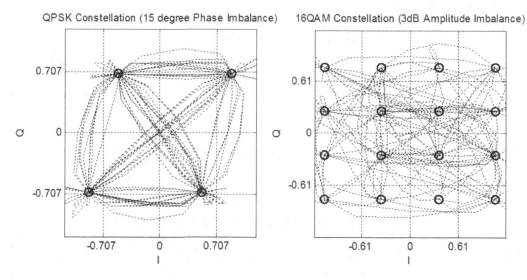

Figure 6-48: Example Constellations and Traces with IQ Phase and Amplitude Imbalance

6.3.7   Sample Timing Offset and Drift

Timing Offset

Timing offset is more a fact of life than an actual problem caused by our communication system hardware. The single-tone QAM transmitter sends pulse-shaped symbols across the wireless channel, and it is up to the receiver to sample these pulses at their peaks to achieve optimal detection. In the absence of linear distortion or after an equalizer has corrected the signal spectrum at the receiver, the peaks represent the transmitted symbols that we must detect.

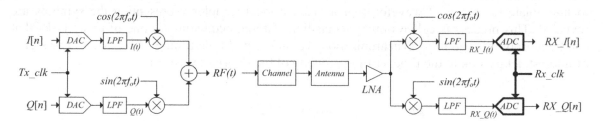

Figure 6-49: Sample Timing Offset and Drift

Unfortunately, as we sample the incoming waveforms into the receive portion of the base band or DSP processor, we have no idea where these perfect peaks are. The timing would be perfect only if the DAC modules in the transmitter and the ADC modules in the receiver were driven by the same physical master clock and the signal transmission time was zero.

The figure below shows the *I* portion of a QPSK signal as it works its way through the transmit chain. The open circles represent the symbols produced by the mapper block, while the black dots represent the sample stream after the symbols have been zero-stuffed and pulse-shaped. The continuous waveform connecting the samples represents the signal after the DAC reconstruction filters in the analog domain.

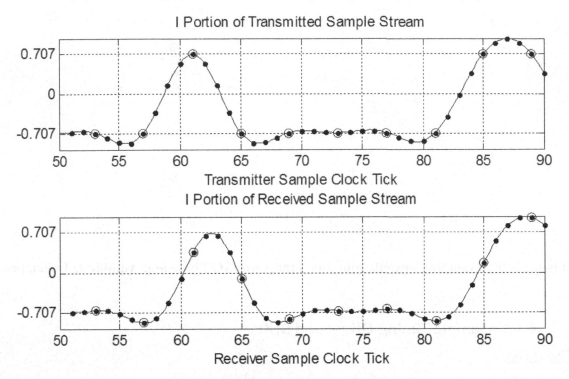

Figure 6-50: Static Timing Offset Example

The signal arrives at the receiver ADC modules with a delay of approximately 1.5 sample periods. If you look at the ADC samples, the black dots, it becomes clear that none of them represents the ideal sample instant that would reveal the desired symbol. It is the receiver's task to interpolate the received waveform such that the ideal samples representing the symbols are revealed. This process is called timing acquisition. Timing acquisition is one of many tasks that can be loosely referred to as synchronization. To understand how timing offset affects our signal of interest, let us look at the time-shifting property of the Fourier transform.

$$given \qquad x(t) \overset{FT}{\leftrightarrow} X(f)$$

$$then \qquad x(t - t_o) \overset{FT}{\leftrightarrow} X(f) \cdot e^{-j2\pi f t_o}$$

It is clear from the equation above that, in the frequency domain, a phase shift occurs, which grows linearly with increasing frequency. We may therefore think of the timing offset as a form of linear distortion that can be removed by an equalizer or interpolator.

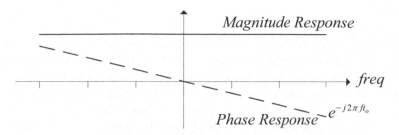

Figure 6-51: The Effect of Timing Offset on the Frequency Response of a Waveform

All interpolation techniques discussed in chapter 3 may be utilized in some way to introduce timing offset or drift. The function below is taken directly from that chapter and uses a FIR filter to time-shift or interpolate the input signal. In the second code segment, we introduce offset using polynomial interpolation via the method of splines.

```
function [Output] = FIR_Interpolator(Input , SampleDelay)


N      = 11;                        %% Easier if odd number of taps
m      = -(floor(N/2)):1:floor(N/2); %% Frequency index m
Mag    = ones(1, N);                %% Magnitude response
Phase  = -2*pi*m*SampleDelay/N;     %% Linear phase response
X      = Mag.*cos(Phase) + j*Mag.*sin(Phase);
                                    %% Total frequency response
DiscreteTime = -(floor(N/2)):1:floor(N/2); %% Time index
h = zeros(1,N);            %% Place holder for impulse response

%% Inverse discrete Fourier transform (compute impulse response)
for i=1:N
    n      = DiscreteTime(1,i);
    h(1,i) = X*exp(j*2*pi*m*n/N).'; %% The correlation equation
end

%% Overlay the Hanning window on top of impulse response
Hanning = hann(N+2)';
FIR_Coefficients = h.*Hanning(1,2:end-1);
FIR_Coefficients = real(FIR_Coefficients ... %% Normalizing
                /sum(FIR_Coefficients) );

%% Filtering operation
Output = filter(FIR_Coefficients, 1, Input);
```

```
function Output = Cause_TimingOffset(Input, SampleDelay)
% Input         --> Input sequence


SampleStep  = 1;
InputIndex  = 1:length(Input);
OutputIndex = 1+ SampleDelay:SampleStep:length(Input);


Output      = interp1(InputIndex, Input, OutputIndex, 'spline');
```

Timing Drift

Timing drift is due to a difference between sample rates at the receiver and transmitter. Master clocks, usually implemented as crystal oscillators, provide references to the synthesizer as well as the digital base band processor chip. Mixed signal (ADC and DAC) modules naturally derive their sample clock from that same master. The crystal oscillator frequency error is specified in parts per million, or ppm. Thus, a 10MHz crystal oscillator with an error specification of ±20 ppm will cause a frequency deviation of up to ±200Hz from the nominally specified value. If we have similarly specified crystal oscillators at both transmit and receive modules then the maximum error may increase to ±400Hz. This difference in master clocks at the two ends of the communication link produces a timing offset that slowly drifts. By slowly tracking and correcting this offset, we eliminate both timing offset and drift. MatLab provides additional interpolation functions that are based on splines and feature convenient input parameters to make our life easier. One such function, interp1(), is used in the code segments below, which introduce timing offset and timing drift into the input sequence. The interp1() function may be reviewed in the online MatLab documentation.

```
function Output = Cause_TimingDrift(Input, Drift)
% Input --> Input sequence
% Drift --> in parts per million (ppm)

SampleStep  = 1 + Drift/1e6;
InputIndex  = 1:length(Input);
OutputIndex = 1:SampleStep:length(Input);

Output      = interp1(InputIndex, Input, OutputIndex, 'spline');
```

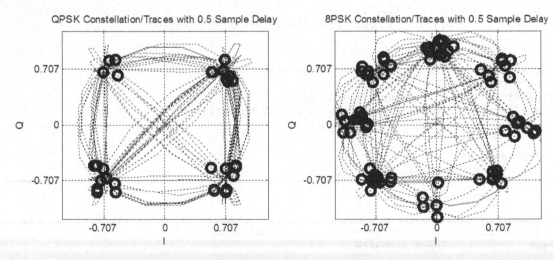

Figure 6-52: QSPK and 8PSK Constellation Positions with Signal Traces at a Delay of 0.5 Samples

6.3.8 MatLab Model Summary

The Defect Model

This first section of code represents the defect model function call that accepts a transmit sample stream and two structures detailing defect conditions. It returns the distorted transmit sample stream, as well as the impulse response of the multipath channel for plotting purposes.

```
function [Output_Waveform FIR_Taps] = Defect_Model(TX_Samples, ...
                                       Settings, Mode)

%% 1. Transferring the mode parameters
Multipath_Select       = Mode.Multipath;  % 0/1  means  exclude/include
AW_GaussianNoise_Select = Mode.ThermalNoise;
PhaseNoise_Select      = Mode.PhaseNoise;
Frequency_Offset_Select = Mode.Freq_Offset;
IQ_Imbalance_Select    = Mode.IQ_Imbalance;
Timing_Offset_Select   = Mode.TimingOffset;
Timing_Drift_Select    = Mode.TimingDrift;


%% 2. Transferring settings
N                      = Settings.NumberOfTaps; % For Multipath Model
DelaySpread            = Settings.DelaySpread;  %
SampleRate             = Settings.SampleRate;
SNR_dB                 = Settings.SNR_dB;        % Thermal Noise
PhaseNoiseProfile1     = Settings.PhaseNoiseProfile;
FrequencyOffset        = Settings.FrequencyOffset;
PhaseImbalance         = Settings.PhaseImbalance;
AmplitudeImbalance_dB  = Settings.AmplitudeImbalance_dB;
Sample_Offset          = Settings.Sample_Offset;
Drift_ppm              = Settings.Drift_ppm;


%% 3. Generating multipath filter
FIR_Taps = GetMultipathFilter(SampleRate, DelaySpread, N);
if(Multipath_Select == 1)
    TX_Samples = filter(FIR_Taps, 1, TX_Samples);
    VarOutput  = var(TX_Samples);
    TX_Samples = TX_Samples./sqrt(VarOutput);
end

%% 4. Generating additive white gaussian noise (thermal noise)
if(AW_GaussianNoise_Select == 1)
    TX_Samples    = TX_Samples + Generate_AWGN(TX_Samples, SNR_dB);
end

%% 5. Adding phase noise
NumberOfSamples      = length(TX_Samples);
[PhNoise Gain Freq RMSN] = PhaseNoiseGenerator(SampleRate, ...
    NumberOfSamples, PhaseNoiseProfile1(2,:), PhaseNoiseProfile1(1,:));
PhNoise              = PhNoise - mean(PhNoise);
PhaseNoise           = exp(j*PhNoise);
```

```
if(PhaseNoise_Select == 1);
    display(['Integrated Phase Noise1: ' num2str(RMSN*57.3)]);
    TX_Samples            = TX_Samples.*PhaseNoise;
end

%% 6. Adding frequency offset
OffsetSignal = ...
        exp(j*2*pi*(1:length(TX_Samples))*FrequencyOffset/SampleRate);
if(Frequency_Offset_Select == 1)
    TX_Samples    = TX_Samples.*OffsetSignal;
end

%% 7. Adding IQ imbalance
IGain                     = 10^(0.5*AmplitudeImbalance_dB/20);
QGain                     = 10^(-0.5*AmplitudeImbalance_dB/20);

if(IQ_Imbalance_Select == 1)
    TX_Samples    = IQImbalance(TX_Samples, PhaseImbalance,  IGain, QGain);
end

%% 8. Add timing offset
if(Timing_Offset_Select == 1)
    TX_Samples     = Cause_TimingOffset(TX_Samples, Sample_Offset);
end

%% 9. Add timing drift
if(Timing_Drift_Select == 1)
    TX_Samples    = Cause_TimingDrift(TX_Samples, Drift_ppm);
end

Output_Waveform = TX_Samples;
```

Exercising the Defect Model

In this section we configure the single tone QAM transmitter and defect model to become familiar with the impact of noise, distortion and offsets on the transmit constellation.

```
%% 1.  Transmitter Setup
%  1a. General setup information
ModulationFormat = 1;     % 0/1/2/3  =  BPSK/QPSK/8PSK/16QAM
NumberOfSymbols  = 300;
Alpha            = 0.5;  % This is the raised cosine index
Delay            = 4;    % Recommended -> Delay = 3 for Alpha > .3
                         % Recommended -> Delay = 6 for .1 > Alpha > .3
                         % Recommended -> Delay > 6 for Alpha < .1
OSR              = 4;    % Recommended --> 4 or 8
SymbolRate       = 10e6;
SampleRate       = SymbolRate*OSR;


% 1b. Generate random bits
InputBits        = round(rand(1, NumberOfSymbols*(ModulationFormat + 1)));
```

```
% 1c. The mapping operation
[MapperSymbols]  = Mapper(InputBits, ModulationFormat);

% 1d. The zero-stuffing and pulse shaping operation
[TX_Samples]    = TX_Waveform_Mapping([MapperSymbols 0 0 0 0 0 0 0 0], ...
                                        OSR, Alpha, Delay);
Range           = 21:OSR:540;  % Waveform indices at which the waveform
                               % sample is equal to the symbol
%% 2. Defect model
% 2a. Configuring the default model
Mode.Multipath          = 0;              % 0/1 = exclude/include defect
Mode.ThermalNoise       = 1;
Mode.PhaseNoise         = 0;
Mode.Freq_Offset        = 0;
Mode.IQ_Imbalance       = 0;
Mode.TimingOffset       = 0;
Mode.TimingDrift        = 0;


Settings.SampleRate     = 40e6;           % Hertz
Settings.NumberOfTaps   = 40;             % For multipath model
Settings.DelaySpread    = 150e-9;         % Seconds
Settings.SNR_dB         = 35;             % AWG thermal noise
Settings.PhaseNoiseProfile = [1e3 10e3 20e3 35e3 50e3 80e3 90e3 100e3 ...
                              120e3 150e3 200e3 300e3 500e3 1e6; ...
                              -70 -72 -72 -74 -76 -85 -90 -95 ...
                              -100 -105   -110 -120 -130 -140];
Settings.FrequencyOffset    = -100e3;     % Hz
Settings.PhaseImbalance     = pi/2000;    % Radians
Settings.AmplitudeImbalance_dB = -.1;     % dB
Settings.Sample_Offset      = -1;         % Samples at the sample rate
Settings.Drift_ppm          = -80;        % Given above in ppm (parts per
                                          % Million
% 2b. Applying the default model
[TX_Output FIR_Taps]    = Defect_Model(TX_Samples, Settings, Mode);
```

6.4 DAC Effects, and Nonlinear Distortion

We end this chapter with defects that don't lend themselves well to integration into the default defect model presented in the last chapter. Techniques for resolving DC offsets, for example, require more elaborate receiver modeling than was undertaken in the last few sections. Sample and hold distortion in digital to analog converter modules is an effect that is best understood and minimized rather than laboriously modeled. Its impact on the constellation is similar to a small amount of linear distortion and becomes negligible when properly choosing sampling rates. Nonlinear distortion is a critically important topic which does not lend itself well to MatLab simulation because it is difficult to find a proper model for it. You may review such models in these texts [2, 3, 4]. Very expensive RF circuit simulators such as Agilent ADS or Cadence Virtuoso Spectre are needed to estimate nonlinear distortion accurately in analog transceivers. In this chapter, we explain the effects of nonlinear distortion on receivers and transmitters but forego modeling them.

6.4.1 Distortion in Digital to Analog Converters

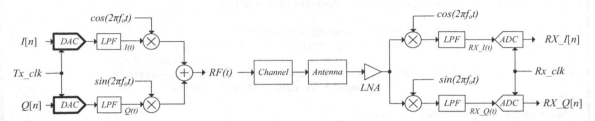

Figure 6-53: Distortion in Digital to Analog Converters

Mixed signal converters (ADC/DAC modules) have come a long way over the years. For most communication systems, there is an ample choice of parts that feature impressive linearity and speed. Although problems such differential and integral nonlinearity errors still exist, their effects are mostly negligible compared to the nonlinear distortion produced by the RF analog components in a typical transceiver. There are, however, linear distortion effects, produced by the sample and hold operation of digital to analog converters, that are worth exploring. As you may know, digital to analog converters hold the output voltage, which they were instructed to produce by the digital input word, steady for one sample period. The figure below expresses the digital sample stream as a sequence of scaled unit impulses, which are converted into a sequence of rectangular pulses at the output by the DAC.

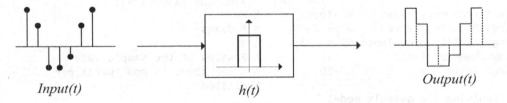

Input(t)   *h(t)*   *Output(t)*

Figure 6-54: Sample and Hold Distortion as a Convolution Process

How does the frequency response of our analog output signal differ from that of the digital input signal? To answer that question, we must realize that the sample and hold effect can be easily modeled using simple convolution. From our work on convolution, we may write the following expressions.

$$Output(t) = convolution(Input(t), H(t))$$
$$Output(f) = Input(f) \cdot H(f)$$

From the convolution property of the Fourier transform, we know that frequency response of the input must be multiplied by that of the system, $H(f)$, to get to the response of the output. The impulse response of the system, which our input signal traverses, is a gate function. This gate function is defined as unity from minus half the sample period to plus half the sample period and zero everywhere else.

$$h(t) = \begin{vmatrix} 1 & for & -\frac{T_S}{2} <= t <= \frac{T_S}{2} \\ 0 & otherwise \end{vmatrix}$$

To get to its frequency response, we simply take the Fourier transform of $h(t)$. The definition of $h(t)$ conveniently constrains the limits of integration of the Fourier transform equations.

$$H(f) = \frac{1}{T}\int_{-\infty}^{\infty} h(t) \cdot e^{-j2\pi ft}\,dt$$

$$= \frac{1}{T}\int_{-T/2}^{T/2} 1 \cdot e^{-j2\pi ft}\,dt$$

$$H(f) = \frac{1}{T}\int_{-T/2}^{T/2} 1 \cdot e^{-j2\pi ft}\,dt$$

$$= \frac{\sin(\pi fT)}{\pi fT}$$

$$= sinc(fT)$$

Below we take a graphical look at the frequency response of the type of signal that we will encounter in the next chapter. The input signal shown is sampled at 40MHz and features a single-sided bandwidth of slightly less than 10MHz. The sample and hold effect causes the main frequency image of our input signal to roll off prematurely. The attenuation toward the edge of the band is less than 1dB, which we may safely consider to be negligible compared to the multipath channel distortion that will affect the signal during transmission. By the time the first alias image appears at 40MHz, the sample and hold distortion has kicked into full gear, significantly attenuating the image's frequency content.

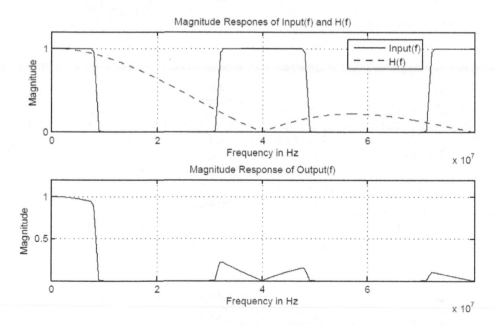

Figure 6-55: The Effect of Sample and Hold Distortion on the Transmit Signal Spectrum

In fact, this phenomenon may be considered somewhat beneficial, since it attenuates the sampling images that are inherent in the samples signal. The requirements on the analog reconstruction filter that follows the DAC module are therefore alleviated. Now, if our input signal had been sampled at 20 instead of 40MHz, its frequency response inside ±10MHz would have been affected significantly more. The figure below takes us through the all too familiar process of mapping, zero-stuffing, and pulse-shaping. The spectrum of the scaled unit impulse train that represents the symbol sequence is flat, and the zero-stuffing operation changes nothing but the sampling rate conditions. It is after the pulse-shaping operation that we first observe the characteristics that will govern the response of the final signal. Note the effect of the sinc function – owing to the sample and hold phenomenon of the DAC – which is multiplied with the periodic spectrum of the sample stream. It is not until the DAC reconstruction filters that all sampling images are removed and the final analog signal – ready for up-conversion – takes shape.

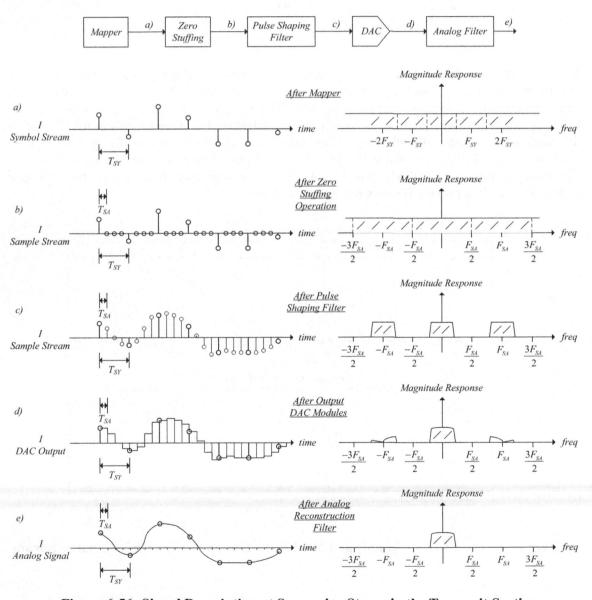

Figure 6-56: Signal Description at Successive Stages in the Transmit Section

References

[1] TS 25.101 (2013) *User Equipment (UE) Radio Transmission and Reception (FDD)*, 3GPP Technical Specification, Version 11.4.0, Section 6.6.2.1.1

[2] Vuolevi, J. and Rahkonen, T. (2003) *Distortion in RF Power Amplifier*, Artech House, London/Boston

[3] Johns, D. and Martin, K. (1997) *Analog Integrated Circuit Design*, John Wiley & Sons, New York, 181-221

[4] Jeruchim, M., Balaban, P., and Shanmugan, K. Sam (2002) *Simulation of Communication Systems: Modeling, Methodology, and Techniques*, Second Edition, Kluwer Academic, New York, 203-289

[5] Razavi, B. (2000) *Design of Analog CMOS Integrated Circuits*, McGraw-Hill Higher Education, New York, 191-233

[6] Chayat, N. (1997) *"Tentative Criteria for Comparison of Modulation Methods"*, Document: IEEE P802.11-97/96, September

[7] Saleh, A.M. and Valenzuela, R.A. (1987) *"A Statistical Model for Indoor Multipath Propagation"*, *IEEE Journal on Selected Areas in Communications*, Vol. SAC-5, no.2, 128-137

[8] Oppenheim, A., Willsky, A., and Young, I. (1983) *Signals and Systems*, Prentice-Hall, Englewood Cliffs, NJ

[9] Lyons, R. (2004) *Understanding Digital Signal Processing*, Prentice-Hall, Upper Saddle River, NJ

PART 4

Modern OFDM Based Communication Systems

7 **OFDM Based Communication Systems**

OFDM, or orthogonal frequency division multiplexing, is a modulation technique that has been utilized in virtually all high throughput wireless communication systems that have been developed over the last twenty years. Except for the 802.11b standard, OFDM is used in all 802.11 WLAN standards, as well as in the 4G LTE and 5G New Radio downlink direction. In addition to straight OFDM modulation, the 4G and 5G uplink and sidelink directions use a related modulation technique called SC-FDMA, or single carrier frequency division multiple access. In this chapter we will study communication systems using these two very popular modulation schemes.

A Preview of this Chapter

Before starting this chapter, ensure that you have read sections 5.4 and 5.5, which introduce us to the basics of both OFDM and SC-FDMA modulation. In addition to some further introductory information, we will cover the following communication technologies.

→ We first introduce the IEEE 802.11a modem, which lays the ground work of all 802.11 WLAN variants that followed. The 802.11a modem is a great way to learn about OFDM as it is straight forward and does without a lot of bells and whistles that can overcomplicate some of the later implementations. We discuss these bells and whistles in chapter 8. The 802.11a physical layer is described in some 50 pages in the specification, which are surprisingly easy to read.

→ The 4G LTE (3GPP) technology introduced OFDM to the world of cellular mobile networks. The 4G physical layer is a definite step up from the 802.11a modem as it must deal with terminals that are in motions and must coordinate communication as the terminal moves from one cell to the next. The physical layer description spans multiple documents and is far more difficult to read than the IEEE 802.11 documentation. This text will strip many of the unnecessary complications of the 4G modem and presents the key signal processing topics that make LTE work.

→ The C-V2X technology is a 3GPP standard that enables communication between different vehicles, traffic structures and pedestrians to create an intelligent traffic system. C-V2X is based on SC-FDMA and provides an excellent case study as it to avoid some of the excessive complication that can be found in its LTE parent.

→ In this text, we only briefly mention 5G as at its core it is based on the same principles as LTE. The study of 5G without working knowledge of 4G will be quite difficult. I will mention the differences between LTE and 5G New Radio in many of the pages that make up this chapter.

Note that the physical layer descriptions provided by these standards specify the construction of the OFDM or SC-FDMA based transmit waveforms. They only indirectly point us to the techniques we need to use in the receiver to recover it. In this chapter, we will spend considerable time with techniques to resolve timing and frequency offsets and drifts as well as methods that estimate the channel response (leading to equalizer coefficients) and the signal to noise ratio of the received waveform.

7.1 OFDM – Orthogonal Frequency Division Multiplexing

7.1.1 A Short Recap

Section 5.4 introduced OFDM modulation and the reader should understand these basics prior to proceeding with this chapter. The figure below provides a high-level overview that summarizes what we have learned previously and adds a few simple yet very useful concepts, which we will use throughout this chapter.

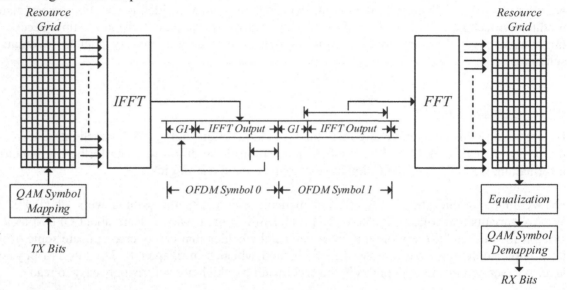

Figure 7-1: Simplified OFDM Modem Structure

An OFDM modem features several basic processing steps, which are illustrated in the figure above. Typically, the data bits that we which to transmit are scrambled and encoded to protect them from errors before being mapped into QAM (BPSK, QPSK, 16QAM, … etc) constellation position. The set of QAM symbols are then organized into a matrix called the resource grid, prior to being mapped into the IFFT input buffer. The QAM symbols located in a single column are simultaneously mapped into the IFFT to produce a time domain sequence, which we naturally refer to as the IFFT output sequence. The last few samples of this IFFT output sequence are copied and appended to the front to form what we call an OFDM symbols. This procedure is repeated for each column in the resource grid to form a time domain sequence that will be read out to the digital to analog converters and send through the RF portion of the transmitter. At the receiver, the incoming waveform is digitized, and a portion of each OFDM symbol is extracted and mapped into the FFT input buffer. Similarly to what happened in the transmitter, the FFT outputs are organized into the columns of a resource grid prior to being extracted for further processing. Notice that the receiver may not extract the exact IFFT output portion of the OFDM symbol as can be seen in the figure above. This is done to avoid intersymbol interference that may result due to the action of the multipath channel. The apparent timing offset as well as the linear distortion produced by the multipath channel are estimated and their effects are removed by the equalizer. After the equalization step, the QAM symbols are demapped to produce receive bits. The convenient resource grid concept is used extensively in LTE and 5G, but is not formally defined in the 802.11 specifications.

7.1.2   The Concepts of Subcarriers and Resource Grids

As mentioned in the last section, OFDM symbols are generated by taking the IFFT (see equation below) of a sequence of complex transmit data values that usually belong to a QAM constellation such as BPSK, QPSK, 16 and 64 QAM. These complex data values are mapped somewhere into the IFFT input vector, $X[m]$, such that the IFFT time domain output sequence forms a signal with a well-defined bandwidth. To better understand how the complex transmit data values are organized, we introduce the concept of subcarriers and resource grids.

$$x[n] = \sum_{m=0}^{N-1} X[m] \cdot e^{\frac{j2\pi nm}{N}} \qquad for\ n = 0, 1...N-1$$

Subcarriers (Tones) and Subcarrier Spacing

A length N IFFT produces a superposition of N complex sinusoids, $exp(j2\pi nm/N)$, which are scaled by the complex data values $X[m]$. The 4G and 5G specifications refer to these complex sinusoids as *subcarriers*, whereas the 802.11 specifications call them *tones*. Given a sample rate of Fs, these subcarriers have the following properties:

→ A subcarrier features N normalized time instances n, where $n = 0, 1 ... N-1$, and actual time instances at n/Fs seconds.
→ A subcarrier can feature one of N normalized frequencies of value m/N, where $m = 0, 1, ... N-1$, which translate to the actual frequencies shown below.

$$F_{actual} = \begin{cases} \dfrac{m}{N}Fs \ \ (Hz) & for\ m = 0...\dfrac{N}{2}-1 \quad (positive\ Frequencies) \\[3mm] \dfrac{m-N}{N}Fs \ (Hz) & for\ m = \dfrac{N}{2}...N-1 \ (aliased\ to\ negative\ Frequencies) \end{cases}$$

→ The spacing between two subcarriers, or subcarrier spacing, is equal to $\dfrac{Fs}{N}$.

The Resource Grid and Resource Elements

The right portion of the next figure illustrates how the index, m, of the IFFT input vector $X[m]$, maps to subcarriers SC. The convention of mapping indices $m = 0, 1, ... N/2-1$ to positively valued SC indices, and $m = N/2, ..., N-1$ to negatively valued SC indices comes from the 802.11 specification. The 802.11A modem, uses an $N = 64$ length IFFT at a sample rate of 20MHz. The tone $SC = -1$, at $m = 63$, represents the subcarrier with the smallest negative frequency equal to $(63-64)/64 \cdot 20e3 = -312.5$KHz, whereas the tone SC = 1 represents the subcarrier with the smallest positive frequency of 312.5KHz. In the 802.11A specification, only subcarriers SC = 1, 2 ... 26 and SC = -1, -2, ..., -26 carry complex data symbols resulting in a signal spanning frequencies from $-26 \cdot 312.5$KHz = -8.125MHz to $26 \cdot 312.5$KHz = 8.125MHz. The remaining subcarriers, including the DC subcarrier at SC = 0, do not carry any information and are loaded with a value equal to 0. The DC subcarrier in the 802.11a/g/n and LTE specifications does not carry information as its content is difficult to detect due to DC offsets in modern direct conversion receivers.

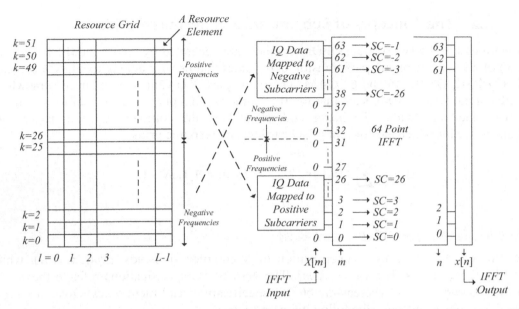

Figure 7-2: The Mapping of Subcarriers into the IFFT and Its Relationship to the Resource Grid

The resource grid, which is shown to the left of the IFFT structure is only defined in the 3GPP specifications that detail the LTE and 5G New Radio technologies. The resource grid is a matrix of complex data values that we wish to transmit using our OFDM modulation mechanism. Each column, represented by the variable l, contains a sequence of complex data values that we want to map into the IFFT input vector, $X[m]$, and subsequently transform into a time domain OFDM symbol. According to the 3GPP specifications (LTE/5G), the complex data value at $k = 0$ maps to the most negative subcarrier, whereas the largest index k maps to the most positive subcarrier. Section 7.5.4 discusses how the index k of the resource grid maps to the IFFT input vector index, m, for all supported LTE bandwidths. The LTE specification further defines a quantity called a **resource element**, which represents a single complex number fed into the IFFT input. The value of the resource element is the orientation and size of a single subcarrier at a single OFDM symbol. The 802.11a specification doesn't directly specify a resource grid or resource element, but we can easily use the one shown in the figure above. The figure below, illustrates how LTE uses the resource grid to plan the location of different downlink signals and physical channels. For a detailed look at these downlink entities, see section 7.5.3.

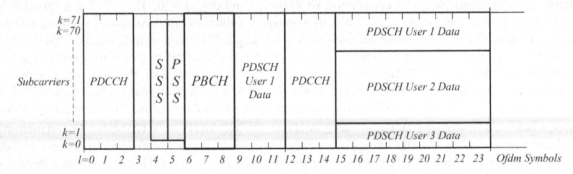

Figure 7-3: Typical Resource Grid for 1.4MHz Lte Signal

7.1.3 Multipath Channel Effects in OFDM Systems

The OFDM modulation and demodulation process described in the last section was pretty straight forward and involved computing the IFFT of a sequence of complex data symbols, transmitting the generated output and computing the FFT at the receiver to recover the original data symbols. The only unusual aspect of the process was the insertion of the guard interval (IEEE nomenclature) or cyclic prefix (3GPP nomenclature), which protects the signal from intersymbol interference. In this section, we will take a look at how the multipath channel affects the transmitted OFDM symbol, how the guard interval protects us them the aforementioned intersymbol interference and how we best repair the received OFDM signal. To start our endevor, let's examine the multipath channel model which we introduced in section 6.3.1.

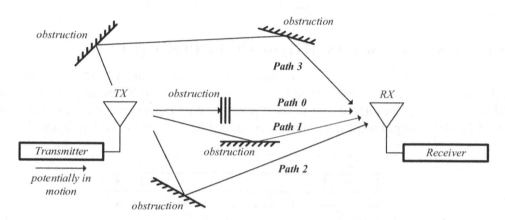

Figure 7-4: Simple Multipath Channel Model

We can make the following generalizations for a multipath channel model where both the transmitter and receiver are stationary.

→ The transmitted signal is generally radiated in all direction and will bounce of any number of obstructions before reaching the receive antenna where the various copies are summed into one superimposed receive signal.

→ As the transmit signal takes different paths to get to the receive antenna, the various copies arrive with different time delay, τ_p.

→ Given the simple concept of path loss, the copies of the transmit signal that require the longest time to arrive, tend to feature more attenuation than those copies that took a more direct path and arrived earlier. This is especially true in cases where there is a direct line of sight. However, the scenario above shows the case where the most direct path also features an obstruction. The path attenuation is embedded in the magnitude of the complex scalar C_p as shown below.

→ If either the transmitter or receiver are in motion, Doppler shift will occur.

→ Due to the different distances that the RF signals traverse, the signals complex envelope will have a random phase when it arrives at the receiver. This is reflected in the phase of the scalar C_p.

At baseband, we model the effect of a single path, p, on a transmit signal $Tx(t)$ via the following expression (ignoring Doppler for now).

$$Rx_p(t) = C_p \cdot Tx(t - \tau_p)$$

The transmit signal experiences a delay, τ_p, and complex scaling operation via C_p, both of which depend on the conditions of path p. If one or both of the terminals are in motion, then Doppler is introduced and the expression changes slightly to include the frequency term f_p.

$$Rx_p(t) = C_p \cdot exp^{j2\pi f_p t} \cdot Tx(t - \tau_p)$$

Review section 6.3.2 and realize that relative motion between two terminals looks like a small frequency offset in the arriving copy of the transmitted signal. The final received signal now takes the following form, where P refers to the number of paths arriving at the receive antenna.

$$Rx(t) = \sum_{p=0}^{P-1} C_p \cdot exp^{j2\pi f_p t} \cdot Tx(t - \tau_p) \tag{7.1}$$

7.1.3.1 The Guard Interval (GI) or Cyclic Prefix (CP)

To appreciate the role of the guard interval (cyclic prefix) we ignore all aspects of our multipath channel model except for the delay. Below, we take a look at the IFFT portions of the transmitted OFDM symbol. As mentioned in our introduction of OFDM (section 5.4.2), the CP is the final section of the IFFT output sequence, which is appended to the start the OFDM symbol.

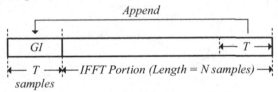

Figure 7-5: Generation of the Guard Interval

An equivalent method of producing the guard interval is highly instructive in explaining how it works. Simply replace the IFFT operation by the IDFT operation below and extend the time instant range at which it is evaluated. Rather than evaluating the IDFT expression at $n = 0, 1 \dots N-1$, which is the normal case, simply evaluate it at $n = -T \dots 0, 1 \dots N-1$, where T represents the guard interval sample length. Because the IDFT expression produces a periodic waveform with period equal to N, any sample rendered before time $n = 0$, will be equal to sample rendered before time $n = N$. Equation 7.2 below represent the normal IDFT expression, whereas expression 7.3 additionally appends the guard interval.

$$x[n] = \sum_{m=0}^{N-1} X[m] \cdot e^{\frac{j2\pi nm}{N}} \to n = 0, 1 \dots N-1 \tag{7.2}$$

$$x[n] = \sum_{m=0}^{N-1} X[m] \cdot e^{\frac{j2\pi nm}{N}} \to n = -T \dots 0, 1 \dots N-1 \tag{7.3}$$

At the receiver, we are at liberty to start the FFT operation at any sample instant, G, inside the guard interval. The information recovered by the FFT operation is identical to the original transmit symbols, $X[m]$, except for a phase shift that we can quantify using the time shifting property of the Fourier transform, and which can be easily undone after the FFT operation.

$$x[n - G] \to DFT \to X[m]e^{\frac{-j2\pi Gm}{N}}$$

So at this point, we have convinced ourselves that the guard interval doesn't cause problems, but how does it help us? The figure depicts the four paths in our channel model just prior to being summed in the receive antenna. With our previous thinking, we would want the take the FFT of the IFFT portion of path 1, which was the one arriving with the highest power. Notice that before the end of the IFFT portion of path 1, the OFDM symbol of path 0 has finished and a new symbol has begun. This intersymbol interference is unacceptable especially when we transmit signals with a large QAM constellation. The figure below illustrates the preferred section of the signal for which we need to compute the FFT to avoid intersymbol interference problems.

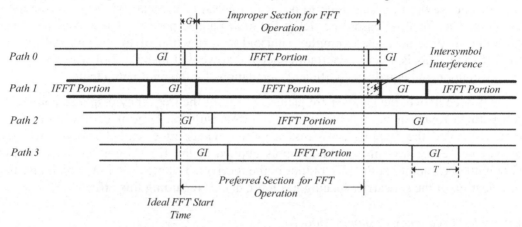

Figure 7-6: Transmit Signal Arriving via Different Paths at Receive Antenna

Via judicious choice of the FFT start instance, we have reduced the multipath channel problem to one of linear distortion only. Let us now understand the impact of each path on the overall frequency response, which characterizes said linear distortion.

7.1.3.2 The Impact of the Multipath Channel on the Frequency Response

To start our task of understanding of how each path impacts the overall magnitude response, we will at first ignore the influence of the Doppler shift and look at how the time shifting property of the Fourier transform provides us with the following frequency domain expression.

$$if \qquad Tx(t) \rightarrow DFT \rightarrow X(f)$$

$$then \qquad C_p \cdot Tx(t - \tau_p) \rightarrow DFT \rightarrow C_p \cdot X(f) \cdot e^{-j2\pi\tau_p f}$$

If the term $X(f)$ represents the Fourier transform of the transmit signal, then the frequency response is defined by those terms that have altered the spectrum $X(f)$ after it has passed through the channel. The following expression represents the frequency response for a single path.

$$FreqResponse_p(f) = C_p \cdot e^{-j2\pi\tau_p f} \qquad (7.4)$$

If we wanted to include the Doppler shift, f_p, then the frequency response changes with time, and technically we would have to take the Fourier transform of equation 7.1, which would introduce a small frequency shift into our last expression. However, without invalidating our model, we will choose an approximated version of the frequency response expression for path p, which we show below. The term $exp(-j2\pi\tau_p f)$ causes the channel response to change over frequency, whereas the term $exp(-j2\pi f_p t)$ causes the response to change over time.

$$FreqResponse_p(f,t) = C_p \cdot e^{-j2\pi\tau_p f} \cdot e^{-j2\pi f_p t}$$

Let's halt here and state some important facts regarding the expression above.

→ The delay, τ_p, is the difference between the FFT start time instant, which we chose, and the actual start time of the IFFT section of the OFDM symbol embedded in each path, p. In the last section, we choose the FFT start time to occur just ahead of the start of the IFFT section of the earliest path, and for this reason the delay for each path is positive. Curiously, the frequency response due to time delay is a complex sinusoid as a function of frequency, rather than as a function of time. This is a very important concept that we will make use of when we look at the channel estimation and equalizer coefficient calculations for the multipath channel.

→ As mentioned above, the Doppler frequency, f_p, causes the frequency response component due to each path to rotate in time. As long as this offset isn't too large it can be corrected at the receiver. However, the Doppler can cause a more devious effect called loss of orthogonality, which will cause the transmit QAM symbols applied at the IFFT input in the transmitter to interfere with one another at the FFT output in the receiver. The next section, which discusses the proper selection of the subcarrier spacing has more details regarding this effect.

The Combined Frequency Response Model

Notwithstanding loss of orthogonality, the total frequency response is simply the superposition of the frequency responses of each path, where P indicates the number of paths.

$$FreqResponse(f,t) = \sum_{p=0}^{P-1} C_p \cdot X(f) \cdot e^{-j2\pi\tau_p f} \cdot e^{-j2\pi f_p t} \tag{7.2}$$

In the MatLab code below, we provide a simple channel model that would be appropriate for a cellular communication system in an urban environment. The code evaluates equation 7.5 for this model, and may be used to plot the frequency response in polar format (magnitude/phase) and rectangular format thus illustrating the real and imaginary terms (see plots that follow).

```
%% 1. Determine Channel Model with Four Paths
              %    Cp        Delay(sec)    Doppler
ChannelModel    = [0.5,        0e-6,         20;  ... % Path 0 (The earliest path)
                   1j,         0.5e-6,      -100;  ... % Path 1 (The strongest path)
                   0.5j,       2.5e-6,        50;  ... % Path 2
                   0.2,        4.5e-6,       150];      % Path 3 (The Latest path)

%% 2. Compute the frequency response
f            = -1.5e6:15000:1.5e6;
Time         = 0.001*ones(1, length(f));
FreqResponse = ComputeFrequencyResponse(ChannelModel(2,:), f, Time); % Of Path 2 only


%% 3. Compute the frequency model according to equation 7.2
function FreqResponse = ComputeFrequencyResponse(...
                        ChannelModel      ... % Multipath Channel Model
                      , f                 ... % The frequencies and the time at which
                      , Time)                 % to evaluate the Frequency Response
```

```
FreqResponse = zeros(size(f));
NumPaths     = size(ChannelModel, 1);      % Get number of rows of the matrix

for p = 1:NumPaths
   Cp      = ChannelModel(p, 1);           % Complex scalar
   Tau_p   = ChannelModel(p, 2);           % The delay
   f_p     = ChannelModel(p, 3);           % The Doppler Shift
   FreqResponse = FreqResponse + Cp * exp(-1j*2*pi*Tau_p*f) .* exp(1j*2*pi*f_p*Time);
end
```

The figure below illustrates the channel response of the second (our strongest) path, which in the model features a delay of 500 nanoseconds. Note the complex sinusoidal behavior of the channel response across frequency.

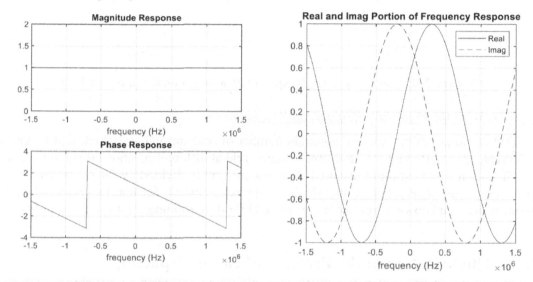

Figure 7-7: The Frequency Response for Path 1 (The Strongest Path)

The next figure illustrates the channel response assuming that all paths are active. We can make the following statements regarding its behavior:

→ The peaks and nulls that occur in the magnitude response, which is defined as |*FreqResponse(f)*|, are formed by the constructive and destructive interferences that are the result of the summation of the constituent frequency responses (the different complex sinusoids) at the receive antenna.

→ Looking back at equation 7.4, we realize that the rate at which the frequency response changes is directly proportional to the path with the largest delay, τ_p, as the associated complex sinusoid $exp(-j2\pi\tau_p f)$ will have the highest rate of change.

→ If one or both terminals are in motion, and Doppler shift exists, then the frequency response will change completely as time progresses.

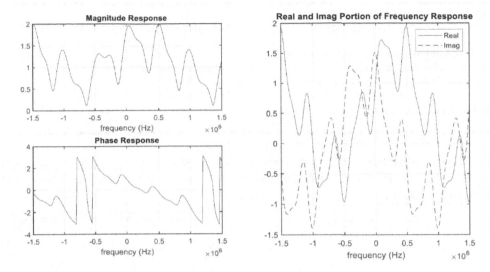

Figure 7-8: The Frequency Response as a Superposition of All Paths

How to Measure the Frequency Response in Real Situations

In OFDM communication systems, a certain number of constant values, which will be known at the receiver, are inserted into the transmit resource grid at different positions (resource elements). The FFT output values at those resource elements are simply divided by the corresponding IFFT input values at the transmitter to provide the frequency response for that OFDM symbol. These constant values are known as reference signals (LTE/5G), or training/pilot information WLAN.

7.1.3.3 The Selection of the Proper Subcarrier Spacing

One of the most fundamental design choices when conceiving of an OFDM communication system is the selection of the subcarrier spacing, *SC*, representing the distance in Hertz between the adjacent orthogonal frequencies produced by the modulation process which is done using the IFFT. The subcarrier spacing is defined as ratio of sample rate and FFT size, *N*.

$$SC = \frac{SampleRate}{N}$$

The subcarrier spacing used in various communication systems is provided below.

Technology	Subcarrier Spacing	Comments
802.11a	20MHz / 64 = 312.5KHz	802.11n/ac use higher sample rate and FFT sizes to achieve the same SC = 312.5KHz
4G LTE	30.72MHz / 2048 = 15KHz	For different LTE bandwidths sample rates and FFT sizes vary to achieve SC = 15KHz
5G New Radio	15/30/60/120/240 KHz	Sample rates and FFT sizes vary to achieve the different bandwidths

Figure 7-9: Subcarrier Spacing for Different OFDM Based Wireless Technologies

The most fundamental advantage of OFDM over other wireless modulation technologies is the fact that the frequency response value, *FreqResponse*[*m*], at a particular subcarrier with index, *m*, can be considered constant, which allows us to employ a single equalization coefficient of $1/FreqResponse[m]$. We thus converted a broad band communication signal, which traditionally must be equalized using a sophisticated time domain filter structure, into a series of parallel narrowband signals, each of which can be equalized by a single constant. This simplification not only eases equalization but also enables other sophisticated technologies such as MIMO, which were far harder to implement when using more traditional wideband signals.

As we mentioned in the last section, it is the path with the largest delay that will provide the frequency response component with the highest rate of change. WLAN is primarily an indoor technology where the differences in flight time between the different paths stays within a few hundred nanoseconds, whereas the need of LTE to operate in large urban areas results in flight time differences of up to several microseconds. The various delays in the receiver, which we defined as the difference between the optimal FFT start time in the receiver and the start time of the IFFT portion of the OFDM symbol, will be significantly smaller in WLAN than in LTE. Therefore, we are able to choose a larger subcarrier spacing for WLAN since we know that the frequency response doesn't change at the high rate from which the LTE channel suffers.

The Advantage of Large Subcarrier Spacing and Loss of Orthogonality

So why not choose very small subcarrier spacings in order to guarantee a constant frequency response around each IFFT tone? The reason to stick with the largest possible subcarrier spacing has to do with the frequency offset between the TX and RX LO signals. The IFFT maps transmit symbols onto complex sinusoids of orthogonal frequencies. At the receiver, the FFT reverses the process by computing the correlation of the received signal against each one of these complex sinusoids. If the received signal suffers from frequency offset or excessive Doppler, then the spectrum of the transmitted waveform is shifted slightly, and the complex subcarriers generated by the IFFT in the transmitter don't line up properly with their FFT counterparts in the receiver. This effect, called loss or orthogonality, worsens as the frequency offset increases. Any one tone in the FFT now recovers information note just from its corresponding tone in the IFFT, but content from all other tones as well resulting in interference. A larger subcarrier spacing provides robustness again this deleterious effect. Notice that 5G new radio can select larger subcarrier spacings than LTE. This flexibility is built in to allow 5G to operate at higher center frequencies than LTE, where frequency offsets and Doppler shifts are more prevalent. This in turn forces the size of the cell to shrink to avoid large delay differences between the paths, and keeps the frequency response constant in the vicinity of each subcarrier. Alas, the shrinking cell size is quite acceptable as path loss at higher frequencies is larger, and we would have had to boost the output power at the cell tower significantly to retain the same cell size. LTE with its relatively small subcarrier spacing thus needs to spend more resources to provide very accurate clocks that feed the LO generators, or synthesizers, and additional features that allow the receiver to resolve the frequency offset and Doppler that remain.

7.1.3.4 The Role and Design of Pilot or Reference Signals

All wireless transmit signals feature embedded symbols that are known to the receiver beforehand and are used to estimate the frequency response of multipath channel at a certain instant of time. Depending on the wireless technology, this information is referred to as reference signals, pilot symbols or training information. In OFDM systems, this training information takes the shape of a certain number of known QAM (usually BPSK or QPSK) symbols at the IFFT input of the transmitter. Let's take a look at the positioning of this information in both some of the more popular OFDM systems.

Training Information in 802.11A

In the case of 802.11a/g, an $N = 64$ point IFFT is used to produce subcarriers that span orthogonal frequencies between $-32 \cdot 312.5KHz = -10MHz$ to $31 \cdot 312.5KHz = 9.6875MHz$. Only those subcarriers located at $(-1:-26) \cdot 312.5KHz$ and $(1:26) \cdot 312.5KHz$ actually carry information thus limiting the frequency content to between -8.125Mhz to 8.125MHz. The subcarrier at DC is zeroed out as any value there would be lost in the direct conversion receivers due to DC offset. The training information is organized as follows:

→ Known QPSK data is transmitted in a long training sequence which spans approximately 2 OFDM symbols These OFDM symbols feature identical QPSK data values on all 52 information carrying subcarriers (SC = -1:-26 and 1:26).

→ In all subsequent OFDM symbols, pilot information (also known QPSK symbols) are transmitted at subcarriers -21, -7, 7, 21.

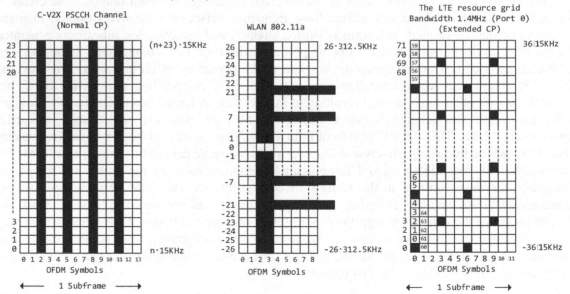

Figure 7-10: 802.11A Resource Grid with Training Information as Block Squares

At the receiver, the FFT outputs at SC = 1:26 and SC = -1:-26 of these two initial training symbols are first averaged, and the frequency response may be found for each subcarrier by dividing these FFT output values by the corresponding IFFT input values (our training information). Although it won't become clear until the next section, the 802.11a/g specification allows us to compute a very accurate frequency response estimate via this long training sequence.

As the terminals are not in motion, the only mechanisms that can change the frequency response later on are as follows.
→ Frequency offset and phase noise (see Section 6.3.4 / 6.3.5).
→ Timing drift (see Section 6.3.7).

Section 7.3 gives a detailed account of how to analyze and remove these effects (using pilot information) to maintain an accurate frequency response of the channel as time increases.

Training Information in the LTE Downlink

The situation in LTE is somewhat different. For the 1.4MHz bandwidth scenario, the information carrying subcarriers are $-1 \cdot 15KHz : -36 \cdot 15KHz$ and $1 \cdot 15KHz : 36 \cdot 15KHz$, and the training information, called cell specific reference signals, or CRS [7], are spread out across the reference grid as illustrated by the black squares in the last figure. This figure only shows CRS for antenna port 0, but those of ports 1, 2, and 3 can be seen in [7]. Unlike in 802.11a/g/n, one terminal, the cell phone, is often in motion and training information must be continuously and evenly supplied across time and bandwidth to track any changes in the multipath channel. As we will discuss in the next section, the spacing of the CRS in frequency determines how much delay can be tolerated, whereas the spacing in time determines how much Doppler or frequency offset may be tolerated. See section 7.5.6 for more information on cell specific reference signals in the LTE downlink.

Training Information in the LTE C-V2X Sidelink

The C-V2X, or cellular vehicle to everything technology, is a 3GPP technology that standardizes communication between automobiles in what is called intelligent traffic systems, or ITS. This technology is based on the LTE sidelink (very similar to the uplink) and uses SC-FDMA modulation, which we saw in section 5.5. There is only one transmit antenna in C-V2X, and after examining the figure above a little more closely, you can tell that there are far more reference signals for this single port, than there are for port 0 of the LTE downlink signal. There are thus fewer resource elements available for transmit user data, but our ability to estimate the channel (frequency response) is significantly improved.

7.1.3.5 Channel Estimation and RS-CINR Estimation

Good channel estimation is critical in high performance OFDM systems. Remember that OFDM systems are so popular, because each subcarrier is narrow enough such that the frequency response across it hardly changes. Therefore, at the FFT output in the receiver, we simply scale or equalize each recovered subcarrier orientation by a complex number to remove the channel effects (the frequency response). Given that we know the location and original complex value of each reference signal (training value), it is a simple matter to compute the frequency response and corresponding equalizer value at the reference signal locations.

$$RawFreqResponse = \frac{RefSignal_{Rx}}{RefSignal_{Tx}} \qquad EqualizerValue = \frac{RefSignal_{Tx}}{RefSignal_{Rx}}$$

Notice that whereas we know the original transmit value of the reference signal exactly, the received version of it will not just have suffered linear distortion, which is what we are trying to measure, but also additive white Gaussian noise. Therefore, the estimate of the linear distortion (the frequency response) will also suffer from noise. For now, we will call the noisy result of the computation above the raw frequency response. The goal of channel estimation in OFDM systems is to find the raw frequency response and remove the Gaussian noise to maximize its signal to noise ratio to subsequently equalize the channel as accurately as we possibly can. All high performance OFDM systems undertake this very interesting exercise. Notice that due to the spacing of the reference signals in the LTE downlink, we do not know the frequency response at every subcarrier, and we must use interpolation to make up for this deficit.

> If you are new to OFDM communication systems and are trying to learn the basic mechanics involved, then you are at liberty to skip the rest of this section and come back to it later. The topic is very interesting and essential for high performance modems, but it can distract a reader that is new to the topic of OFDM. If you are ready to tackle this exiting topic, then ensure that you have a good grasp of the principle component analysis by reviewing it in chapter 1.3.

To understand how we can remove Gaussian noise from the raw frequency estimate, let's take another look at the model of the frequency response that we developed in section 7.1.3.2.

$$FreqResponse(f) = \sum_{p=0}^{P-1} C_p \cdot X(f) \cdot e^{-j2\pi\tau_p f} \cdot e^{-j2\pi f_p t}$$

The code segment below, provides a typical WLAN channel that features delays, τ_p, of no more than +/- 300 nanoseconds and no Doppler as the terminals are not in motion. Remember that 802.11a/g modems feature a tone or subcarrier spacing of 312.5KHz, and that there are 26 positive and 26 negative frequency subcarriers that carry information. Even though the transmit waveform carries no information at DC, the frequency response is obviously defined there.

```
%% 1. Determine Channel Model with Four Paths
               %    Cp        Delay(sec)   Doppler
ChannelCondition = [  1,           0,         0;  ... % Path 0 (The strongest path)
                   -0.7j,       -20e-9,       0;  ... % Path 1 (The earliest path)
            0.3 - 0.7j,        80e-9,        0;  ... % Path 2
                     1j,       120e-9,       0;       % Path 3
                    0.4,       200e-9,       0;       % Path 4
            0.1 - 0.1j,       250e-9,       0];       % Path 5 (The Latest path)

%% 2. Compute the frequency response (See code from Section 7.1.3.1)
SubcarrierSpacing = 312.5e3; %KHz;
Subcarriers       = -26:1:26;    % Data carrying and the DC subcarrier
f                 = Subcarriers*SubcarrierSpacing;
Time              = 0;
FreqResponse      = ComputeFrequencyResponse(ChannelCondition(:,:), f, Time);
```

The real and imaginary portions of the frequency response are shown in the next figure. The thing to note here is that the frequency response is an complex valued waveform that is not changing terribly fast as we move along subcarriers. However, the random noise that would be part of the raw frequency response would have instantaneously changing values as we move from one subcarrier to the next. The astute reader will rightly assume that some type of filtering may be in order. The second plot takes the inverse Fourier transform yielding the impulse response of the

---

transversal filter that models the multipath condition and approximately provides us with the coefficients and delays of each path. I say approximately, as the waveform is of finite length and FFT spectrum leakage makes it impossible to read off the path coefficients exactly. However, it shows us that even for a channel with reasonable delays (for indoor conditions), there is a lot of empty space to the right and left of the peaks in the channel impulse response thus giving us a good idea of how much of a benefit filtering can provide.

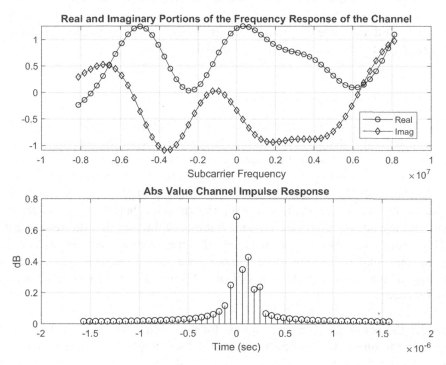

Figure 7-11: Frequency Response of Typical WLAN Channel

Filtering in the Frequency Domain

The idea of filtering the frequency response is a bizarre one, but simply takes some getting used to. We have a few choices when trying to filter the frequency response. We can take the IFFT of the frequency response, set those bins to zero where we know that only noise and no content can exist, and then take the FFT back to return to the filtered frequency response. This method has a drawback as this type of nulling will also removes valid content. This valid content is due to FFT leakage that appears and it can be seen in the figure above. The delays in the model are restricted to -20 to 250 nanoseconds, but we can clearly see content decaying away as we look at times outside that range. Simply nulling these bins and taking the FFT to get back to a cleaner frequency response introduces some distortion. Thus, a different method of noise reduction will be used in this chapter. Note that the time extent and time step in the impulse response above can be figured out via the Nyquist rules which indicate the following time range and step.

→ Time range: $\pm(1/SubcarrierSpacing)/2 = \pm(1/312.5KHz)/2 = \pm1.6e\text{-}6$ seconds
→ Time step: $(1/SubcarrierSpacing)/N = (1/312.5KHz)/53 = 60.4$ nanoseconds

Filtering in the Time Domain

Let's now take a look at the reference signal arrangement for the LTE downlink as shown in the last section. Notice how reference signal locations are staggered in the time domain. For antenna ports 0, reference signals are available once per slot for those subcarriers that actually feature reference signals. A slot period equals 0.5 milliseconds which translates to a sample rate of 2KHz. Thus, Doppler or frequency offset could be detected within the Nyquist range of -1KHz to +1KHz. However, realistically speaking Doppler frequencies of more than +/-300Hz are the exception as you can deduce from section 6.3.2. It should therefore be possible to reject noise in the frequency range between +/-300Hz and +/-1000Hz.

The Use of the Principle Component Analysis

When looking at the reference signal positions of the LTE resource grid, it should become clear to the reader that filtering the reference signals to reject noise in the traditional sense is a lot of work and a bit messy. To generate the cleaned-up frequency response of one subframe would require filtering across frequency, in the vertical direction of the resource gird, for each column containing reference signals as well as somehow filtering in the time domain, the horizontal direction, afterwards. To make matters worse, the distance between reference signals in the vertical direction may be six resource elements, but due to the fact that the resource grid does not show the DC subcarrier, there is a distance of seven subcarriers as we cross DC. This inconvenient spacing makes filtering more cumbersome. Using the principle component analysis allows us to use all reference signals in a slot or subframe all at once without having to worry about their positioning.

Step 1: Consider the set of all reference signals for a particular antenna port in the desired time span starting with the one at the earliest time and the most negative frequency (at the bottom left of the resource grid) and ending with the one at the latest time and most positive frequency (at the top right of the resource grid). For the LTE resource grid (1.4MHz bandwidth), we would have 12 reference signals at OFDM symbol indices l = 0, 3, 6, and 9 for a total of 48. Considering only antenna port 0 for now, we start our exercise by compiling the coordinates of the reference signals via two vectors:

→ Frequency vector: Contains the frequency position in Hz of each reference signal.
→ Time vector: Contains the time position in seconds of each reference signal.

Step 2: In our earlier example that used the principle component analysis in chapter one, we were presented with a large set of observations, where each observation consisted of two numbers, the current and the voltage. In our case, each observation is a vector of raw frequency response values (48 for the LTE example, 52 for WLAN, and 24·4 for C-V2X) at our reference signal coordinates for a particular channel condition. Also, this time we are not presented with a large set of observations, and we thus generate them ourselves for all reasonable channel conditions that we may encounter. Each wireless communication system will have a set of boundary conditions that specify the maximum and minimum reasonable time delays and Doppler values. The table below illustrates these sets of boundary conditions, which we will call a channel model.

Table 7-1: Channel Models for Different OFDM Communication Systems

	Min Delay (sec)	Max Delay (sec)	Min Doppler (Hz)	Max Doppler (Hz)
LTE	-1e-6	+4e-6	-300	+300
WLAN	-50e-9	+250e-9	0	0
C-V2X	-400e-9	+2e-6	-750	750

Our first instinct would be to generate a large number, N, of channel conditions each with some number of paths that would feature scaling factors C_p, delays t_p, and Dopplers f_p with values randomly assigned within the range given in the last table. We would end up with N observation vectors featuring raw frequency response values for each reference signal location we have available. Interestingly enough, this turn out to be unnecessary. It is sufficient to pick a large number of channel conditions, where each channel condition consists of a single path, which features a complex scalar C_p with unit magnitude and a random phase, as well as time delay and Doppler values that evenly span the boundary conditions set out by our channel model. The following code defines the reference signal coordinates and channel model for the 1.4MHz LTE downlink signal, the C-V2X PSCCH channel and the WLAN signal.

```
%% Step 1. Set the reference signal coordinates and channel model given the standard
Standard = 'WLAN';    % LTE / V2X / WLAN
if     (strcmp(Standard, 'LTE') == 1)
    SubcarrierSpacing = 15e3;
    OfdmSymbolPeriod  = 0.001/12; % Extended Cyclic Prefix (12 symbols per subframe)
    Frequencies  = [-36:6:-1,  1:6:36, ...        % Symbol l = 0
                    -33:6:-1,  4:6:36, ...        % Symbol l = 3
                    -36:6:-1,  1:6:36, ...        % Symbol l = 6
                    -33:6:-1,  4:6:36] * SubcarrierSpacing;       % Symbol l = 9
    Times  = [zeros(1,12), 3*ones(1,12), 6*ones(1,12), 9*ones(1,12)]*OfdmSymbolPeriod;
    DelayRange   = [-1e-6, 4e-6];                 % Channel Model
    DopplerRange = [-300 300];                    % Channel Model
elseif (strcmp(Standard, 'V2X') == 1)
    SubcarrierSpacing = 15e3;
    OfdmSymbolPeriod  = 0.001/14; % Normal Cyclic Prefix (14 symbols per subframe)
    Frequencies = [0:23, ...                      % Symbol l = 2
                   0:23, ...                      % Symbol l = 5
                   0:23, ...                      % Symbol l = 8
                   0:23]* SubcarrierSpacing;      % Symbol l = 11
    Times = [2*ones(1,24), 5*ones(1,24), 8*ones(1,24), 11*ones(1,24)]*OfdmSymbolPeriod;
    DelayRange   = [-0.4e-6, 2e-6];               % Channel Model
    DopplerRange = [-750 750];                    % Channel Model
elseif (strcmp(Standard, 'WLAN') == 1)
    SubcarrierSpacing = 312.5e3; % Hz
    OfdmSymbolPeriod  = 4e-6;    % seconds
    Frequencies = [-26:1:-1,  1:1:26]* SubcarrierSpacing;
    Times       = zeros(1,52)*OfdmSymbolPeriod;
    DelayRange   = [-50e-9 250e-9];               % Channel Model
    DopplerRange = [10 10];                       % Channel Model
else
    assert(false, 'The selection is unsupported.');
end
```

The following code generates a large number of possible frequency responses given the channel model of the desired standard. Each frequency response features $K = NumCoordinates$ values (52 for WLAN), and we say that each frequency response is a single point in K^{th} dimensional space.

```
%% Step 2. Compute the Frequency response for a large set of channel conditions
DelayStep         = (DelayRange(1,2)   - DelayRange(1,1))   / 100;
DopplerStep       = (DopplerRange(1,2) - DopplerRange(1,1)) / 50;
DelayVector       = DelayRange(1,1):DelayStep:DelayRange(1,2);
DopplerVector     = DopplerRange(1,1):DopplerStep:DopplerRange(1,2);
NumChannels       = length(DelayVector)*length(DopplerVector);
NumCoordinates    = length(Frequencies);    % The number reference signals

FreqResponseMatrix  = zeros(NumChannels, NumCoordinates);
NumChannelCondition = 0;
for A = 1:length(DelayVector)
    Delay = DelayVector(1,A);               % Grab the delay from source vector
    for B = 1:length(DopplerVector)
        NumChannelCondition = NumChannelCondition + 1;
        Doppler = DopplerVector(1,B);       % Grab the doppler from the source vector
        % The time delay will cause the frequency response to rotate over frequency.
        % The doppler will cause the frequency response to rotate over time.
        % These rotations need to start at a random angle in order for the
        % NumCoordinates dimensional space to be covered properly. If we don't do this,
        % the eigenvalues and eigenvectors produced by the PCA will not work properly.
        StartAngle         = rand(1,1)*2*pi;
        for Position = 1:NumCoordinates
            Frequency      = Frequencies(1, Position);
            Time           = Times(1, Position);
            % The Complex scalar of the path
            C_Scalar       = exp(1j*StartAngle);
            % Impact on the frequency response at this ref signal due to time delay.
            AngleDueToDelay   = -2*pi*Frequency * Delay;
            C_Delay           = exp(1j*AngleDueToDelay);
            % Impact on the frequency response at this reference signal due to Doppler.
            AngleDueToDoppler = 2*pi*Doppler* Time;
            C_Doppler         = exp(1j*AngleDueToDoppler);
            % The final frequency response at the current reference signal coordinate.
            C_FreqResponse    = C_Scalar * C_Delay * C_Doppler;
            % Hold the total frequency response of each channel condition in the rows of
            % the matrix below.
            FreqResponseMatrix(NumChannelCondition, Position) = C_FreqResponse;
        end
    end
end
```

Step 3: Form the Covariance matrix and compute its eigenvalue decomposition. The eigenvectors are the principle components. Just as we saw in our treatment of the principle component analysis in chapter 1, the data extends along certain principle components (eigenvectors) more than along others. It is the eigenvalues that indicate to which degree the data (the set of frequency responses computed in step 2) extends along each principle component. The eigenvectors are stored in the columns of the matrix V, and the nth eigenvector is associated with the nth eigenvalue. The next figure shows a plot of the eigenvalues, which indicates that the data only extends along six of the 52 eigenvectors (WLAN scenario). It is because the individual values of the frequency response are correlated (they change slowly as you move across frequency) that the Kth dimensional space is not completely occupied. However, the noise component of the raw frequency response changes randomly thus extending evenly along each principle component (eigenvector).

```
%% 2. Compute the Covariance Matrix and the Eigen value decomposition.
CovarianceM = cov(conj(FreqResponseMatrix)); % NumCoordinates by NumCoordinates matrix
[V, D]      = eig(CovarianceM);              % The eigenvalue decomposition
% Grab the Eigenvectors, which are in the columns of V
EigenVectors = V;
% Grab the Eigenvalues, which are in the diagonal of D
EigenValues = diag(D);
figure(1);
stem(EigenValues, 'k'); grid on;
title('EigenValues - Data Extent Along each Principle Component');
```

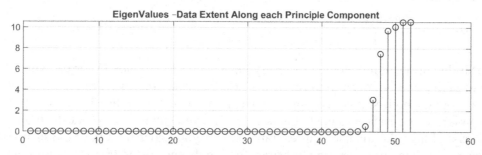

Figure 7-12: The Set of Eigenvalues

Note that each one of the K values in the raw frequency response scales a basis vector, where the set of basis vectors are orthogonal. The set of original basis vectors are the columns in the matrix shown below. The eigenvectors, or principle components (the column in matrix V), are simply a different set of orthogonal basis vectors along which the data may be expressed more conveniently. See the example in chapter one.

$$OrginalBasisVectors = \begin{bmatrix} 1 & 0 & \cdots & 0 \\ 0 & 1 & \cdots & 0 \\ \vdots & \vdots & \ddots & \vdots \\ 0 & 0 & \cdots & 1 \end{bmatrix} \quad NewBasisVectors = columns\ of\ V$$

Step 4: At this point, we need to process an RX waveform compliant with the standard that we chose and generate the raw frequency response. In this WLAN example, we use the same channel shown earlier and corrupt the resulting frequency response with enough white Gaussian noise to produce an *RS-CINR* = 5dB (reference signal carrier and interference to noise ratio).

```
%% Step 4. Generate a typical frequency response and force a particular CINR.
                % Cp        Delay(sec)   Doppler
ChannelCondition = [   1,         0,        0; ... % Path 0 (The strongest path)
                    -0.7j,    -20e-9,       0; ... % Path 1 (The earliest path)
                0.3 - 0.7j,    80e-9,       0; ... % Path 2
                       1j,    120e-9,       0;     % Path 3
                      0.4,    200e-9,       0;     % Path 4
                0.1 - 0.1j,   250e-9,       0];    % Path 5 (The Latest path)

%   Compute the frequency response (See code from Section 7.1.3.2)
FreqResponse    = ComputeFrequencyResponse(ChannelCondition(:,:), Frequencies, Times);

%   Force a particular reference signal CINR (carrier to interference and noise ratio)
FreqResponsePower = MeanSquare(FreqResponse);
RsCinrdB        = 5;                      % dB
RsCinrLinear    = 10^(RsCinrdB/10);
```

```
NoisePower          = FreqResponsePower/RsCinrLinear;
GaussianNoise       = 0.7071*sqrt(NoisePower)*(randn(1, length(FreqResponse)) + ...
                                         + 1j*randn(1, length(FreqResponse)) );
RawFreqResponse     = (FreqResponse + GaussianNoise).';  %% Convert to column vector
```

Step 5: Whereas the noise power in the frequency responses is spread across all principle components, the signal power is spread across only six, and we execute a type of process that is akin to filtering. Our task is to determine the extent to which this particular raw frequency response extends along the new basis vectors, the principle components. We do this by projecting the raw frequency response vector against each one of the principle components to produce a norm, or vector length.

$$Norm_k = \frac{dot(RawFrequencyResponse, Eigenvector_k)}{dot(Eigenvector_k, Eigenvector_k)} \ where\ k = 0, 1, ...K - 1$$

We repeat this projection operation for every eigenvector to produce K norms. Note, this process is akin to taking the Fourier transform of the raw frequency response. Some of lower frequency bins will contain both signal and noise, whereas most of the higher frequency bins will only contain noise. To go back into the frequency domain, you can now zero out the higher frequency bin values and take the inverse Fourier transform. If you understand the Fourier transform, you will also realize that you can simply retransform only the lower frequency bins into the time domain and simply ignore those tones associated with the higher frequency bins. This is exactly what we will do here. In order to transform back to the original basis vector set, we execute the following simple expression.

$$Waveform = \sum_{k=Start}^{Stop} Norm_k \cdot EigenVector_k$$

Computing the Channel Estimate (The Improved Accuracy Frequency Response)

→ If we set *Start* = 0 and *Stop* = K-1 (K = 52 for WLAN), then the generated waveform is simply the original raw frequency response, and we have gained nothing via the exercise.

$$RawFrequencyResonse = \sum_{k=0}^{K-1} Norm_k \cdot EigenVector_k$$

→ If we set *Start* = K-L and *Stop* = K-1, then the generated waveform represents the channel estimation we have been after. L is the number of significantly sized eigenvalues, which was equal to 6 for this WLAN example.

$$ChannelEstimate = \sum_{k=K-L}^{K-1} Norm_k \cdot EigenVector_k$$

→ If we set *Start* = 0 and *Stop* = K-L-1, then the generated waveform represents noise only. We can now compute the variance of that noise and divide by K-L yielding the mean noise power along each principle component. Simply multiply by K produces the total estimated noise power in the raw frequency response.

$$NoisePower = \frac{K}{K-L} \cdot var(\sum_{k=0}^{K-L-1} Norm_k \cdot EigenVector_k)$$

The signal power in the channel estimate is equal to the variance of the channel estimate minus the amount of noise power residing along the last N principle components.

$$SignalPower = MeanSquare(ChannelEstimate) - \frac{L}{K} \cdot NoisePower$$

The reduction in noise power residing in the frequency response and thus the SNR benefit can now be calculated as follows.

$$ChangeInNoisePower = 10 \cdot log_{10}\left(\frac{L}{K}\right) = 10 \cdot log_{10}\left(\frac{6}{52}\right) = -9.4dB$$

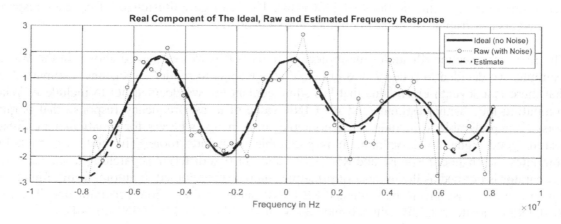

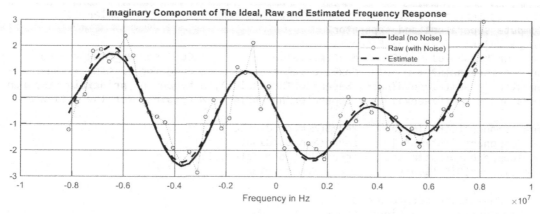

Figure 7-13: The Channel Estimate Versus the Ideal and Raw Frequency Responses

The figure above illustrates just how much closer the channel estimate is to the ideal frequency response compared to the raw frequency response. For this example, the gain in signal to noise ratio was around 9.4dB, but there are situations where we can restrict the delay and Doppler ranges even further and achieve more than 15dB of improvement.

Computing the RS-CINR or Reference Signal – Carrier to Interference and Noise Ratio

Cellular systems are not just corrupted by noise but by interference from neighboring cells as well. The LTE and 5G standard go through great pains to ensure that this interference from signals of neighboring cells appears noise-like to your cell phone, and the algorithm developed above will not be able to tell the difference. The term carrier to interference and noise ratio stems from this fact that interference and noise add. WLAN is different in that usually only a single transmit signal is on-the-air at a time and interference is absent. Measuring RS-CINR is important in cellular system as the cell phone must constantly measure the signal quality of several nearby cell towers in order to connect to the one with the best reception in case that the user starts a call.

$$RsCinrdB = 10 \cdot log_{10}(\frac{SignalPower}{NoisePower})$$

Thus computing a cleaned up version of the frequency response features the added benefit of being able to easily compute the RS-CINR value. The last figure illustrated a frequency response that had been corrupted by noise to produce a RS-CINR of 5 dB.

The final code section illustrates the projection process that was explained above. In the WLAN example, we decided to retain $L = 6$ principle components to render the channel estimate. Note that there is a seventh eigenvalue that has some content but we decided not to include it. When computing the channel estimate and RS-CINR value of a raw frequency response that is quite noisy, it is preferable to include fewer principle components, whereas for situations that already feature good signal to noise ratio, it is preferable to include more. The code section below computes the channel response and RS-CINR value for two different separators, L values, where a separator is defined as the number of principle components retained. It automatically figures out which results to use given the estimated RS-CINR value resulting from separator one. The final RS-CINR estimate is 4.7dB, which is incredibly close to the actual RS-CINR of 5dB.

```
%% Step 5. Generate a typical frequency response and force a particular CINR
% Compute seperator1 and seperator2
MaxEigenValue       = max(EigenValues);
SeperatorThreshold1 = 0.02  * MaxEigenValue;          % Good for RS_CINR < 20dB
SeperatorThreshold2 = 0.001 * MaxEigenValue;          % Good for RS-CINR > 15dB
Seperator1 = sum(EigenValues > SeperatorThreshold1); % Number of principle components
Seperator2 = sum(EigenValues > SeperatorThreshold2); % retained for the estimate

SignalComponentOfEstimate1  = zeros(NumCoordinates, 1);
NoiseComponentOfEstimate1   = zeros(NumCoordinates, 1);
SignalComponentOfEstimate2  = zeros(NumCoordinates, 1);
NoiseComponentOfEstimate2   = zeros(NumCoordinates, 1);

for EigIndex = 1:NumCoordinates
    EigenVector     = EigenVectors(:, EigIndex);
    Norm            = sum(RawFreqResponse .* conj(EigenVector)) / ...
                      (sum(conj(EigenVector).*EigenVector));
    if(EigIndex > NumCoordinates - Seperator1)
    % Build the signal component of the frequency response. Note that the eigenvectors
    % belonging to the signal components will have some noise as well.
        SignalComponentOfEstimate1 = SignalComponentOfEstimate1 + Norm * EigenVector;
    else
    % Compute the noise belonging to those eigenvectors that do not belong to the signal
    % component of the noisy raw frequency response. We use this to compute the RS-CINR.
```

```
                 NoiseComponentOfEstimate1  = NoiseComponentOfEstimate1  + Norm * EigenVector;
        end

        if(EigIndex >  NumCoordinates - Seperator2)
            SignalComponentOfEstimate2 = SignalComponentOfEstimate2 + Norm * EigenVector;
        else
            NoiseComponentOfEstimate2  = NoiseComponentOfEstimate2  + Norm * EigenVector;
        end
end

% Compute RS-CINR for Seperator1
TotalNoisePower1  = MeanSquare(NoiseComponentOfEstimate1)  ...
                    * NumCoordinates / (NumCoordinates - Seperator1);
TotalSignalPower1 = MeanSquare(SignalComponentOfEstimate1) ...
                    - TotalNoisePower1 * Seperator2/NumCoordinates;
RsCinrdB1         = 10*log10(TotalSignalPower1/TotalNoisePower1);

% Compute RS-CINR for Seperator2
TotalNoisePower2  = MeanSquare(NoiseComponentOfEstimate2)  ...
                    * NumCoordinates / (NumCoordinates - Seperator2);
TotalSignalPower2 = MeanSquare(SignalComponentOfEstimate1) ...
                    - TotalNoisePower1 * Seperator2/NumCoordinates;
RsCinrdB2         = 10*log10(TotalSignalPower2/TotalNoisePower2);

% Only choose the results of seperator2 only if its RS-CINR results is truely better
if(real(RsCinrdB2) > (real(RsCinrdB1) + 1))
   RsCinrdB        = real(RsCinrdB2);    % Results for seperator1
   ChannelEstimate = SignalComponentOfEstimate2;
else
   RsCinrdB        = real(RsCinrdB1);    % Results for seperator2
   ChannelEstimate = SignalComponentOfEstimate1;
end
```

This method of channel and RS-CINR estimation is quite nice compared to using filtering as much of the data can be precomputed. Only the projection process has to be repeated every time an estimate is desired, making this method computationally very efficient.

7.2 The OFDM 802.11a Transmitter

7.2.1 Packet Structure and Synchronization

Many introductory communication courses will not present much beyond the basic modulation/demodulation process of OFDM. And, if those basics were all there was to OFDM communication, then most undergraduates with a basic knowledge of the Fourier transform and some digital circuit know-how could build a modem. Alas, this is not the case. By now, however, we all know that the digital demodulator must solve a host of problems that have nothing whatsoever to do with the core IDFT/DFT process. Modern modems must deal with the offsets, distortions, noises, and imbalances that we examined in chapter 5 prior to presenting the demapper with complex symbols that can be turned into received bits. The process of altering the received signal so that the detector can make optimal decisions as to the transmitted values is called synchronization. The physical layer description of the WLAN 802.11a standard [1] describes a packet structure that yields a lot of interesting information on how the digital demodulator should go about the process of synchronization.

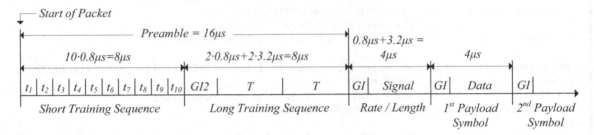

Figure 7-14: WLAN 802.11a Packet Structure

Short Training Sequence

The first part of the preamble is the short training sequence, which extends over an eight-microsecond period of time and is used by the receiver to facilitate the following synchronization steps.

→ AGC – Automatic gain control
→ Packet detection
→ Coarse frequency correction

While the receiver is listening for packet traffic on a vacant channel, it generally sets the analog RX gain to maximum amplification. In the absence of any traffic, the only signal that fills the input dynamic range of the ADC modules is amplified thermal noise due to the antenna resistance and noise figure of the receiver. As soon as an on-channel packet arrives at the antenna, the maximum gain setting will cause the signal at the output of the analog receiver portion to completely overwhelm the ADC inputs of the baseband processor. The AGC module in the baseband processor now reduces the analog gain in the receiver until the signal fits nicely into the dynamic input range of its ADC. This closed loop process needs to finish well before the eight-microsecond limit. Once the signal is beginning to stabilize at the ADC inputs, the frequency detector will analyze the remaining portion of the short training sequence to determine the validity of the 802.11a packet (packet detection), as well as the coarse frequency offset between the transmit signal and the receiver LO. The short training sequence features a structure that is specifically crafted to facilitate the detection and subsequent correction of the frequency offset.

Long Training Sequence

The long training sequence features a guard interval, GI2, followed by two long training symbols. This second part of the preamble facilitates the following synchronization processes.

→ Fine frequency correction
→ FFT timing reference acquisition
→ Channel estimation

Based on the result of the coarse frequency detector, a complex NCO (numerically controlled oscillator) is programmed to frequency shift the incoming signal toward zero Hz. This step decreases the offset to within the detection range of the fine frequency offset estimator. Once the fine offset is calculated, the NCO is reprogrammed to the optimized frequency. Although the final remaining offset does not have to be zero, it must be small enough such for its time varying phase shift to be essentially constant over a four microsecond OFDM symbol.

The FFT timing reference can best be understood by taking another look at Figure 6-7 describing the basic modulation/demodulation process. It shows that the received sample stream fills a shift register prior to the FFT execution. Well, just when is this shift register filled with the proper information? The long training symbols provide a beacon that allows us to determine this timing reference, which indicates the start of the FFT calculation.

Channel estimation is the process of modeling the multipath channel such that we may program the equalizer – following the FFT – to remove the associated linear distortion. We will show how the equalizer is programmed later on in this chapter.

Ongoing Synchronization during the Payload

The early payload symbols enjoy the benefits of proper synchronization provided by the tasks mentioned above. However, synchronization must be maintained over the entire packet which requires the following additional tracking operations.

→ Phase drift correction
→ Timing drift correction

Phase drift occurs due to phase noise and the small frequency offset remaining after the fine frequency acquisition process. Timing drift is due to the unavoidable difference in the master clock frequencies at the transmitter and receiver. Thus, while our transmitter and its associated DAC modules are producing output signals at a certain clock rate, the receiver ADC modules work off a rate that is not exactly the same. The receiver will use a priori known training information riding on the pilot tones of each OFDM symbol to correct both types of drift.

A Brief Summary of the Transmit Specifications

The WLAN 802.11a standard calls for the use of 52 total orthogonal tones or subcarriers of which four carry a priori known pilot information used for symbol by symbol tracking of phase and timing variation. The 52 tones, which are symmetrically spread out between positive (tones 1 through 26) and negative (tones -1 through -26) frequencies, contain pilots at subcarriers -21, -7, 7, and 21, as illustrated in the figure below. The specification requires that all tones be spaced at intervals of 312.5KHz and produce an IFFT output length of 3.2 microseconds. To this end, it suggests the use of a 64-point IFFT outputting samples at 20MSPS. The additional 12 inputs available for the IFFT calculation are to be set to zero, including tones 0, 27 through 31, and -27 through -32.

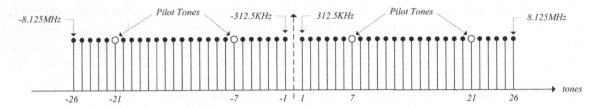

Figure 7-15: Distribution of Information Carrying and Pilot Tones

The standard provides for payload data rates of 6, 9, 12, 18, 24, 46, 48, and 52MBPS and uses BPSK, QPSK, 16QAM, and 64QAM modulation to that end.

Data Rate (MBits/sec)	Modulation	Coding Rate	Coded Bits per Subcarrier (N_{CBSC})	Coded Bits per OFDM Symbol (N_{CBPS})	Data Bits per OFDM Symbol (N_{DBPS})
6	BPSK	½	1	48	24
9	BPSK	¾	1	48	36
12	QPSK	½	2	96	48
18	QPSK	¾	2	96	72
24	16QAM	½	4	192	96
36	16QAM	¾	4	192	144
48	64QAM	⅔	6	288	192
54	64QAM	¾	6	288	216

Figure 7-16: Rate Configuration for the 802.11a Standard

The figure above presents an overview of the different available data rates and their corresponding modulation formats, coding rates and bit configurations. Note that convolutional encoding is used as a means of forward error correction which increases the effective rate at which bits must be transmitted. You can easily compute the encoded data rate by dividing the coded bits per OFDM symbol by the symbol length of 4 microseconds.

7.2.2   An Overview of the Transmit Architecture

We have earlier seen the modem's packet construction, which dictates that a 16 microsecond header section precedes the payload data. This payload data is formed by consecutive four-microsecond OFDM symbols which are composed of a 0.8-microsecond guard interval and an IFFT output sequence of 3.2-microseconds length. Parts of the header sequence, like the long training symbols can by formed via a real-time IFFT operation, but since the header never changes, it is more convenient to calculate it a priori and store the complex samples in a read only memory module, as shown below. Thus, when a new packet needs to be transmitted, the multiplexer, shown before the upsampler, first passes the header ROM content before switching over to yield access to the payload through the bottom path.

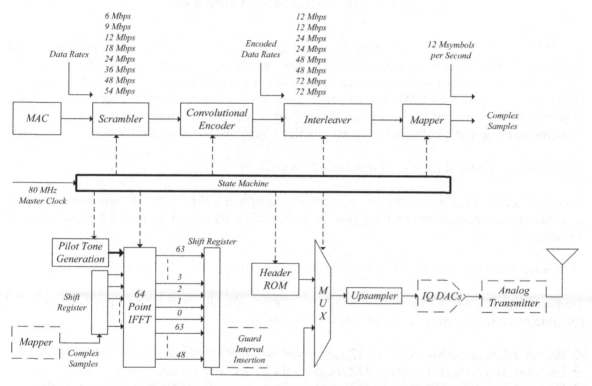

Figure 7-17: Overview of 802.11a Transmitter

The modem chain that generates the payload data is subdivided into three portions: the bit manipulation portion at the top of the figure, the complex value manipulation portion at the bottom, and the state machine in between. The bit manipulating portion consists of the scrambler, convolutional encoder, interleaver, and mapper input. The mapper assembles groups of bits and maps them into one of the reference positions belonging to our current modulation format (BPSK, QPSK, 16QAM, or 64QAM). At this point, the information enters the complex value manipulation portion of the modem, including the IFFT and the upsampling filter. The final portion of the modem is the state machine, which acts like the conductor of an orchestra telling each block in the system when to process and when not to.

Modem Rate Management

You may have noticed the large number of information rates that have to be supported at different positions in the transmitter. To understand the timing structure of the modem we start by considering the output of the IFFT block, which is stitched into the OFDM symbols as follows.

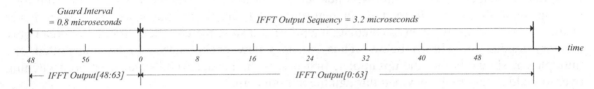

Figure 7-18: OFDM Symbol Structure

The IFFT takes in 64 numbers representing the orientation of 64 complex sinusoids, each oscillating at a different orthogonal frequency. The summation of all these sinusoids results in an IFFT output sequence of 64 values representing complex samples in time. These 64 samples are placed at the end of the OFDM symbol, whereas the last 16 samples of that same IFFT output sequence are appended at the beginning of the symbol. The result is a set of 80 samples within a 4-microsecond period yielding a rate of 80 samples / 4-microseconds or 20MSPS.

→ Sample rate (IFFT shift register output) = 20MSPS

Therefore, every 4 microseconds, the mapper must supply the IFFT with 48 information bearing symbols, which are combined with 4 pilot symbols and 12 zeros to form an IFFT input vector of 64 entries.

→ Symbol rate (mapper output) = 48 symbols / 4 microseconds = 12 Mega symbols/second.

Transmission of a 64QAM symbol requires 6 encoded data bits, which translates to an encoded data rate of 12 mega symbols/second times 6 bits/symbol.

→ Encoded data rates (64QAM) = (12 mega symbols / second) · 6 bits/symbol = 72MBPS
→ Encoded data rates (16QAM) = (12 mega symbols / second) · 4 bits/symbol = 48MBPS
→ Encoded data rates (QPSK) = (12 mega symbols / second) · 2 bits/symbol = 24MBPS
→ Encoded data rates (BPSK) = (12 mega symbols / second) · 1 bit/symbol = 12MBPS

When we then consider the different convolutional encoding rates of ½, ⅔, and ¾, we can easily backtrack to the original data rates of 6, 9, 12, 18, 24, 36, 48, and 54MBPS.

How does the state machine manage all these rates, given a single master clock of 80MHz?

As mentioned above, we must feed the IFFT shift register 48 complex symbols every four microseconds, which, at an 80MHz clock rate, corresponds to 320 clock cycles. Therefore, the mapper is allotted no more than 320/48 = 6.67 clock cycles to produce one complex symbol. We consequently allot six clock cycles per mapping conversion.

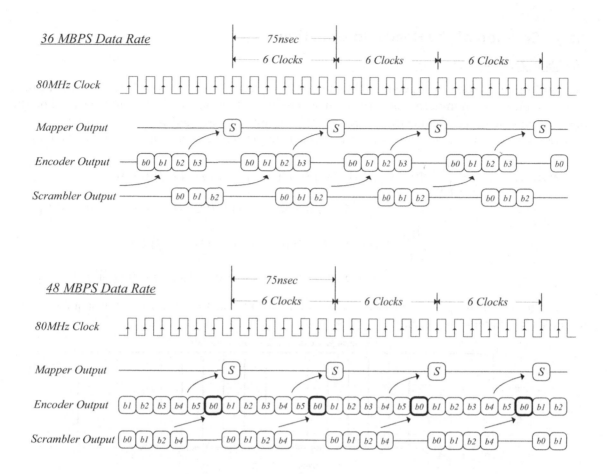

Figure 7-19: Timing Diagram for the 36MBPS and 48MBPS Scenarios

The timing diagrams above show the information rates at the mapper, encoder, and scrambler outputs for the 36 and 48 MBPS scenarios. Note that no matter which of the eight possible data rates we select, the mapper will always produce one symbol every six clock cycles. During these six clock cycles the encoder will provide the mapper with 1, 2, 4, or 6 encoded bits depending on the modulation format of BPSK, QPSK, 16QAM, or 64QAM. In the 36 and 48 MBPS cases, the coding rate turns out to be ¾ and ⅔, respectively, which yields convenient sets of three and four bits (within six clocks) at the scrambler output. The state machine controls the computation of each module by enabling or disabling its clocked logic elements. Depending on the rate, the scrambler combination may not always produce the same number of bits during successive six clock periods. It really does not matter. The state machine will enable the MAC output and scrambler often enough for the convolutional encoder to produce 1, 2, 4, or 6 encoded output bits within the six clock period. Note that it take 48·6 = 288 clock cycles for the mapper to fill the IFFT shift register. During the 32 clocks that remain in the four-microsecond interval, the state machine will disable the various blocks and the bit manipulation portion becomes idle.

7.2.3 Creation of the Header Information

The Short Training Sequence

The 802.11a specification dictates that the short training symbol should be derived from a combination of 12 non-zero subcarriers (of a possible 64 provided for by our IFFT). For the positive and negative frequency indices of $m = 0$ through 31 and $m = -32$ through -1, the orientation of the subcarriers are as follows. (The actual frequencies are equal to $m \cdot 312.5 KHz$)

$$Positive_{m=0,1...31} = \sqrt{\frac{13}{6}} \cdot [0\ 0\ 0\ 0\ (-1-j)\ 0\ 0\ 0\ (-1-j)\ 0\ 0\ 0\ (1+j)\ 0\ 0\ 0$$

$$(1+j)\ 0\ 0\ 0\ (1+j)\ 0\ 0\ 0\ (1+j)\ 0\ 0\ 0\ \ \ 0\ 0\ 0\ 0]$$

$$Negative_{m=-32,-31...-1} = \sqrt{\frac{13}{6}} \cdot [0\ 0\ 0\ 0\ \ \ 0\ 0\ 0\ 0\ \ (1+j)\ 0\ 0\ 0\ (-1-j)\ 0\ 0\ 0$$

$$(1+j)\ 0\ 0\ 0\ (-1-j)\ 0\ 0\ 0\ (-1-j)\ 0\ 0\ 0\ \ (1+j)\ 0\ 0\ 0]$$

The scaling factor of $sqrt(13/6)$ ahead of the vectors normalizes the variance of the sequence to be in line with the rest of the packet.

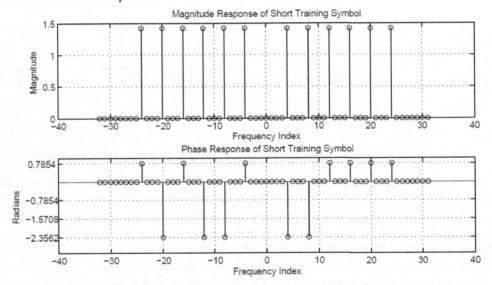

Figure 7-20: Frequency Response for the Short Training Symbol as Specified in the Standard

We now concatenate the frequency orientations above into a vector *Total* = [*Negative Positive*], and compute the short training sequence as follows.

$$ShortTraining(t) = \sum_{m=-32}^{31} Total_m \cdot e^{j2\pi t\left(\frac{m}{64}\right)} \quad for\ t = 0, 1, ...159$$

The equation above is in its normalized state, where the normalized frequency $m/64$ corresponds to an actual frequency of 20MHz $\cdot m/64$. Furthermore, the normalized time variable t refers to an actual time of $t/20$MHz. The numbers $Total_{-32}$ and $Total_{31}$ represent the first and last entries in the vector and correspond to frequencies at $-32 \cdot 312.5KHz = -10$MHz and $31 \cdot 312.5KHz = 9.6875$MHz.

```
function ShortTrainingSequence = Get_ShortTrainingSequence(Step)
                        %% Step = 1 (for 20MHz) or 1/2 (for 40MHz)


Positive = [0 0 0 0  -1-j 0 0 0  -1-j 0 0 0   1+j 0 0 0  ...
            1+j 0 0 0   1+j 0 0 0   1+j 0 0 0    0 0 0 0];
Negative = [0 0 0 0    0 0 0 0   1+j 0 0 0  -1-j 0 0 0  ...
            1+j 0 0 0  -1-j 0 0 0  -1-j 0 0 0   1+j 0 0 0];
Total    = sqrt(13/6)*[Negative Positive];
m        = -32:31;


N                     = 64;                      %% Number of frequency tones
ShortTrainingSequence = zeros(1, 160/Step);
for n = 0:length(ShortTrainingSequence)-1   %% The IDFT operation
    t                         = n*Step;
    E                         = exp(j*2*pi*t*m/N);
    ShortTrainingSequence(1,n+1) = Total*E.';
end
```

Notice that we evaluated the equation for $t = 0, 1 \ldots 159$, which yields the correct short training sequence for a sample period of 1/20MHz. To find out what the short training sequence is for a rate of 1/40MHz, we simply evaluate the equation for $t = 0, \frac{1}{2}, 1 \ldots 159.5$. The code above lets us switch between these two rates (the most likely to be used in the 802.11a modem) by selecting a step size of 1 or 0.5.

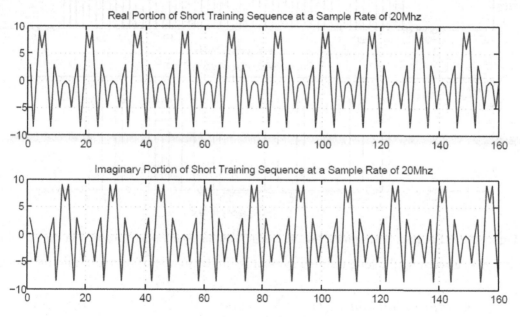

Figure 7-21: Short Training Sequence at 20MHz Sample Rate

The short training sequence was crafted in this fashion to yield a waveform whose periodicity could be exploited by a frequency detection algorithm. We will see this algorithm in later sections that discuss the digital demodulator.

The Long Training Sequence

The 802.11a specification dictates that the long training symbol should be derived from a combination of 52 non-zero subcarriers (of 64 possible tones provided in our IFFT).

$$Positive_{m=0,1...31} = \quad [0 \ 1 \ -1 \ -1 \ \ 1 \ 1 \ -1 \ 1 \ \ -1 \ 1 \ -1 \ -1 \ \ -1 \ -1 \ -1 \ 1$$
$$1 \ -1 \ -1 \ 1 \ \ -1 \ 1 \ -1 \ 1 \ \ 1 \ 1 \ 1 \ 0 \ \ \ 0 \ 0 \ 0 \ 0]$$

$$Negative_{m=-32,-31...-1} = [0 \ 0 \ 0 \ 0 \ \ 0 \ 0 \ \ 1 \ 1 \ \ -1 \ -1 \ 1 \ 1 \ \ -1 \ 1 \ -1 \ 1$$
$$1 \ \ 1 \ 1 \ \ 1 \ \ 1 \ -1 \ -1 \ 1 \ \ 1 \ -1 \ 1 \ -1 \ \ 1 \ 1 \ 1 \ 1]$$

Once again, we create a vector called *Total*, which concatenates the negative and positive subcarrier arrangement as *Total* = [*Negative Positive*]. The figure below illustrates the orientation of the different subcarriers required for the creation of the long training symbol.

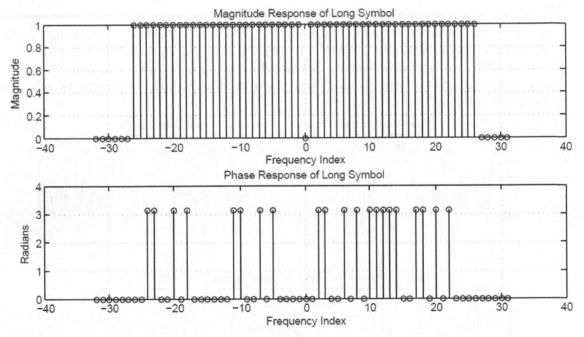

Figure 7-22: Frequency Response for the Long Training Symbol as Specified in the Standard

The equation governing the long training symbol is similar to that of its short training counterpart.

$$LongTrainingSymbol(t) = \sum_{m=-32}^{31} Total_m \cdot e^{j2\pi t\left(\frac{m}{64}\right)} \quad for \ t = 0, 1, ...63$$

To form the long training sequence, two long training symbols of 3.2 microseconds each are concatenated and preceded by a guard interval of 1.6 microseconds. This guard interval is identical to the second half of a regular long training symbol. Once again, the function below allows us to generate a long training symbol appropriate for a sample rate of 20MHz (where $t = 0$, 1 ... 63) and 40 MHz ($t = 0$, ½, 1 ... 159.5) by choosing a step size of 1 and ½, respectively.

```
function [LongTrainingSequence AllTones] = Get_LongTrainingSequence(Step)
                        %% Step = 1 (for 20MHz) or 1/2 (for 40MHz)
Positive = [0  1 -1 -1   1  1 -1  1   -1  1 -1 -1   -1 -1 -1  1 ...
            1 -1 -1  1  -1  1 -1  1    1  1  1  0    0  0  0  0];
Negative = [0  0  0  0   0  0  1  1   -1 -1  1  1   -1  1 -1  1 ...
            1  1  1  1   1 -1 -1  1    1 -1  1 -1    1  1  1  1];
AllTones = [Negative Positive];
m        = -32:31;  %% Frequency index (actual frequency = m * 3125.KHz)
N        = 64;          %% Number of frequency tones
LongTrainingSymbol = zeros(1, 64/Step);


for n = 0:length(LongTrainingSymbol)-1    %% The IDFT computation
    t = n*Step;
    E = exp(j*2*pi*t*m/N);
    LongTrainingSymbol(1,n+1) = AllTones*E.';
end


if(Step == 1)
    LongTrainingSequence = [LongTrainingSymbol(1, 33:64)  ...
                            LongTrainingSymbol LongTrainingSymbol];
else
    LongTrainingSequence = [LongTrainingSymbol(1, 65:128)  ...
                            LongTrainingSymbol LongTrainingSymbol];
end
```

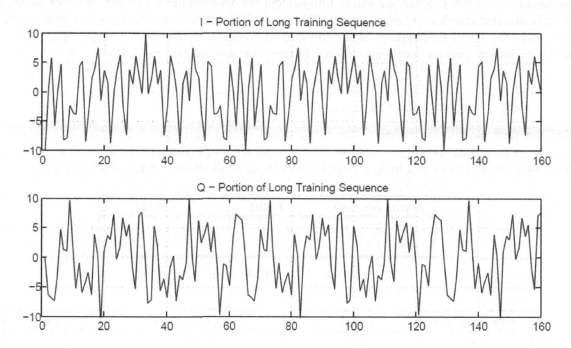

Figure 7-23: Long Training Sequence at 20MHz Sample Rate

As a curious reader, you may ask yourself why the standards committee decided to go with this particular combination of 1 and -1 values for the orientation of the 52 tones. Why not just make all tones equal to 1? This combination, likely one of many, gives the long training sequence a well-behaved peak to average ratio. A waveform featuring a low peak to average ratio will experience less nonlinear distortion when passing through the analog transmit chain. Modify the code above by changing all orientations to unity and then plot the revised output sequence. You will see huge peak excursions that would make life very hard for your RF power amplifier.

7.2.4 The Bit Manipulation Chain: MAC, Scrambler, and FEC

We start off with a serial bit stream that is provided by the media access controller (MAC), which is really nothing more than a microprocessor. The MAC accepts data from the host processor, usually in a PC or laptop, and serially transmits that data to the OFDM baseband processor (digital modem) at rates of 6, 9, 12, 18, 24, 36, 48, and 54 MBPS. The bit stream is then scrambled for security purposes. The convolutional encoder augments the data stream with information that allows the receiver to correct errors that occur. The encoding rates that are available are ½ (one input/two output bits), 2/3 (two input/three output bits) and ¾ (three input/four output bits). The interleaver takes the sequential bits fed to it by the convolutional encoder and rearranges them in time such that no consecutive encoded bits are grouped by the mapper. The reason for this has to do with the multipath effect. As the multipath channel becomes more severe, frequency notches will appear in the spectrum, which tend to completely attenuate isolated tones. It is imperative that consecutive encoded bits are not impacted by this notching effect, since the convolutional decoder has more difficulty correcting consecutive errors than isolated ones. In this chapter, we will not implement the descrambler/scrambler, decoder/encoder, or deinterleaver/interleaver, since we are, for now, only interested in developing algorithms that maximize the EVM that appears at the input of the receiver's demapper/detector. The 802.11a specification presents an excellent overview of these bit manipulation modules.

7.2.5 The Mapping Process

For the four modulation formats, the mapper will process groups of 1, 2, 4, or 6 bits and transform them into complex reference positions in the respective constellation. The mapping process is shown below as a lookup table followed by a modulation dependent scaling factor.

Input Bit (b0)	I Out	Q Out
0	-1	0
1	1	0

Figure 7-24: BPSK Map

Input Bit (b0)	I Out	Input Bit (b1)	Q Out
0	-1	0	-1
1	1	1	1

Figure 7-25: QPSK Map

Input Bits (b0 b1)	I Out	Input Bits (b2 b3)	Q Out
00	-3	00	-3
01	-1	01	-1
11	1	11	1
10	3	10	3

Figure 7-26: 16QAM Map

Input Bits (b0 b1 b2)	I Out	Input Bits (b3 b4 b5)	Q Out
000	-7	000	-7
001	-5	001	-5
011	-3	011	-3
010	-1	010	-1
110	1	110	1
111	3	111	3
101	5	101	5
100	7	100	7

Figure 7-27: 64QAM Map

After the mapping process, the complex output values are scaled, depending on the chosen constellation. The BPSK, QPSK, 16QAM, and 64QAM output values are multiplied by the factors 1, 1/*sqrt*(2), 1/*sqrt*(10), and 1/*sqrt*(42), respectively. This is done so that all constellations will have the same *rms* value, thus the same average output power. Although the process is shown as a two-step task, the hardware implementation can combine them into one lookup table per modulation format.

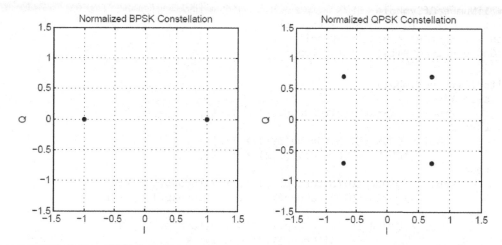

Figure 7-28: Normalized BPSK and QPSK Constellations for WLAN 802.11a

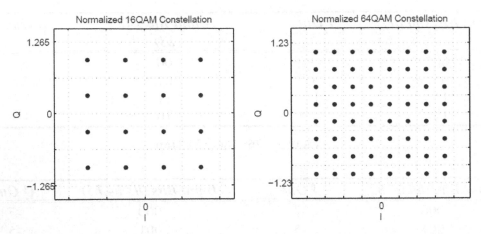

Figure 7-29: Normalized 16 and 64 QAM Constellations for WLAN 802.11a

```matlab
function [OutputSymbols] = Mapper_OFDM(InputBits, BitsPerSymbol)
% BitsPerSymbol: 1, 2, 4, 6 --> BPSK, QPSK, 16QAM, 64QAM

persistent BPSK_LUT QPSK_LUT QAM16_LUT QAM64_LUT
if(isempty(BPSK_LUT))
    BPSK_LUT  = [-1;  1];
    QPSK_LUT  = [-1;  1] / sqrt(2);
    QAM16_LUT = [-3; -1; 3; 1] / sqrt(10);        % Careful Gray Coding
    QAM64_LUT = [-7; -5; -1; -3;  7;  5;  1;  3] / sqrt(42);
end                                               % Careful Gray Coding

NumberOfSymbols = floor( length(InputBits)/BitsPerSymbol );
OutputSymbols   = zeros(1,NumberOfSymbols);


for i = 1:NumberOfSymbols
    Start    = 1 + (i - 1)*BitsPerSymbol;
    Stop     = Start + BitsPerSymbol - 1;
    BitGroup = InputBits(1,Start:Stop);


    switch(BitsPerSymbol)
        case 1
            Symbol = BPSK_LUT(BitGroup(1,1) + 1, 1);
        case 2
            Symbol = QPSK_LUT(BitGroup(1,1) + 1,1) + ...
                   j*QPSK_LUT(BitGroup(1,2) + 1, 1);
        case 4
            Symbol = QAM16_LUT(BitGroup(1,1)*2 + BitGroup(1,2)+1,1)+  ...
                   j*QAM16_LUT(BitGroup(1,3)*2 + BitGroup(1,4) + 1,1);
        case 6
            Symbol = QAM64_LUT(BitGroup(1,1)*4 + BitGroup(1,2)*2 + ...
                               BitGroup(1,3) + 1,1)  + ...
                   j*QAM64_LUT(BitGroup(1,4)*4 + BitGroup(1,5)*2 + ...
                               BitGroup(1,6) + 1,1);
    end
    OutputSymbols(1,i) = Symbol;
end
```

7.2.6 The 64 Point IFFT and Guard Interval Insertion

This section illustrates how the mapper symbols and pilot tone information are applied to the IFFT module. Review the transmit summary at the start of section 6.2 to complement the information in the figure below.

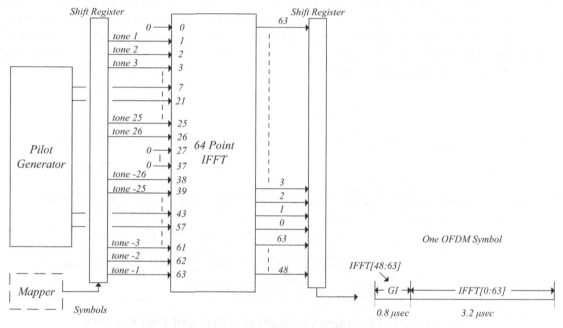

Figure 7-30: IFFT and Guard Interval Insertion

This guard interval is inserted to prevent pre-cursor and post-cursor copies of an OFDM symbol from overlaying their neighbors. In textbooks, these overlays are referred to as inter-symbol interference, or ISI. ISI in single-tone modems is corrected using sophisticated time domain equalizers such as those that we studied in chapter 4. In OFDM systems, a guard interval is used in concert with a far simpler frequency domain equalizer. The figure below illustrates a simple multipath scenario we will use to demonstrate the need for the guard interval. The figure shows multiple images of the signal arriving at the antenna, with the strongest synchronized to a normalized time of 0 seconds. Two weaker paths, called post-cursors, appear later at 200 and 600 nanoseconds, respectively, and the pre-cursor arrives 100 nanoseconds ahead of the main path.

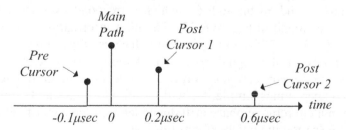

Figure 7-31: Simple Multipath Example

If we call the OFDM symbol that we are currently processing in the receiver the main symbol, then it is the post-cursors of the previous symbol and the pre-cursor of the following symbol that interfere with the main symbol. As you can see from the figure below, the post-cursors of the previous symbols overlap with the guard interval of the main symbol. This luckily does not cause any damage to the main IFFT output portion of the main symbol. If a pre-cursor is present, as is the case in here, then the next OFDM symbol will interfere with the main IFFT output of the main symbol, causing interference. However, a close look at the figure below shows that the main symbol does feature a 3.2-microsecond interval that is free of any interference. This interval includes the last 0.2 microseconds of the guard interval. In the receiver, therefore, we will want the FFT operation to process the region of the received OFDM symbol that is most likely to be clear of any pre- and post-cursor interference.

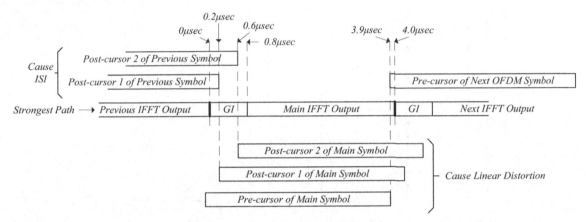

Figure 7-32: The Impact of Multipath Pre- and Post-Cursors

Additionally, pre- and post-cursor images of the main symbol interfere with the strongest copy that we wish to process in the receiver. This type of interference leads to simple linear distortion, including selective frequency fading, and can be remedied by a frequency domain equalizer that follows the FFT operation, as we will see soon.

Simplifying the IFFT Input and Pilot Tone Organization

Whereas the 802.11a specification describes a unique procedure in which symbols are assigned to subcarriers, we will simplify the process and map consecutive mapper output symbols to IFFT input ports 1 through 26 and 38 through 63, which correspond to tones 1 to 26 and -26 to -1, respectively (just as illustrated in Figure 6-25). The pilot information at tones 7, 21, -7, and -21 (FFT inputs 7, 21, 57, 43) is BPSK modulated and changes every frame in a random manner to avoid potential artifacts in the signal spectrum. Although beneficial in some aspects, this randomization adds unnecessary complication to our code and a modem that must already appear complex enough. To keep things simple, we will fix all pilot tones at $1 + j0$. Remember that the pilot tones help the modem remove phase drift due to latent frequency offset and phase noise. The simplification in pilot scaling in no way affects this task.

```
function Payload = IFFT_GI(Symbol_Stream)

NumberOfSymbols = floor(length(Symbol_Stream) / 48);
Payload         = zeros(1, NumberOfSymbols * 80);
                          % 80 samples per symbols 64 (IFFT) + 16 (GI)
IFFT_Input      = zeros(1,64);
for i = 1:NumberOfSymbols
    Start                       = (i-1)*48 + 1;     Stop = (i-1)*48 + 48;
    CurrentInput                = Symbol_Stream(1,Start : Stop);
    IFFT_Index                  = [1:6 8:20 22:26 38:42 44:56 58:63];
    IFFT_Input(1,1+IFFT_Index)  = CurrentInput;  %% 48 data  symbols
    IFFT_Input(1,1+[7 21 43 57]) = [1 1 1 1];  %% inserting  4 pilot symbols
    IFFT_Output   = ifft(IFFT_Input);   %% The IFFT operation
    Guard_Interval   = IFFT_Output(1, 1 + (48:63)); %% Guard interval
                                                    %% Definition
    Start                       = (i-1)*80 + 1;     Stop = (i-1)*80 + 80;
    Payload(1, Start:Stop)      = 64*[Guard_Interval IFFT_Output];
end
```

7.2.7   The Upsampling Process

To achieve a spacing of 312.5 KHz between tones using a 64 point IFFT requires a processing or sample rate of 20MHz. If we were to go straight to the DAC modules at that rate, we would be faced with two undesirable side effects. First, the sinc distortion inherent in the DAC module's sample and hold action would cause drooping in the magnitude spectrum of our signal. More importantly, a significant amount of frequency content from aliased images starts to appear just beyond the ±10MHz Nyquist rate. These images, even if attenuated by the sample and hold action of the DAC, must be filtered in the analog transmitter lest they cause adjacent channel interference. A filter capable of passing up to ±9MHz and significantly attenuating content beyond ±11MHz is very hardware intensive. It is therefore advisable to first upsample the waveform to 40MHz, such that the signal, arriving at the DAC modules, features no signal content between ±10MHz and ±30MHz. This will significantly reduce the sinc droop and facilitate the DAC reconstruction filter design in the analog transmitter. Please review the previous chapter for additional information regarding this phenomenon.

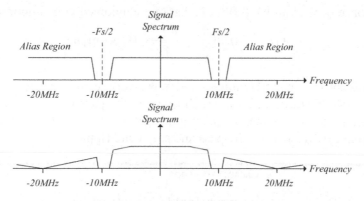

Figure 7-33: 20MSPS Signal Spectrum before and after the DAC Modules

Designing the Upsampling Filter

The upsampling structure is a concatenation of a zero-stuffing element, which inserts zero samples between those that exist at the 20MHz rate, and a half-band FIR filter. The zero-stuffing operation does not change the periodically repeating spectrum of the 20MHz sample stream at all. It simply moves the Nyquist rate from 10MHz to 20MHz (see section 2.10 of chapter 5). The figure below illustrates the signal path through the zero-stuffing element and half band filter (see chapter 2 for additional details).

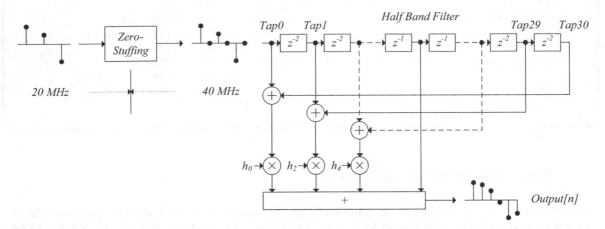

Figure 7-34: The Complete Upsampling Structure

The half band filter has the wonderful feature that almost every second sample is equal to zero, thus reducing the number of hardware multipliers needed. Further, given that the impulse response is an even function, symmetrically located tap values may be added and then share a hardware multiplier to further reduce needed resources. For this implementation, we choose a 31 tap half-band FIR filter that in fact requires only eight multipliers and, given that we'll need two instantiations, one for the *I* and one for the *Q* path, the total number of multiplication units rises to a humble 16.

Example 6.1: *Half Band Filter Design Example for 802.11a Application*

The impulse response of the half band filter is that of a windowed *sinc* function, as shown below.

$$h[n] = sinc(\frac{n}{2} - \frac{N-1}{4}) \cdot Window[n]$$

In this example, we chose to use a Hanning window whose coefficients are raised to the ½ power. The basic *sinc* function features a bit too much ripple in the pass band but excellent sharpness in its transition from pass to stop band. Overlaying different variants of the Hanning window improves the response by decreasing the pass and stop band ripple.

```
N    = 31;                      % Number of Taps
n    = 0:N-1;
Arg  = n/2 - (N-1)/4;           % Argument inside sinc function
Hann = hann(length(n) + 2)';    % The Hanning window as a row vector
```

```
%% Half Band Filter Impulse Response
h    = sinc(Arg).*(Hann(1,2:end-1) .^0.5); % Hanning window exponent = 0.5


%% Computing the Frequency Response
FResolution = 0.002;
Frequencies = -0.5:FResolution:0.5;
FrequencyResponse = zeros(1,length(Frequencies));

for i = 1:length(Frequencies)                    %% The DTFT Operation
    f                    = Frequencies(1,i);
    AnalysisTone         = exp(j*2*pi*n*f);
    FrequencyResponse(1,i) = (1/N)*h*AnalysisTone';
end
LogResponse = 20*log10(abs(FrequencyResponse));

figure(1);
subplot(1,2,1)
stem(n, h, 'k'); title('Impulse Response of Halfband Filter'); xlabel('n');
axis([0 N-1 -.25 1.1])
subplot(1,2,2);
plot(Frequencies*40e6, LogResponse - max(LogResponse), 'k'); grid on;
title('Normalized Magnitude Response of Half Band filter');
set(gca, 'XTick', 40e6*[-.5 -.25 0 .25 .5]);
xlabel('Normalized Frequency');
ylabel('dB');axis([SampleRate*([-.5 .5]) -60 5]);
```

Try modifying the Hanning window's exponent in the code above and watch the normalized magnitude response change – to your liking or dismay.

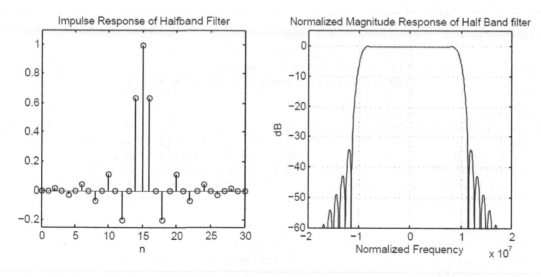

Figure 7-35: Half Band Filter Impulse and Magnitude Response

7.2.8 The 128 Point IFFT and Guard Interval Insertion

Whereas the transition from a guard interval to the IFFT output is smooth and continuous, the boundary samples between OFDM symbols are not. After all, their IFFT output and associated guard intervals are generated from completely different sets of transmit symbols. During the upsampling process, we remove digital images using the half band filter discussed in the last section. The filtering process will smear the boundaries between symbols, causing some information from one to leak into the other. A closer look at the half-band filter's impulse response makes the extent of that phenomenon clear. This smearing and the pre-cursor images generated during multipath events are the primary reasons that we will want to advance the timing instant of the FFT in the receiver, thereby preventing information of the next symbol from interfering with the current calculation.

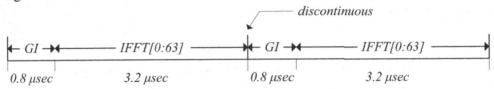

Figure 7-36: Discontinuous OFDM Symbol Boundary

However, rather than using an upsampling filter to change the transmit sample rate from 20 to 40MHz, it is possible to double the size of the IFFT from 64 to 128 points. This alternate method of upsampling, though more hardware intensive, renders the upsampled symbols directly and sidesteps any smearing associated with the filtering operation. The first 32 input symbols destined for the 64 point IFFT are placed at the same input positions 0 through 31 of the 128 point IFFT. The upper input symbols 32 through 63 at the input of the 64 point IFFT are placed at input positions 96 through 127, as seen in the code below.

```
function Payload = IFFT128_GI(Symbol_Stream)

NumberOfSymbols = floor(length(Symbol_Stream) / 48);
Payload         = zeros(1, NumberOfSymbols * 160);
                  % 160 samples per symbols 128 (IFFT) + 32 (GI)

IFFT64_Input        = zeros(1,64);
for i = 1:NumberOfSymbols
    Start                            = (i-1)*48 + 1; Stop = (i-1)*48 + 48;
    CurrentInput                     = Symbol_Stream(1,Start : Stop);
    IFFT64_Index                     = [1:6 8:20 22:26 38:42 44:56 58:63];
    IFFT64_Input(1,1+IFFT64_Index)   = CurrentInput;%% Place 48 data symbols
    IFFT64_Input(1,1+[7 21 43 57])   = [1 1 1 1];   %% Place 4 pilot symbols

    IFFT128_Input                    = zeros(1,128);
    IFFT128_Input(1,[1:32 97:128])   = IFFT64_Input;
    IFFT_Output                      = ifft(IFFT128_Input,128); %% The IFFT
    Guard_Interval                   = IFFT_Output(1, 1 + (96:127));

    Start                            = (i-1)*160 + 1; Stop = (i-1)*160 + 160;
    Payload(1, Start:Stop)           = 128*[Guard_Interval IFFT_Output];
end
```

7.2.9 The Complete OFDM (802.11a) Transmitter Code

The transmit function below requires two input parameters that determine the number of OFDM symbols to be transmitted, as well as the type of modulation expressed in terms of bits per symbol. The function returns both the ideal 40MHz sample stream of the packet as well as the symbol stream that was produced by the mapping function. This information will later be compared to the symbol stream at the input of the receive demapper for EVM computation purposes.

```
function [Sample_Output Symbol_Stream] = OFDM_TX_802_11(...
                        NumberOf_OFDM_Symbols, BitsPerSymbol)
           %% BitsPerSymbol = 1,2,4,6 = BPSK, QPSK, 16QAM, 64QAM


%% 1. Compute short and long training sequences at 20MHz (Step = 1/2)
Step                 = 1;
ShortTrainingSequence = Get_ShortTrainingSequence(Step);
LongTrainingSequence  = Get_LongTrainingSequence(Step);



%% 2. Generating random bits for mapping operation
rand('seed', 0);        %% Fixing the seed of the random number generator
NumberOfBits = (48*BitsPerSymbol)*NumberOf_OFDM_Symbols;
Source_Bits  = round(rand(1,NumberOfBits)); %% Creating random input bits
Symbol_Stream = Mapper_OFDM(Source_Bits, BitsPerSymbol);


%% 3. Generating the payload
Payload = IFFT_GI(Symbol_Stream);


%% 4. Zero-Stuffing operation
Packet_20MHz                    = [ShortTrainingSequence ...
                                   LongTrainingSequence Payload];
Packet_Zero_Stuffed             = zeros(1,2*length(Packet_20MHz));
Packet_Zero_Stuffed(1,1:2:end)  = Packet_20MHz;


%% 5. Computing Halfband filter coefficients
N   = 31;                   % Number of taps
n   = 0:N-1;
Arg = n/2 - (N-1)/4;        % Argument inside sinc function
Hann = hann(length(n) + 2)'; % The Hanning window as a row vector
h   = sinc(Arg).*(Hann(1,2:end-1) .^1);


%% 6. The halfband filtering operation
Sample_Output     = filter(h, 1, [Packet_Zero_Stuffed zeros(1,100)]);
```

To change the code above to the 128 point IFFT version, which, instead of using an upsampling filter, directly produces the samples at 40 MHz, we make the following changes. Change the step variable in section one of the code from 1 to ½. Replace the function IFFT_GI() with IFFT128_GI() in section three of the code. Finally, concatenate the short and long training sequences with the payload to form the sample output array, which is returned by function. Just replace Packet_20MHz with Sample_Output in section four of the code and delete all else.

7.3   The 802.11a Receiver

7.3.1   An Overview of the Receiver Architecture

The FFT algorithm, which embodies the core processing block of the OFDM demodulator, is supported by a rich collection of synchronization algorithms, ensure that the final symbol detector (the demapper) is presented with the most pristine receive symbols possible. The figure below illustrates the positioning of the FFT and synchronization algorithms in the portion of the digital receiver that deals with complex information. After the phase and timing correction module, the complex receive symbols are presented to the demapping block, which changes them into a serial bit stream. The discussion in this chapter will limit itself to algorithms up to the demapping block. Based on the information given in the figure below, let us introduce the main synchronization and signal conditioning tasks required for proper OFDM operation.

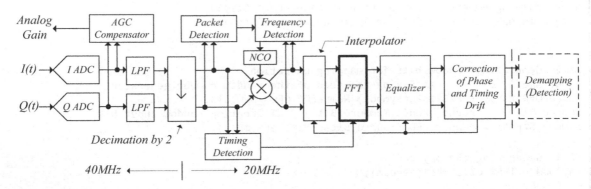

Figure 7-37: Portion of the OFDM Demodulator that Processes Complex Samples

Low Pass Filtering and Sample Rate Decimation

The discrete low pass filtering and decimation processes serve a similar purpose as the upsampling algorithm ahead of the DAC modules in the transmitter. We assume, as seen in part *a*) of the figure below, that our signal of interest is accompanied by additional traffic content on a neighboring channel located at an offset of 20MHz. If we were to sample the signal at 20MHz, an analog anti-aliasing with the ability to pass frequency content below 9 MHz and significantly suppress it beyond 11MHz would have to be designed. Such a filter would be unnecessarily expensive in terms of complexity and current consumption.

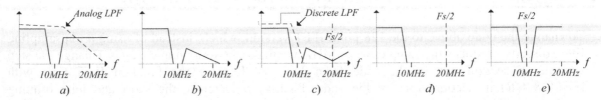

Figure 7-38: Signal Progression before, during and after the Analog to Digital Conversion Process

By sampling the signal at 40MHz rather than 20MHz, content between 10MHz and 30MHz can no longer alias into our band of interest, and the specifications of the analog anti-aliasing filter may be relaxed. The function of the analog filter therefore changes to the task of attenuating the neighboring traffic such that it cannot saturate the ADC input and rejecting content beyond 30MHz, which would otherwise alias into our signal band. By relaxing the analog filter requirements, some content between 10 and 30MHz will likely remain and interfere with our receive signal, as shown in part *b*) of the last figure. However, once the signal has been sampled (part *c*), this content is trivially removed via a digital filter (part *d*) and the waveform is ready for the decimation process that follows (part *e*).

AGC Compensation

The AGC compensator forms part of an overall control loop that ensures that the received signal at the inputs of the analog to digital converters is properly sized. The module, in fact, compensates or stabilizes the loop. The AGC compensator measures the magnitude of the signal directly after the ADC rather than after the LPF modules, for two reasons. The received waveform consists of both the desired in-band signal and the filtered content in neighboring channels, and it is this compound signal that must be correctly scaled at the ADC inputs. The second reason to choose the ADC output signal rather than its low pass filtered counterpart is the additional time delay that would be introduced into the loop. Filtering takes time, and time delay is the enemy of fast tracking performance in closed loop control applications.

Packet Detection

While waiting for the arrival of a packet, the receiver's AGC compensator will keep the analog gain at a maximum to ensure that even low-power signals are properly detected. This condition is what we might call the idle or waiting state of the demodulator. When a packet does arrive, the large gain of the analog receiver usually causes the receive signal to completely saturate the ADC modules. The magnitude detector inside the AGC compensator is therefore the first circuit to notice the arrival of this signal. Once the AGC loop has returned the ADC input signal to normal levels, the actual packet detector begins to search for the unique signature of the short training sequence. Only when the packet detector has recognized the short training sequence can we say that the waveform is indeed a valid 802.11a signal, rather than interference from the microwave oven where a coworker is heating up his lunch. Remember that many WLAN systems use an unlicensed band which they share with other applications.

Frequency Detection and Correction

Frequency offsets in the local oscillator signals must be detected and corrected for the OFDM demodulator to function correctly. If uncorrected, each FFT output symbol represents not only the orientation and magnitude of a single subcarrier, but contains trace information of all other subcarriers as well. We refer to this effect as loss of orthogonality. The task of frequency offset correction is subdivided into coarse and fine frequency acquisition; the former algorithm uses the short training sequence, whereas the latter uses the long version. So, just how much of a frequency offset can we expect? The transmit and receive LO signals that up and down convert the base band waveforms are created by synthesizer modules. Each synthesizer uses a reference clock, usually from a TCXO, which oscillates near its rated frequency. The error in reference frequency is specified in terms of parts per million, or ppm. Given a reference clock error of ±20ppm, an LO signal oscillating at a channel frequency of 2.4GHz may deviate by as much as 2.4·(±20e-6) GHz, or ±48KHz. To make matters worse, the frequency error of the reference crystals at the transmitter and receiver may be at opposite extremes, thus causing a maximum possible frequency offset in the received signal of ±96KHz. Some small amount of frequency error will remain after the coarse and fine frequency offset correction is done. Note that frequency error is equivalent to a moving phase offset. The error must be small enough so that the resulting change in phase over one OFDM symbol is negligible. Moving phase offsets that remain at low levels may be reliably calculated based on information provided by the pilot tones and subsequently removed.

Timing Detection and Correction

Timing detection is the process of determining the correct instant at which the FFT block's shift register is filled with proper information. This timing reference represents the start signal for the FFT algorithm and repeats every 4 microseconds. The first instant when this timing reference must assert is the moment at which the FFT input shift register contains the long training symbol, providing us with an estimation of the multipath channel. While the correct timing instant is established during the reception of the training symbols, additional timing drift will occur due to the same error in the crystal oscillator references that caused the frequency offset mentioned above. These slightly different reference clocks provide the sampling instant for the DAC and ADC modules and, for that matter, for the entire digital transmitter and receiver, respectively. Timing drift will be calculated from pilot tone information and corrected prior to the demapping module. Timing drift in the system will require us to update the equalizer coefficients and probably counteract the problem using an interpolator.

Equalization

One of the features that have driven the adoption of OFDM is the simplified way of equalizing the received information. Unlike single-tone, wide-band signals that suffer selective fading at the hands of the multipath channel, the multi-tone nature of OFDM permits us to treat multipath effects as flat fading on a subcarrier by subcarrier basis. Thus, the FFT result of each narrow band subcarrier may be rescaled by a single complex value. Given that the subcarrier orientations of the long symbols are known a priori, it is a simple matter of taking the FFT of a long symbol and comparing the ideal version originally sent by the transmitter. The reciprocal for each subcarrier is then applied during the equalization step for all subsequent OFDM symbols.

Phase Drift Correction

Owing to unresolved frequency offset and phase noise, some amount of slow phase drift will continue to disturb the 802.11a receive packet, as OFDM symbols successively enter the receiver. This phase drift will be corrected via the use of pilot tones that are embedded in each symbol. The orientation of each pilot tone is known ahead of time, which allows us to compare them to the pilot orientations actually calculated by the FFT. The four phase differences are averaged and the result used to derotate all received subcarriers.

Bit Manipulation

Although our interest in the demodulator ends with the calculation of the error vector magnitude (EVM), of the symbols handed to the demapping module, there naturally exist additional blocks that process the resulting serial bit stream. These blocks may be seen in the figure below. While the deinterleaver and descrambler modules are of low to medium sophistication and may easily be derived from the 802.11a specifications, the Viterbi decoder is a far more complex module and its inner working will not be presented until a future edition of this text.

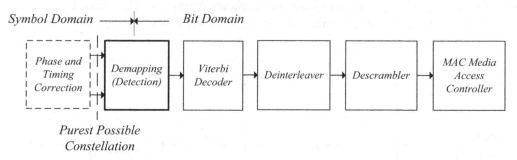

Figure 7-39: Portion of the OFDM Demodulator that Processes Bits

7.3.2   Packet Detection

The packet detector, running at 20MHz, exploits the fact that the short training symbols repeat every 16 samples. Therefore, if the short training sequence is present, the average of $x[n] \cdot x[n-16]^*$ should be approximately equal to the average of $x[n] \cdot x[n]^*$. These two averages are in fact estimates of the autocorrelation of $x[n]$ for a delay of 16 samples and the variance of $x[n]$.

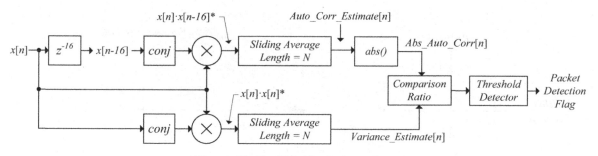

Figure 7-40: 802.11a Packet Detector

$$AutoCorrEstimate[n] \rightarrow \frac{1}{N} \sum_{k=0}^{N-1} x[n-(N-1)+k] \cdot x[n-(N-1)+k-16]^*$$

$$VarienceEstimate[n] \rightarrow \frac{1}{N} \sum_{k=0}^{N-1} x[n-(N-1)+k] \cdot x[n-(N-1)+k]^*$$

Comparing these two quantities is our primary method of indicating the presence of a valid 802.11a packet.

Valid packet → if autocorrelation ≈ variance

Invalid packet → if autocorrelation << variance

The last and next figures illustrate the information flow through the packet detector as well as the corresponding graphical interpretation of its results. We may create a normalized comparison by calculating the ratio of the autocorrelation and variance estimates. This ratio will hover around a value of 1 during a valid short preamble and be significantly less otherwise.

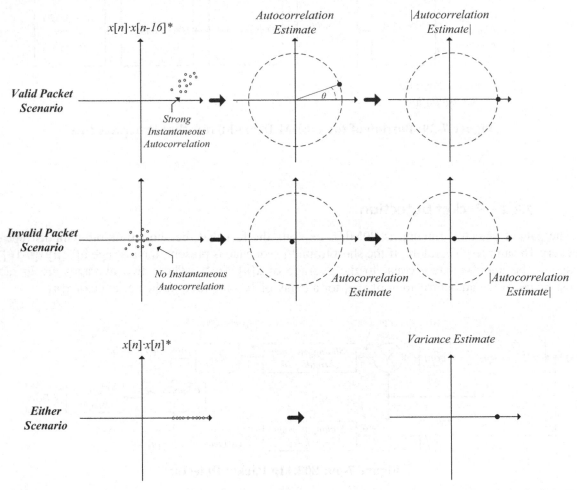

Figure 7-41: Graphical Interpretation of the Autocorrelation and Variance Results

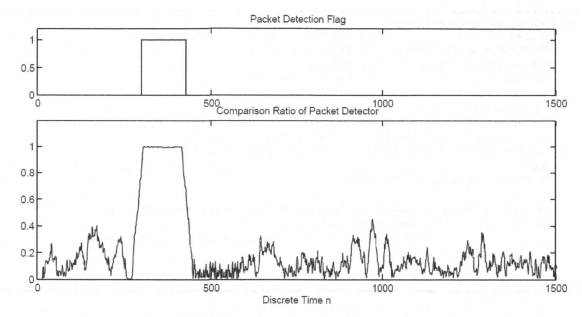

Figure 7-42: Comparison Result as Autocorrelation Estimate/Variance

$$ComparisonRatio[n] = \frac{AutoCorrEstimate[n]}{VarianceEstimate[n]}$$

The packet detection flag is asserted when the comparison result exceeds 0.85 and deasserts as it drops below 0.65. The figure also provides a crude timing reference (FallingEdgePosition), which indicates the end of the short training sequence and will be used in subsequent modules.

```
%% Packet detector and coarse frequency detector
function [Comparison_Ratio PacketDetFlag FallingEdgePosition ...
         AutoCorr_Est]  = Packet_Detector(RX_Input)
persistent             Detection_Flag;

AutoCorr_Est           = zeros(1, length(RX_Input));
Comparison_Ratio       = zeros(1, length(RX_Input));
PacketDetFlag          = zeros(1, length(RX_Input));
Delay16                = zeros(1,16);
SlidingAverage1        = zeros(1,32);  %% For the AutoCorr estimate
SlidingAverage2        = zeros(1,32);  %% For the variance estimate
Detection_Flag         = 0;  %% Short training symbol flag


for i = 1:length(RX_Input)
    RX_Input_16        = Delay16(1,16);   % The 16 sample delay
    Delay16(1,2:16)    = Delay16(1,1:15);
    Delay16(1, 1)      = RX_Input(1,i);
         %% Compute absolute value of autocorrelation estimate
    Temp               = RX_Input(1,i)*conj(RX_Input_16);
    SlidingAverage1(1,2:32) = SlidingAverage1(1,1:31);
    SlidingAverage1(1,1)    = Temp;
    AutoCorr_Est(1,i)       = sum(SlidingAverage1)/32;
    AbsAutoCorr_Est         = abs(AutoCorr_Est(1,i));
```

```
%% Compute variance estimate
    InstPower               = RX_Input(1,i)*conj(RX_Input(1,i));
    SlidingAverage2(1,2:32) = SlidingAverage2(1,1:31);
    SlidingAverage2(1,1)    = InstPower;
    Variance_Est            = sum(SlidingAverage2)/32;

    Comparison_Ratio(1,i) = AbsAutoCorr_Est/Variance_Est;
    %% PacketDetFlag with hysteresis window
    if    (Comparison_Ratio(1,i) > 0.85); Detection_Flag = 1;
    elseif(Comparison_Ratio(1,i) < 0.65); Detection_Flag = 0;
    end
    PacketDetFlag(1,i) = Detection_Flag;

    %% Initial timing reference by looking at deasserting PacketDetFlag
    if( (i > 1) && (i < length(RX_Input)/2) && ...
                    (PacketDetFlag(1,i) - PacketDetFlag(1,i-1)) == -1)
        FallingEdgePosition = i;
    end
end
```

7.3.3 Frequency Offset Detection

Both the coarse and fine frequency offset detectors employ the same autocorrelation approach that was used in the packet detector. This time, however, we are not interested in the absolute value of the autocorrelation estimate but in its angle. This angle represents the average phase difference between the current sample $x[n]$ and the sample $x[n\text{-Period}]$ that was valid one period ago. Given that the short and long training sequences repeat after a period of 16 and 64 samples, respectively, we formulate the autocorrelation functions as follows.

$$AutoCorrEstimate_{Short} \rightarrow \frac{1}{N}\sum_{k=0}^{N-1} x[n-(N-1)+k] \cdot x[n-(N-1)+k-16]^*$$

$$AutoCorrEstimate_{Long} \rightarrow \frac{1}{N}\sum_{k=0}^{N-1} x[n-(N-1)+k] \cdot x[n-(N-1)+k-64]^*$$

For this application, we set the length of the averaging operation for the short and long angle calculation to $N = 32$ and 64, respectively.

$$\theta_{Short}[n] = atan2(AutCorrEstimate_{Short})$$

$$\theta_{Long}[n] = atan2(AutCorrEstimate_{Long})$$

The phase difference, θ, reflects the angle traveled by the frequency offset signal over a period of 16 or 64 samples.

$$\theta = 2\pi \cdot FrequencyOffset \cdot Period \cdot SamplingPeriod = \frac{2\pi \cdot FrequencyOffset \cdot Period}{SampleRate}$$

$$FrequencyOffset_{coarse} = \theta_{Short}\frac{20e6Hz}{2\pi \cdot 16}$$

$$FrequencyOffset_{fine} = \theta_{Short}\frac{20e6Hz}{2\pi \cdot 64}$$

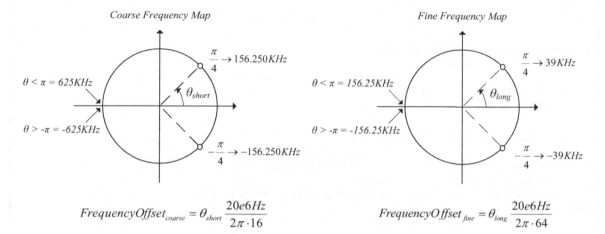

Figure 7-43: Correlation Output to Frequency Offset Conversion

The figure above matches the frequency offsets to the $-\pi < \theta < \pi$ argument. The coarse frequency process is meant to detect offsets of up to ±625KHz and correct them such that they fall into the smaller capture range of the fine frequency process, which features better sensitivity.

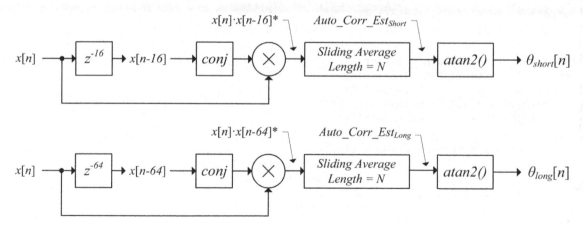

Figure 7-44: Coarse and Fine Frequency Detectors

The *atan2* function shown in the autocorrelation modules above may be conveniently implemented using a cordic algorithm [3]. Each module uses timing references such as the packet detection flag to schedule its computation process. Based on the calculation of the coarse frequency detector, a numerically controlled oscillator, or NCO, is programmed to generate a complex sinusoid with the offset frequency but opposite direction. The fine offset calculation then updates the NCO output signal.

```matlab
function [FrequencyOffsets] = Detect_FrequencyOffsets(RX_Input, ...
                             FallingEdgePosition)

FrequencyOffsets    = zeros(1,2);      % 1- coarse and 2- fine offsets
AutoCorr_Est        = zeros(1, length(RX_Input));
Delay16             = zeros(1,16);
SlidingAverage1     = zeros(1,32);     % For the AutoCorr estimate

%% Compute Coarse Frequency Offset
for i = 1:length(RX_Input)
    RX_Input_16     = Delay16(1,16);   % The 16 sample delay
    Delay16(1,2:16) = Delay16(1,1:15);
    Delay16(1, 1)   = RX_Input(1,i);

    %% Compute autocorrelation estimate
    Temp                   = RX_Input(1,i)*conj(RX_Input_16);
    SlidingAverage1(1,2:32) = SlidingAverage1(1,1:31);
    SlidingAverage1(1,1)    = Temp;
    AutoCorr_Est(1,i)       = sum(SlidingAverage1)/32;
end

Theta               = angle(AutoCorr_Est(1, FallingEdgePosition - 50));
FrequencyOffsets(1,1) = Theta*20e6/(2*pi*16);

figure(14);
plot(real(AutoCorr_Est), 'r'); grid on; hold on;
plot(imag(AutoCorr_Est), 'b'); stem(FallingEdgePosition - 50, 1, 'k')

AutoCorr_Est_Fine   = zeros(1, length(RX_Input));
Delay64             = zeros(1,64);
SlidingAverage2     = zeros(1,64);     % For the AutoCorr estimate

%% Compute fine frequency offset
for i = 1:length(RX_Input)
    RX_Input_64     = Delay64(1,64);
    Delay64(1,2:64) = Delay64(1,1:63);
    Delay64(1, 1)   = RX_Input(1,i);

    %% Compute autocorrelation estimate
    Temp                   = RX_Input(1,i)*conj(RX_Input_64);
    SlidingAverage2(1,2:64) = SlidingAverage2(1,1:63);
    SlidingAverage2(1,1)    = Temp;
    AutoCorr_Est_Fine(1,i)  = sum(SlidingAverage2)/64;
end

Theta               = angle(AutoCorr_Est_Fine(1, FallingEdgePosition +125));
FrequencyOffsets(1,2) = Theta*20e6/(2*pi*64);

figure(15);
plot(real(AutoCorr_Est_Fine), 'r'); grid on; hold on;
plot(imag(AutoCorr_Est_Fine), 'b'); stem(FallingEdgePosition +125, 1, 'k')
```

7.3.4 Timing Acquisition

The packet detection flag provided us with a crude timing reference that we could use to start our frequency offset detection processes. The receiver, however, requires a higher accuracy timing reference in order to synchronize the start of the FFT operation with flow of the complex sample stream. In other words, we must know when the last 64 samples of an OFDM symbol have arrived in the FFT's input shift register to start the transform calculation. To create a timing reference that is accurate to within one sample, we cross correlate the sample stream with a local copy of the long training symbol. When the cross correlator produces its maximum peak, the long training symbol of the strongest path is properly aligned in the correlator shift register. The mathematic expression of the cross correlation function we need to implement is shown below.

$$CrossCorrelation[n] = \frac{1}{64}\sum_{k=0}^{63} x[n+k-63] \cdot LongTraining[k]^*$$

Correlation using a transversal filter as we do in this application can be somewhat confusing because of the awkward notation. Note that for $k = 0$, the earliest input sample, $x[n-63]$, is multiplied by the conjugate of the long training symbol's first sample, $L[0]^*$.

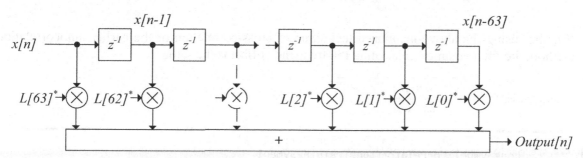

Figure 7-45: Sliding Cross-correlator Implementation for Long Training Sequence

The sliding cross-correlator features 64 complex multiplications, along with a division-by-64 operation, which reduces to a bit shifting operation to the right by six positions. Unfortunately, though, the remaining 64 complex multiplications, or 4·64 real multiplications, make the module very hardware intensive. To get around this issue, we replace the coefficients of the long training sequence by their sign equivalent, called $L[k]$ in the figure above, thus quantizing each of them to ±1.

$$QuasiCrossCorrelation[n] = \frac{1}{64}\sum_{k=0}^{63} x[n+k-63] \cdot L[k]^*$$

$$= \frac{1}{64}\sum_{k=0}^{63} x[n+k-63] \cdot (sign(real(LongTraining[k])) + j \cdot sign(imag(LongTraining[k])))^*$$

This simplification works surprisingly well for our purposes and eliminates all multiplication elements. The next figure shows three separate cross correlation peaks, the first indicating the guard interval position, while the others locate the first and second long training symbol. We may use the crude timing reference provided by the packet detection flag to establish two timing windows within which each of the two cross-correlation peaks would fall. The two peaks provide us with a timing reference that is accurate to within one sample, as well as a second method of calculating the fine frequency offset. Multiplying the complex value of the second cross-correlation peak by the complex conjugate of the first produces a number whose angle, θ, is proportional to the frequency offset.

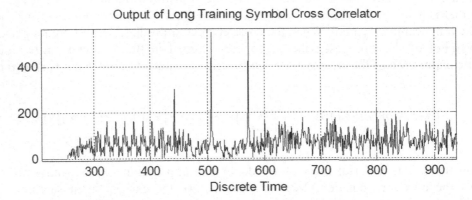

Figure 7-46: Output of Long Training Symbol Cross-correlator

Whether using the above-mentioned technique of cross-correlation or the earlier autocorrelation method, the final equation calculating the frequency offset is the same.

$$FrequencyOffset = \theta\frac{20e6\ Hz}{2\pi \cdot 64}$$

```
function [LTPeak_Value1 LTPeak_Position OutputLong] = ...
        LongSymbol_Correlator(LongTrainingSymbol, ...
                              RX_Waveform, FallingEdgePosition)


L               = sign(real(LongTrainingSymbol)) + ...
                    j*sign(imag(LongTrainingSymbol));
OutputLong      = zeros(1, length(RX_Waveform));
LTPeak_Value1   = 0;
LTPeak_Position = 0;
Cross_Correlator = zeros(1,64);
for i=1:length(RX_Waveform)-64
    Output          = Cross_Correlator*fliplr(L)'; %% The crosscorrelation
    OutputLong(1,i) = Output;                      %% Operation
    Cross_Correlator(1,2:64) = Cross_Correlator(1,1:63);
    Cross_Correlator(1,1)    = RX_Waveform(1,i);
    %% Search for first long training symbol
    if((i > FallingEdgePosition + 54)&&(i < FallingEdgePosition + 54 + 64))
        if( abs(Output) > abs(LTPeak_Value1) )
            LTPeak_Value1   = Output;
            LTPeak_Position = i;
        end
    end
end
```

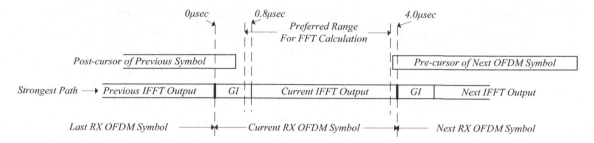

Figure 7-47: The Preferred Range for the FFT Calculation

The FFT Start Time

In an 802.11a communication link that is distorted by multiple paths, the two largest peaks will indicate the time instant when the long symbols of the strongest path are properly aligned within the sliding cross correlator. This timing reference allows us to estimate the time instant when the last 64 samples of the received OFDM symbol are properly aligned with the FFT input shift register. Is this the instant at which the state machine should light off the FFT calculation? The figure above shows three OFDM symbols that belong to the strongest path, as well as post- and pre-cursor symbol information from weaker paths. The current RX OFDM symbol of the strongest path is bound by the 0µsec to 4.0µsec time stamps. We can clearly see the symbols' composition as the cascade of the guard interval and the IFFT output. While initially counterintuitive, we don't want the FFT to operate on the exact IFFT output range (0.8µsec to 4.0µsec), which encompasses the last 64 samples of the current OFDM symbol. The figure above indicates a preferred FFT range residing slightly earlier in time than the IFFT output information. The preferred range ensures that possible pre-cursor information of the next symbol does not slip into the FFT calculation of the current OFDM symbol. The fact that the preferred range will contain some amount of the guard interval is fine, given that the GI is a cyclic extension of the IFFT output information. This time advanced version of the IFFT information causes the subcarriers' orientations calculated by the FFT to be rotated according to the time-shifting property of the Fourier transform. The equalizer, which was trained by a slightly time-advanced long symbol, will be aware of this rotation and correct it for all symbols. Therefore, the start instant of the FFT should be a few samples advanced in time compared to IFFT output sequence of the strongest path.

7.3.5 Channel Estimation and Equalizer Setup

The channel estimation process determines the impact of the multipath channel on the frequency response of the original signal. Owing to the multipath channel's linear distortion effects, the receiver will find the signal spectrum to be bent out of shape in both its magnitude and phase domain. Since we have full knowledge of the long training symbol's subcarrier orientations, it is a simple matter of comparing the received subcarriers (FFT outputs) of the long symbols with the orientation with which they were sent. As a matter of fact, since there are two long training symbols, we can use averaging to improve the signal to noise ratio of the estimate. Let us define the orientation of all long symbol subcarriers, as was shown earlier in section 6.2.1.

$$Positive_{m=0,1...31} = \quad [0 \quad 1 \; -1 \; -1 \quad 1 \; 1 \; -1 \; 1 \quad -1 \; 1 \; -1 \; -1 \quad -1 \; -1 \; -1 \; 1$$

$$1 \; -1 \; -1 \; 1 \quad -1 \; 1 \; -1 \; 1 \quad 1 \; 1 \; 1 \; 0 \quad 0 \; 0 \; 0 \; 0]$$

$$Negative_{m=-32,-31...-1} = [0 \; 0 \; 0 \; 0 \quad 0 \; 0 \quad 1 \; 1 \quad -1 \; -1 \; 1 \; 1 \quad -1 \; 1 \; -1 \; 1$$

$$1 \; 1 \; 1 \; 1 \quad 1 \; -1 \; -1 \; 1 \quad 1 \; -1 \; 1 \; -1 \quad 1 \; 1 \; 1 \; 1]$$

$$Ideal\_Tones[m] = [Negative \; Positive]$$

We construct the expression for the estimate of the frequency response of the channel as follows.

$$RX\_Tones[m] = Frequency\_Response[m] \cdot Ideal\_Tones[m]$$

$$Frequency\_Response[m] = \frac{RX\_Tones[m]}{Ideal\_Tones[m]}$$

Undoing the effects of the channel is the duty of the equalizer, which multiplies the orientation of each received subcarrier by the reciprocal of the corresponding frequency response subcarrier orientation.

$$Equalizer[m] = \frac{1}{Frequency\_Response[m]} = \frac{Ideal\_Tones[m]}{RX\_Tones[m]}$$

This simple frequency domain explanation is given in most elementary communication text books, but it was not until the advent of frequency domain modems, such as OFDM architectures, that equalizer hardware implementation was able to benefit from this simplicity. Most time domain modems still need to find the equalizer coefficients via sophisticated linear algebra steps or adaptive algorithms. The topic of time domain equalization is discussed in detail in chapter 4, and after reviewing it, you may draw your own conclusion about which equalization approach is easier to understand and implement.

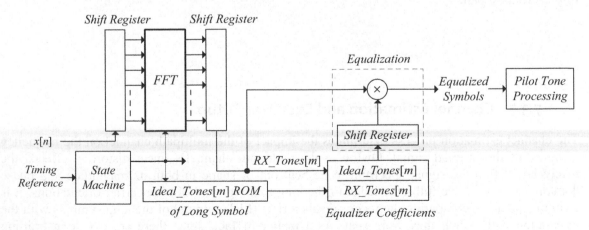

Figure 7-48: Equalizer Positioning in the OFDM Receiver

The last figure illustrates the basic interconnection between the different processing algorithms of the equalization process. While receiving the long training sequence, the ideal long symbol subcarrier orientations, Ideal_Tones[m], stored in read only memory, are divided by the received long symbol subcarriers values, RX_Tones[m], to form the equalizer coefficients. The coefficients are stored in a shift register and serially applied as the received tones of payload OFDM symbols arrive. The state machine uses the earlier gained timing reference to coordinate and schedule execution of all modules. After the equalization process, the symbol set of 52 complex values is ready for pilot tone processing.

```
%% 3. Channel estimate and equalizer setup
  % --> Extract long training symbols from rx waveform and average them
First_LongSymbol = TX_Output20MHz(1,LTPeak_Position-64:LTPeak_Position - 1);
Second_LongSymbol= TX_Output20MHz(1,LTPeak_Position   :LTPeak_Position +63);
Averaged_LongTraining_Symbol = First_LongSymbol*.5 + Second_LongSymbol*.5;


  % --> Take its FFT and extract the positive and negative tone orientations
FFT_Of_LongTraining_Symbol = ifft(Averaged_LongTraining_Symbol);
RX_Positive_Tones          = FFT_Of_LongTraining_Symbol(1, 2:27);
RX_Negative_Tones          = FFT_Of_LongTraining_Symbol(1, 39:64);
RX_Tones                   = [RX_Negative_Tones RX_Positive_Tones];


  % --> Establish ideal tone orientations
[Ignore AllTones]    = Get_LongTrainingSequence(1);
Ideal_Positive_Tones = AllTones(1, 34:59);
Ideal_Negative_Tobes = AllTones(1,  7:32);
Ideal_Tones          = [Ideal_Negative_Tones Ideal_Positive_Tones];


  % --> Channel estimate and equalizer coefficient calculation
Channel_Estimate      = RX_Tones ./ Ideal_Tones;
Equalizer_Coefficients = 1./Channel_Estimate;
```

What we have computed above is called the raw channel estimate, which can be corrupted by quite a bit of noise. Use the technique described in 7.1.3.5 to refine the estimate by removing large amounts of noise thereby increasing the estimates signal to noise ratio and thus its accuracy. Every high performance OFDM receiver does this.

7.3.6   Removing Phase and Timing Drift

Detecting and Removing Phase Drift

OFDM demodulation is a coherent process, meaning that any residual phase offset (resulting from phase drift) must be corrected to place the received symbols onto their intended constellation position. Phase drift is primarily caused by latent frequency offset and phase noise. If we assume that the frequency offset has been corrected to within a small residual error, then the subsequent channel estimation/equalizer combination will identify and correct any remaining phase offset. The residual frequency offset must be small enough to ensure a negligible change in phase over one OFDM symbol. For packets with large numbers of symbols, however, this small phase change will add over time to become significant. To analyze and correct slowly increasing phase errors, pilot tones are embedded into each OFDM symbol. The following expression shows the phase deviation $\theta[n]$ altering our signal of interest $x[n]$.

$$y[n] = x[n] \cdot e^{j\theta[n]}$$

If the change in phase is indeed minimal over a single 4.0 μsec OFDM symbol, as we assume it is, then all tones, including the pilots, generated by the FFT will be rotated away from their nominal values by that quasi constant phase, θ.

$$if \ \ x[n] \rightarrow DFT \rightarrow X[m]$$

$$then \ x[n] \cdot e^{j\theta} \rightarrow DFT \rightarrow X[m] \cdot e^{j\theta}$$

Since we know what the values of the pilot tones are ahead of time, we can compare how far, on average they have moved away from their ideal positions and then correct the phase of all 52 information bearing tones accordingly. For this OFDM implementation, we have simplified all pilot orientations to $1 + j0$, which of course is not the case for the full 802.11a specifications, where the receiver would first need to derotate all pilots to $1+j0$ before taking the average and determining the phase offset. The phase error of the current symbol is therefore equal to the average of the actually received pilot tone angles minus the intended pilot angles of 0 radians. To set things straight, the phase error is simply subtracted from all FFT outputs.

$$PhaseError = \frac{1}{4} \sum_{k=0}^{3} angle(Pilot[k])$$

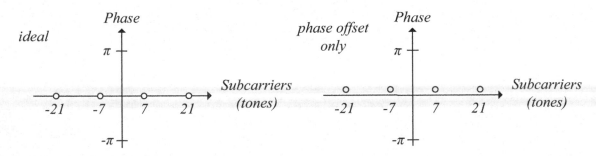

Figure 7-49: Pilot Tone Phases for Varying Scenarios of Phase Offsets

We discussed the properties of phase noise and its methods of generation in chapter 5. From that discussion, we learned that phase noise is characterized by its profile in the frequency domain, as well as its total *rms* phase deviation. Just from a casual observation of any phase noise profile, it is clear that a good amount of its content resides at low frequencies. These low frequency components will not cause the overall phase noise waveform to change over any one single OFDM symbol and are subsequently removed during phase offset detection and correction. The problem lies in the higher frequency components, which do change during a single OFDM frame. These components can neither be detected nor corrected and cause a loss of orthogonality during the FFT process. The correction mechanism therefore acts as a low pass filter removing low frequency phase noise components but leaving the higher rate counterparts intact. The proper phase noise profile for the receive and transmit synthesizers must therefore avoid content at higher frequencies. We will show the impact of various profiles on the constellation and EVM performance of the receiver later in this chapter.

Detecting and Removing Timing Drift

As was discussed in detail in chapter 5, timing drift occurs because of the difference in reference clocks at the transmit and receive portions of the communication link. These reference clocks are provided by temperature compensated crystal oscillators, which feature frequency uncertainties specified in parts per million. For errors of ±20ppm, the digital transmit/DAC combination may be working at $40 \cdot (1 + 20e\text{-}6)\text{MHZ} = 40.0008\text{MHz}$, while the receiver/ADC combination samples and processes the signal at $40 \cdot (1 - 20e\text{-}6)\text{MHZ} = 39.9992\text{MHz}$. As we illustrated in chapter 5, timing error introduces a linear phase shift into the spectrum of the signal we are processing. This phase shift is revealed by the time shifting property of the Fourier transform.

$$given \qquad x(t) \overset{FT}{\leftrightarrow} X(f)$$

$$then \qquad x(t - t_o) \overset{FT}{\leftrightarrow} X(f) \cdot e^{-j2\pi f t_o}$$

A positive timing offset, $t_o > 0$, indicates that we are late in observing or sampling the waveform $x(t)$, thereby causing a phase shift $e^{-j2\pi f t_o}$ to occur in the signal spectrum. This phase shift is directly proportional to the frequency, f, and causes negative phase change for positive frequencies and positive change for negative frequencies. The two bottom plots of the figure below illustrate the effects of a negative timing offset, $t_o < 0$, indicating that we are early in our observation. To detect the offset, we compute the slope of the straight line described by the pilot phases as shown in the figure. Note that the plot at the bottom right shows pilot orientations that indicate both timing and phase offsets.

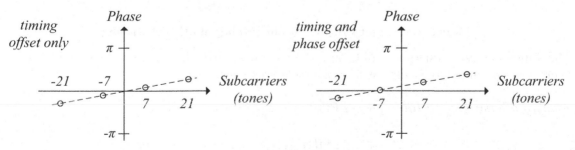

Figure 7-50: Pilot Phases for Negative Timing Offset ($t_o < 0$)

The primary method of removing the timing offset from the FFT result is simply to modify the phases of all information bearing subcarriers in accordance with the following formula.

$$PhaseCorrection(f,t) = e^{j2\pi f t_o}$$

If the packet is very long, the timing drift may cause the sampling point to wander over several samples periods during transmission. In this case, simple post FFT phase correction of the subcarriers is no longer possible. For the clock inaccuracies given above, a total reference frequency offset of 40.0008 MHz – 39.9992 MHz = 1.6 KHz will cause the sampling point to wander past the 25nsec sampling period within less than 1msec. For long packets, therefore, an interpolator will have to be designed that features enough past and future samples to time-correct the waveform. These techniques are explained in the section on interpolation in chapter 3. Here, we assume that the packets are short enough and the timing drift benign enough for post FFT modification of the subcarrier phases to be sufficient.

Implementation of the Phase and Timing Tracking Algorithms

Remember that the initial timing reference provided by the long symbol cross-correlator was accurate to within one sample only. Both the initial phase and timing inaccuracy, or slant in subcarrier phase, are revealed during the channel estimation and corrected in the equalizer at the beginning of the packet. Although both phase and timing offset may have been addressed during the equalizer set up, phase and timing drift require continuous correction. The correction happens in two steps. First, the average phase offset is calculated, and all equalized subcarrier results are corrected accordingly. Next, the average slope is calculated, and the subcarrier results modified once again. In the second step, we take the averaged slope and phase offset results to update some or all of the following: the equalizer coefficients, an interpolator, and the numerically controlled oscillator. If we were not to update these modules, the pilot phases would increase in magnitude until cross-over between ±π would make proper average phase and slope calculations impractical.

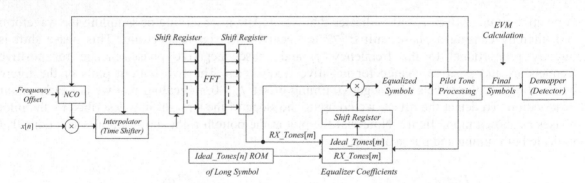

Figure 7-51: Pilot Processing Positioning in OFDM Receiver

The following two equations guide the average phase and slope calculations. Note that the average slope value represents the change in phase per subcarrier.

$$AveragePhase = c_1 \angle Pilot_{-21} + c_2 \angle Pilot_{-7} + c_3 \angle Pilot_7 + c_4 \angle Pilot_{21}$$

$$AverageSlope = -c_1 \frac{\angle Pilot_{-21}}{21} - c_2 \frac{\angle Pilot_{-7}}{7} + c_3 \frac{\angle Pilot_7}{7} + c_4 \frac{\angle Pilot_{21}}{21}$$

While the coefficients c_1 through c_4 may each be set to ¼ for an equal gain combining average, we may also use maximum ratio combining to optimize the signal to noise ratio of the final result. In this case, the coefficients are set in accordance with their signal strength as indicated by the channel estimate. Maximum ratio combining allows us to deemphasize pilot information that has been suppressed by selective fading, and therefore features poor signal to noise ratio.

7.4 802.11a Performance Evaluation and Transceiver Code

To evaluate the performance of the ODFM receiver, we introduce two very important test metrics. The first, called EVM versus frequency, allows us to view the average EVM performance over one packet as a function of subcarrier, where each of these subcarriers is located at $[-26:26] \cdot 312.5 KHz$. The second metric, called EVM versus time, averages EVM over all subcarriers and records it on an OFDM symbol by symbol basis. As a debugging tool, these two test metrics, shown as plots, are invaluable as we turn on different defects in our model. Both multipath and improper timing control will prominently appear in the EVM versus frequency plot, whereas effects such as improper phase tracking will be clearly visible in the EVM versus time graph. These graphs are available on all modern VSA (vector signal analyzer) software packages such as the Agilent 89600 suite.

The Impact of White Gaussian Noise on Performance

The two plots below show the best possible performance of the receiver for a signal to noise ratio of 25dB and a packet length of 128 OFDM symbols. Beyond the Gaussian noise, no other effects were included and the pilot based timing and phase tracking algorithms were turned off. This performance is independent of the chosen QAM modulation format.

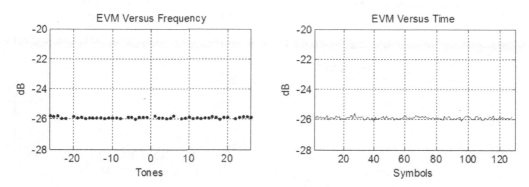

Figure 7-52: EVM Performance for SNR = 25dB (no Additional Defects / no Tracking)

One curiosity is the fact that the output EVM $\approx$ -26dB, which is more or less a reformulation of the signal to noise ratio expression, is in fact superior to the SNR conditions = 25dB of the input signal. Why would this be? Note that we applied thermal white Gaussian noise while the signal was sampled at 40MHz. Although this noise is flatly dispersed over the range of -20 MHz to 20MHz, the signal content is in fact constrained to approximately -8.3 to 8.3MHz. In the receiver code, we forgo the half band filtering process, which would have cut the noise power in half, before decimating. Decimation without filtering causes the noise power outside the legal Nyquist range to alias into the band. The noise power is therefore preserved and the signal to noise ratio continues to be as we had intended it. However, the noise power is still flatly distributed across the ±10MHz Nyquist range, while the signal is not. Inside the signal's frequency band, the signal to noise ratio per subcarrier is improved by approximately $10 \cdot log_{10}(10/8.3) = 0.8$dB. This idealized EVM performance quickly degrades as we turn on different tracking features required in the modem. Enabling phase tracking reduces performance slightly, since the pilot information, on which the tracking algorithms are based, is corrupted by Gaussian noise.

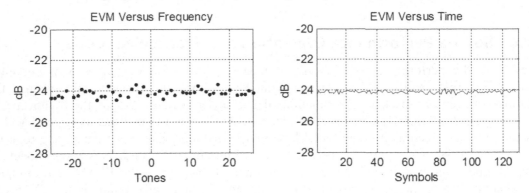

Figure 7-53: EVM Performance for SNR = 25dB (no Additional Defects/Phase Tracking Enabled)

The first really interesting result occurs, however, when timing acquisition based on pilot tone information is enabled. Since the pilot tones are noisy, as determined by the current SNR condition, the calculated timing offset value and subsequent correction will be suboptimal. From the time-shifting property of the Fourier transform, we know that the effect of timing offset becomes larger as frequency increases. The resulting concave tendency of the EVM versus frequency graph will, however, be drowned out by the selective fading effects that are introduced after the multipath effects are turned on.

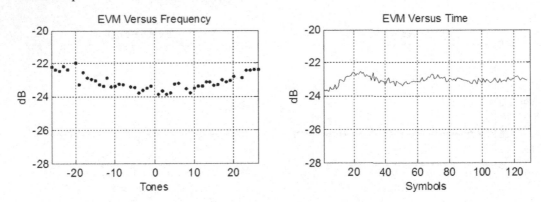

Figure 7-54: EVM Performance for SNR = 25dB (no Additional Defects / all Tracking Enabled)

As a rule of thumb, robust detection (demapping) of a 64 QAM constellation requires an error vector magnitude of less than or equal to -25dB. Owing to the loss in modulation accuracy caused by the noise affected tracking operation, the input signal to noise ratio will need to be at least 27dB. The two graphs below show constellation information at the demapper input for input signal to noise ratios of 25 and 30dB. Notice that constellation positions that are further removed from the center are slightly more degraded than the close-in ones. This effect is primarily due to noise impaired equalizer coefficients whose errors cause larger deviations for higher magnitude position than for smaller ones. Phase tracking inaccuracies will also appear more prominently at higher magnitude constellation, but phase noise will not.

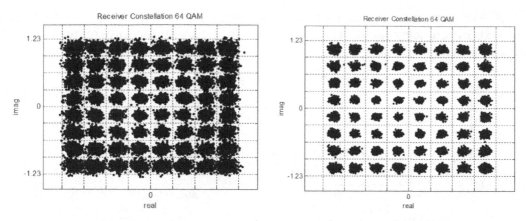

Figure 7-55: Receive 64 QAM Constellation Plots for Input SNR = 25dB (Left)/30dB (Right)

Adding Multipath Effects

The next figure shows the receiver performance as we introduce a harmless multipath channel featuring an *rms* delay spread of 10 nanoseconds. The bottom graphs present the absolute value of the multipath filter's impulse response (FIR taps), as well as its accompanying magnitude behavior in the frequency domain. Since the multipath filter runs at 40MHz, the normalized frequency span between ±0.21Hz ($\pm0.205\cdot40$MHz $= \pm8.2$MHz) properly covers the 52 tones at $\pm26\cdot312.5$ KHz $= \pm8.125$MHz. The multiple paths accentuate certain frequencies while attenuating others, as can be clearly seen by the EVM versus frequency plot. Note that 802.11a is somewhat rigid in its assignment of data to subcarriers, which must all operate using the same constellation (BPSK, QPSK …). Newer OFDM systems use adaptive modulation, which inserts more data bits (higher constellation) at subcarriers that feature favorable SNR ratios and fewer at those that don't.

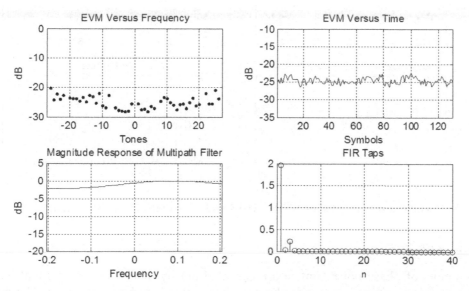

Figure 7-56: Performance Graphs for SNR = 25dB and Multipath Delay Spread of 10 Nanoseconds

The next two figures measure the receiver performance for multipath conditions of 50 and 100 nanoseconds of *rms* delay spread. The EVM versus frequency plots clearly mirror the magnitude response of the multipath filter.

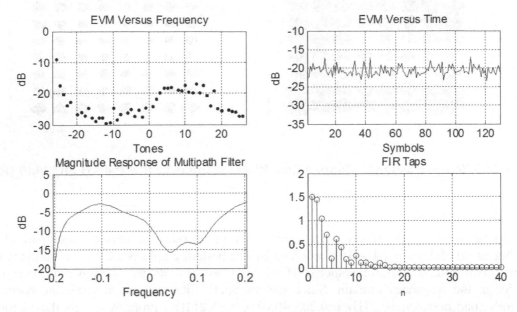

Figure 7-57: Performance Graphs for SNR = 25dB and Multipath Delay Spread of 50 Nanoseconds

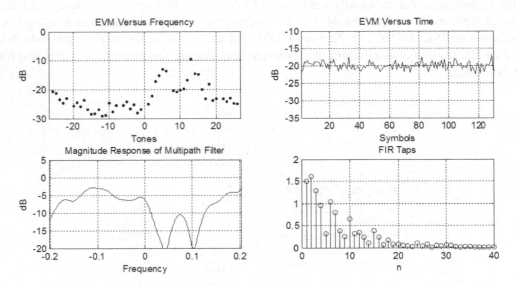

Figure 7-58: Performance Graphs for SNR = 25dB and Multipath Delay Spread of 100 Nanoseconds

The Impact of Phase Noise

Certain amounts of phase noise content are canceled out by the symbol by symbol phase offset tracking mechanism, which acts as a high pass filter to the disturbance in angle. The phase noise profile shown below features a total *rms* phase error of approximately 4.4 degrees, but harnesses its content at lower frequencies. The signal to noise ratio used in this simulation was 30dB, leading us to a best possible EVM metric of approximately -28dB given the unavoidable performance penalty of the pilot aided tracking operations. And indeed, the constellation below features an error vector magnitude of -28dB, indicating that the performance did not deteriorate due to the presence of phase noise. Outside the white Gaussian and phase noises, no other defects were injected in the results below.

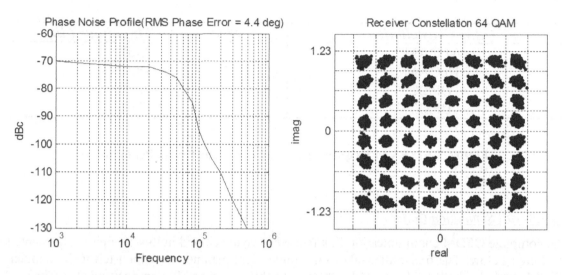

Figure 7-59: Negligible Impact of Phase Noise on 64QAM Constellation at 30dB Input SNR

The second phase noise profile also features a 4.4 degree *rms* error but distributes its content over a wider frequency range, thus allowing more of the disturbances to escape the phase offset tracking mechanism. The difference in profiles is obvious to the naked eye and so is the deterioration of the constellation's EVM, which now hovers around -24dB. Simulations such as these provide system engineers with valuable insight that can be passed on to RF IC designers in charge of synthesizer development.

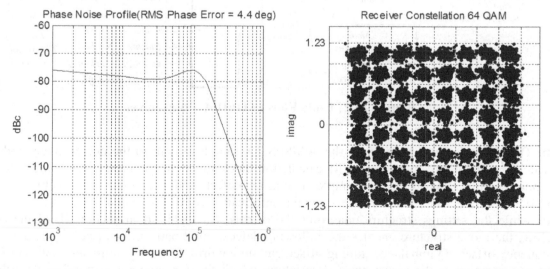

Figure 7-60: Significant Impact of Phase Noise on 64QAM Constellation at 30dB Input SNR

EVM versus QAM Constellation

The table below presents a list of approximate EVM thresholds below which the 4 QAM modes should be decoded with few or no errors. These numbers ignore multipath effects which, could produce an average EVM metric that appears good but includes some low signal to noise ratio subcarriers that would cause errors.

QAM Constellation	Error Vector Magnitude (dB)
64 QAM	< -25dB
16 QAM	< -18dB
QPSK	< -12dB
BPSK	< -7dB

Figure 7-61: EVM Requirements for Excellent Error Performance

The Overall Simulation Code

The complete OFDM communication link that we have developed in this chapter is represented in the figure below. The transmitter takes only three input parameters, which tell it the number of OFDM symbols, the IFFT length (64/128), and the bits per constellation position (1 = BPSK / 2 = QPSK / 4 = 16QAM / 6 = 64QAM).

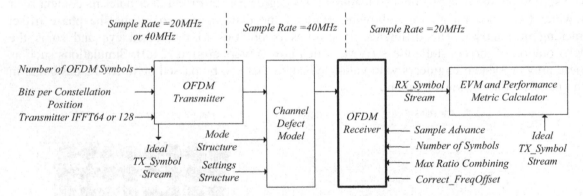

Figure 7-62: Code Flow of OFDM Transceiver

The OFDM transmitter hands over the 40MSPS TX sample stream to the defect model and the ideal TX symbol stream to the EVM calculation block for later comparison with the receiver output symbols. Beyond the TX sample stream, the defect model requires two structures called mode and setting. The mode structure contains information regarding which defect to turn on and off, while the settings structure quantifies the defects. As specified in section 1b of the code below, the mode structure enables the following defects: multipath, thermal noise, phase noise, frequency offset, IQ imbalance, timing offset, and timing drift. The only confusing portion of the settings structure is the phase noise description, which features two rows, the first describing frequency positions and the second matching them to dBc values. Otherwise, the configuration of the defect model should be self-explanatory and follows the format introduced in chapter 5.

The miscellaneous controls allow several options associated with the receiver and the simulation in general. In the receiver, it allows us to turn off frequency offset correction (for debugging purposes), switches in the use of maximum ration combining for pilot processing, and advances the FFT timing reference in order to avoid processing samples that are located too close to the next OFDM symbol. By holding the random seed generator at a static value, we are able to repeat the simulation using different modes and options but identical noise values. Therefore, the multipath filter, phase noise, and thermal Gaussian noise remain unchanged from simulation to simulation.

```matlab
%%                      1. Overall simulation setup
%% 1a.Configuring the transmitter
NumberOf_OFDMSymbols = 1500;
BitsPerSymbol        = 6;                % BitsPerSymbol = 1,2,4,6 =
                                         % BPSK,QPSK,16QAM,64QAM
TransmitterChoice    = 1;                % 0/1 = upsampled / IFFT128


%% 1b. Configuring defect model code
Mode.Multipath       = 0;                % 0/1 = exclude/include defect
Mode.ThermalNoise    = 1;
Mode.PhaseNoise      = 0;
Mode.Freq_Offset     = 0;
Mode.IQ_Imbalance    = 0;
Mode.TimingOffset    = 0;
Mode.TimingDrift     = 0;


Settings.SampleRate      = 40e6;         % Hertz
Settings.NumberOfTaps    = 40;           % Of multipath FIR model
Settings.DelaySpread     = 150e-9;       % RMS delay spread in seconds
Settings.SNR_dB          = 35;           % Signal to noise ratio in dB
Settings.PhaseNoiseProfile = [1e3 10e3 20e3 35e3 50e3 80e3 90e3 100e3 ...
                      120e3 150e3 200e3 300e3 500e3 1e6; ...
                     -70 -72 -72 -74 -76 -85 -90 -95 ...
                     -100 -105    -110 -120 -130 -140];
Settings.FrequencyOffset     = -100e3;   % Hz
Settings.PhaseImbalance      = pi/2000;  % radians
Settings.AmplitudeImbalance_dB = -.1;    % dB
Settings.Sample_Offset       = -1;       % samples at the sample rate
Settings.Drift_ppm           = -80;      % given above in ppm (parts per
                                         % million
%% 1c. Configuring miscellaneous simulation controls
CorrectFrequencyOffset   = 1;            % 0/1 = no/yes
UseMaxRatio_Combining    = 1;            % 0/1 = don't use / do use
StaticNoise              = 0;            % 0/1 = different / same noise
NoiseSeed                = 600;          % Any positive integer
SampleAdvance            = 3;            % Min = 0 / Max = 15

if(StaticNoise); randn('state',NoiseSeed); end


%% 2. Transmitter setup      --> See code at the beginning of this chapter
if(TransmitterChoice == 1)
    [Sample_Output TX_Symbol_Stream] = ...
          OFDM_TX_802_11_128(NumberOf_OFDMSymbols, BitsPerSymbol);
else
    [Sample_Output TX_Symbol_Stream] = ...
          OFDM_TX_802_11(NumberOf_OFDMSymbols, BitsPerSymbol);
```

```
end
TX_Output_Clean        = [zeros(1,10) Sample_Output zeros(1,10)];

%% 3. Defect model      --> See code in chapter 5
[TX_Output FIR_Taps]   = Defect_Model(TX_Output_Clean, Settings, Mode);

%% 4. Receiver code     --> See code at the end of this chapter
[Corrected_Symbols]    = OFDM_Receiver(TX_Output, ...
                                       SampleAdvance, ...
                                       CorrectFrequencyOffset, ...
                                       NumberOf_OFDMSymbols,...
                                       UseMaxRatio_Combining);

%% 5. Performance evaluation
ErrorVectors              = TX_Symbol_Stream - Corrected_Symbols;
Average_ErrorVectorPower = ...
        (1/length(ErrorVectors))*(ErrorVectors*ErrorVectors');

EVM = 10*log10(Average_ErrorVectorPower/1);
display(['EVM = ' num2str(EVM) 'dB']);

% EVM versus rime calculation
Error_Time = zeros(1, NumberOf_OFDMSymbols);
for i = 1:NumberOf_OFDMSymbols
    Start            = (i-1)*48 +  1;
    Stop             = (i-1)*48 + 48;
    CurrentSymbol    = Corrected_Symbols(1, Start:Stop);
    IdealSymbol      = TX_Symbol_Stream(1, Start:Stop);
    ErrorVector      = IdealSymbol - CurrentSymbol;
    Error_Time(1,i)  = (1/48)*(ErrorVector*ErrorVector');
end

% EVM versus frequency calculation
Error_Frequency = zeros(1,48);
for i = 1:NumberOf_OFDMSymbols
    Start            = (i-1)*48 +  1;
    Stop             = (i-1)*48 + 48;
    CurrentSymbol    = Corrected_Symbols(1, Start:Stop);
    IdealSymbol      = TX_Symbol_Stream(1, Start:Stop);
    ErrorVector      = IdealSymbol - CurrentSymbol;
    Error_Frequency = Error_Frequency + ...
                   (ErrorVector.*conj(ErrorVector))/NumberOf_OFDMSymbols;
end


EVM_Frequency      = 10*log10(Error_Frequency);
EVM_Time           = 10*log10(Error_Time);
PosIndex           = [1:6 8:20 22:26];
NegIndex           = [38:42 44:56 58:63] - 64;

f = -.5:0.001:.5;      %% Use the DFT to find multipath magnitude response
Response = zeros(1,length(f));
n = 0:1:length(FIR_Taps)-1;
for d=1:length(f)
    E = exp(j*2*pi*n*f(1,d)/1);
    Response(1,d) = FIR_Taps*E';
end
```

```
MagResponse        = 20*log10(abs(Response));
MagResponse_norm = MagResponse - max(MagResponse);


figure(1);
hold off;
subplot(2,2,1); hold off;
plot([PosIndex NegIndex], EVM_Frequency,'k.'); grid on; ylabel('dB')
title('EVM Versus Frequency'); xlabel('Tones'); axis([-27 27 -35 0]);
subplot(2,2,2); hold off;
plot(EVM_Time,'k'); grid on; title('EVM Versus Time');
xlabel('Symbols'); ylabel('dB'); axis([1 130 -35 -10])
subplot(2,2,3); hold off;
plot(f, MagResponse_norm,'k'); grid on; xlabel('Frequency'); hold on;
title('Magnitude Response of Multipath Filter');
axis([-13/64 13/64 -20 5]); ylabel('dB');
subplot(2,2,4); hold off;
stem(abs(FIR_Taps),'k'); grid on; title('FIR Taps'); xlabel('Symbols');


figure(2);
plot(real(Corrected_Symbols), imag(Corrected_Symbols),'k.', 'Markersize', 10);
grid on;xlabel('real'); ylabel('imag'); axis([-1.5 1.5 -1.5 1.5]);
```

The OFDM Receiver Code

The receiver code does not filter the signal prior to the decimation process, as no signal content in neighboring channels was produced by the transmitter. The filter is similar to the one used for upsampling in the transmitter and is skipped for simplicity. The packet detector's main function is to supply the falling edge position indicator that serves as a first crude timing reference used by the frequency detection function.

The frequency detection function returns an array called frequency offset, which contains both the coarse and the fine estimate. We therefore run it once, correct the coarse offset, and then run it again to correct the fine frequency offset. Note that disabling the frequency correction in the receiver and setting the frequency offset and phase noise to zero in the defect model facilitates the testing of the phase offset tracking algorithm. If no phase drift exists, then the four pilot tone orientations must average to zero.

In section five of the code, we establish the IFFT timing reference, which we advance in time to establish a small distance (buffer region) from the discontinuous boundary, which exists between OFDM symbols. Sections six and seven in the code compute the equalizer coefficients in the exact fashion spelled out in section 6.3.5, as well as the maximum ratio combining averaging factors.

Section eight of the code contains the main loop that processes every payload OFDM symbol by first equalizing it and tracking the phase and timing drift defects. The loop goes on to store all corrected IFFT output symbols for later comparison against the original transmitted values. The code contains an adjustable length smoothing filter to reduce the error in the calculated slope that forms part of the timing drift correction process. The variable L allows control over the filter length and can be set as low as L= 2.

```
function [Corrected_Symbols] = OFDM_Receiver(TX_Output, ...
                                   SampleAdvance, ...
                                   CorrectFrequencyOffset, ...
                                   NumberOf_OFDMSymbols,...
                                   UseMaxRatio_Combining
%% 1. The receiver code : decimation by 2
RX_Waveform_20MHz = TX_Output(1,1:2:end);


%% 2. Packet detector
[Comparison_Result PacketDetFlag FallingEdgePosition AutoCorr_Est] ...
       = Packet_Detector(RX_Waveform_20MHz);


%% 3. Detect and correcting the coarse frequency offset estimate
[FreqOffset]       = Detect_FrequencyOffsets(RX_Waveform_20MHz, ...
                        FallingEdgePosition);
CoarseOffset       = FreqOffset(1,1);
n                  = 1:length(RX_Waveform_20MHz);
NCO_Signal         = exp(-j*2*pi*n*CoarseOffset/20e6);
if(CorrectFrequencyOffset == 1)
    RX_Waveform_20MHz = RX_Waveform_20MHz.*NCO_Signal;
end


%% 4. Detect and correcting the fine frequency offset estimate
[FreqOffset]       = Detect_FrequencyOffsets(RX_Waveform_20MHz, ...
                        FallingEdgePosition);
FineOffset         = FreqOffset(1,2);
n                  = 1:length(RX_Waveform_20MHz);
NCO_Signal         = exp(-j*2*pi*n*FineOffset/20e6);
if(CorrectFrequencyOffset == 1)
    RX_Waveform_20MHz = RX_Waveform_20MHz.*NCO_Signal;
end


%% 5. Long training symbol correlator
LongTrainingSequence = Get_LongTrainingSequence(1);
LongTrainingSymbol   = LongTrainingSequence(1,33:96);
[LTPeak_Value1 LTPeak_Position OutputLong] = LongSymbol_Correlator( ...
       LongTrainingSymbol, RX_Waveform_20MHz, FallingEdgePosition);


LTPeak_Position = LTPeak_Position - SampleAdvance; % Insert time advance


%% 6. Channel estimate and equalizer setup
  % --> Extract and average both long training symbols from RX waveform
First_LongSymbol =  RX_Waveform_20MHz(1,LTPeak_Position - 64: LTPeak_Position - 1);
Second_LongSymbol = RX_Waveform_20MHz(1,LTPeak_Position    : LTPeak_Position +63);
Averaged_LongTraining_Symbol = First_LongSymbol*.5 + Second_LongSymbol*.5;


  % --> Take its FFT and extract the positive and negative tone orientations
FFT_Of_LongTraining_Symbol = (1/64)*fft(Averaged_LongTraining_Symbol);
RX_Positive_Tones          = FFT_Of_LongTraining_Symbol(1, 2:27);
RX_Negative_Tones          = FFT_Of_LongTraining_Symbol(1, 39:64);
  % --> Establish ideal tone orientations
[Ignore AllTones]          = Get_LongTrainingSequence(1);
Ideal_Positive_Tones       = AllTones(1, 34:59);
```

```
Ideal_Negative_Tones        = AllTones(1,  7:32);
   % --> Channel estimate and equalizer coefficient calculation
Channel_Estimate_Pos        = RX_Positive_Tones ./ Ideal_Positive_Tones;
Channel_Estimate_Neg        = RX_Negative_Tones ./ Ideal_Negative_Tones;
Channel_Estimate            = [0 Channel_Estimate_Pos zeros(1,11) ...
                               Channel_Estimate_Neg];


Equalizer_Positive          = 1./Channel_Estimate_Pos;
Equalizer_Negative          = 1./Channel_Estimate_Neg;
Equalizer_Coefficients      = [0 Equalizer_Positive zeros(1,11) ...
                               Equalizer_Negative];


%% 7. Examine pilot tones for subsequent maximum ratio combining
Pilot_Minus21_Strength      = abs(Channel_Estimate(1, 1 + 43) );
Pilot_Minus07_Strength      = abs(Channel_Estimate(1, 1 + 57) );
Pilot_Plus_07_Strength      = abs(Channel_Estimate(1, 1 +  7) );


Pilot_Plus_21_Strength      = abs(Channel_Estimate(1, 1 + 21) );


SumIt = Pilot_Minus21_Strength + Pilot_Minus07_Strength + ...
        Pilot_Plus_07_Strength + Pilot_Plus_21_Strength;


if(UseMaxRatio_Combining == 0)
    C1      = 1/4;
    C2      = 1/4;
    C3      = 1/4;
    C4      = 1/4;
else
    C1      = Pilot_Minus21_Strength/SumIt;
    C2      = Pilot_Minus07_Strength/SumIt;
    C3      = Pilot_Plus_07_Strength/SumIt;
    C4      = Pilot_Plus_21_Strength/SumIt;
end



%% 8. Process all received symbols
L = 8;
AverageSlopeFilter = zeros(1,L);


Corrected_Symbols = zeros(1, 48*NumberOf_OFDMSymbols);
for i = 1:NumberOf_OFDMSymbols
    Start               = (i-1)*80 + LTPeak_Position + 64 + 16;
    Stop                = (i-1)*80 + LTPeak_Position + 127 + 16;
    CurrentOFDMSymbol   =  RX_Waveform_20MHz(1, Start: Stop);
    CurrentFFTOutput    = (1/64)*fft(CurrentOFDMSymbol);
    EqualizedSymbol     = CurrentFFTOutput .* Equalizer_Coefficients;

   % --> Processing the phase drift
    Pilot_Minus21       = EqualizedSymbol(1, 1 + 43);
    Pilot_Minus07       = EqualizedSymbol(1, 1 + 57);
    Pilot_Plus_07       = EqualizedSymbol(1, 1 +  7);
    Pilot_Plus_21       = EqualizedSymbol(1, 1 + 21);


    Averaged_Pilot      = C1*Pilot_Minus21 + C2*Pilot_Minus07 + ...
                          C3*Pilot_Plus_07 + C4*Pilot_Plus_21;
```

```
    Theta               = angle(Averaged_Pilot);
    DerotationScalar    = exp(-j*Theta);
    CorrectedSymbol1    = EqualizedSymbol * DerotationScalar;
    Equalizer_Coefficients = Equalizer_Coefficients  * exp(-j*Theta/L);


    % --> Remove average phase offset from pilots
    Pilot_Minus21       = EqualizedSymbol(1, 1 + 43)*conj(Averaged_Pilot);
    Pilot_Minus07       = EqualizedSymbol(1, 1 + 57)*conj(Averaged_Pilot);
    Pilot_Plus_07       = EqualizedSymbol(1, 1 +  7)*conj(Averaged_Pilot);
    Pilot_Plus_21       = EqualizedSymbol(1, 1 + 21)*conj(Averaged_Pilot);


    Slope               = - C1*angle(Pilot_Minus21)/21  ...
                          - C2*angle(Pilot_Minus07)/7   ...
                          + C3*angle(Pilot_Plus_07)/7   ...
                          + C4*angle(Pilot_Plus_21)/21;


    AverageSlopeFilter(1,2:L)  = AverageSlopeFilter(1,1:L-1);
    AverageSlopeFilter(1,1)    = Slope;
    AverageSlope               = sum(AverageSlopeFilter)/(L);


    StepPlus         = (  0:31)*AverageSlope;
    StepMinus        = (-32:-1)*AverageSlope;
    AppliedCorrection = [StepPlus StepMinus];
    CorrectedSymbol2 = CorrectedSymbol1 .* exp(-j*AppliedCorrection);
    Equalizer_Coefficients = Equalizer_Coefficients .* ...
                                    exp(-j*AppliedCorrection/L);


    Start                      = (i-1)*48 +  1;
    Stop                       = (i-1)*48 + 48;
    IFFT_Index                 = [1:6 8:20 22:26 38:42 44:56 58:63];
    Corrected_Symbols(1, Start:Stop) = CorrectedSymbol2(1, 1 + IFFT_Index);
end
```

7.5 The LTE Transmitter

The 802.11a modem that we covered in the last two sections is a wonderful introduction to OFDM as its design is not overly complicated. As a matter of fact, the IEEE standard [1] covers the entire physical layer in just over 50 pages and a single, motivated engineer can read and get a complete overview of the transmitter architecture within a reasonably short time. Unfortunately, the 3GPP physical layer description of the LTE modem spans several documents each boasting hundreds of pages. Fortunately for us, a great deal of this material is not related to digital signal processing and but rather with the allocation of information of different physical channels into the OFDM grid. Before we start to dive into the material, let's examine what makes the LTE modem different from the 802.11A modem.

Mobility

802.11a transceivers operate in a physically stationary environment, meaning that they don't move. A client transceiver is not meant to be integrated into a moving car and communicate with an access point sitting on the side of the road. As a consequence, the 802.11a signal provides reference signals at the start of the packet allowing the receiver to fully train its equalizer only at that time. A few reference signals are provided later on during the packet to detect phase noise and timing drift, but all in all, the magnitude response is thought to be stable as the multipath channel remains static with time. However, since the LTE mobile is designed to be used in moving cars, the multipath channel is constantly changing thus requiring reference/pilot signals to be present throughout time.

Packetization

Wifi information is sent back and forth within self-contained packets. The receiver listens to find out whether the channel is being used by an access point or client unit via simple RF power detection and is free to start transmission if all is clear. An access point and client unit transmit information at the same frequency but not at the same time. This scheme is called time division duplexing, or TDD. Furthermore, there is no clear schedule during which each transceiver maintains transmission. This is very different from the LTE system where both time and frequency division duplexing, or FDD, are supported. In FDD, the base station is continuously transmitting at one frequency, whereas all mobile phone talk back on a different frequency. Even in the TDD mode, which is heavily used in China, the base station and mobile phones have very precise schedules when they are allowed to transmit. Furthermore, the information that the mobile phone needs to receive from the base station in order communicate properly is spread out over time and not confined to a single packet.

Interference

LTE mobile phones are designed from the get-go to operate in the presence of interference from multiple cell towers. The coverage area of different LTE base stations must obviously overlap in order for the mobile phone to terminate its link with one and establish a link with another cell tower. For this to work, the LTE link has to be maintained at signal to noise ratios of 0 dB or less, and significantly more redundancy and error protection is provided compared to the 802.11a standard.

7.5.1 An Overview of LTE Topics

In this text, we will focus on two major signal processing activities undertaken by mobile phones.

Scanning

An LTE mobile phone is continuously scanning its environment to ascertain the cell identity, the signal quality as CINR (carrier to interference and noise ratio) and timing synchronization of each cell it can hear. The mobile phone will keep and continually update a list of base stations with these parameters in case it needs to establish a link with one of them, or execute a hand-over from one cell to the next during a call. Determining the cell identity and timing reference requires the processing of synchronization signals, whereas determination of CINR is best done using the synchronization or the abundant reference signals available in the downlink (cell tower to mobile) signal.

Demodulation and Information Decoding

The mobile will further receive and demodulate broadcast or system information, which is periodically available from each base station and provides more detailed information such as LTE bandwidth, the network operator name and access rights to the cell. The demodulation and decoding steps involved are more sophisticated than those present in the 802.11a modem. Only when this broadcast information has been determined can the mobile phone establish two-way communication with the cell tower.

Table 7-2: Useful 3GPP Specifications for LTE

3GPP Specification	Topics
TS36.211 Section 6.2	Resource Grids, Resource Elements and Resource Blocks
TS36.211 Section 6.3	General Physical Channel Scrambling, Modulation, Layer Mapping, Precoding and Resource Element Mapping
TS36.211 Section 6.6	Physical Downlink Broadcast Channel Scrambling, Modulation, Layer Mapping, Precoding and RE Mapping
TS36.211 Section 6.8	Physical Downlink Control Channel Scrambling, Modulation, Layer Mapping, Precoding and RE Mapping
TS36.211 Section 6.10	Explanation of Reference Signals
TS36.211 Section 6.11	Explanation of Synchronization Signals
TS36.211 Section 6.12	OFDM Baseband Signal Generation
TS36.212 Section 5.1.1	CRC Encoding
TS36.121 Section 5.1.3	Tail Biting Convolutional Encoding / Turbo Encoding
TS36.121 Section 5.1.4	Rate Matching Explanation
TS36.331 Section 6.2.2	Definition of MIB (Master Information Block) and SIB1 (System Information Block 1)

7.5.2 LTE Frequency Allocation and Frame Structure

The LTE base station signal is OFDM modulated and features bandwidths between 1.4MHz to 20MHz. The frequency domain is organized into **resource blocks**, or RBs, consisting of 12 subcarriers, where subcarriers are spaced at 15 KHz resulting in a fundamental RB bandwidth of 180 KHz. A resource block extends over a time period of one 0.5msec slot in the time domain and consists of either 6 or 7 OFDM symbols. The resource block is the smallest resource unit that is subject to information scheduling in LTE. Thus, any block of control or user data scheduled in the LTE grid must be in integer multiples of a resource block. Notice that a resource block contains 12·7 = 84 or 12·6 = 72 positions where complex valued information may be placed. A position such as this is called a **resource element**, or RE. The figure below illustrates the resource grid of a 1.4 MHz and 3 MHz LTE signal.

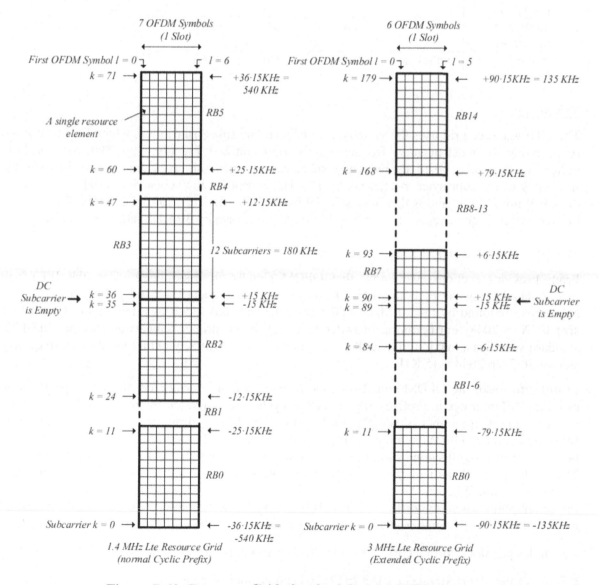

Figure 7-63: Resource Grids for the 1.4 and 3.0MHz Signals

LTE Signal Bandwidth

The last figure illustrates the LTE resource grid organization of a 1.4 and 3MHz signal featuring 6 and 15 resource blocks respectively. It is clearly indicated that the subcarriers for the two bandwidths extent from -540 to 540 KHz = 1.08 MHz and -1.35 to 1.35 MHz = 2.7 MHz, which fall short of the official figures of 1.4 and 3 MHz. The official bandwidths include a guard band, which is necessary as the OFDM signal spectrum leaks some content beyond its outer subcarriers. We know this phenomenon as FFT leakage, and the guard band ensures that leakage from one LTE carrier doesn't leak into the bandwidth of a neighboring LTE carrier.

Table 7-3: Bandwidth Organization of LTE Signals

	1.4MHz	3MHz	5MHz	10MHz	15MHz	20MHz
Number of RBs (NDLRB)	6	15	25	50	75	100
Number of Subcarriers	72	180	300	600	900	1200
Subcarrier Extent	1.08MHz	2.7MHz	4.5MHz	9MHz	6.75MHz	9MHz

Subcarriers

The LTE resource grid uses the variable k to indicate the subcarrier index, which begins at $k = 0$ representing the most negative frequency subcarrier and $k = 71, 179, 299, 599, 899$, or 1199 indicating the most positive frequency subcarrier. In general, as k increments by one the frequency of the subcarrier increments by 15 KHz, except for the subcarrier at 0Hz, which does not carry information and is therefore skipped by the index k. Thus, for the 1.4 MHz LTE signal, the subcarrier frequencies for indices $k = 35$ and 36 increment by 30 KHz rather than 15 KHz.

The OFDM Symbol and its Cyclic Prefix

From previous sections, we know that OFDM signals are modulated using the IFFT and demodulated using the FFT algorithm. In order to provide space for 1200 information carrying subcarriers required by the 20 MHz LTE signal, the standard specifies a maximum default FFT size of $N = 2048$. Furthermore, in order to establish the proper subcarrier spacing, the LTE standard specifies a maximum default sample rate of 30.72 MSPS, yielding the desired frequency step of 30.72e6/2048 = 15 KHz.

In order to make the OFDM signal immune to intersymbol interference, the LTE specification provides different cyclic prefixes. To create the cyclic prefix (called guard interval in 802.11a), we copy a certain number of samples at the end of the IFFT output sequence and prepend it to form the complete OFDM symbol. LTE provides a normal and extended cyclic prefix, where the normal version is 160 samples long in the first symbol of a slot and 144 samples long otherwise. The cyclic prefix length of the extended cyclic prefix has a length of 512 samples for every symbol in a slot. Whereas the extended CP is hardly used, it does avoid intersymbol interference for geographies where the difference in delays is very large. Whether the cyclic prefix is normal or extended, the total number of samples in a slot must be the same as we show below.

\# Samples per slot (Normal CP)   → (160+2048)·1 symbols + (144+2048)·6 symbols = 15360

\# Samples per slot (Extended CP) → (512+2048)·6 symbols = 15360

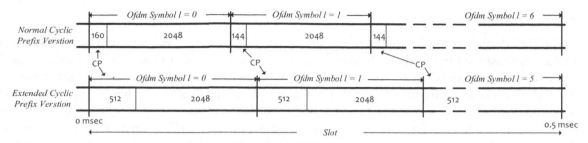

Figure 7-64: Number of Samples for OFDM Symbols with Normal and Extended CP

Notice that the FFT size of 2048 and the associated sample rate of 30.72 MHz are maximum default values for the 20MHz LTE signal. Clearly, we don't need to operate a huge $N = 2048$ I/FFT to process the 1.4 MHz signal, which only features 72 subcarriers. A 128-point FFT should be just fine to accommodate the smaller bandwidth signal and save processing power in the baseband processors. In the table below, we show both the different FFT sizes that may be used to handle the different bandwidth LTE signals, and how the number of samples per OFDM consequently scale.

	1.4MHz	3MHz	5MHz	10MHz	15MHz	20MHz
Number of RBs	6	15	25	50	75	100
Number of Subcarriers	72	180	300	600	900	1200
FFT Size	128	256	512	1024	1408	2048
Sample Rate (MSPS)	1.92	3.84	7.68	15.36	21.12	30.72
Normal CP Length (Samples)	10/9	20/18	40/36	80/72	110/99	160/144
Extended CP Length (Samples)	32	64	128	256	352	512
Samples Per Slot	960	1920	3840	7680	10560	15360

Figure 7-65: Sample Parameters for Different LTE Bandwidths

Given the number of samples for the IFFT output ($N = 2048$) and that of the various cyclic prefix variants (160/144/512), we can compute the quantities shown in the table below by dividing by the sample rate of 30.72 MSPS.

	Normal CP	Extended CP
CP Time (µsec)	5.21 / 4.69	16.67
IFFT (µsec)	66.6667	66.6667
Total Symbol Time (µsec)	71.875 / 71.3542	83.3334
OFDM Symbols Per Slot	7	6

Figure 7-66: Symbol Times in Seconds

Frame Structure

In the time domain, the LTE signal is subdivided not just into slots, but into subframes and radio frames as illustrated in the figure below. As mentioned above, a slot may consist of either 6 or 7 OFDM symbols depending on the cyclic prefix type and features a length of 0.5 milliseconds. Two slots form a subframe of length 1 millisecond, which features either 12 or 14 OFDM symbols. Ten subframes, or 20 slots, make up 10 millisecond radio frames, which are numbered from 0 to 1023 via a variable called system frame numbers, or SFN, which wrap at a value of 1024. Lte refers to a group of 1024 radio frames as a hyper frame. The LTE mobile phone is always scanning and demodulating broadcast messages from the base station. One of the broadcast messages contains information regarding the system frame number of the current radio frame being transmitted. The LTE mobile uses this information to locate additional broadcast messages, which are located in frames with specific system frame numbers.

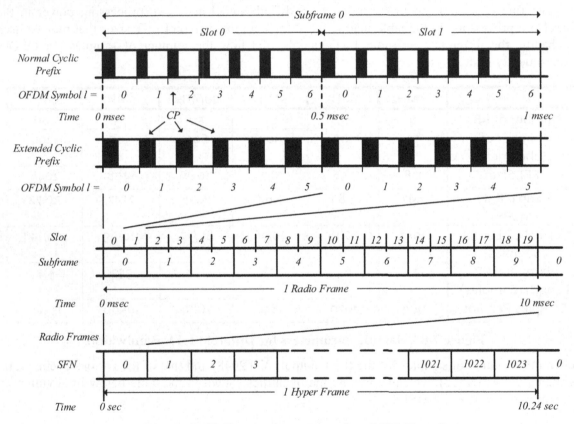

Figure 7-67: Frame Structure of an LTE Signal

7.5.3 Modulation of LTE Base Station Signals

The LTE base station signal is an OFDM modulated waveform featuring a different number of subcarriers that depends on the bandwidth of the signal. The signal is generated by loading the complex data symbols of the resource elements belonging to an OFDM symbol, l, into the input buffer of an IFFT. The OFDM time domain symbol is composed of the output of the IFFT and the prepended cyclic prefix. The figure below illustrates the process of loading the IFFT for the various LTE bandwidths.

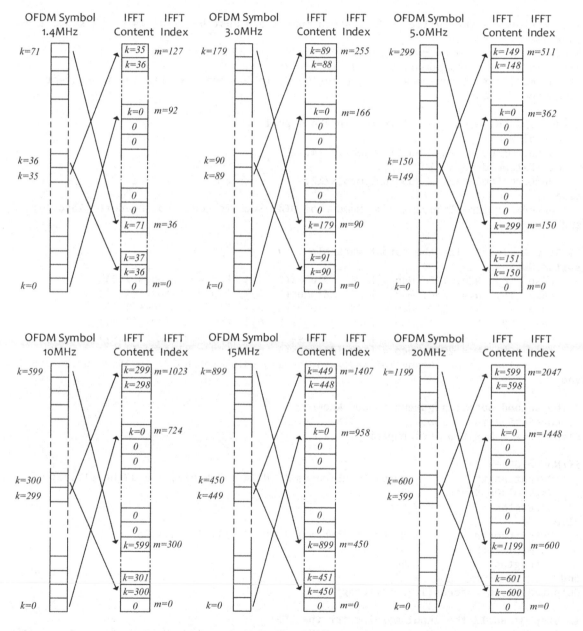

Figure 7-68: IFFT Input Buffer Loading Process for LTE Signals of Different Bandwidths

```
function [ OutputWaveform, SampleRate ] = RenderToTimeDomain(ResourceGrid)

%% Step 1: Check input argument for proper size
%  1a. Find number of resource blocks in signal
[NumSubcarriers, NumOfdmSymbols]   = size(ResourceGrid);

% 1b. Determine number of downlink resource blocks (or NDLRB)
NDLRB          = NumSubcarriers/12;
if(mod(NumSubcarriers, 12) ~= 0); error('RenderToTimeDomain.m: Bad input argument
dimension.'); end

% 1c. Determine the cyclic prefix, given that there must be eithe 14 or 12 symbols per
subframe.
NormalCP       = mod(NumOfdmSymbols, 14) == 0;
ExtendedCP     = mod(NumOfdmSymbols, 12) == 0;
if(~NormalCP && ~ExtendedCP);      error('RenderToTimeDomain.m: Bad input argument
dimension.'); end

% 1d. Find the number of subframes in the signal.
if(NormalCP == 1);
    NumberOfSubframes = NumOfdmSymbols/14;
elseif(ExtendedCP == 1)
    NumberOfSubframes = NumOfdmSymbols/12;
else
    error('RenderToTimeDomain.m: Number of Ofdm Symbols could not be determined.');
end

%% Step 2: Initialize the output variable
switch(NDLRB)
    case   6;  N =  128; Ncp = [ 10   9*ones(1, 6)]; Ecp =  32*ones(1,6);
    case  15;  N =  256; Ncp = [ 20  18*ones(1, 6)]; Ecp =  64*ones(1,6);
    case  25;  N =  512; Ncp = [ 40  36*ones(1, 6)]; Ecp = 128*ones(1,6);
    case  50;  N = 1024; Ncp = [ 80  72*ones(1, 6)]; Ecp = 256*ones(1,6);
    case  75;  N = 1408; Ncp = [110  99*ones(1, 6)]; Ecp = 352*ones(1,6);
    case 100;  N = 2048; Ncp = [160 144*ones(1, 6)]; Ecp = 512*ones(1,6);
    otherwise; error('RenderToTimeDomain:Invalid value for NDLRB (6,15,25,50,75,100)');
end

% The second output argument is now known
SubcarrierSpacing = 15000;        % Hz
SampleRate = N * SubcarrierSpacing;

if(NormalCP == 1)
    OutputLength        = NumberOfSubframes * (N*14 + 2*Ncp(1,1) + 12*Ncp(1,2));
    NsymbolsPerSlot     = 7;
    CPLengths           = Ncp;
else
    OutputLength        = NumberOfSubframes * (N*12 + 12*Ecp(1,1));
    NsymbolsPerSlot     = 6;
    CPLengths           = Ecp;
end
OutputWaveform = zeros(1, OutputLength);

%% Step 3: Build the input mapping for the IFFT
PositiveSubcarriers = NumSubcarriers/2:(NumSubcarriers - 1);  % Index in resource grid
NegativeSubcarriers = 0:(NumSubcarriers/2 - 1);               % Index in resource grid
```

```
PosFreqIndicesIFFT   = 1:NumSubcarriers/2;                % Index in IFFT
NegFreqIndicesIFFT   = (N - NumSubcarriers/2):(N-1);      % Index in IFFT

StartIndex = 0;
for SymbolIndex = 0:NumOfdmSymbols-1
    % Compute IFFT output
    IFFT_Input                           = zeros(N, 1);
    IFFT_Input(PosFreqIndicesIFFT + 1,1) = ResourceGrid(PositiveSubcarriers+1, SymbolIndex+1);
    IFFT_Input(NegFreqIndicesIFFT + 1,1) = ResourceGrid(NegativeSubcarriers+1, SymbolIndex+1);
    IFFT_Output                          = N*ifft(IFFT_Input, N);

    % Generate cyclic prefix
    SymbolIndexInSlot                    = mod(SymbolIndex, NsymbolsPerSlot);
    CurrentCpLength                      = CPLengths(1, SymbolIndexInSlot + 1);
    CyclicPrefix                         = IFFT_Output(N-CurrentCpLength+1:N, 1);

    % Generate Output symbol
    OutputSymbol                         = [CyclicPrefix; IFFT_Output];

    % Place output symbols into output waveform array
    Range                   = StartIndex+1:StartIndex+length(OutputSymbol);
    OutputWaveform(1, Range) = OutputSymbol.';
    StartIndex              = StartIndex + length(OutputSymbol);
end
assert(StartIndex == OutputLength, 'The output waveform array was not completely
filled');
end
```

7.5.4   Basic Construction of the LTE Base Station Signals

An LTE downlink, base station to mobile phone, signal is composed of many signals and physical channels that are computed, placed into the resource grid and then OFDM modulated prior to being transmitted of 1, 2, or 4 different TX antennas. The figure below illustrates the FDD resource grid of a 1.4MHz LTE signal featuring 6 resource blocks ($6 \cdot 12 = 72$ subcarriers) and 12 subframes.

Synchronization Signals

The LTE downlink provides a primary (PSS) and secondary (SSS) synchronization signal. Each one of these signals is confined to a single OFDM symbol and the 6 center resource blocks in the LTE resource grid. In the frequency division duplex case, these signals appear in the last and second last OFDM symbol of slot 0 and slot 10 of each radio frame. When a mobile phone comes within the reach of a new cell, it will process the synchronization signals of the base station in order to acquire the frequency offset, the timing reference and the physical cell ID. The physical cell ID is a number between 0 and 503 and is used to process additional information sent on the remaining physical channels transmitted by the base station. As explained in section 6.3.2, it is beneficial that there be a single crystal oscillator providing the timing reference to both baseband processor and synthesizer. This eases frequency detection using the correlation peaks of the PSS. We shall see more details regarding the synchronization signals in the next section.

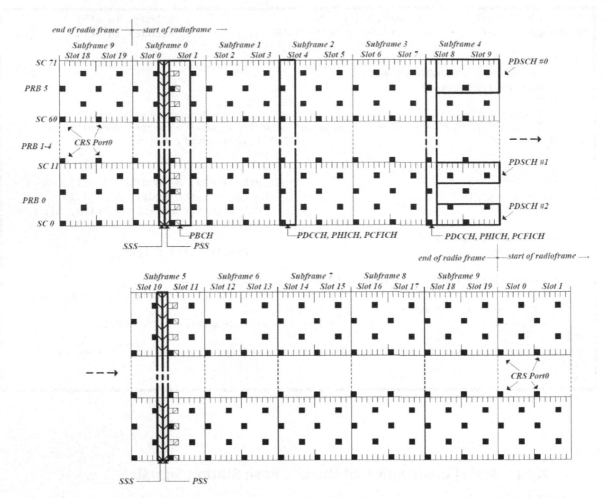

Figure 7-69: Construction of a Single Radio Frame Consisting of 10 Subframes

Reference Signals in LTE

Reference signals are identified by the black squares in the resource grid above. At these locations, the OFDM downlink signal inserts QPSK symbols that are known ahead of time at the receiver. By comparing the received and ideal values of these reference signals, the mobile phone can compute the channel estimation, the subsequent equalizer coefficients and the reference signal carrier no noise and interference ratio, or RS-CINR. If the RS-CINR is good enough, the mobile phone will attempt to decode the physical broadcast and shared channels to extract broadcast information about the base station such as the network operator residing on the tower and the actual bandwidth of the downlink signal. This all happens without the user being aware of it and without the user initiating two-way communication. The QPSK values of these reference signals are randomized across the resource grid. The randomization seed depends directly on the physical cell identity that we found via the synchronization signals. Therefore, the reference signals are unique for each base station. It is for this reason, that these reference values are actual called, cell specific reference signals, or CRS. (For more details, see section 7.5.4.1).

Physical Channels in LTE

It is only after the cell phone has processed the synchronization signals and cell specific reference signals that is will turn its attention to the demodulation and decoding of the physical channels located in the downlink signal. A physical channel in LTE is defined via three aspects:

→ The allocation in the resource grid in terms of number and position of resource blocks.
→ The encoding definition in terms of forward error correction, scrambling and interleaving.
→ The nature of the information that it carries.

Physical Broadcast Channel (PBCH)

The physical broadcast channel occupies the first four OFDM symbols of the second slot of subframe 0. It is therefore available in the first subframe of every radio frame, or every 10 milliseconds. The PBCH carries the master information block, or MIB, which contains very basic information such as the LTE signal bandwidth. You must decode the physical broadcast channel before any other physical channel can be decoded, and it is for this reason that the bit encoding definition for the PBCH provide the highest degree of error protection in the downlink. The physical layer encoding of the PBCH is presented in great detail in section 7.5.7. Note that the encoding of the PDCCH and PSSCH is very similar to that of the PBCH.

Physical Downlink Control Channel (PDCCH)

The physical downlink control channel occupies the first L OFDM symbols of every subframe in a radio frame. It carries one or more downlink control information messages, or DCIs, which identify the physical location of one or more physical downlink shared channels, or PDSCH. The base station uses the PDCCH to tell each cell phone where both broadcast and individual user information is located in the resource grid.

Physical Control Format Indicator Channel (PCFICH)

The Physical Control Format Indicator Channel specifies how many OFDM symbols L the PDCCH occupies. The more cell phones a base station is servicing, the larger the number of DCI messages are in the PDCCH and therefore more OFDM symbols might be required. The PDFICH channel takes up only a few resource elements in the first OFDM symbols of every subframe. It is collocated with the PDCCH.

Physical channel HybridARQ Indicator Channel (PHICH)

Just as the PCFICH, this channel is also collocated with the PDCCH and provides information regarding acknowledgements. The base station and cell phone acknowledge the successful reception of each other's messages in order to determine whether retransmission is necessary. This channel is involved in this process.

Physical Downlink Shared Channel (PDSCH)

The physical downlink shared channel carries additional broadcast information as well as user information such as speech and data traffic. The broadcast information, intended to be decoded by all mobiles in the cell, provides a large amount of additional information about the cell tower such as the identity of the network provider being hosted. In addition, the PDSCH provides all user information dedicated for individual mobile phones.

7.5.5 Synchronization Signals

The LTE downlink signal features a primary, PSS, and a secondary, SSS, synchronization signal that aids the cell to acquire the following:

→ Timing synchronization: By correlating the incoming downlink signal against the PSS, we are able to gain knowledge about the frame, subframe and OFDM symbol start time.

→ Timing drift and frequency offset: By observing the magnitude peaks of the PSS correlation, we are able to detect timing drift due to the difference in timing base (master clock and Doppler) as well as the associated frequency offset that comes along with it.

→ Physical cell ID: After processing the PSS, we decode the secondary synchronization signal and the combination of the two provide us with a physical cell identification number, whose value is between 0 and 503. This physical layer ID allows the mobile to gain information regarding the orientation of the reference signals and scrambler initializations used for various steps in the encoding of the physical channels.

The Physical Cell Identity

There are 504 unique physical-layer cell identities. They are grouped into 168 unique groups, each containing three unique identities. The physical cell ID is expressed as the sum of two sub IDs:

$$N_{ID}^{cell} = 3N_{ID}^{(1)} + N_{ID}^{(2)}$$

$N_{ID}^{(1)}$ is extracted from the secondary synchronization signal and represents the physical layer identity group, which ranges from 0 to 167. $N_{ID}^{(2)}$ is extracted from the primary synchronization signal to provide the sector identity ranging from 0 to 2 within a group.

7.5.5.1 The Primary Synchronization Signal (PSS)

In order to build the primary synchronization signal, we use a sequence $d(n)$, which features 62 entries. The sequence is generated from a frequency domain Zadoff-Chu sequence according to the following expression.

$$d_u(n) = \begin{cases} e^{-j\frac{\pi u(n+0)(n+1)}{63}} & n = 0,1,...30 \\ e^{-j\frac{\pi u(n+1)(n+2)}{63}} & n = 31,32,...61 \end{cases}$$

The factor u is related to $N_{ID}^{(2)}$ as indicated in the table below.

Table 7-4: Zadoff-Chu Root Index Table

$N_{ID}^{(2)}$	*Root Index u*
0	25
1	29
2	34

The PSS is located in the last OFDM symbol of slot 0 and 10 no matter what type of CP is valid. Remember that the resource grid of the LTE downlink signal consists of $12 \cdot NDLRB$ subcarriers (not counting the subcarrier at DC). Review table 7-3 to determine the number of downlink resource blocks, $NDLRB$, for each LTE bandwidth. The PSS is mapped into 62 resource elements sitting at the center of the resource grid. The PSS occupies the following subcarrier indices, k:

$$k = n - 31 + \frac{NDLRB \cdot 12}{2} \quad for\ n = 0,1...61$$

Remember that the smallest bandwidth LTE signal at 1.4MHz extends over 72 subcarriers (not including DC), and the PSS fits neatly inside this bandwidth. Note that the PSS covers 62 resource elements, not the entire 72 of the 1.4MHz LTE downlink signal. The bottom five and top five resource elements with index k_{bottom} and k_{top} are set to zero.

$$k_{bottom} = n - 31 + \frac{NDLRB \cdot 12}{2} \quad for\ n = -5, -4,... -1$$

$$k_{top} = n - 31 + \frac{NDLRB \cdot 12}{2} \quad for\ n = 62,63,...66$$

If the LTE downlink signal did in fact feature a bandwidth of 1.4MHz ($NDLRB = 6$), then the indices k would look as follows.

→ k $\quad = 5, 6, \ldots 66$

→ $k_{bottom} = 0, 1, \ldots 4$

→ k_{top} $\quad = 67, 68, \ldots 71$

Clearly, the LTE downlink signal can feature six different bandwidths, and when the mobile phone processes the PSS, it must always filter the received waveform to expose the center 72 subcarriers. At is as this point, that the mobile will run a sliding correlation of the received waveform and the three possible Zadoff-Chu time domain sequences with root indices of 25, 29, and 34. The following MatLab code generates the PSS.

```
function d = GeneratePSS_Sequence(PCI)

%% Step 0: Figure out N1 and N2
N1 = floor(PCI/3);
N2 = PCI - 3*N1;

switch(N2)        % See table 6.11.1.1-1
    case 0;  u = 25;
    case 1;  u = 29;
    case 2;  u = 34;
end

d = zeros(1,62);
% Generate first portion from n = 0:30
n = 0:1:30;
d(1,n+1) = exp(-1j*pi*u*n.*(n+1)/63);

% Generate second portion from n = 31:61
n = 31:1:61;
d(1,n+1) = exp(-1j*pi*u*(n+1).*(n+2)/63);
```

7.5.6   Cell Specific Reference Signals

The primary purpose of the cell specific reference signals, or CRS, is to provide the receiver with a method of estimating the frequency response of the multiply path channel at any instance of time. Furthermore, the cell specific reference signals allow us to compute the LTE downlink signal's carrier to interference and noise ratio, which we rely on to select the proper cell to access. LTE supports either 1, 2 or 4 transmit antennas and each antenna port provides its own reference signals. The figures below illustrate the resource grid for the 1.4MHz LTE signal along with the locations of the CRS for each port.

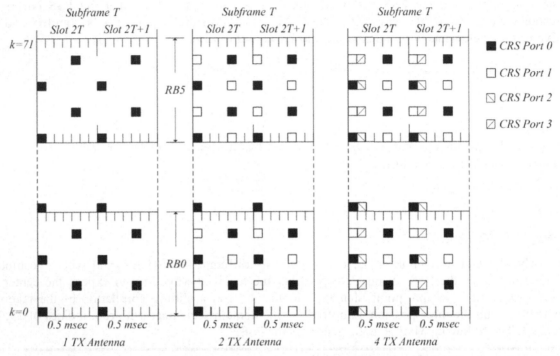

Figure 7-70: LTE Cell Specific Reference Signal Locations for Antenna Ports 0, 1, 2 and 3

The specification of the values and locations of the cell specific reference signals is described in Section 6.10 of TS36.211. If the base station decides to use only one transmit antenna port, then only CRS for port 0, which are identified by the black squares, are transmitted. The other resource elements are available for physical channel signals. If the base station decides to use two antennas, then then the CRS locations indicated by the white squares are populated on antenna port 1. Note that on port 0, the CRS locations of port 1 must remain empty just as on port 1, the CRS locations of port 0 must remain empty. This allows the mobile phone to receive and measure the CRS for each transmit antenna port separately without interference. This same idea is extended to the four transmit antenna port as well. We can make the following statement regarding the cell specific reference signals.

→ The cell specific reference signal values and locations are different depending on the physical cell identity (0 – 503), which we determined by processing synchronization signals. The CRS values are QPSK modulated and their orientation is scrambled where the scrambling seed depends on the physical cell identity. Further, the locations are shifted in frequency by some offset in k, which also depends on the physical cell identity. Just imagine the entire CRS pattern being moved up by some amount. The idea is to ensure that CRS of neighboring cells don't overlay, and if they do overlay, that their orientations are completely uncorrelated. In this manner, the overlaid CRS value from a neighboring cell appear like a noise contributor to the CRS values of the current cell. Under all circumstances, the 4G specifications attempt to avoid that there is any correlation in CRS orientation between cells as this quickly leads to detection problems.

→ For each cell, the values of the CRS will repeat where the repetition period is equal to one radio frame. Thus, the CRS values are only unique for 10 subframes or 20 slots.

→ The raw channel estimate can be easily computed by dividing the received CRS values by the ideal transmit values of the CRS. The raw channel estimate will only be known at the CRS position and will be affected by the noise inherent in the receiver. Using interpolation, the channel estimate must be extended to the other resource elements that carrier data symbols. We may significantly improve the signal to noise ratio of the frequency response estimate by using the techniques discussed in section 7.3.1.5.

$$RawChannelEstimate = \frac{CRS\_Rx}{CRS\_Tx}$$

→ Note that CRS for port 0 and 1 are available once per slot in those subcarriers that carry reference signals. Given that a slot is of length 0.5e-3 seconds, the sample rate at which CRS are available in the time domain is equal to 1/0.5e-3 seconds = 2000 Hz. This yield a Nyquist rate of between -1KHz and +1KHz, which bounds the maximum frequency offset and Doppler shift combination that the system can handle. Taking another look at section 6.3.2, we can conclude that and an LTE link at 2GHz can accommodate Doppler frequencies of +/- 1KHz, which occurs when a vehicle travels at a speed of approximately 250Kph directly toward a base station.

→ Note further that the CRS are available twice per resource blocks representing a step in frequency of 15000 KHz times 6 subcarriers equaling 90KHz. The sample rate at which CRS are available in the frequency domain is equal to 1/90KHz = 11.1 microseconds, which yield a Nyquist range of between -5.55 and +5.55 microseconds. If we take another look at section 7.1.3.2, we realize that frequency response due to the time delay, τ_p, of a path, p, in relation to the FFT start time is a complex sinusoid in the frequency domain.

$$FreqResponse_p(f) = C_p \cdot e^{-j2\pi\tau_p f}$$

The Nyquist range indicates the absolute time shift, τ_p, that can be resolved in LTE. The resolvable time shift must obviously be greater than the cyclic prefix range.

→ Notice that given the offset pattern of CRS in symbols 0 and 4 of each slot, the ranges over which Doppler and time delay may be resolved can be increased by some clever processing. Section 7.1.3.5 illustrates a clever method of computing the channel estimate without burdensome preinterpolation or preprocessing of cell specific reference signals.

The information above provides a nice overview of the purpose and features of the CRS. The physical layer definition will not mention any of these facts, it simply describes their location and value as a function of physical resource block and slot timing.

7.5.7 Encoding of the Physical Broadcast Channel (PBCH)

Showing how the LTE standard encodes the PBCH serves as an excellent exercise that illustrates the general procedure used to encode the remaining physical channels such as the PDCCH and PDSCH. The PBCH carries the master information block, or MIB, which is the first broadcast message that must be demodulated and decoded by the mobile phone in order to find out basic transmission parameters associated with the cell. Once the MIB is demodulated, the phone will proceed to demodulate additional broadcast information messages known as system information blocks, or SIBs, for information regarding available network providers hosted in the cell and access rights to the cell. The MIB information is confined to an area in the LTE resource grid called the physical broadcast channel, which takes up the first four symbols of slot 1 in every radio frame.

Information Contained in the MIB

The master information block features 24 bits containing three important pieces of information.
→ 3 bits for the system bandwidth indicating the NDLRB (number of downlink resource blocks)
→ 3 bits describing the physical HARQ indicator channel, or PHICH
→ The 8 most significant bits of the 10 bit system frame number of the current radio frame
→ 10 bits of value [0 0 0 0 0 0 0 0 0 0] reserved for future use.

The three most significant bits in the MIB indicate the number of downlink resource blocks, NDLRB, of the base station's LTE signal.

Bit Pattern	NDLRB	Bandwidth (MHz)
0 0 0	6	1.4
0 0 1	15	3
0 1 0	25	5
0 1 1	50	10
1 0 0	75	15
1 0 1	100	20

Figure 7-71: Bit Pattern Indicating the LTE Downlink Bandwidth

The three PHICH bits provide information to the mobile phone regarding repeat request in case the base station was unable to decode a message from the phone. Having nothing to do with signal processing, we will not consider these bits in detail.

Each LTE radio frame is assigned a system frame number, which ranges from 0 to 1023 and then rolls over. Thus the 1024th radio frame sent by the base station will have a system frame number, or SFN, equal to 0. Additional broadcast messages are located in frames with precisely defined system frame numbers making it necessary for the mobile phone to discover the SFN of the current frame that it is processing before proceeding to decode the additional messages. As the MIB contains the 8MSB of the SFN, the MIB content changes at radio frames that obey mod(SFN, 4) = 0. As a matter of fact, via quite a few processing stops, we encode the MIB into 1920 bits which we spread across four successive PBCH occurrences (one in each radio frame). In one of the processing steps, the 1920 error encoded MIB bits are scrambled via a bit sequence of length 1920. The final 1920 MIB bits are then divided up into 4 sets of 480 bits each and mapped into the physical broadcast channels at the frames obeying mod(SFN, 4) = 0, mod(SFN, 4) = 1, mod(SFN, 4) = 2, and mod(SFN, 4) = 3. It is because of the fact that the scrambling sequence is unique in each set of 480 PBCH bits that the receiver will be able to recognize which of the four frames it is looking at, and we it can finally retrieve the 2 LSBs of the system frame number. The encoding of the MIB is a bit tricky, which gives the reader a small taste of some of the odd physical layer procedures that are defined in the 4G standard. The 4G specification actually expects the mobile to undertake trial and error type algorithms to determine the properly decoded MIB information. This leads us to an interesting specification approach instigated by the 3GPP group. The 3GPP specifications assume that the mobile phone has tremendous processing power.

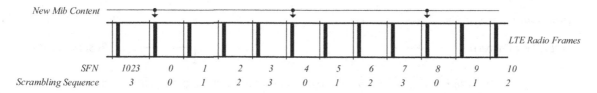

Figure 7-72: Location of PBCH in Ten Consecutive Radio Frames

PBCH Encoding Steps

Processing Step	Explanation
CRC Attachment	A 16 bit CRC, cyclic redundancy check, word is computed and appended to the 24 bit MIB to form a 40 bits message. The CRC, which acts like to a parity check, is used to detect errors in the original 24 MIB bits. When the decoding process in the mobile phones receiver reaches the point that reveals the 40 bit word, it recomputes the CRC of the first 24 bits and compares them to the last 16 of the received word. The CRC process is explained in great detail in section 5.2.1 of this text.
CRC Mask (Antenna Check)	Once the 16 bit CRC word has been computed, we XOR it with one of 3 unique 16 bit mask indicating the number of transmit antennas (1, 2, or 4) at the base station. The receiver must therefore repeat the CRC check three times to test for each antenna configuration.
FEC Encoding	The 40 bit message consisting of the MIB and modified CRC bits are FEC encoded using a rate 1/3 tail biting convolutional encoder to produce 120 encoded bits.
Rate Matching	The 120 encoded bits are first interleaved and then repeated to form a sequence of 1920 bits (normal CP) or 1728 bits (extended CP).
Scrambling	The rate matched bits are scrambled using a 1920 or 1728 bit scrambling sequence which is unique for each physical layer cell identity, PCI. Remember that the PCI is readily available to us as it was found after detecting the primary and secondary synchronization sequence.
QPSK Mapping	The 1920 or 1728 bits are mapped to either 960 or 864 QPSK symbols.
Layer Mapping Precoding	The QPSK symbols are layer mapped and precoded such that they can be transmitted from either 1, 2, or 4 antennas.
RE Mapping	The precoded information is inserted into the PBCH area of the LTE resource grid of each antenna. There can be up to four resource grids.

Figure 7-73: The PBCH Encoding Steps

The 24 MIB bits are encoded into a physical broadcast channel, PBCH, message with either 1920 (normal cyclic prefix) or 1728 (extended cyclic prefix) bits. The encoding process consist of many steps and embeds information such as the 2 LSB of the SFN and the number of transmit antennas. The last tables and figures below illustrate the original MIB construction and the encoding chain that leads us to the physical broadcast channel message.

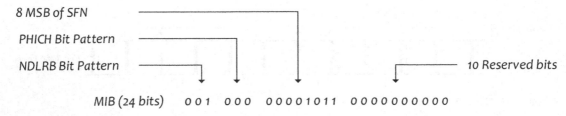

Figure 7-74: MIB Content Indicating a Bandwidth of 1.4MHz (NDLRB = 6) and 8MSB of SNF = 11

As the reader can see in the figure below, there are quite a few encoding steps, which we will cover one by one in the following sections, except for the CRC computation which is covered in section 5.2 of this text.

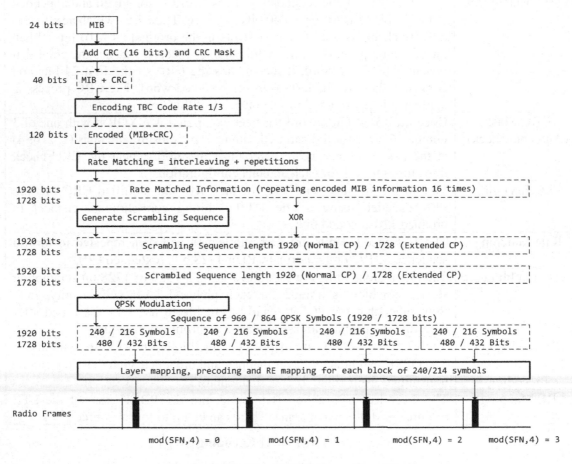

Figure 7-75: The PBCH Encoding Chain

7.5.7.1 Generating the Cyclic Redundancy Check Bits

We take the current 24 MIB bits and from them compute 16 CRC bits that we will append at the end to form a sequence of 40 total bits. The LTE demodulator will look at the 40 incoming bits and split them into 24 received MIB and 16 received CRC bits. It uses the 24 received MIB bits to compute 16 CRC bits and then ensures that they are equal to the received CRC bits. If they agree, then the 24 received MIB bit are deemed to be correct. This process is explained in detail in section 5.6.1 for the LTE specification.

7.5.7.2 Adding the CRC Mask

The 16 bit cyclic redundancy check, or CRC, is XOR using a CRC mask, whose value changes depending on the TX antenna configuration of the LTE base station. Modern LTE base stations have either 1, 2 or 4 TX antennas, and the associated mask it shown in the figure below.

Number of transmit antenna ports at eNodeB	PBCH CRC mask $< x_{ant,0}, x_{ant,1},...., x_{ant,15} >$
1	<0, 0, 0, 0, 0, 0, 0, 0, 0, 0, 0, 0, 0, 0, 0, 0>
2	<1, 1, 1, 1, 1, 1, 1, 1, 1, 1, 1, 1, 1, 1, 1, 1>
4	<0, 1, 0, 1, 0, 1, 0, 1, 0, 1, 0, 1, 0, 1, 0, 1>

Figure 7-76: TX Antenna Configuration CRC Mask

```
function Output = AddCrcMask(MIB_Bits, CRC, NumberOfTxAntennas)

switch(NumberOfTxAntennas)
 case 1; Crc_Modified = mod(CRC + zeros(1,16), 2);   % Can be omitted
 case 2; Crc_Modified = mod(CRC + ones(1,16), 2);    % Invert all bits
 case 4; Crc_Modified = mod(CRC + [0, 1, 0, 1, 0, 1, 0, 1, 0, 1, 0, 1, 0, 1, 0, 1], 2);
                                       % Invert every second bit
 otherwise; assert(false, 'Illegal TX antenna configuration');
end
Output = [MIB_Bits, Crc_Modified];
```

7.5.7.3 Tail Biting Convolutional Encoder

The LTE transmitter uses a tail-biting convolutional encoder with a constraint length of 7 and generator polynomials g_1, g_2, and g_3, with values 133oct = 1'011'011bin, 171oct = 1'111'001bin, and 165oct = 1'110'101bin respectively. The structure below reveals how the binary equivalent determines the positioning of the modulo-two additions along the shift register. Note that there are constraint length = 7 positions along the shift register, at which bits are processed. The input position of first register as well as all six register output positions. The convolutional encoder is defined in document TS36.212 Section 5.1.3.1 of the 3GPP specifications. The 3GPP architects decided that adding the usual 6 zero valued tail bits to a small 40 bit Mib/Crc message seemed like an unnecessary amount of overhead, and decided to use tail biting convolutional coding instead. Tale a look at section 5.6.4 for a full discussion on tail-biting in binary convolutional encoding. At the encoder side, we simply need to initialize the initial state of the encoder to the state in which the encoder must finish once all 40 MIB bits have been processed. The MatLab code below shows how to do this using native MatLab function calls as well as the encoder we have coded ourselves in section 5.6.4.

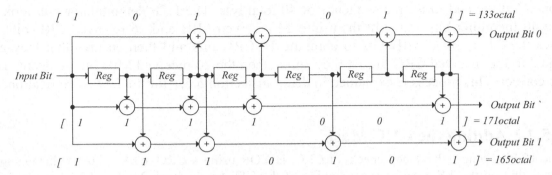

Figure 7-77: LTE Convolution Encoder Structure

```
MibCrcBits         = randi([0,1], 1, 40); % For now we just use random bits
PolynomialsOct     = [133,171,165];
ConstraintLength   = 7;
trellis            = poly2trellis(ConstraintLength, PolynomialsOct);


%% Native MatLab Version. initstate = last 6 bits of input stream
initstate = Input(length(Input)-5:length(Input)) * [1 2 4 8 16 32]';
Output1   = convenc(Input, trellis, initstate);


%% Encoder Developed in Section 5.6.4
TailBitingFlag = true;
Output2 = ConvolutionalEncoder(MibInputBits,   ConstraintLength,  PolynomialsOct, ...
                          TailBitingFlag, 'Hard');
```

7.5.7.4 Rate Matching

Rate matching is a process that consists of a bit interleaver step and a bit repetition step. Interleaving is a process that rearranges the position of the input bits, in order to prevent errors in successive bit positions. Assume for a moment, that interleaving was skipped, and 4 successive transmit bits were transformed into a single 16QAM constellation position and placed at a particular resource element. If the frequency response of the channel at that resource element is particular terrible (high amounts of attenuation, thus poor SNR), it is likely that all or many of these 4 bits would be in error when the 16QAM position is demapped into bits at the receiver. The Viterbi decoder does not handle successive errors very well, but has no problem with those that are spread out. Another scenario that should be avoid is the placement of successive bits in the same resource element of neighboring symbols. If the frequency response of a particular resource element is bad in one of the symbols, then it is very likely that the channel response is similarly bad in the neighboring symbols as the overall multipath channel configuration doesn't change all that quickly. Interleaving aims to avoid the scenario of consecutive errors and can be customized based on how the information is placed into the resource grids. The LTE interleavers are somewhat complicated but for basic applications one can simply read in the source bits into a matrix, one row at a time, and then read them out one column at a time to get some initial positional separation.

Bit repetition is another way to add redundancy to the bit stream. In the LTE PBCH, physical broadcast channel, the 120 interleaved bits are repeated 16 and 14.4 times to produce either 1920 or 1728 rate matched bits depending on whether we are transmitting using the normal or extended cyclic prefix (more on this later). For the extended prefix we repeat the interleaved bits 14 times and then concatenate the first 48 bits of the original 120 interleaved bits.

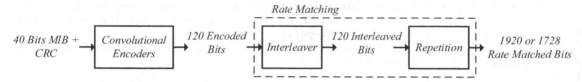

Figure 7-78: Rate Matching in the LTE PBCH Channel

In the receiver, the encoding process is undone until we get to the rate dematching step were we need to collapse 1920 or 1728 noise corrupted bits back into 120. We do this by simply adding them all up. There is a substantial SNR benefit due to this averaging process. Remember that if you add two identical signal values that have been corrupted with uncorrelated noise, then the signal power increases by a factor of 4, whereas the noise power of the sum increased by a factor of 2. The SNR benefit for each bit due to the averaging is equal to $10 \cdot log_{10}(16) = 12$ dB and $10 \cdot log_{10}(14.4) \approx 11.5$ dB respectively. The rate matching step is explained in the 3GPP document TS36.211 section 5.1.3.2. The MatLab code below generates a set of either 1920 or 1728 indices into the 120 bit error encoded output vector. Interleaving and repetition are thus achieved simultaneously.

```matlab
function InterleaverTableTbc = GenInterleaverTableTbc(NumCodedBits, ...
                                      DesiredNumberOfOutputBits)
%% 0. Subblock Interleaving
CodeRate   = 1/3;
% The permutation pattern according to table 5.1.4-2
PermPattern = [1, 17, 9, 25, 5, 21, 13, 29, 3, 19, 11, 27, 7, 23, 15, 31, ...
               0, 16, 8, 24, 4, 20, 12, 28, 2, 18, 10, 26, 6, 22, 14, 30];
% The input index of the bit sequence entering the rate matcher is as follows
EncodedIndex = 0:NumCodedBits-1;
% We reshape the index to isolate the indices of d0(i), d1(i), d2(i) to different rows
ReshapedIndex = reshape(EncodedIndex, 3, []);  % [0, 3, 6,  9 ...]
                                               % [1, 4, 7, 10 ...]
                                               % [2, 5, 8, 11 ...]
% The number of input bits into each interleaver
NumInputBits = NumCodedBits*CodeRate;

% Step 1: The interleaver table has 32 columns and enough rows to fit the number of
% input bits
NumColumns              = 32;
% Setp 2: Determine Number of Rows
NumRows                 = ceil(NumInputBits/NumColumns);
TotalNumberOfTableEntries = NumColumns * NumRows;      % Kpi in the specification
wk   = zeros(3*TotalNumberOfTableEntries,1);           % Final index sequence

for i = 1:3     % 1 through number of interleavers (there are 3)
    % Step 3: Clearly, the NumInputBits will not always be an integer multiple of 32,
    % and dummy elements are inserted at the start of the interleaver table. In the
    % pruning process later on, these dummy elements are eliminated
    NumDummyBits = TotalNumberOfTableEntries - NumInputBits; % ND in the specification
```

```
    InputIndex    = ReshapedIndex(i,:);
    LinearIndex   = [-ones(1,NumDummyBits) InputIndex];
    InterleaverMatrixStep3    = reshape(LinearIndex,NumColumns,NumRows)';

    % Step 4: Change the columns according to the PermPattern
    InterleaverMatrixStep4    = InterleaverMatrixStep3(:, PermPattern + 1);

    % Step 5: The output of the block interleaver are the Interleaver Matrix4 bits
    %         read out column by column. In our case, the matrix is filled with indices.
    OutputIndex   = InterleaverMatrixStep4(:);
    Range         = (i-1)*TotalNumberOfTableEntries : i*TotalNumberOfTableEntries - 1;
    wk(Range+1, 1)    = OutputIndex;
end

%% 2. Bit collection, selection and transmission
E = DesiredNumberOfOutputBits;
Kw = length(wk);
InterleaverTableTbc = zeros(E, 1);
k = 0;
j = 0;
while (k < E)
    if(wk(mod(j, Kw) + 1, 1) ~= -1)
        InterleaverTableTbc(k+1, 1) = wk(mod(j, Kw) + 1, 1);
        k = k + 1;
    end
    j=j+1;
end
```

7.5.7.5 Scrambling

Scrambling is a very important concept is LTE, and it is employed during many signal processing steps in the generation of the transmit signal. Imagine for a moment that your cell phone is listening to and is in contact with the closest cell, but also receives signals from other neighboring cell towers. The signals sent by all towers are almost always time aligned, meaning they all send their own PBCH at exactly the same time and exactly the same location in frequency. To avoid the scenario in which several neighboring cells transmit similar information, which can constructively interfere at your cell phones antenna, scrambling that differs from one cell tower to the next is introduced to make otherwise similar information look totally random. Remember the interleaving step of the last section. Undoing it meant the averaging, or summation, of repeated bits, a process as a result of which the power of the bit values increases faster than that of their noise components. Without the randomization effect of scrambling the interfering signals from neighboring cells would might appear correlated to your cell phone and add constructively to cause reception problems. The scrambler consists of two maximum length sequences, called m-sequences, which are combined into a single output, $c[n]$, using an XOR gate. To make sequences in adjacent cells feature the lowest possible cross-correlation, the generator is clocked 1600 times just after initialization. The first m-sequence will always be initialized with $x_1[0] = 1$, $x_1[n]=0$ for $n = 1,2 \dots, 30$. The initialization of the second m-sequence depends on the LTE information stream that is to be scrambled. For the PBCH channel, the second initialization word called c_{init} in the specification is equal to the physical layer cell ID of the base station. Physical layer cell identifications range from value of 0 to 503. The scrambling process for the PBCH channel is explained in the 3GPP document TS36.211 Section 6.6.1, but the diagram below is of far more help.

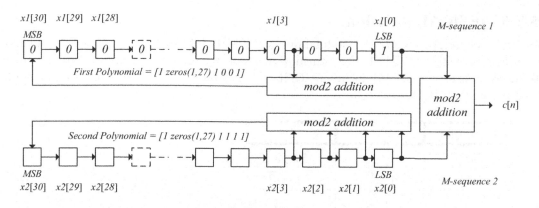

```
function Output = GoldCodeGenerator(c_init, NumberOfOutputBits)

% Here are some initial conditions that never change and don't need to be passed as
arguments.
FirstPolynomial    = [1 zeros(1,27) 1 0 0 1];
SecondPolynomial   = [1 zeros(1,27) 1 1 1 1];
FirstInitialCond   = [0 zeros(1,29) 1];
Shift              = 1600;

% convert c_init to SecondInitialCond
Init = zeros(31,1);
for i = 1:31   % convert number to bit vector
    if(bitand(c_init, 2^(i-1)) ~= 0); Init(i,1) = 1; end
end
SecondInitialCond = fliplr(Init.');

%% Initializing and executing the shift of the Gold Code Generator
Output = zeros(NumberOfOutputBits, 1);
TotalLength = Shift + NumberOfOutputBits;          %% This is the faster version
Reg1       = [zeros(1, TotalLength) FirstInitialCond];
Reg2       = [zeros(1, TotalLength) SecondInitialCond];
EndLocation = TotalLength;

for i = 1:Shift                    %% Run the fist 1600 itertations
    Mod1 = mod( Reg1(1, EndLocation+28) + Reg1(1, EndLocation+31), 2);
    Mod2 = mod( Reg2(1, EndLocation+28) + Reg2(1, EndLocation+29) + ...
                Reg2(1, EndLocation+30) + Reg2(1, EndLocation+31), 2);
    Reg1(1, EndLocation)  = Mod1;
    Reg2(1, EndLocation)  = Mod2;
    EndLocation = EndLocation - 1;
end

for i = 1:NumberOfOutputBits       %% Produce actual scrambling bits
    Output(i, 1) = mod(Reg1(1, EndLocation + 31) + Reg2(1, EndLocation + 31), 2);
    Mod1 = mod( Reg1(1, EndLocation+28) + Reg1(1, EndLocation+31), 2);
    Mod2 = mod( Reg2(1, EndLocation+28) + Reg2(1, EndLocation+29) + ...
                Reg2(1, EndLocation+30) + Reg2(1, EndLocation+31), 2);
    Reg1(1, EndLocation)  = Mod1;
    Reg2(1, EndLocation)  = Mod2;
    EndLocation = EndLocation - 1;
end
```

7.5.7.6 QPSK Modulation

The PBCH channel uses QPSK modulation as the MIB information is important and must be transmitted robustly. In QPSK we transmit two bits per complex symbol with the mapping definitions appearing in 3GPP document TS36.211 Section 7.1.2. The mapping of an even length bit sequence $b(i)$ is defined as follows.

bit0, bit1	I	Q
00	$1/\sqrt{2}$	$1/\sqrt{2}$
01	$1/\sqrt{2}$	$-1/\sqrt{2}$
10	$-1/\sqrt{2}$	$1/\sqrt{2}$
11	$-1/\sqrt{2}$	$-1/\sqrt{2}$

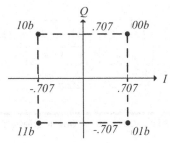

Figure 7-79: QPSK Mapping Definition

The 1920 or 1728 rate matched bits are scrambled converted into 960 or 864 QPSK symbols.

```
function ModulatedSymbols_Tx = QPSK_Mapping(MIB_Scrambled)
% This simple step is spelled out in to clause 6.6.2 and 7.1.2 of TS36.211.
NumModulatedSymbols   = length(MIB_Scrambled)/2;   % => 1920/2 = 960   or   1728/2 =
864
ModulatedSymbols_Tx   = zeros(1, NumModulatedSymbols);
for i = 0:(E/2)-1
    Bit0  = MIB_Scrambled(2*i + 0 + 1, 1);
    Bit1  = MIB_Scrambled(2*i + 1 + 1, 1);
    if(Bit0 == 0); I = 1/sqrt(2);
    else;          I = -1/sqrt(2); end
    if(Bit1 == 0); Q = 1/sqrt(2);
    else;          Q = -1/sqrt(2); end
    ModulatedSymbols_Tx(1,i + 1) = I + 1j*Q;
end
```

7.5.7.7 Layer Mapping and Precoding

Layer mapping and precoding are steps necessary to map the QPSK symbols to the different transmit antennas. A base station transmits its signal from either 1, 2 or 4 of its antennas (it also receives on those same antennas). Because the PBCH content is very important, the transmit methodology is based on SISO (single input single output), where one transmit antenna communicates with one receive antenna, or the very robust MISO (multiple input single output) method called transmit diversity method. In transmit diversity we spread the information across several transmit antennas, which provides better processing gain at a single and multiple receive antennas. The transmit diversity method uses a technique called space frequency block coding, which is described in great detail in the next section 8.2. It is a fascinating technique that increases the robustness of the channel by exploiting the diversity provided when the base station transmits to the mobile phone over 2 or 4 transmit antennas rather than just one. The 960/864 QPSK symbols are broken up into four sets of 240/216 symbols each. The four sets are mapped to subframe 0 of four consecutive frames as shown in the section on resource element mapping.

Layer Mapping

Layer mapping is a simple step that divides the vector of modulated symbols into either 1, 2 or 4 new sequences. Assume the QPSK mapping step produces a complex number sequence, QPSK[n], of N symbols. These symbols are mapped to one, two or four layers given that one, two or four antennas will transmit the information. (See TS36.211 Section 6.8)

1. For the single transmit antenna case, the symbols in layer 1 are identical to those of the QPSK[n] vector.

$$Layer0[i] = QPSK[i] \; for \; i = 0, 1, ...N - 1$$

2. For the two transmit antenna case, the QPSK symbols are mapped to two layers as follows:

$$Layer0[i] = QPSK[2i] \quad for \; i = 0, 1, ...N/2 - 1$$

$$Layer1[i] = QPSK[2i + 1] \; for \; i = 0, 1, ...N/2 - 1$$

3. For the four transmit antenna case, the QPSK symbols are mapped to four layers as follows:

$$Layer0[i] = QPSK[4i] \quad for \; i = 0, 1, ...N/4 - 1$$

$$Layer1[i] = QPSK[4i + 1] \; for \; i = 0, 1, ...N/4 - 1$$

$$Layer2[i] = QPSK[4i + 2] \; for \; i = 0, 1, ...N/4 - 1$$

$$Layer3[i] = QPSK[4i + 3] \; for \; i = 0, 1, ...N/4 - 1$$

In case that the length, N, of the QPSK sequence is not divisible by four, meaning $mod(N, 4) \neq 0$, then 2 values $= 0 + j0$ are appended to the QPSK sequence and are mapped into the final positions of layer 2 and 3.

```
function [Layer0, Layer1, Layer2, Layer3] = LayerMapping(ModulatedSymbols_Tx)
Layer1 = []; Layer2 = []; Layer3 = [];

if(NumLtePorts == 1)                    % Single Antenna port. There are 960
   Layer0  = ModulatedSymbols_Tx;       % or 864 modulated symbols in this layer
elseif(NumLtePorts == 2)                % Two Antenna ports
   NumSymbolsPerLayer = NumModulatedSymbols/2;
   ModSymb         = reshape(ModulatedSymbols_Tx, [2 NumSymbolsPerLayer]);
   Layer0          = ModSymb(1, :);             % Either 480 or 432 bits in each layer
   Layer1          = ModSymb(2, :);
else                                    % Four Antenna ports
   if(mod(NumModulatedSymbols,4) == 0)
       NumSymbolsPerLayer = NumModulatedSymbols/4;
       ModSymb            = reshape(ModulatedSymbols_Tx, [4 NumSymbolsPerLayer]);
   else
       NumSymbolsPerLayer        = (NumModulatedSymbols+2)/4;
       ModulatedSymbols_Tx_Extended = [ModulatedSymbols_Tx 0 0]; % Appending 2 zeros
       ModSymb = reshape(ModulatedSymbols_Tx_Extended, [4 NumSymbolsPerLayer]);
   end
   Layer0  = ModSymb(1, :);             % Either 240 or 216 bits in each layer
   Layer1  = ModSymb(2, :);
   Layer2  = ModSymb(3, :);
   Layer3  = ModSymb(4, :);
end
```

Precoding

In the precoding step, we map the QPSK symbols residing in layers 0, 1, 2, and 3 into new vectors that we shall call PrecodedP0, PrecodedP1, PrecodedP2, and PrecodedP3, whose content will be directly mapped onto antenna 0, 1, 2, and 3.

1. Precoding for the single antenna case is the simple act of copying those symbols residing in layer 0 in to PrecodedP0.

$$PrecodedP0[i] = Layer0[i] \; for \; i = 0, 1, ...N - 1$$

2. Precoding for the two-antenna case used the transmit diversity technique first pioneered by Siavash Alamouti in 1998, and provides a similar performance to maximum ratio combining, which can be used for the case that we have a single transmit antenna but two receive antennas. As mentioned earlier, please review this fascinating techniques of STBC and SFBC in section 8.2. The mapping proceeds as follows:

$$PrecodedP0[2i + 0] = \; 0.7071 \cdot Layer0[i] \qquad for \; i = 0, 1, ...N/2 - 1$$

$$PrecodedP0[2i + 1] = \; 0.7071 \cdot Layer1[i] \qquad for \; i = 0, 1, ...N/2 - 1$$

$$PrecodedP1[2i + 0] = - \, 0.7071 \cdot conj(Layer1[i]) \; for \; i = 0, 1, ...N/2 - 1$$

$$PrecodedP1[2i + 1] = \; 0.7071 \cdot conj(Layer0[i]) \; for \; i = 0, 1, ...N/2 - 1$$

3. Precoding for the four-antenna case uses the identical transmit diversity technique shown above for the 2 transmit antenna case using the following:

$$PrecodedP0[4i + 0] = 0.7071 \cdot Layer0[i] \qquad for \; i = 0, 1, ...N/4 - 1$$

$$PrecodedP0[4i + 1] = 0.7071 \cdot Layer1[i] \qquad for \; i = 0, 1, ...N/4 - 1$$

$$PrecodedP0[4i + 2] = PrecodedP0[4i + 3] = 0$$

$$PrecodedP2[4i + 0] = - \, 0.7071 \cdot conj(Layer1[i]) \; for \; i = 0, 1, ...N/4 - 1$$

$$PrecodedP2[4i + 1] = \; 0.7071 \cdot conj(Layer0[i]) \; for \; i = 0, 1, ...N/4 - 1$$

$$PrecodedP2[4i + 2] = PrecodedP2[4i + 3] = 0$$

$$PrecodedP1[4i + 0] = PrecodedP1[4i + 1] = 0$$

$$PrecodedP1[4i + 2] = 0.7071 \cdot Layer2[i] \qquad for \; i = 0, 1, ...N/4 - 1$$

$$PrecodedP1[4i + 3] = 0.7071 \cdot Layer3[i] \qquad for \; i = 0, 1, ...N/4 - 1$$

$$PrecodedP3[4i + 0] = PrecodedP2[4i + 1] = 0$$

$$PrecodedP3[4i + 2] = - \, 0.7071 \cdot conj(Layer3[i]) \quad for \; i = 0, 1, ...N/4 - 1$$

$$PrecodedP3[4i + 3] = \; 0.7071 \cdot conj(Layer2[i]) \quad for \; i = 0, 1, ...N/4 - 1$$

Notice how entries 2, 3, 6, 7, 10, 11 … etc in PrecodedP0 and PrecodedP1 remain unpopulated as 0 + j0 values. The same goes for entries 0, 1, 4, 5, 8, 9 … etc in PrecodedP2 and PrecodedP3. Thus for the four antenna case, the same space frequency block coding technique is used as for the two antenna cases, except we are alternating the transmit information between the antenna pairs 0, 1 and 2, 3, allowing us additional path diversity.

```
function [PrecodedP0, PrecodedP1, PrecodedP2, PrecodedP3] =
                                PBCH_Precoding(Layer0, Layer1, Layer2, Layer3)
NumSymbolsPerLayer = length(Layer0);
PrecodedP0 = zeros(1, NumSymbolsPerLayer);  % Invalid content
PrecodedP1 = PrecodedP0; PrecodedP2 = PrecodedP0; PrecodedP3 = PrecodedP0;
if(NumLtePorts == 1)
        PrecodedP0 = Layer0;
elseif(NumLtePorts == 2)  % The SFBC precoding step for the 2 Tx antennas
    for i = 0:NumSymbolsPerLayer - 1
        PrecodedP0(1, 2*i + 0 + 1, 1) = (1/sqrt(2))*Layer0(1, i+1);
        PrecodedP0(1, 2*i + 1 + 1, 1) = (1/sqrt(2))*Layer1(1, i+1);
        PrecodedP1(1, 2*i + 0 + 1, 1) = (1/sqrt(2))*conj(-Layer1(1, i+1));
        PrecodedP1(1, 2*i + 1 + 1, 1) = (1/sqrt(2))*conj( Layer0(1, i+1));
    end
elseif(NumLtePorts == 4)  % The staggered SFBC precoding step for the 4 TX Antennas
    for i = 0:NumSymbolsPerLayer - 1
        PrecodedP0(1, 4*i + 0 + 1, 1) = (1/sqrt(2))*Layer0(1, i+1);
        PrecodedP0(1, 4*i + 1 + 1, 1) = (1/sqrt(2))*Layer1(1, i+1);
        PrecodedP2(1, 4*i + 0 + 1, 1) = (1/sqrt(2))*conj(-Layer1(1, i+1));
        PrecodedP2(1, 4*i + 1 + 1, 1) = (1/sqrt(2))*conj( Layer0(1, i+1));
        PrecodedP1(1, 4*i + 2 + 1, 1) = (1/sqrt(2))*Layer2(1, i+1);
        PrecodedP1(1, 4*i + 3 + 1, 1) = (1/sqrt(2))*Layer3(1, i+1);
        PrecodedP3(1, 4*i + 2 + 1, 1) = (1/sqrt(2))*conj(-Layer2(1, i+1));
        PrecodedP3(1, 4*i + 3 + 1, 1) = (1/sqrt(2))*conj( Layer3(1, i+1));
    end
end
```

7.5.7.8 Resource Element Mapping

The complex symbols in the precoded vectors are now mapped into the first, second, third and fourth OFDM symbols of the second slot in subframe 0 of the current frame that we are generating. Each precoded vector contains either 240 or 216 complex symbols depending on whether we are operating in normal or extended cyclic prefix. The specification says, that the first complex number in the precoded vector should be mapped to the lowest available subcarrier index, k, and the first OFDM symbol in the second slot. The second complex number in the precoded vector is mapped to next highest subcarrier index, k, and the same OFDM symbol. Once you reach subcarrier index $k = Max$ (where $Max = NDLRB \cdot 12 - 1$), we move back to subcarrier position $k = 0$ of the next OFDM symbol. We work our way through each subcarrier position and each symbol. Note the presence of the cell specific reference signals, or CRS. When mapping the precoded numbers, the resource element that contain CRS must be skipped. Note that no matter into which antenna port we are mapping, we must skip the CRS resource element of all ports. Any single receive antenna must be able to read the CRS of every port in order to estimate the frequency response (channel condition) from all transmit antennas. All in all, there are exactly 240 or 216 available resource elements located in the four OFDM symbols dedicated to the PBCH. The next figure illustrates the resource grid for subframe 0 and clearly shows the cell specific reference signal positions that are reserved. If the LTE base station features 4 antennas, which is very common nowadays, then the four precoded vectors are mapped to four different resource grids. The PBCH channel is only one of many channels and signals that need to be mapped into the resource grid. The synchronization signals, the reference signals, the PDCCH (physical downlink control channel), the PDSCH (physical downlink shared channel), and other information must be mapped into the resource grid. Once all information is mapped, each resource grid is separately OFDM modulated and its time domain signals transmitted via each antenna.

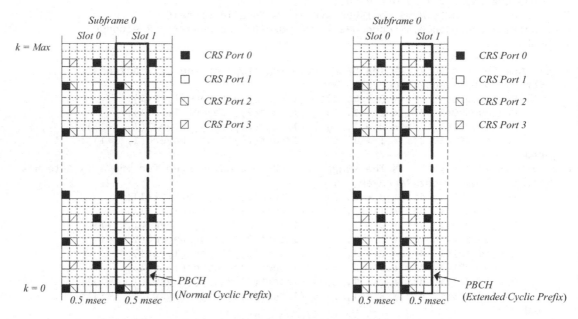

Figure 7-80: The PBCH Location in Subframe 0

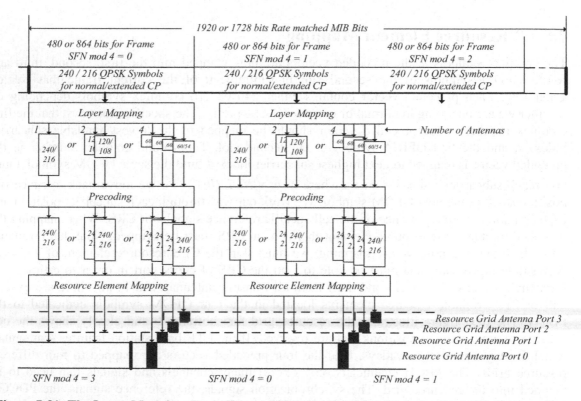

Figure 7-81: The Layer Mapping, Precoding and Element Mapping Steps for the 4 TX Antenna Case

7.6 The LTE Receiver

Most signal processing based communication tasks are run by the LTE's mobile phone receiver while it is simply listening to the base station. Remember that LTE has several different bandwidth options, which the smallest of the at 1.4MHz occupying 6 resource blocks, whereas the largest at 20MHz occupying 100 resource blocks. The next couple of tracking and synchronization algorithms act on information that is embedded in the center 6 resource blocks of the LTE bandwidths no matter what the actual bandwidth is. At this point the mobile simply does not know the LTE signal's bandwidth, which will be revealed in the Master Information Block, or MIB, and these synchronization signals are therefore conveniently placed such that their existence is guaranteed no matter the bandwidth. The receiver filters out all signal content except for the center 6 resource blocks before starting it synchronization steps. In chronological order, the mobile phone executes the following tasks.

Determine the PCI (Physical Cell ID) of a base station

→ The mobile checks for the existence of primary synchronization signals by running a time domain sliding correlation of the received IQ signal against all three possible PSS variants.

→ Once it has detected a sizable peak representing the presence of a PSS, the mobile will OFDM demodulate the OFDM symbol containing the PSS. Given that the receiver knows the exact frequency domain construction of the PSS Zadoff-Chu sequence, it will compute a channel estimate (the frequency response) at that OFDM symbol.

→ The mobile now demodulates the OFDM symbol containing secondary synchronization sequence and equalizes its content. The OFDM symbol containing the SSS is the one just prior to the one featuring the PSS. We can equalizer the SSS based on the channel estimate of the PSS because the multipath channel hardly changes over such a small period of time.

→ The mobile now runs a cross correlation of the content of the received SSS against all 168 possible SSS variants. The cross correlation with the largest score indicates the received SSS variant.

Thus, the mobile has detected both the PSS and SSS variants, which allows it to compute the physical cell identity. In the expression below, $N_{ID}^{(2)}$ (ranging from 0 to 2) indicates the PSS index, whereas the $N_{ID}^{(1)}$ (ranging from 0 to 167) indicates the SSS variant.

$$N_{ID}^{cell} = 3N_{ID}^{(1)} + N_{ID}^{(2)}$$

The physical cell identity (range = 0 to 503) is required as many scrambling operations in the physical layer use the PCI as part of their random generator's starting seed.

Detecting Timing and Frequency Drift

Once the mobile has determined the variant of the SSS it has received, it will start to run time domain sliding correlations of the received IQ signal against a perfect copy of the detected SSS variant. As there are only three possible PSS sequences, the likelihood that at one point the mobile will receive the same PSS variant from different base stations is rather high. For this reason, to track the signal of a particular base station, it is preferable to run the sliding correlation against the SSS, which features 168 possible variants. The mobile phone determines both the timing and frequency drift by observing the occurrence of the SSS correlation peaks over time. This process is explained in detail in the next section.

Computing the FFT Timing Instance

The cross-correlation peaks of the PSS and SSS indicate to the mobile when OFDM signals start. The mobile will track the peaks in order to predict the start of any OFDM symbol it is receiving. Once this timing is known, the mobile will know of which section of the signal to take the FFT in order to produce the frequency content of the OFDM symbol.

Computing Quality Metrics such as RS-CINR and RSRP

The mobile phone continually scans reference signal transmissions from different base stations in order to evaluate the signal quality and rank the base stations accordingly. Two important quality metrics that it keeps are RS-CINR, or reference signal carrier to interference and noise ratio, and RSRP, or reference signal receive power. These quality metrics together with the decoded Master Information Block and System Information Blocks discussed below form a record that described signal quality and access information for each base station. The mobile phone maintains a list of these records and when the user decides to start a call, the mobile will chose the base station to contact based on this list. See additional information on how RS-CINR and RSRP are determined in section 7.6.3.

Decoding the Master Information Block

In section 7.5.7, we have already covered the physical broadcast channel, which holds the MIB. Primarily, the MIB contains information regarding the bandwidth of the LTE signal and approximately indicates which radio frame range we are currently processing. Once the bandwidth is known, the mobile changes its channel bandwidth and ADC sample rate to capture the entire LTE signal, not just the center 6 resource block. The MIB also tells us which radio frame we are currently processing, a fact which is very important as other broadcast information is present in certain radio frames only. We cover the decoding of the PBCH in detail in section 7.6.4.

Decoding of System Information Block 1

The LTE mobile phone now has FFT timing, radio frame timing and knows the proper bandwidth of the base station signal. It's job now is to decode the system information block 1, or SIB1, which indicates the networks operator available on the base station tower (let's hope the one you are subscribed to is on the list) as well as the schedule for further system information messages which contain yet more broadcast information. The SIB1 appears in a fixed radio frame schedule which the mobile is aware of. Inside the appropriate radio frame, the LTE cell phone now checks the PDCCH, which appears at the start of every subframe that carries user information, to determine in which resource blocks of the current subframe the SIB1 is located.

Decoding of the Remaining System Information Blocks

Once the mobile has decoded the SIB1, it reads the schedule for the remaining SIBs and decodes them as well. At this point, all broadcast information is known and the phone is at liberty to contact the base station directly to start a call or data upload.

7.6.1 Timing and Frequency Offset Detection using Synchronization Signals

The synchronization signals provided by the LTE base station are used to acquire the following:

→ FFT timing – FFT timing refers to the knowledge of the receiver start samples of all OFDM symbols in the incoming LTE signal.

→ Timing drift – The analog to digital converters in the receiver and the digital to analog converters in the transmitter will run at a sampling rate dictated by separate master reference clocks usually in the form of a temperature compensate crystal oscillators. Since these oscillators are different physical modules and are located at physically different location, they will never produce exactly the same reference frequency. Therefore, the rate at which IQ samples are generated at the transmitter will differ slightly from the rate at which they are read in at the receiver. This is called timing drift. It is important to keep track of timing drift because it slowly changes the FFT timing instant and it allows us to compute frequency offset as mentioned below.

→ Frequency offset – In LTE communication equipment, the master reference clocks are used not only to generate and read in IQ data, they are also used to generate the TX and RX local oscillator sinusoids, which upconvert and downconvert the IQ data stream to and from the cellular frequency bands. Therefore, the timing drift between the transmit and receive IQ streams mentioned above must be accompanied by a corresponding frequency offset between the two LO sinusoids. Therefore, if you have found the timing drift, then you can immediately calculate the frequency offset. Take a look at the end of section 6.3.2, which explains that time drift and frequency offset are also produced due to the motion of the transmitter relative to the receiver. This phenomenon, called the Doppler effect, is indistinguishable from the error between the TX and RX reference clocks. Thus, detecting and fixing timing drift and frequency offset solves the effects of Doppler and timing base error simultaneously.

→ The synchronization signals are used to find the physical cell identity, which is used to scramble and descramble the values of the reference signals in the LTE waveform and the data bits carried by the different physical channels. Check the start of section 7.6 for a good explanation of how the physical cell identity is found.

To compute timing drift and FFT timing, we pass the received IQ sequence into a sliding correlator, which compares shifting sections of the received waveform with one or more ideal copies of the synchronization signals. The good autocorrelation properties of the synchronization signals allow the production of sharp and distinct peaks when the received synchronization signals are aligned inside the correlator. The peaks indicate the start position of the OFDM symbol. Due to noise and timing drift, there will be some deviation from the expected timing instant at which the peaks occur. Therefore, we must track the arrival instant of the correlation peak over time. Take a look at the next figure and summarize what we can say about it.

→ The transmitter produces a synchronization signal (either PSS or SSS) every 500 microseconds based on the transmitters reference clock.

→ The 'o' symbols on the graph are the coordinates that pair transmit time at which a synchronization signal must have been sent to receive time of the observed correlation peak.

→ If the receiver clock is identical to the transmitter reference clock and there is no Doppler, then the receiver will see the correlation peak every 500 microseconds. The two '+' symbols on the graph indicate when the receiver should see peaks due to the synchronization signal at 4 and 8 milliseconds TX time.

→ It is clear from the slope of the graph before a transmit time of four milliseconds that the receive time was slower than the transmit time and the slope of the dashed line is less than 1.0. Let us assume that this is the default time base error between the transmit and receive reference clocks. The accompanying frequency offset seen at the receiver is positive as the transmit LO is at a higher frequency then the receiver LO.

→ After four milliseconds, the slope increases above 1.0 indicating that the transmit clock has decelerated compared to the receive clock. This can happen when the transmitter is starting to move away from the receiver causing a dilation in time leading to a negative frequency offset seen at the receiver. (See section 6.3.2)

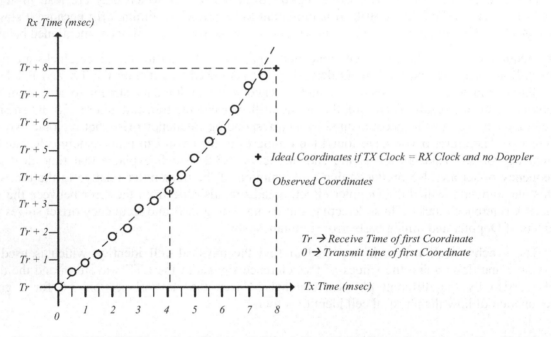

Figure 7-82: TX / RX Time Coordinates of Correlation Peaks due to Synchronization Signals

As time progresses, it is our goal to continually recompute the linear regression of the time coordinates, thus producing a line representing the true relationship between the transmit and receive time basis. We will use the weighted least squares algorithm which will emphasize accuracy for later time coordinates. This curve fitting exercise will allow us to predict time coordinates for which we do not have data. As a matter of fact, it is our intent to compute receiver time instances in between the observation to determine the FFT start time of every OFDM symbol that must be demodulated. Note that time at the receiver and transmitter is tracked by a counter incrementing by one for every sample that the ADC and DAC process. The quantity Tr is the receiver sample time at which the first correlation peak was detected. Without loss of generality, we can assume that the original transmit time of the first peak observation was zero. We can later add an offset that represents the PSS or SSS OFDM symbol start time within a radio frame.

Understanding Timing Error

Time base error of a reference clock is quantified in parts per million, or ppm. A 10MHz crystal oscillator with positive error of 1 part per million, will produce 10e6 clock pulses /second · 1e-6 · 1second = 10 additional clock pulses in one second. Compared to a perfect clock, time referenced by this crystal oscillator runs too fast by a factor of 1 microsecond every second. The overall time base error or timing drift seen at the receiver will be equal to the following.

$$TotalTimeBaseErrorPpm \approx RxErrorPpm - TxErrorPpm$$

In the figure above, the slope of the line relating the time coordinates will be equal to 1 if both the receive and transmit time references are perfectly aligned and no time base error exists. If they are not, then the slope of the line will obey the following expression.

$$Slope = 1 + (TotalTimeBaseErrorPpm) \cdot 1e - 6$$

Understanding Frequency Error

The transmitter and receiver upconvert and down convert the IQ information to and from the RF center frequency. If the transmit reference clock error is equal to 1ppm, and the desired RF center frequency is 2GHz, then the transmit IQ stream will actually be upconverted to the following frequency.

$$ActualRfFreq = DesiredRfFreq * (1 + \frac{TimingError_{ppm}}{1e6})$$
$$= 2e9Hz * (1 + \frac{1}{1e6})$$
$$= 2e9Hz + 2000Hz$$

The total corresponding frequency offset seen at the receiver is equal to the following expression.

$$FrequencyOffset = DesiredRfFreq(TxErrorPpm - RxErrorPpm) \cdot 1e - 6$$
$$= - DesiredRfFreq(TotalTimeBaseErrorPpm) \cdot 1e - 6$$

The following MatLab code simulates cross correlation peak measurements featuring time coordinates matching up transmit time, TxTime, to receive peak observation time, or RxTime. The simulation allows you control over the transmitter and receiver time base errors as well as the carrier frequency, observation period and maximum receive time inaccuracy that incurs when we record the receive time of the incoming correlation peak. The accuracy of the frequency offset estimate and receive OFDM symbol start times increases with increasing observation time and decreasing maximum receive time inaccuracy. The FFT start times are simply equal to the sum of the OFDM symbol start times and the cyclic prefix time.

```
% -------------------------------------------------------------------------
% Define the time base error
TxErrorPpm            = 2;                    % Transmit time base error
RxErrorPpm            = 0;                    % Receiver time base error
TotalTimeBaseErrorPpm = RxErrorPpm - TxErrorPpm;
```

```
% Define the actual frequency offset produced due to the total time base error.
CarrierFreq          = 2e9;                                      % Hertz
ActualFreqOffset     = -CarrierFreq * TotalTimeBaseErrorPpm * 1e-6;   % Hertz


% --------------------------------------------------------------------------
% Approximate the TxTime and RxTime coordinates of the cross-correlation peaks
% produced by the total time base error configuration above. Experiment with the
% ObservationPeriod.
Slope                = 1 + TotalTimeBaseErrorPpm * 1e-6;
ObservationPeriod    = 200e-3;
TxTime               = (0:0.0005:ObservationPeriod)'; % Seconds

% The RxTime coordinates have an intial value at TxTime = 0. We add this offset
% as well as inaccuracies when measuring the receive time, RxTime, at the
% correlation peaks. Experiment with the size of the MaxRxTimeInaccuracy.
RxTimeOffset         = 0.2341;                        % The intercept point.
MaxRxTimeInaccuracy  = 100e-9;                        % Seconds
RxTimeNoise          = MaxRxTimeInaccuracy * (2*rand(length(TxTime), 1) - 1);
RxTime               = TxTime * Slope + RxTimeOffset + RxTimeNoise;


% --------------------------------------------------------------------------
% We have the observed correlation peak time coordinates now as (TxTime, RxTime).
% It is our goal to compute the linear regression to estimate the intercept point
% and slope of the line describing the actual correlation peak time coordinates.
F      = [ones(size(TxTime)), TxTime];  % See Section 3.1.2 for more details
Weight = 1:length(TxTime); % Weigh the later coordinates more than the eariler ones.
W      = diag(Weight);
AW     = inv(F' * W * F) * F' * W * RxTime;  % F'*W*F is a simple 2x2 matrix

% AW represents the intercept point and the slope of a line
DetectedInterceptPoint    = AW(1,1);
DetectedSlope             = AW(2,1);

% We can now estimte the frequency offset that must exist in the signal
EstimatedTotalBaseErrorPpm  = (DetectedSlope - 1) * 1e6;
EstimatedFreqOffset         = -CarrierFreq * EstimatedTotalBaseErrorPpm * 1e-6;

% We can also estimate the receive time at which OFDM symbols starts.
% For the extended CP, the LTE OFDM symbol period is 1e-3/ 12 = 83.33333e-6 seconds
% The receive times estimates will achieve the highest accuracy toward the end.
OfdmSymbolPeriod  = 1e-3 / 12;
TxTimeSymbolStart = 0:OfdmSymbolPeriod:ObservationPeriod;
RxTimeSymbolStart = DetectedInterceptPoint + DetectedSlope * TxTimeSymbolStart;

disp(['Actual    Frequency Offset (Hz): ' num2str(ActualFreqOffset) ]);
disp(['Estimated Frequency Offset (Hz): ' num2str(EstimatedFreqOffset) ]);
disp(['UnresolvedFrequency Offset (Hz): ' num2str(ActualFreqOffset -
EstimatedFreqOffset)]);
```

7.6.2 OFDM Demodulation in LTE

At its core, OFDM demodulation is simply the application of the FFT operation on the section of the sampled waveform that represent the IFFT portion of the OFDM symbols. But where does this IFFT portion start? Take a look at the diagram below, which illustrates the arrival at the receive antenna of an OFDM signal via four paths. Path 2 turns out to be the strongest, whereas path 1 (also called a precursor) arrived 600 nanoseconds earlier and paths 3 and 4 (also known as post cursors) arrived 1 and 2 microseconds later. The LTE modem has determined the proper start time of the strongest path by correlating the incoming waveform against the synchronization signals. Therefore, the start time of the IFFT portion of the OFDM symbol for the strongest path (path 2 in figure below) is known with high accuracy. However, is it really our goal to take the FFT of the IFFT portion of the OFDM symbol of path two? In fact, it is not, and the reason can be seen at the end of the OFDM symbol of path two. Notice how the cyclic prefix portion of the next symbol of path one has started before the end of path two's symbol. As all the paths are added into a single waveform at the receive antenna, intersymbol interference occurs at the end of path two's OFDM symbol. To avoid this problem, we take the FFT at an arrival time of approximately -1 microseconds, well ahead of the precursor. However, we must take care not to move the FFT start time any further back, otherwise the last symbol of the post cursors (paths 3 and 4) will interfere with the section of the signal that we want to provide to the FFT. As the cyclic prefix is a cyclic extension of the IFFT portion of the OFDM symbol, it is completely fine to move the FFT start time inside the CP area. Using the time shifting property of the Fourier transform, we simply compensate the FFT output values.

All in all, advancing the FFT start time by about 1 microsecond is a good compromise as in most cellular environments, precursors don't usually exceed 1 microsecond of time advance and post cursors don't exceed 4 microseconds of time delay. This is based on LTE field trial experiments and the cyclic prefix length specified by the LTE standard is thus just shy of 5 microseconds.

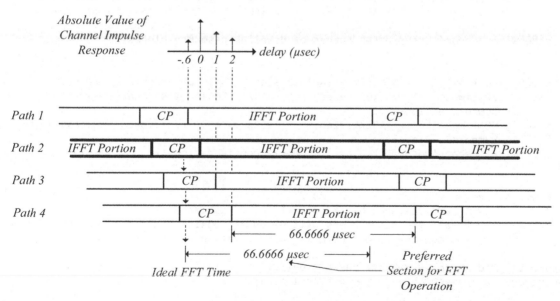

Figure 7-83: Typical Multipath Scenario in Cellular Communication Systems

Before we provide MatLab code that provides the OFDM demodulation, we need to ask ourselves whether it really is necessary to explicitly compensate FFT output for the 1 microsecond of advance upon which we had agreed. And once again, the answer is 'no'. Once we have the FFT outputs of several symbols, we examine the embedded reference signals to estimate the frequency response of the channel. Based on this information we compute equalizer coefficients that will undo the linear distortion of the multipath channel. The time advance embedded in out FFT start time is simply another linear distortion that is discovered during channel estimation and is corrected when applying the equalizer coefficients.

The following MatLab code demodulates the OFDM symbols of a waveform given the number of downlink resource blocks (which leads to the sample rate and FFT size), the cyclic prefix type, the start time and the sequence of complex waveform samples. Note, as the start time can indicate an instant in-between available samples. The start sample used by the FFT will be the sample that occurs just ahead of the start time. After the FFT is computed, we use the time shifting property of the Fourier transform to correct the effect on the phase due to the apparent delay that occurred.

```matlab
function ResourceGrid = DemodulateLteSignal(...
              Waveform    ... % Vector of complex samples
            , NumSubframes... % Number subframes to demodulate
            , NDLRB       ... % Number download resource blocks
            , CP_Selection... % Cyclic Prefix 1/2 - normal/ext
            , StartTime)

% Set parameters and some error checking
SubcarrierSpacing = 15000;   % Hz
TimeIfftSection  = 1/15000; % Length (sec) of IFFT portion of OFDM symbol = 66.667usec

switch(NDLRB)
    case   6; SampleRate =  1.92e6; FFT_Size =  128;  % Corresponds to BW =  1.4MHz
    case  15; SampleRate =  3.84e6; FFT_Size =  256;  % Corresponds to BW =  3.0MHz
    case  25; SampleRate =  7.68e6; FFT_Size =  512;  % Corresponds to BW =  5.0MHz
    case  50; SampleRate = 15.36e6; FFT_Size = 1024;  % Corresponds to BW = 10.0MHz
    case  75; SampleRate = 21.12e6; FFT_Size = 1408;  % Corresponds to BW = 15.0MHz
    case 100; SampleRate = 30.72e6; FFT_Size = 2048;  % Corresponds to BW = 20.0MHz
    otherwise; error('Unsupported number of downlink resource blocks.');
end

switch(CP_Selection)        % First Symbol in slot  -  All other symbols in slot]
    case 1; CP_Times      = [10/1.92e6,       9/1.92e6]; % seconds - Normal CP
            SymbolsPerSlot = 7;
    case 2; CP_Times      = [32/1.92e6,      32/1.92e6]; % seconds - Extended CP
            SymbolsPerSlot = 6;
    otherwise; error('Unsupported cyclic prefix configuration.');
end

NumberOfPositiveSubcarriers = 12 * NDLRB / 2;
                % Each resource block consists of 12 subcarriers
                % Half the subcarriers will be on the positive and half on the negative
                % frequency side. There is no information at the DC subcarrier.

% Resource grid where we deposit the FFT output
ResourceGrid = zeros(NDLRB*12, SymbolsPerSlot*2);

FFT_Time            = StartTime;
TotalNumberOfSymbols = NumSubframes*2*SymbolsPerSlot; % There are two slots / subframe
for Symbol = 0:TotalNumberOfSymbols-1
    if(mod(Symbol, SymbolsPerSlot) == 0);  CP_Time = CP_Times(1,1);
```

```
    else;                                    CP_Time = CP_Times(1,2);
    end
    FFT_Time      = FFT_Time + CP_Time;
    EarlySample   = floor(FFT_Time * SampleRate);

    % We are taking the FFT of samples that happen earlier in time.
    % To the FFT it looks like we are transforming a delayed signal.
    ApparentDelay = FFT_Time - EarlySample/SampleRate;

    % Take the following samples and FFT transform them
    Range         = EarlySample:(EarlySample + FFT_Size - 1);
    Samples       = Waveform(1, Range + 1);
    FFT_Output    = fft(Samples, FFT_Size)/FFT_Size;

    % Adjust the FFT output phases given the Apparent Delay
    Pos_Frequencies = (1:1:NumberOfPositiveSubcarriers)*15000;  % Remember -> skip DC
    Pos_Index_Range =  1:1:NumberOfPositiveSubcarriers;
    FFT_Output(1, Pos_Index_Range + 1) = ...
    FFT_Output(1, Pos_Index_Range + 1) .* exp(1j*2*pi*ApparentDelay*Pos_Frequencies);

    Neg_Frequencies = (-NumberOfPositiveSubcarriers:1:-1) *15000;
    Neg_Index_Range = FFT_Size - NumberOfPositiveSubcarriers:1:FFT_Size - 1;
    FFT_Output(1, Neg_Index_Range + 1) = ...
    FFT_Output(1, Neg_Index_Range + 1) .* exp(1j*2*pi*ApparentDelay*Neg_Frequencies);

    % Map the FFT output into the resource grid
    Pos_Range = NumberOfPositiveSubcarriers:2*NumberOfPositiveSubcarriers - 1;
    Neg_Range = 0:1:NumberOfPositiveSubcarriers-1;
    ResourceGrid(Pos_Range + 1, Symbol + 1) = FFT_Output(1, Pos_Index_Range + 1);
    ResourceGrid(Neg_Range + 1, Symbol + 1) = FFT_Output(1, Neg_Index_Range + 1);

    % Increment FFT timing instant
    FFT_Time      = FFT_Time + TimeIfftSection;
end
```

7.6.3 Signal Quality Metrics – RS-CINR, RSRP and RS-RSSI

The mobile phone continually evaluates reference signals transmitted from different base stations in order to rank them in order of signal quality. The mobile phone will use this information to select the base station to contact in case the user commences a call. Reference signals are QPSK modulated and different base stations will use different scrambling sequences to determine their orientation. Because reference signals from different cells feature different orientation they are called actually called cell specific reference signals, or CRS. The following quality metrics are computed given the CRS of different base stations.

Reference Signal – Carrier to Intereference and Noise Ratio (RS-CINR)

The RS-CINR is the ratio of the average reference signal power and the averge noise/interference power within the resource elements assigned to the CRS. Remember that the base station whose reference signals we are trying to measure experiences interference from signal transmitted by other base station. The interference from these base stations may be CRS as well or due to generic information traffic. As this interference appears noise like, thermal noise coming from the receiver of the mobile phone and interference from other base station signals can appear as a single quantity.

$$RS\_CINR = 10 \cdot log_{10}\left(\frac{Average\ Reference\ Signal\ Power}{Average\ Noise\&Interference\ Power}\right) dB$$

Reference signals and the computation of RS-CINR were discussed in detail in sections 7.5.6, 7.1.3.4 and 7.1.3.5.

Reference Signal Receive Power (RSRP)

RSRP represents the average power of the CRS of the cell that we are currently evaluating. The resource elements (in the resource grid) that contain CRS will also feature noise and interference. During the RS-CINR measurement, the dimensionless reference signal power, or mean square, has to be computed separately from that of the noise and interference. This reference signal mean square has to be adjusted via a scalar that is proportional to the gain of the receiver. This scalar is a function of the receiver gain and will likely be available in tabular form from the chip set manufacturer. The RSRP is computed in terms of dBm.

$$RSRP = 10 \cdot log_{10}(Reference\ Signal\ Mean\ Square) + Adjustement(ReceiverGain)\quad dBm$$

Reference Signal Receive Strength Signal Indicator (RS-RSSI)

The RS-RSSI is the total average power of OFDM symbols that carry cell specific reference signals. The OFDM symbols must carry CRS but may also feature LTE traffic as well as featuring noise and interference. A simple way to compute the RSSI is to OFDM demodulate the time domain sequence, fill in the resource grid based on the FFT outputs and compute the mean square of all resource elements that fall within a group of closely collocated OFDM symbols. Remember that the columns of the resource grid represent the frequency domain information of each OFDM symbol. You therefore average the products of each resource element value with its complex conjugate. A single OFDM symbol features a number of subcarriers equal to $NDLRB \cdot$ 12 subcarriers/resource block, where $NDLRB$ represents the number of downlink resource blocks. The mean square of a single OFDM symbols is computed as follows.

$$MeanSquare = \frac{1}{NDLRB \cdot 12} \sum_{k=0}^{NDLRB \cdot 12 - 1} ResourceElement[k] \cdot conj(ResourceElement[k])$$

$$RS\_RSSI = 10 \cdot log_{10}(Mean\ Square) + Adjustement(ReceiverGain)\quad dBm$$

7.7 C-V2X (Vehicle to Everything Communication)

Cellular vehicle to everything technology, or C-V2X, is a 3GPP communication standard that specifies communication between vehicles, road side units such as traffic lights, and pedestrians. This communication is meant to form the back bone of intelligent traffic systems, ITS, which are meant to increase safety on our roads. In its basic form, V2X stations such as vehicles, road side units, and cellular phones carried by pedestrians will periodically broadcast data such as position, velocity, heading, and traffic light status. Each V2X station continually listens to these broadcasts thereby becoming aware of the local traffic conditions, allowing it to take action to avoid risky situations. The most common scenario would be that of a vehicle braking autonomously if it approaches another car in front of it too quickly due to negligence of the driver. C-V2X exists as a 4G (3GPP Release 14) type technology, which is being deployed as of the year 2020/2021, and a 5G (3GPP Release 16) technology that will likely see deployment in 2023/2024. In this section, we introduce LTE C-V2X and discuss how it solves common signal processing and physical layer tasks. To better understand this material, the reader should be familiar with basic LTE concepts such as OFDM demodulation and the resource grid that was introduced in section 7.5.2.

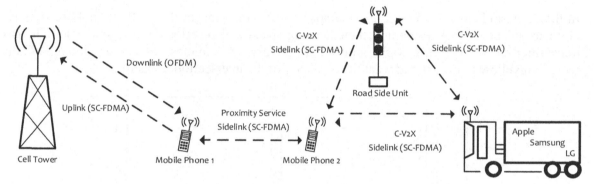

Figure 7-84: Downlink, Uplink and Sidelink Communication Provided by the LTE Standard

Virtually all discussions from section 7.5 and 7.6 in this text have focused on the **downlink** that provides OFDM based information transmission from the cell tower to the mobile phone. The **uplink** direction from the mobile phone to the cell tower uses SC-FDMA, which we have met in section 5.5 and is implemented here to allow for the use of more efficient RF power amplifiers in the mobile. The 3GPP specification also provides for a **sidelink**, which enables two-way communication between mobile phones via a feature called proximity service. This service enables emergency communication for police, paramedics and firefighter in areas with poor cell coverage. The cellular vehicle to everything, C-V2X, technology borrows the proximity service sidelink technology, improves it to allow for communication with fast moving vehicles and moves the transmit frequency from the popular cellular bands up to 5.9GHz. At this higher transmit frequency, signal coverage will be limited to a range of no more than 1Km. Compared to lower frequency cellular communication, you can also expect the frequency offset (Doppler) problem to worsen but the multipath channel to be somewhat better behaved given that transmission paths will arrive more densely spaced in time (see section 6.3.1, 6.3.2 and 7.1.3.2).

C-V2X is an interesting use case as it will soon become quite common place in our communication landscape. It shares a lot of features with the uplink, which people usually don't cover in books, and does without many of the complicated physical layer procedures of the down and uplinks.

V2X Communication in the Time and Frequency Domain

The figure below shows the transmissions of five separate V2X stations each featuring a period of 100 milliseconds. C-V2X only allows two separate bandwidths including 10MHz, or 50 sidelink resource blocks (shown below), and 20MHz, or 100 sidelink resource blocks, where each resource blocks occupies $12 \cdot 15\text{KHz} = 180\text{KHz}$. Unlike the litany of physical channels in LTE, C-V2X transmission is limited to two physical channels, the PSCCH and PSSCH, which are responsible for organizing and conveying the traffic messages. The C-V2X receiver continually scans the incoming RF signal to detect the presence of one or more PSCCH transmission in the each incoming subframe. The decoded content of the PSCCH points the receiver to the position and encoding scheme of the PSSCH, which holds the traffic message. The C-V2X transceiver gathers information about the location of physical channels it sees and based on this information determines where it should transmit its own V2X content. The idea is somewhat similar to clear channel assessment in 802.11a/b/g/n, where a station transmits only when no other signal is on the air. Communication is therefore ad-hoc as compared to the LTE downlink and uplink, where the location in time and frequency of each transmission is precisely determined by instructions from base station.

In the scenario below, we could, for example, assume that transmitters 1 through 4 are vehicles close to an intersection providing their heading speed and position to everyone else, whereas transmitter 5 is a traffic light informing all nearby vehicles what the timing of the red/green/yellow lights are and what the geometry of the intersection looks like.

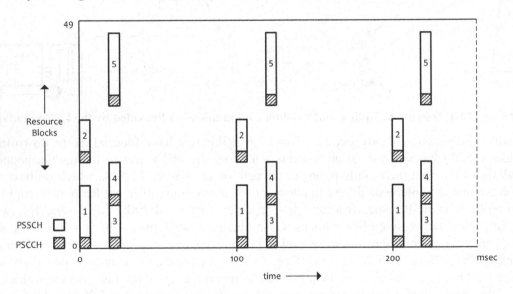

Figure 7-85: Resource Grid Showing C-V2X Traffic For Five V2X Transmitters

7.7.1   C-V2X Physical Channels

The figure above introduces us to two physical channels that carry information that we need to demodulate. The first is the physical sidelink control channel, or PSCCH, which contains information regarding the physical sidelink shared channel, or PSSCH. The PSSCH contains the actual traffic related information sent by the vehicle, pedestrian or road side unit.

The Physical Sidelink Control Channel – PSCCH

The PSCCH hold 32 bits of control information providing facts about the physical sidelink shared channel that holds the V2X traffic message.

Table 7-5: Information in the PSCCH

Field	Number of Bits	Description
Priority	3	The priority field enables quality of service features.
Resource Reservation	4	Indicates when this V2X station will transmit a new message. Usually 100 milliseconds in the future.
Frequency Resource Allocation	Varies	This field indicates the resource blocks that are occupied by the PSSCH. The PSSCH will always occur in the same subframe as the PSCCH as the last figure shows.
Time Gap	4	A C-V2X message usually occurs as a pair of identical transmissions giving the receiver two chances to decode the same message. This time gap indicates the distance in subframes to the other transmission.
MCS	5	The modulation and coding scheme of the PSSCH channel tells the receiver what constellation the PSSCH information is using and how much rate matching is being applied. Information that is very important will likely have a lot of rate matching (bit repetition) and use QPSK rather than the less robust 16 or 64QAM.
Retransmission Index	1	Tells us whether the current transmission is the early original transmission or the later retransmission.
Pad bits	Varies	To ensure that the entire PSCCH message consists of 32 bits, we add zeros at the end.

The 32 bit sequence shown in the table above is called sidelink control information, or SCI, format 1. These bits are encoded and processed until complex numbers that represent these bits are mapped into the resource grid and OFDM modulated. The next figure illustrates the encoding process of the SCI format 1 bits, and it becomes readily apparent that it is quite involved. The following list provides details regarding each one of these processing blocks.

→ CRC Attachment: The 16 bit cyclic redundancy check of the 32 bit SCI is computed and appended to form a 48 bit message. The CRC bits are used at the receiver to detect bit errors. The CRC process is described in section 5.6.1. (See TS36.212 Section 5.4.3, 5.3.3.2, and 5.1.1).
→ Convolutional Encoding: The 1/3 rate tail biting convolutional encoder adds 96 redundancy bits that are used at the receiver to correct errors. This process is described in detail in section 5.6.3. (See TS36.211 Sections 5.3.3.3 and 5.1.3.1)
→ Rate Matching: The rate matching step simply repeats the 144 convolutionally encoded bits (some of them 4 times, other only 3 times) to produce 480 rate matched bits. It also includes some interleaving. At the receiver, the repeated bits are simply added to increase the signal to noise ratio prior to being Viterbi decoded. The more the 144 bits are repeated, the higher the SNR advantage due to the accumulation process at the receiver. (TS36.212 Section 5.3.3.4 / 5.1.4.2)
→ PUSCH Interleaving: This additional interleaving step is something that the sidelink shares with the uplink and simply increases the separation of closely located bits. (TS36.212 Section 5.4.3)

→ Scrambling: Scrambling is traditionally used to avoid long bit sequences of 1 or 0s. In the mobile phone environment, it is used to make an information sequence appear random and noisy like to receivers not intended to understand it. (TS36.211 Section 9.4.1)

→ QPSK Mapping: The bits are mapped into QPSK symbols according to section 7.5.7.6 in this text. We always use the most robust constellation here in order to ensure that the message is understood. (TS36.211 Section 9.4.2 and 5.3.2)

→ Transform precoding: This is the step required to implement SC-FMA, or single carrier – frequency division multiple access. In this step, the 24 complex numbers that are destined to go into the resource elements of a single OFDM symbol are first FFT transformed before being mapped into the resource grid. (TS36.211 Section 9.4.6)

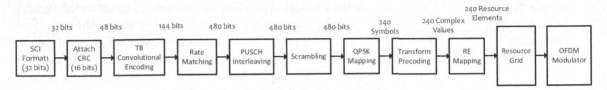

Figure 7-86: The PSCCH Encoding Chain

The figure below shows the 14 OFDM symbols belonging to a PSCCH instance. Symbols 2, 5, 8, and 11 are reserved for demodulation reference signals that are known to the receiver and will be used there to detect frequency offset as well as the frequency response of the multipath channel. As shown above, the 480 scrambled data bits are reduced to 240 QPSK symbols, which after transform precoding are placed into the resource elements of OFDM symbols 0, 1, 3, 4, 6, 7, 9, 10, 12, and 13. However, the information in the 13th OFDM symbol is subsequently zeroed out before OFDM modulation and transmission. Some of the QPSK symbols are therefore lost, and OFDM symbol 13 will serve as a guard period. This loss is easily made up given the extensive amount of rate matching which inflates the 144 encoded bits to 480 rate matched (repeated) bits.

Why the 144 encoded bits are not simply rate-matched to $9 \cdot 48 = 432$ bits and therefore to $9 \cdot 24 = 216$ QPSK symbols to nicely fill up 9 data OFDM symbols isn't completely clear but it may have to do with staying compatible with some of the other sidelink and uplink procedures in LTE.

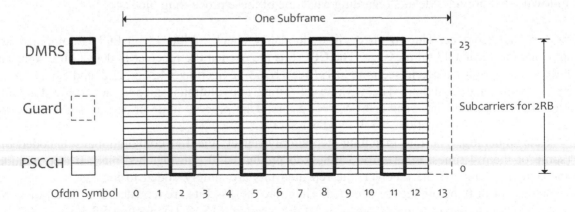

Figure 7-87: The PSCCH Information Spread Across OFDM Symbols 0, 1, 3, 4, 6, 7, 9, 10, and 12

The Physical Sidelink Shared Channel – PSSCH

There are only a few differences between the encoding process of the PSCCH and the PSSCH.

→ The number of bits in the transport block (the PSSCH message) is set by the modulation and coding rate and the number of physical resource blocks. You can derive both of these pieces of information from the SCI along with a few tables from the specification.

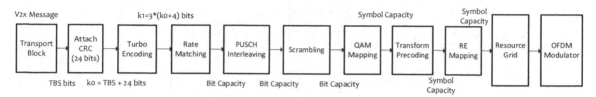

Figure 7-88: The PSSCH Encoding Chain

The biggest differences between the control channel and shared channel encoding are as follows:

→ The PSCCH features a transport block of 32 bits, whereas the PSSCH can feature a large number of different sized bit streams.
→ The shared channel features 24 bits of CRC protection for a potentially long bitstream, whereas the control channel uses 16 bits for a transport block of only 32 bits. The control channel protection is far more robust.
→ The PSSCH uses turbo coding as compared to tail biting binary convolutional coding used in the PSCCH. Turbo coding is superior to BCC for long messages.
→ The PSCCH uses QPSK, 16QAM and 64QAM constellations in the QAM mapper, whereas the control channel uses only QPSK.

7.7.2 SC-FDMA Modulation and Demodulation in C-V2X

The modulation scheme for C-V2X is single carrier frequency division multiple access, or SC-FDMA, which we covered in detail in section 5.5 of this text. Take a look at V2X station 1, which produces two bit payloads. One is the SCI (sidelink control information) destined for the PSCCH, and one is the actual V2X traffic message destined for the PSSCH. After the PSCCH and PSSCH encoding steps, the resulting QAM symbols streams are DFT precoded, which converts OFDM modulation into SC-FMDA modulation. The PSCCH QAM symbol stream is divided into 10 24 QPSK symbol sections, which are each DFT precoded before being mapped into the 10 data carrying OFDM symbols positions in the resource grid. Remember, the PSCCH and PSSCH map into OFDM symbols $l = 0, 1, 3, 4, 6, 7, 9, 10, 12,$ and 13, whereas DMRS are mapped into $l = 2, 5, 8,$ and 11. The insertion of the DMRS is not shown in the figure below as this process is not SC-FDMA modulation but plain OFDM modulation. The resource grid covers a single subframe with 14 OFDM symbols and $RB = 0 \ldots 99$ resource blocks indicating a bandwidth of 20MHz. The variable k to the right of the resource grid indicates the subcarrier index. Remember that each resource block consists of 12 subcarriers for a total of 1200 subcarrier in our C-V2X signal. V2X station 2 is not co-located with V2X station 1 and their physical channels don't map into the same resource grid but their own separate grids. The figure simply suggests the difference in resource block positions that the V2X stations must obey in order to avoid interfering with one another.

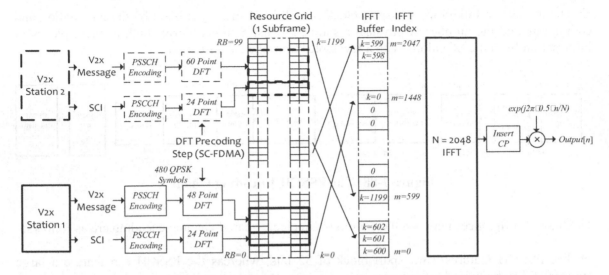

Figure 7-89: SC-FDMA Modulation Process in LTE C-V2X

The remaining OFDM modulation process is very similar to that of the downlink (see section 7.5.3) except for one very important difference. In the downlink, the content of the resource grid is not mapped to the DC carrier ($m = 0$) of the IFFT input buffer. Here, subcarrier $k = 600$ is indeed mapped into the DC position of the IFFT. In the downlink, the DC position was set to zero as the mobile phone receiver is likely of the zero-IF type. This type of receiver directly down converts the RF signal to baseband. The zero-IF receiver features a lot of DC offsets, which overwhelm the OFDM content there, especially if the receive signal is small. Unlike the mobile phone, the receiver at the base station can use a more sophisticated super heterodyne structure, which does not suffer from the same problem as the zero-IF structure (which is cheaper). We therefore have a value at DC, 599 at positive frequencies but 600 at negative frequencies. In order to properly center the signal bandwidth, a 7500Hz frequency offset is introduced after the OFDM symbol has been generated. Thus, the first positive frequency and negative frequency subcarriers are now located at +7500 and -7500Hz respectively. The code below OFDM modulates a resource grid pertaining to either a 10MHz or 20MHz bandwidth C-V2X signal and adds the 7500Hz shift (see 3d).

```
% Brief:    CV2xModulate() function converts the ResourceGridArray into an IQ sequence.
% Inputs:   ResourceGridArray -> The resource grid to be OFDM modulated.
%           FftSize           -> Usually (N = 1024 -> 10MHz | 1408/1536/2048 -> 20MHz)
% Outputs:  SampleRate        -> Self Explanatory
%           TxSequence        -> The Ofdm modulated output sequence.
% Specs:  See TS36.211 Section 5.6
function [SampleRate, TxSequence] =  CV2xModulate(ResourceGridArray, FftSize)

    [NumSubcarriers, NumOfdmSymbols]  = size(ResourceGridArray);
    ProperNumSubcarriers = NumSubcarriers == 600 || NumSubcarriers == 1200;
    ProperNumSymbols     = mod(NumOfdmSymbols, 14) == 0;
    assert(ProperNumSubcarriers, 'We need to have either 600 or 1200 subcarriers.');
```

```
    assert(ProperNumSymbols,      'We to have an integer number of subframes.');

    % 1. Setup parameters
    N_FFT_Target = FftSize;
    SampleRate   = FftSize * 15000;      % 15000Hz is the subcarrier spacing

    % The number of CP samples of the first and remaining symbols in a slot
    Ncp_Target   = [10, 9, 9, 9, 9, 9, 9] * FftSize/128; % Correct of Normal CP

    NumSubframes  = NumOfdmSymbols/14;    % There are 14 Ofdm Symbols per subframe
    NumTxSamples  = NumSubframes * ...
                  (N_FFT_Target*14 + 2*Ncp_Target(1,1) + 12*Ncp_Target(1,2));
    TxSequence    = zeros(1, NumTxSamples);

    % 2. Build the input mapping for the IFFT (Index in resource grid)
    PositiveSubcarriers = NumSubcarriers/2 : (NumSubcarriers - 1);
    NegativeSubcarriers = 0:(NumSubcarriers/2 - 1);

    % Index in IFFT (Different than the downlink -> 1:NumSubcarriers/2)
    PosFreqIndicesIFFT  = 0:NumSubcarriers/2 -1;
    NegFreqIndicesIFFT  = (N_FFT_Target - NumSubcarriers/2):(N_FFT_Target-1);

    % 3. Ofdm Modulate each symbol (See TS36.211 Section 5.6)
    StartIndex = 0;
    for SymbolIndex = 0:NumOfdmSymbols-1
        % 3a. Fill the  IFFT input buffer
        IFFT_Input                        = zeros(N_FFT_Target, 1);
        IFFT_Input(PosFreqIndicesIFFT + 1, 1) = ...
                ResourceGridArray(PositiveSubcarriers + 1, SymbolIndex + 1);
        IFFT_Input(NegFreqIndicesIFFT + 1, 1) = ...
                ResourceGridArray(NegativeSubcarriers + 1, SymbolIndex + 1);

        % 3b. Take the IFFT
        IFFT_Output               = N_FFT_Target*ifft(IFFT_Input, N_FFT_Target);
                                    % Scaling by N_FFT_Target guarantees same
                                    % output scaling for all FFT sizes
        % 3c. Generate cyclic prefix
        SymbolIndexInSlot  = mod(SymbolIndex, 7);  % CP - normal (7 Symbols per slot)
        CurrentCpLength = Ncp_Target(1, SymbolIndexInSlot + 1);
        CyclicPrefix     = IFFT_Output(N_FFT_Target-CurrentCpLength+1:N_FFT_Target, 1);

        % 3d.Generate time domain OFDM symbol including CP and 7500Hz shift
        OutputSymbol             = [CyclicPrefix; IFFT_Output];
        n                        = (-CurrentCpLength:N_FFT_Target-1).';
        f                        = 0.5 * (1/N_FFT_Target);
        OutputSymbol             = OutputSymbol.*exp(1j*2*pi*n*f);

        % 3e. Place output symbols into output waveform array
        Range                    = StartIndex+1:StartIndex+length(OutputSymbol);
        TxSequence(1, Range)     = OutputSymbol.';
        StartIndex               = StartIndex + length(OutputSymbol);
    end
end
```

7.7.3   Timing Synchronization in C-V2X

Just like LTE, the C-V2X signal is subdivided into the same radio frames, subframes and OFDM symbols, lasting 10 milliseconds, 1 millisecond, and approximately 1/14 milliseconds respectively (see section 7.5.2). The LTE downlink signal provided synchronization signals that were used by the receiver to time align the signal and demodulate the OFDM symbols. However, the C-V2X technology must function in areas without network coverage, and for this reason, the designers decided to use GPS timing data to define the start of a radio frame. A GPS receiver, which must be part of the C-V2X station, provides two pieces of important timing information.

→ A very accurate pulse per second, PPS, signal that features either an abrupt falling or an abrupt rising edge exacty once per second. The signal edge represents the start of a new second based on UTC, or coordinated universal time, which is the successor to GMT, or Greenwich mean time.
→ The GPS receiver will also provide the number of seconds that have passed since the Unix epoch at midnight on the 1st of January, 1970.

Given these timing references, the C-V2X station can determine the following.

→ Given the PPS signal, the C-V2X station can – with great accuracy – determine the time at which each sample is either written to the digital to analog converters in the transmitters or read into the analog to digital converters in the receiver.
→ Given the Unix time, the number of seconds since 1970, the C-V2X station computes the number of seconds that have passed since midnight on the 1st of January, 1900. The 3GPP specification defines the theoretical start time of the first radio frame at midnight at the change form the 19th to the 20th century.

At this point, the C-V2X station knows the exact time, T_{V2X}, in seconds since midmight on January 1st of 1900 and computes the direct subframe number, or DFN, (comprable to the system frame number, or SFN, in LTE). Remember that the SFN, and also the DFN, are 10 bit unsigned integers that wrap around at a value of 1024.

$$DFN \ (Direct \ Frame \ Number) = \bigl(floor(100 * T_{V2X})\bigr) mod \, 1024$$

$$DSFN \ (Direct \ Subframe \ Number) = \bigl(floor(1000 * T_{V2X})\bigr) mod \, 10$$

Knowledge of the current DSFN is important as it partakes in the scrambling of the physical sidelink shared channel information.

7.7.4   Frequency Offset Acquisition in C-V2X

According to the specification, each C-V2X station must feature a timing reference, a temperature compensated crystal osciallator, with an accuracy of 0.1 parts per million. Thus, at a center frequency of 5.9GHz, a maximum frequency error of $\pm 0.1e\text{-}6 \cdot 5.9GHz = \pm 590Hz$ can be observed in the local oscillator signals that up and down converts the IQ information. From section 6.3.2 in this text, we can also determine the Doppler frequency that would occur for a maximum reasonable driving speed of 250Kph = 70 m/sec.

$$F_{Doppler} = F_{Tx} \frac{Velocity}{SpeedOfLight}$$

$$= 5.9GHz \frac{70 \, m/s}{300e6 \, m/s} = 1307Hz$$

Therefore, in reasonable traffic conditions, we can expect to see a maximum frequency offset, the sum of the Doppler and LO frequency error, of no more than ±2KHz. Our goal is to determine an algorithm that can realiably detect this total frequency offset seen at the receiver.

Taking another look at the figure showing the DMRS, or demodulation reference signals, in OFDM symbols 2, 5, 8, and 11, we realize that the distance between them is equal to 3 symbols or 3 · 0.001 seconds / 14 = 214 microseconds. Take a look at the figure below, which represents a condensed version of the image we have seen a few pages ago. Note that the 24 resource elements placed in OFDM symbol 2 feature the same values as those placed in symbol 5, 8, and 11.

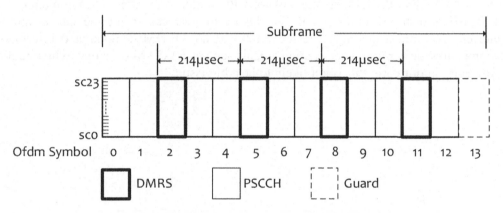

Figure 7-90: C-V2X Subframe Layout of a PSCCH Occurance

Therefore, we can compute the frequency offset by observing the change in the 24 DMRS resource elements values, *REs*, from OFDM symbol 2 to 5, 5 to 8 and 8 to 11. We compute the frequency offset by first computing the mean complex rotation as follows.

$$Rotation = \frac{1}{72} \left(\sum_{k=0}^{23} RE_5[k] \cdot conj(RE_2[k]) + RE_8[k] \cdot conj(RE_5[k]) + RE_{11}[k] \cdot conj(RE_8[k]) \right)$$

$$RotationInCycles = \frac{angle(Rotation)}{2\pi}$$

$$FreqOffsetEstimate = \frac{RotationInCycles}{214e-6 \, seconds}$$

Note that we can only find a reliable frequency offset estimate if the rotation in cycles is less that ±½ as rotations larger than that will look like rotations in the opposite direction. The maximum frequency offset that can reliable be detected is thus within the range of ±0.5 / 214e-6 seconds = ±2336Hz.

This method is actually quite good and provides good estimates even down at signal to noise ratios of -5dB. Once we have found the offset we have to decide how to remove it. There are two choices, one requires more effort and one less. The easier way is to derotate all REs belonging to the PSCCH resource grid from which we have used the DMRS. The more laborious path is to remove the frequency offset from the captured IQ sequence by multiplying it by a complex sinusoid rotating at the frequency offset but in the opposite direction and finally OFDM demodulation again to produce an updated resource grid. Note that the original resource grid that we used to compute the frequency offset contains inter-subcarrier interference due to loss of orthogonality between the IFFT in the transmitter and FFT algorithm in the receiver. Simple derotating the resource element values in the resource grid will not change that fact. Remember that the FFT reverses the IFFT operation only if the set of subcarriers at each end features the same frequencies. If the subcarriers of the received signal feature a frequency offset, then the subcarriers in the FFT block will no longer be orthogonal to those of the received signal and the information embedded within them can no longer be recovered perfectly. The set of output values produces by the FFT in the OFDM demodulator will contain some amount of content that resided on all subcarriers in the received signal. For this reason, we call it inter-subcarrier interference. This interference reduces the signal to noise ratio of the OFDM demodulated information and may become an issue for high constellation signals such as 64 QAM or above, which require high SNR. However, in our case, this is not likely to be a big issue.

7.8 Numerology and Frequency Ranges in 5G New Radio

5G New Radio is the 5th generation cellular communication standard and will over time replace the currently 4G LTE technology. The most important change introduced in the new standard is the much higher RF frequency range to which the 5G signal has access. Whereas 4G signals were primarily used at frequencies between 600MHz and 2.5GHz, the new standard allows communication in the following to frequency ranges.

7.8.1 Frequency Ranges FR1 and FR2

Table 7-6: Frequency Ranges in 5G NR

Name	Frequency Range	Allowed Bandwidths (MHz)	Frame Structure Type
FR1	410 to 7125 MHz	5, 10, 15, 20, 25, 30, 40, 50, 60, 70, 80, 90, 100	FDD or TDD
FR2	24250 to 52600 MHz	50, 100, 200, 400	TDD only

The maximum available bandwidth for an LTE cell was 20MHz (100MHz with carrier aggregation), which compares the maximum 5G bandwidth of 400MHz at frequency range 2. It is not the signal processing or bit encoding schemes that give 5G the large throughput that is being advertised, but to a large extend the bigger bandwidth. Had the FR2 frequency bands been made available for 4G networks, then LTE would have enjoyed higher throughput as well. However, the high throughputs achievable at FR2 are not free as signal propagation is significantly reduced at these higher frequencies, thus forcing network operators to install many base stations to get complete coverage. There is thus no reason to expect 5G signals to operate at FR2 along highways, but rather in high density urban environments.

7.8.2 Numerology

The 5G New Radio standard, encapsulated in the 3GPP TS38 series documents, uses the term numerology to refer to the five different subcarrier spacings and their associated parameters, which can be seen in the figure below.

Table 7-7: Subcarrier Spacings in 5G New Radio

Subcarrier Spacing (KHz)	15 (FR1)	30 (FR1)	60 (FR1/2)	120 (FR2)	240 (FR2)
Numerology μ	0	1	2	3	4
Symbol Length (no CP)	66.66μsec	33.33μsec	16.66μsec	8.33μsec	4.166μsec
Normal CP Period (μsec)	5.2/4.69	2.86/2.34	1.69/1.17	1.10 / 0.586	0.814/2.93
Max Bandwidth (MHz)	49.5	99	198	369	379.44
Max/Min Number of Resource Blocks (RB)	24/275	24/275	24/275	24/275	24/138
Bandwidth Per RB (KHz)	180	360	720	1440	2880
Symbols per Slot	14	28	48/56	112	224
Slots Per Subframe	1	2	4	8	16

The 4G LTE specification only implemented the 15KHz subcarrier spacing, which was chosen such that for typical urban and suburban multipath conditions, the frequency response of the channel would be relatively flat over the narrow 15KHz range. This, of course, is one of the fundamental requirements that enables proper operation and equalization in OFDM. So why did the 5G New Radio specification allow this additional flexibility?

FFT Size

4G LTE called out for a maximum FFT size of 2048 to cover the maximum bandwidth of 20MHz. Now that we can cover up to 400MHz, the reader can imagine the unreasonably large FFT sizes that would be required if the subcarrier spacing were to remain at 15KHz.

Multipath Channel

The larger subcarrier spacings are primarily used at higher RF frequencies where RF propagation is much more limited in range as compared to the lower frequencies. This limited range also implies that different RF paths arriving at the receiver feature smaller delays as the paths themselves are shorter. Smaller time delays in between the paths imply a better behaved frequency response, which changes more slowly and remains flat over a larger subcarrier spacing.

Frequency Offset and Phase Noise

The most important reason for using the larger subcarrier spacing is the increased frequency offset (see section 6.3.5), Doppler shift (see section 6.3.2) and phase noise (see section 6.3.4) at higher local oscillator frequencies. Increased phase noise causes a deterioration of the demodulation quality (EVM – error vector magnitude), whereas the unresolved frequency offset, whether due to Doppler or inaccuracies in the local oscillator signal, which up and down converts the IQ waveform, causes loss of orthogonality during OFDM demodulation (see section 7.1.3.3). Whereas frequency offset can be acquired and largely removed, phase noise must be detected using a new type of reference signal. These new reference signals must be spaced tightly in time, a feature which the larger subcarrier spacing makes possible via reduces OFDM symbol lengths.

7.8.3   Signal Processing in 5G

With the exception of some minor details, the 5G standard retains the same modulation techniques that were introduced to the cellular network ecosystem via the 4G standard. The downlink continues the use of OFDM, or orthogonal frequency multiplexing, whereas the uplink uses SC-FDMA (see section 5.5), or single carrier frequency division multiple access. OFDM is used in the downlink due its high versatility, and SC-FDMA once again finds employment in the uplink as it eases the linearity requirements of the RF power amplifiers in the mobile phone. 5G thus shares the big signal processing features of LTE such a modulation, demodulation, and channel estimation, which we have spent significant amounts of time covering in this text. Furthermore, control channel and shared channel processing (see section 7.5.7) has changed little except for the use of algebraic rather than convolutional forward error correction techniques. Thus, the control channels use **polar** coding rather than BCC, whereas the shared channel moves from Turbo to low density parity check, or **LDPC**, coding. Note that the use of algebraic codes does not provide 5G with any appreciable gain in performance, and they were likely chosen for very subtle reasons such as slight advantages in some specific use cases. More than anything, the beam forming aspects of 5G that have been embedded in the physical layer of the standard are the most obvious changes that are interesting for signal processing engineers. Unfortunately, we will not cover this topic in this edition.

7.8.4   Why is 5G so Much Faster than LTE?

In the countless commercials on television, network providers like to point out the tremendous transmission speeds that 5G can provide the customer. Is 5G NR (new radio) really that much faster than 4G LTE? The speed advantage is not due to the fact that the 5G signal processing is better, but the fact that various federal telecommunication agencies in the world are providing much higher bandwidths (primarily at FR2) than those allotted to LTE. In simple terms, 5G takes advantage of the much larger highways that it provided to achieve very large throughput at the expense of smaller cell sizes as the higher frequencies don't travel as far as the lower frequencies.

7.8.5   5G New Radio Use Cases

The LTE 4G specification was originally focused on delivering higher throughput than the 3G technology before it. Additional use cases, such as machine-to-machine communication, were later added in the form of NB-IOT, or narrow band internet of things, and LTE-M, which provide dedicated low data rate information exchange between the base station and machines such as utility meters and a large host of other industrial applications. Unlike this eveolution in 4G, the 5G specification strived to address three different use cases from the get go.

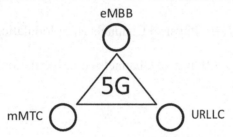

Figure 7-95: Use Cases in 5G

→ *Enhanced Mobile Broadband (eMBB)* - Enhanced mobility broadband addresses the desire for ever higher throughput in cellular systems given the ever-changing data intensive entertainment applications that are continually introduced into the market place. Clearly, the larger bandwidth and usage of dedicated beams at FR2 have the ability to provide unheard of advances in data throughput.

→ *Massive Machine Type Communication (mMTC)* - The 4G NB-IOT and LTE-M technologies will continue to be employed in 5G as well. Using FR2 and beam forming will enable to servicing of yet larger numbers of industrial devices.

→ *Ultra Reliable Low Latency Communication (URLLC)* - URLLC is a new use case that attempts to apply 5G technologies to industrial applications such as communication in automated production settings. Lower latency data exchange is primarily provided by the higher bandwidth at FR2 and the shorter OFDM symbols therefore allowing information to be conveyed within smaller time periods.

References

[1] IEEE Std. 802.11a-1999 (1999), *Wireless LAN Medium Access Control (MAC) and Physical Layer (PHY) Specifications: High-speed Physical Layer in 5 GHz Band*, New York, September

[2] Heiskala, J. and Terry, J. (2001) *OFDM Wireless LANs: A Theoretical and Practical Guide*, Sams Publishing, Indianapolis, IN

[3] IEEE 802.11n-2009—Amendment 5: Enhancements for Higher Throughput. IEEE-SA. 29 October

[4] Fazel, K. and Kaiser, S. (2008) *Multi-Carrier and Spread Spectrum Systems: From OFDM and MC-CDMA to LTE and WiMAX,* Third Edition, John Wiley & Sons, New York

[5] Ergen, M. (2009) *Mobile Broadband – Including WiMAX and LTE*, Springer, NY

[6] Sesia, S., Toufik, I. and Baker, M. (2009) *LTE – The UMTS Long Term Evolution*, John Wiley & Sons, New York

[7] 3GPP TS36.211 Section 6.10 'Physical Channels and Modulation' Version 14.2.0 Release 14

[8] 3GPP TS36.211 Section 9.8 'Physical Channels and Modulation' Version 14.2.0 Release 14

8   An Introduction to Multiple Antenna Systems

Multiple antenna techniques have been developed to mitigate the deleterious effects of multipath distortion and to increase data throughput over wireless communication links. Some of these techniques have been incorporated into recent OFDM based communication systems such as WiMAX [1], WLAN 802.11n/ac [2], and the 3GPP 4G long term evolution [3], and 5G New Radio cellular standard. Modern MIMO (multiple input/multiple output) implementations have favored OFDM systems thanks to the per-subcarrier flat fading model, which may be utilized when equalizing and combining the signal streams arriving at multiple antennas. In this chapter, we will limit our discussion to three multiple antenna techniques that find utility in modern OFDM based communication systems. The techniques in question are receive antenna diversity, transmit antenna diversity, and spatial multiplexing. In the strictest sense, MIMO refers to the concept of spatial multiplexing or a combination of both transmit and receive diversity.

Receive Antenna Diversity (SIMO – single input/multiple output)

Receive antenna diversity utilizes a single transmit antenna and multiple receive antennas to mitigate the selective and flat fading effects of the multipath channel. Channel conditions at one receive antenna will be different from those at the other, and methods of optimally combining the received information will be presented in this chapter. The diversity combining effect produces an overall superior channel condition, which allows the link to use less forward error correction and/or higher symbol constellations, resulting in increased data throughput.

Transmit Diversity (MISO – multiple input/single output)

Transmit diversity utilizes multiple transmit antennas and a single receive antenna to achieve the same multipath benefit as is provided by receive antenna diversity. The original technique was pioneered by Siavash Alamouti [1] and focused on two transmit and one receive antennas. The goal, once again, is to establish a better-behaved multipath channel that allows the use of higher transmit constellations and/or less redundancy in the forward error correction algorithms. The two transmit diversity techniques we will get know are Alamouti's original space time block coding and space frequency block coding, which is used in LTE.

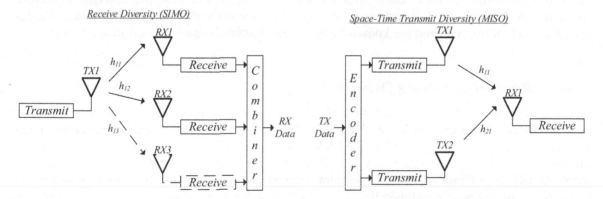

Figure 8-1: Basic Receive and Space-Time Transmit Diversity Scenarios

Spatial Multiplexing (MIMO – multiple input/multiple output)

This technique uses both multiple transmit and multiple receive antennas to send several data streams over the same RF channel. The resulting interference between these streams is removed via digital signal processing techniques at the receiver to yield the desired increase in data throughput. Strictly speaking, spetial multiplexing requires the same number of antennas at the transmitter and at the receiver. However, receive diversity may be easily added to the system to provide a better-behaved multipath channel that results in different numbers of transmit and receive antennas.

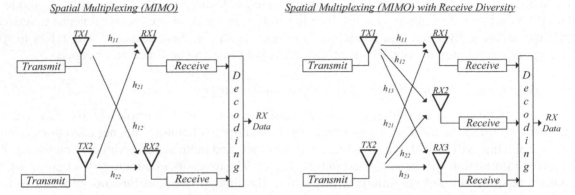

Figure 8-2: Spatial Multiplexing with and without Added Receive Diversity

The Selective Versus Flat Fading Model

The various antenna diversity techniques introduced above feature channel coefficients quantified as simple complex scalars, h_{ab}, where the variables a and b identify the transmit and receive antenna respectively. Channels that may be characterized by simple scalars are represented by the flat fading model, which assumes that the effect of the channel is uniform across the entire signal bandwidth. In chapter 6, we witnessed the harsh impact that selective fading conditions have on wide band signals. Luckily, although OFDM does feature a large bandwidth, it is subdivided into narrowband subcarriers, which may be individually modeled using the flat fading approach. It is for this reason that OFDM lends itself particularly well to the multiple antenna techniques discussed in this chapter. The techniques discussed here assume that all multipath channels, h_{ab}, are of the flat fading type and are known by the receiver after channel estimation is completed.

8.1   Receive Antenna Diversity

Receive antenna diversity has been around much longer than space-time transmit diversity or spatial multiplexing. For decades, we have seen the multiple receive antenna arrangements on cell towers, the roofs of buildings, and on the back of police vehicles. To introduce the idea of receive antenna diversity, let's assume that we have two observations, $y_1[n]$ and $y_2[n]$, of a transmitted symbol, $x[n]$, each featuring different amounts of noise, $v_1[n]$ and $v_2[n]$. It is our goal to find the coefficients of a linear combiner that will result in a final estimate of $x[n]$ that features the maximum possible signal to noise ratio.

$$y_1[n] = x[n] + v_1[n]$$

$$y_2[n] = x[n] + v_2[n]$$

The linear combiner, our estimator, is shown in the figure below and reduces to the following simple expression.

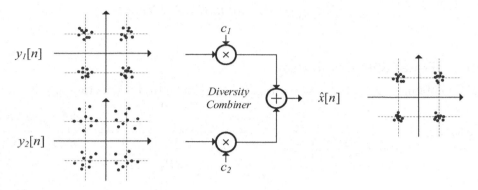

$$\hat{x}[n] = C_1 \cdot y_1[n] + C_2 \cdot y_2[n]$$

$$= C_1 \cdot (x[n] + v_1[n]) + C_2 \cdot (x[n] + v_2[n])$$

$$= (C_1 + C_2)x[n] + C_1 \cdot v_1[n] + C_2 \cdot v_2[n]$$

Assuming that our coefficients are real-valued, the signal to noise ratio of the estimate of x is as follows.

$$Signal\ Power = E\big(\ (c_1 + c_2)x[n] \cdot ((c_1 + c_2)x[n])^*\big) = |(c_1 + c_2)^2|\,\sigma_x^2$$

$$Noise\ Power = E\big(\ (c_1 v_1[n] + c_2 v_2[n]) \cdot (c_1 v_1[n] + c_2 v_2[n])^*\ \big)$$

$$= E\big(\ c_1^2\, v_1[n] \cdot v_1[n]^*\big) + E(c_1 c_2\, v_1[n] \cdot v_2[n]^*)$$

$$+\ E\big(c_2 c_1\, v_2[n] \cdot v_1[n]^*\big) + E(c_2^2\, v_2[n] \cdot v_2[n]^*)$$

$$= c_1^2 \sigma_{v1}^2 + c_2^2 \sigma_{v2}^2$$

$$SNR_{final} = \frac{(c_1 + c_2)^2 \sigma_x^2}{c_1^2 \sigma_{v1}^2 + c_2^2 \sigma_{v2}^2}$$

Because $v_1[n]$ and $v_2[n]$ are uncorrelated, the terms $E(c_1 c_2 \cdot v_1[n] \cdot v_2[n]^*)$ and $E(c_2 c_1 \cdot v_2[n] \cdot v_1[n]^*)$ average toward zero, thus limiting the increase in noise power resulting from the diversity combination. This stands in stark contrast to the signal power calculation, in which all terms will add constructively to yield to positive values. Let's take a look at three combining schemes that have seen use in receive diversity combining applications.

8.1.1   Maximum Output Combining

The simplest of the three schemes, maximum output combining, picks the information at the antenna with the largest signal to noise ratio and ignores the other. If the information on antenna one features a larger signal to noise ratio, then $c_1 = 1$ and $c_2 = 0$. The technique is easy to implement and has been heavily used in older communication systems and situations where sophisticated DSP hardware isn't available.

$$SNR_{final} = maximum(SNR_1, SNR_2)$$

8.1.2 Equal Gain Combining

Equal gain combining averages the two estimates $y_1[n]$ and $y_2[n]$ by selecting $c_1 = c_2 = 0.5$.

$$SNR_{final} = \frac{\sigma_x^2}{0.25\sigma_{v1}^2 + 0.25\sigma_{v2}^2} = \frac{4}{\frac{\sigma_{v1}^2}{\sigma_x^2} + \frac{\sigma_{v2}^2}{\sigma_x^2}} = \frac{4}{\frac{1}{SNR_1} + \frac{1}{SNR_2}} = 4\frac{SNR_1 \cdot SNR_2}{SNR_1 + SNR_2}$$

Equal gain combining shines in cases where the two signal to noise ratios are similar to each other. When they are the same, the signal to noise ratio is twice that of either one of the constituent estimates, thus easily surpassing the performance of maximum output combining.

$$SNR_{final} = 4\frac{SNR_1^2}{2SNR_1} = 2SNR_1$$

Unfortunately, in the case where one of the estimates features a clearly inferior signal to noise ratio, the equal gain technique falls far short of maximum output combining.

8.1.3 Maximum Ratio Combining

Maximum ratio combining merges the best features of the two previous techniques by scaling each estimate according to its signal to noise ratio. The factor $a = 1/(SNR_1 + SNR_2)$ is selected such that $c_1 + c_2 = 1.0$.

$$c_1 = a \cdot SNR_1 = a\frac{\sigma_x^2}{\sigma_{v1}^2} \quad and \quad c_2 = a \cdot SNR_2 = a\frac{\sigma_x^2}{\sigma_{v2}^2}$$

$$SNR_{final} = \frac{(c_1 + c_2)^2\sigma_x^2}{c_1^2\sigma_{v1}^2 + c_2^2\sigma_{v2}^2} = \frac{\left(a\frac{\sigma_x^2}{\sigma_{v1}^2} + a\frac{\sigma_x^2}{\sigma_{v2}^2}\right)^2\sigma_x^2}{a^2\frac{\sigma_x^4}{\sigma_{v1}^4}\sigma_{v1}^2 + a^2\frac{\sigma_x^4}{\sigma_{v2}^4}\sigma_{v2}^2}$$

$$SNR_{final} = \frac{a^2(\frac{\sigma_x^2}{\sigma_{v1}^2} + \frac{\sigma_x^2}{\sigma_{v2}^2})^2\sigma_x^2}{a^2\sigma_x^2(\frac{\sigma_x^2}{\sigma_{v1}^2} + \frac{\sigma_x^2}{\sigma_{v2}^2})} = \frac{(\frac{\sigma_x^2}{\sigma_{v1}^2} + \frac{\sigma_x^2}{\sigma_{v2}^2})^2}{(\frac{\sigma_x^2}{\sigma_{v1}^2} + \frac{\sigma_x^2}{\sigma_{v2}^2})} = \frac{\sigma_x^2}{\sigma_{v1}^2} + \frac{\sigma_x^2}{\sigma_{v2}^2}$$

$$SNR_{final} = SNR_1 + SNR_2$$

Maximum Ratio Combining for Many Inputs

For N observations $y_i[n] = x[n] + v_i[n]$, where $i = 1$ to N, the final signal to noise ratio of the combined estimate of $x_i[n]$ will be equal to the sum of the signal to noise ratios of all individual observations.

$$SNR_{final} = \sum_{i=1}^{N} SNR_i$$

The idea that the final *SNR* is the sum of the individual *SNR* values is true for any number of estimates, not just two. Take care not to add signal to noise ratios expressed in *dB*. Two signal to noise ratios of value $2 = 3dB$ and $4 = 6dB$ add to $6 = 7.8dB$, not $9dB$.

The Real World

In the real world, observations $y[n]$ that feature an undistorted component of $x[n]$, are not directly available. The next figure illustrates a more realistic scenario where the transmit symbol, $x[n]$, is first scaled by the complex channel coefficient, h_b, before being corrupted by Gaussian noise at the receiver to form the observation, $y_i[n]$. The receiver then produces the estimate of the original symbol, $x[n]$, by undoing or equalizing the channel effect. Thus, before applying the maximum ratio equation, we will first equalize each observation to arrive at the individual estimates of $x_i[n]$, such that each estimate features an undistorted component of $x[n]$.

$$the\ observation \rightarrow\ \ y_i[n] = h_i \cdot x[n] + v_i[n]$$

$$the\ estimate \rightarrow \hat{x}_i[n] = \frac{1}{h_i} \cdot y_i[n] = x[n] + \frac{v_i[n]}{h_i}$$

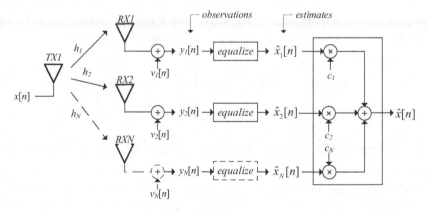

Figure 8-3: Receive Diversity for *N* Observations of Transmitted Signal

8.1.4 The MMSE or Minimum Mean Square Estimator

Whereas the last section and the implementation of the maximum ratio combiner may have been insightful in its own right, what we really need is a single unified approach that can optimally estimate the original transmit symbol, $x[n]$, directly from the observations $y_i[n]$, under variety of different antenna configuration, not just the simple SIMO case. This estimator, called the minimum mean square error estimator, was developed in chapter four and is based on the Wiener-Hopf equation. Before we state the equation for the estimator, let's recast a receive diversity scenario, akin to what we have seen in the last figure, with three receive antennas into matrix form.

$$y_1[n] = h_1 \cdot x[n] + v_1[n]$$

$$y_2[n] = h_2 \cdot x[n] + v_2[n]$$

$$y_3[n] = h_3 \cdot x[n] + v_3[n]$$

$$Y[n] \;=\; H \cdot x[n] \;+\; V[n]$$

$$\begin{bmatrix} y_1[n] \\ y_2[n] \\ y_3[n] \end{bmatrix} = \begin{bmatrix} h_1 \\ h_2 \\ h_3 \end{bmatrix} x[n] + \begin{bmatrix} v_1[n] \\ v_2[n] \\ v_3[n] \end{bmatrix}$$

The estimate of $x[n]$ is formed by the linear combination of the estimator coefficients $C^T = [c_1 \; c_2 \; c_3]$ and the individual observations as follows.

$$\hat{x}[n] = C^T \cdot Y = \begin{bmatrix} c_1 & c_2 & c_3 \end{bmatrix} \cdot \begin{bmatrix} y_1[n] \\ y_2[n] \\ y_3[n] \end{bmatrix}$$

The minimum mean squared error estimator computes the coefficient vector, C^T, using the expression given below.

$$C^T = R_x H^T \cdot inv(H \cdot R_x H^T + R_v)$$

The quantities $R_x = E(X[n] \cdot X[n]^T)$ and $R_v = E(V[n] \cdot V[n]^T)$ refer to the input and noise autocorrelation matrices. Given that noise quantities $v_1[n]$, $v_2[n]$, and $v_3[n]$ are generated at different antennas, they are guaranteed to be uncorrelated, yielding the following expressions.

$$R_x = \sigma_x^2 \quad and \quad R_v = \begin{bmatrix} \sigma_{v1}^2 & 0 & 0 \\ 0 & \sigma_{v1}^2 & 0 \\ 0 & 0 & \sigma_{v3}^2 \end{bmatrix}$$

Example 8.1*: Minimum Mean Square Error Estimator for 1TX by 3RX Antenna Configuration*

The following example computes the optimal estimator coefficients c_1, c_2, and c_3 in the minimum mean square error sense for a receive diversity scenario with three RX antennas. In input symbol stream $x[n]$ is QPSK modulated with unity variance, and the noise variance is equal to 0.01 for all receive antennas. Furthermore, the channel coefficients h_1, h_2, and h_3 are complex with the following values.

$$H = \begin{bmatrix} h_1 \\ h_2 \\ h_3 \end{bmatrix} = \begin{bmatrix} 1+j \\ -0.2 - 1.3j \\ -0.5j \end{bmatrix} \qquad R_x = 1 \qquad R_v = \begin{bmatrix} 0.01 & 0 & 0 \\ 0 & 0.01 & 0 \\ 0 & 0 & 0.01 \end{bmatrix}$$

The MatLab code below doesn't just compute the optimal coefficients in the MMSE sense but proves that the signal to noise ratio at the final estimate is the sum of the individual signal to noise ratios at the three antenna ports. It does this by generating a QPSK input sequence, computing the three observation vectors using simulated Gaussian noise and then producing the estimates using the MMSE coefficients shown next.

$$C^T = [0.25 - j0.25 \quad -005 + j0.326 \quad j0.125]$$

The linear signal to noise ratio of the final estimate $\hat{x}[n]$ is equal to the sum of the signal to noise ratios at each observation. Run the MatLab scripts and it will produce these exact number.

$$SNR(\hat{x}[n]) = \sum_{i=0}^{2} SNR(y_i[n]) = 200 + 173 + 25 = 398$$

```
% Start out by defining the transmit symbol power and noise powers at each antenna.
SigmaX  =    1;       % The signal power (we will use a simple QPSK symbol sequence)
SigmaV1 = 0.01;       % The noise power at antenna1
SigmaV2 = 0.01;       % The noise power at antenna2
SigmaV3 = 0.01;       % The noise power at antenna3

% The channel coefficients describing the frequency response for each antenna path.
h1      = 1 + 1j;     % Channel coefficients for path 1
h2      = -0.2 - 1.3j; % Channel coefficients for path 2
h3      = -1j*0.5;    % Channel coefficients for path 3

% The linear signal to noise ratios at each observation are as follows.
SNR1    = SigmaX*h1*conj(h1)/SigmaV1;  disp(['Linear SNR at y1 = ' num2str(SNR1)]);
SNR2    = SigmaX*h2*conj(h2)/SigmaV2;  disp(['Linear SNR at y2 = ' num2str(SNR2)]);
SNR3    = SigmaX*h3*conj(h3)/SigmaV3;  disp(['Linear SNR at y3 = ' num2str(SNR3)]);

% Autocorrelation matrices needed to determine coefficients of our MMSE estimator.
Rx = SigmaX;                          % This is the signal power
Rv = [SigmaV1,      0,        0; ...   % Autocorrelation matrix for the noise sources
      0,        SigmaV2,      0; ...
      0,            0,  SigmaV3];

% Our channel matrix is nothing but a simple column vector.
H   = [h1; h2; h3];

% Solve for the MMSE Coefficients (The Wiener Hopf equation)
CT = Rx*H' * inv(H*Rx*H' + Rv);

% Transmitting a QPSK symbol stream
N   = 200000;        % Number of TX Symbols
X   = sqrt(SigmaX)*(0.7071*(2*randi([0,1],1,N)-1) + 1j*0.7071*(2*randi([0,1],1,N)-1));

% Generate complex Gaussian noise with noise variance SigmaV1, SigmaV2, and SigmaV3
V1  = sqrt(SigmaV1) * sqrt(1/2) * (randn(1, N) + 1j*randn(1, N));
V2  = sqrt(SigmaV2) * sqrt(1/2) * (randn(1, N) + 1j*randn(1, N));
V3  = sqrt(SigmaV3) * sqrt(1/2) * (randn(1, N) + 1j*randn(1, N));

% The observation vectors
Y   = [X * h1 + V1; ...  % At receive antenna 1
       X * h2 + V2; ...  % At receive antenna 2
       X * h3 + V3];     % At receive antenna 3

% Estimate of the TX symbol sequence x[n], as well as the associated error.
X_Estimate      = CT * Y;
```

```
Error          = X_Estimate - X;                    % The estimate of x[n] minus x[n]
SNR_X_Estimate = var(X_Estimate)/var(X_Estimate - X);
disp(['The linear SNR of the estimate of X is equal to ' num2str(SNR_X_Estimate) ]);
```

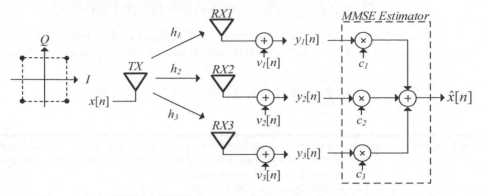

Figure 8-4: 1TX by 3RX Antenna Diversity Combining Scenario

The figure below illustrates the observation vectors at the three receive antenna ports. The SNR of the final estimate is entirely unaffected by the miserable SNR at observation $y_3[n]$. Further, even if it doesn't appear obvious, the SNR of the final estimate is twice that of the SNR at observation $y_1[n]$. Imagine if the noise, or deviation from the mean position, of the constellation of observation $y_1[n]$ where scaled by a factor of $0.7071 = 1/sqrt(2)$. Visually, the deviation away from the mean positions wouldn't change dramatically, but the SNR would in fact double as a result.

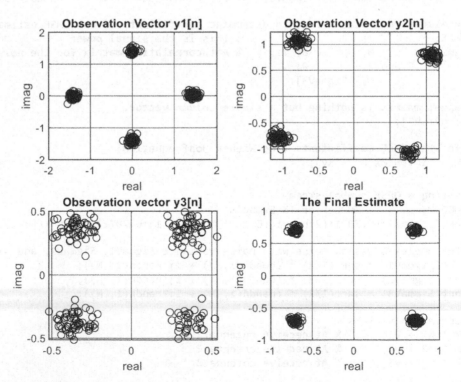

Figure 8-5: Constellation Diagram of the Observation Vectors and the Final Estimate of x[n]

8.2   Transmit Antenna Diversity

Multiple receive antennas can conveniently be fielded at the base station side of a communication link, since cost and hardware resources are not limited there. Base station reception has benefited significantly from the implementation of receive antenna diversity systems. However, mobile stations, like cell phones or mobile wireless LAN devices, are constrained in terms of cost and physical dimensioning, which can make multiple receive antennas impractical. A way had to be found to reap the benefits of diversity gain on a down link (base to mobile station) that is limited to just one RX antenna.

8.2.1   Space-Time Block Coding

In 1998, Siavash Alamouti[4] proposed a transmit antenna diversity scheme that uses two transmit antennas but only a single receive antenna to produce benefits similar to those provided by the maximum ratio combining version of receive antenna diversity. His approach uses a technique called space-time block coding, or STBC, which takes a pair of input symbols, $x[n]$ and $x[n+1]$, and encodes them into two separate data pairs to be transmitted over *TX* antennas one and two. Other efforts have expanded space-time block codes for use with larger numbers of transmit antennas [5]. However, the simple *2TX* by *1RX* antenna configuration, see figure below, turns out to be very advantages to the point where LTE cell towers with 4 transmit antennas will in fact use a staggered *2TX* by *1RX* configuration rather than a straight *4TX* by *1RX* configuration. You can see the exact definitions for the LTE transmit diversity configuration in 3GPP TS36.211 Section 6.3.4.3 [9].

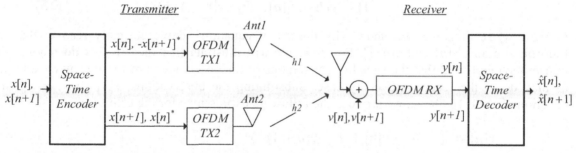

Figure 8-6: Transmit Diversity Communication Link

One of the reasons why receive diversity signals can be combined to achieve SNR values that are superior to those found at the individual receive antennas is that the useful signal components of the two observations add constructively, whereas the uncorrelated noise terms do not (see section 8.1). Given that this time we have only one receive antenna and therefore one noise signal, it appears that uncorrelated noise from multiple sources is no longer available. For this reason, space-time block coding inserts the symbols $x[n]$ and $x[n+1]$ at different times. Assuming that the receiver noise source is white and Gaussian, we know that the contribution, $v[n]$, to the current symbol is uncorrelated to the contribution, $v[n+1]$, of the next symbol. The symbol $x[n]$, for example, is transmitted via antenna one and superimposed with $v[n]$ at the receiver. $x[n]^*$ is transmitted one symbol time later on antenna two and sums with $v[n+1]$. Note that in the figure, we assume that the channel responses h_1 and h_2 don't change significantly over the period of two successive transmit symbols.

The Space-Time Encoder

The space-time encoder takes blocks of two consecutive symbols, $x[n]$ and $x[n+1]$, and reassembles them according to the following simple rules.

Source Block: $[x[n], x[n+1]]$ → Transmit Path 1: $[x[n], -x[n+1]^*]$
→ Transmit Path 2: $[x[n+1], x[n]^*]$

The observations, $y[n]$ and $y[n+1]$, at the receiver are straight-forward formulations of the symbols sent by the two transmit paths.

$$y[n] \quad = \quad h_1 \cdot x[n] \quad\quad + h_2 \cdot x[n+1] + v[n]$$
$$y[n+1] = -h_1 \cdot x[n+1]^* \quad + h_2 \cdot x[n]^* \quad\quad + v[n+1]$$

Space-Time Decoder

The function of the space-time decoder is to sum up the observations into two estimates. The summation process that forms the first estimate must be such that components containing $x[n+1]$ cancel to zero, while those containing $x[n]$ add constructively. Similarly, the summation process yielding the second estimate must retrieve information related to $x[n+1]$ and ignore content related to $x[n]$. The summation method used by the space-time decoder to reassemble the observations, $y[n]$ and $y[n+1]$, into two estimates is shown below.

$$\bar{x}[n] \quad = \quad c(h_1^* \cdot y[n] + h_2 \cdot y[n+1]^*) \quad\quad (1a)$$
$$\bar{x}[n+1] = c(h_2^* \cdot y[n] - h_1 \cdot y[n+1]^*) \quad\quad (1b)$$

In the next section, we will show why the set of equations above correctly recover the two transmit symbols $x[n]$ and $x[n+1]$. For now, we will illustrate that they in fact do recover the transmit symbols and that the signal to noise ratios of the observations of these symbols add just as they do in the receiver diversity case. Substituting the observations, $y[n]$ and $y[n+1]$, into equations 1a and 1b above, we arrive at the following expressions for the estimates.

$$\bar{x}[n] = c(\quad h_1^* \cdot y[n] + h_2 \cdot y[n+1]^*)$$

$$= c(\quad h_1^*(\quad h_1 \cdot x[n] \quad\quad + h_2 \cdot x[n+1] + v[n])$$

$$+ \quad h_2\big(-h_1 \cdot x[n+1]^* + h_2 \cdot x[n]^* \quad\quad + v[n+1]\big)^*)$$

$$= c(\quad h_1^* \cdot h_1 \cdot x[n] \quad\quad + h_1^* \cdot h_2 \cdot x[n+1] + h_1^* \cdot v[n]$$

$$- \quad h_2 \cdot h_1^* \cdot x[n+1] + h_2 \cdot h_2^* \cdot x[n] \quad\quad + h_2 \cdot v[n+1]^*)$$

$$= c(\quad (h_1^* \cdot h_1 + h_2 \cdot h_2^*)x[n] + h_1^* \cdot v[n] + h_2 \cdot v[n+1]^*)$$

Setting $c = 1/(h_1^* \cdot h_1 + h_2 \cdot h_2^*)$ isolates $x[n]$ nicely and the estimate reduces as follows.

$$\bar{x}[n] = x[n] + \frac{h_1^*}{h_1^* \cdot h_1 + h_2 \cdot h_2^*} \cdot v[n] + \frac{h_2}{h_1^* \cdot h_1 + h_2 \cdot h_2^*} \cdot v[n+1]^* \quad\quad (2a)$$

$$\hat{x}[n+1] = c_1(\quad h_2^* \cdot y[n] - h_1 \cdot y[n+1]^*)$$

$$= c_1(\quad h_2^*(\quad h_1 \cdot x[n] \quad + h_2 \cdot x[n+1] + v[n])$$

$$- \quad h_1(-h_1 \cdot x[n+1]^* + h_2 \cdot x[n]^* \quad + v[n+1])^*)$$

$$= c_1(\quad h_2^* \cdot h_1 \cdot x[n] \quad + h_2^* \cdot h_2 \cdot x[n+1] + h_2^* \cdot v[n]$$

$$+ \quad h_1 \cdot h_1^* \cdot x[n+1] - h_1 \cdot h_2^* \cdot x[n] \quad - h_1 \cdot v[n+1]^*)$$

$$= c_1(\quad (h_1^* \cdot h_1 + h_2 \cdot h_2^*)x[n] + h_2^* \cdot v[n] - h_1 \cdot v[n+1]^*)$$

Again, $c = 1/(h_1^* \cdot h_1 + h_2 \cdot h_2^*)$ isolates $x[n+1]$ nicely and the estimate reduces as follows.

$$\hat{x}[n+1] = x[n+1] + \frac{h_2^*}{h_1^* \cdot h_1 + h_2 \cdot h_2^*} \cdot v[n] - \frac{h_1}{h_1^* \cdot h_1 + h_2 \cdot h_2^*} \cdot v[n+1]^* \qquad (2b)$$

At this point, we have proven that equations 1a and 1b will recover the input symbols $x[n]$ and $x[n+1]$. Our big question now is whether the signal to noise ratio benefit of this scheme is similar to that of maximum ratio combining used in receiver diversity, where the signal to noise ratios of the observations at the two antennas added linearly after the estimation process. In that scenario, the observation $y_1[n]$ at antenna 1 and $y_2[n]$ at antenna 2 of $x[n]$ would have been as follows.

$$y_1[n] = h_1 \cdot x[n] + v_1[n]$$

$$y_2[n] = h_2 \cdot x[n] + v_2[n]$$

As mentioned above, using maximum ratio combining, the signal to noise ratio of the final estimate is the sum of the signal two noise ratios of the individual observations $y_1[n]$ and $y_2[n]$.

$$SNR(\hat{x}[n]) = SNR(y_1[n]) + SNR(y_2[n]) = \frac{|h_1|^2 \sigma_x^2}{\sigma_{v1}^2} + \frac{|h_2|^2 \sigma_x^2}{\sigma_{v2}^2}$$

Will the transmit diversity case yield a similar result? Let's take a look at the signal to noise ratio of the estimation $\hat{x}[n]$ of equation 2a.

$$SignalPower(\hat{x}[n]) = \sigma_x^2$$

$$NoisePower(\hat{x}[n]) = \frac{|h_1|^2 \sigma_v^2 + |h_2|^2 \sigma_v^2}{(|h_1|^2 + |h_2|^2)^2} = \frac{(|h_1|^2 + |h_2|^2)\sigma_v^2}{(|h_1|^2 + |h_2|^2)^2} = \frac{\sigma_v^2}{|h_1|^2 + |h_2|^2}$$

The signal to noise ratio of the estimate $\hat{x}[n]$ now reduces to the following expression proving that it takes the same form as in the maximum ratio combining case. Note that the signal to noise ratio for $\hat{x}[n+1]$ of equation 2b is identical.

$$SNR(\hat{x}[n]) = \frac{SignalPower(\hat{x}[n])}{NoisePower(\hat{x}[n])} = \frac{\sigma_x^2(|h_1|^2 + |h_2|^2)}{\sigma_v^2} = \frac{|h_1|^2 \sigma_x^2}{\sigma_v^2} + \frac{|h_2|^2 \sigma_x^2}{\sigma_v^2}$$

Although the theoretical SNR advantage of space-time transmit diversity (two antennas) is the same as that of maximum ratio combining, there are some practical issues that must be considered with the former technique. In the receive diversity case, each receiver will examine the transmitted training sequence to estimate the channel characteristics. This time, however, there is only one receiver, which must estimate the channel characteristics of both transmit paths. Each transmitter must therefore send training sequences that are separated in time, such that the receiver can properly compute both channel estimates. Therefore, the hardware savings at the mobile radio station and improved link robustness come at the cost of slightly more overhead in the overall communication protocol. Given the huge popularity of this scheme in practice, we can safely conclude that it is well worth it. The code below shows the encoding / decoding process.

```
%% The Space time block code example
x0 =  1.0 + 1.0j;          % Used as x[n]
x1 = -0.5 - 2.0j;          % Used as x[n+1]

% Defining the Channel
h1 =  1 + 0.2j;    % Channel coefficient (from TX Antenna1)
h2 =  1 - 1j;      % Channel coefficient (from TX Antenna2)

% The observations are defined via the following encoding scheme
y0 =  h1*x0       + h2*x1;          % Represents the observation y[n]
y1 = -h1*conj(x1) + h2*conj(x0);    % Represents the observation y[n+1]

% The estimates x_0 and x_1 nicely reveal the original transmit symbols
c   = 1/(conj(h1)*h1 + conj(h2)*h2);
x_0 = c*(conj(h1)*y0 + h2*conj(y1));    % Estimate of x[n]
x_1 = c*(conj(h2)*y0 - h1*conj(y1));    % Estimate of x[n+1]
```

Adding Receive Diversity

We can improve the performance of space-time transmit diversity by adding further receivers.

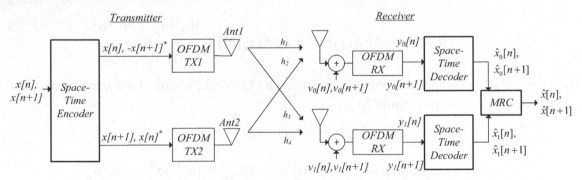

Figure 8-7: A 2x2 Space-Time Transmit Diversity Configuration

By adding a second receive antenna, we have created a MIMO configuration that further improves the channel robustness of the link via receive diversity. The maximum ratio combining algorithm will cause the signal to noise ratios of the individual estimates to add once again.

$$SNR(\hat{x}[n]) \quad = SNR(\hat{x}_0[n]) \quad + SNR(\hat{x}_1[n])$$

$$SNR(\hat{x}[n+1]) = SNR(\hat{x}_0[n+1]) + SNR(\hat{x}_1[n+1])$$

8.2.2   Space-Frequency Block Coding

In this section, we discuss a variation of Alamouti's space time encoder technique that is used in LTE (TS36.211 Section 6.3.4.3). This variation, called space frequency block coding, or SFBC, arranges successive symbols not along the time axis but along the frequency axis. The figure below illustrates our intent to transmit two symbols, $x[0]$ and $x[1]$, which we place in OFDM symbols 0 at successive subcarrier positions in the resource grid of transmit antenna 0. Notice the placement of the conjugated and negative conjugated symbols at the same positions in the resource grid of transmit antenna 1.

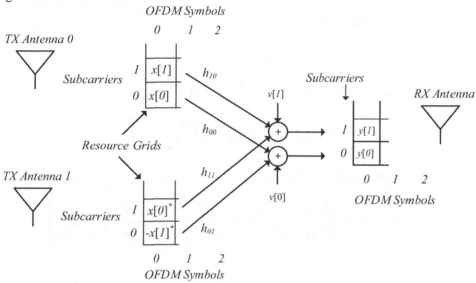

Figure 8-8: Space Frequency Block Coding as Implemented in LTE

For reasons that will become apparent shortly, we will not only show the expressions for the observations $y[0]$ and $y[1]$, but also their complex conjugates.

$$y[0] = h_{00}x[0] - h_{01}x[1]^* + v[0] \;\rightarrow\; y[0]^* = h_{00}{}^*x[0]^* - h_{01}{}^*x[1] + v[0]^*$$

$$y[1] = h_{11}x[0]^* + h_{10}x[1] + v[1] \;\rightarrow\; y[1]^* = h_{11}{}^*x[0] + h_{10}{}^*x[1]^* + v[1]^*$$

Finding the Decoder Equations

In the last section, we provided the space time block code decoder equations for the Alamouti technique without any explanation from where they came. This time around, we want to explain how to get them using the space frequency block code arrangement of the LTE specifications. Our first goal is to arrange the observations in matrix form as follows. Note that H is our channel matrix.

$$\begin{bmatrix} y[0] \\ y[1] \end{bmatrix} = H \cdot \begin{bmatrix} x[0] \\ x[1] \end{bmatrix} + \begin{bmatrix} v[0] \\ v[1] \end{bmatrix}$$

If you take a closer look at the equations of $y[0]$ and $y[1]$, we readily come to the conclusion that the above matrix expression cannot be constructed. However, we can generate a similar matrix expression if we use the equations for $y[0]$ and $y[1]^*$.

$$\begin{bmatrix} y[0] \\ y[1]^* \end{bmatrix} = \begin{bmatrix} h_{00} & -h_{01} \\ h_{11}^* & h_{10}^* \end{bmatrix} \cdot \begin{bmatrix} x[0] \\ x[1]^* \end{bmatrix} + \begin{bmatrix} v[0] \\ v[1]^* \end{bmatrix}$$

$$Y \quad = \quad H \quad \cdot \quad X \quad + \quad V$$

We now try to find the estimate of X.

$$\hat{X} = inv(H) \cdot Y$$

Remember the simple formula for the inverse of a 2x2 matrix.

$$inv\left(\begin{bmatrix} a & b \\ c & d \end{bmatrix}\right) = \frac{1}{ad - cb}\begin{bmatrix} d & -b \\ -c & a \end{bmatrix}$$

Therefore, the inverse of H of becomes

$$inv(H) = inv\left(\begin{bmatrix} h_{00} & -h_{01} \\ h_{11}^* & h_{10}^* \end{bmatrix}\right) = \frac{1}{h_{00}h_{10}^* + h_{11}^*h_{01}}\begin{bmatrix} h_{10}^* & h_{01} \\ -h_{11}^* & h_{00} \end{bmatrix}$$

and

$$\begin{bmatrix} \hat{x}[0] \\ \hat{x}[1]^* \end{bmatrix} = \frac{1}{h_{00}h_{10}^* + h_{11}^*h_{01}}\begin{bmatrix} h_{10}^* & h_{01} \\ -h_{11}^* & h_{00} \end{bmatrix} \cdot \begin{bmatrix} y[0] \\ y[1]^* \end{bmatrix} = c\begin{bmatrix} h_{10}^* & h_{01} \\ -h_{11}^* & h_{00} \end{bmatrix} \cdot \begin{bmatrix} y[0] \\ y[1]^* \end{bmatrix}$$

The estimates of $x[0]$ and $x[1]$ can now be found as follows.

$$\hat{x}[0] = \qquad\qquad\qquad c\left(h_{10}^*y[0] + h_{01}y[1]^*\right)$$

$$\hat{x}[1] = conj\left(c\left(-h_{11}^*y[0] + h_{00}y[1]^*\right)\right) = c^*\left(-h_{11}y[0]^* + h_{00}^*y[1]\right)$$

The variable c stands for the following:

$$c = \frac{1}{h_{00}h_{10}^* + h_{11}^*h_{01}}$$

The following MatLab code steps us through the SFBC encoding and decoding processes. The script illustrates how the original symbols are successfully recovered by the estimator.

```
%% An Example of SFBC as used in LTE
%  Let's define the following input symbols
x0 =  1.0 + 1.0j;       % The source symbol x[n]
x1 = -0.5 - 2.0j;       % The source symbol x[n+1]

% Defining the Channel
% hsymbol,Ant - so h00 and h10 come from the same antenna but different subcarriers
h00 =  1 + 0.2j;    % Channel coefficient (symbol0, Antenna0)
h10 =  1 - 1j;      % Channel coefficient (symbol1, Antenna0)
h01 =  2 + 1j;      % Channel coefficient (symbol0, Antenna1)
h11 = -3 - 1.2j;    % Channel coefficient (symbol1, Antenna1)
% The observations are defined as follows
y0 = h00*x0        - h01*conj(x1);      % The observation y[0]
y1 = h11*conj(x0) + h10*x1;             % The observation y[1]


% The observation vector, Y, is defined as follows
Y  = [y0; conj(y1)];

% The quasi channel matrix
H  = [      h00         -h01; ...
        conj(h11)   conj(h10)];

% Method 1: Finding the Estimates of x0 and x1 using the inv(H) function
X_Estimate1  = inv(H)*Y;
x0_Estimate1 = X_Estimate1(1,1);                % The estimate of x[0]
x1_Estimate1 = conj(X_Estimate1(2,1));          % The estimate of x[1]

% Method 2: Or using the explicit formula for the 2x2 inverse
c            =  1/(h00*conj(h10) + conj(h11)*h01);
X_Estimate2  =  c * [conj(h10) h01; -conj(h11) h00] * Y;
x0_Estimate2 = X_Estimate2(1,1);                % The estimate of x[0]
x1_Estimate2 = conj(X_Estimate2(2,1));          % The estimate of x[1]
```

8.3   Closed Loop Spatial Multiplexing

Spatial multiplexing is a technique that uses two or more transmit antennas to simultaneously send two or more independent data streams, $x_1[n]$ and $x_2[n]$, over a single frequency channel. For the spatial multiplexing technique to operate properly, the number of antennas at the receiver should be equal to or greater than the number of antennas at the transmitter. The figure below illustrates a typical scenario of two transmit antennas communicating with two receive antennas.

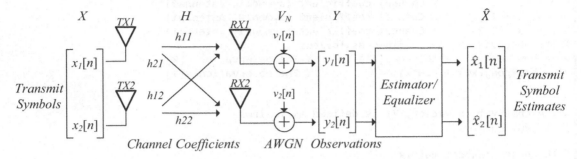

Figure 8-9: Spatial Multiplexing 2x2 Configuration

At first glance, it doesn't appear possible that interference of multiple data streams within the same channel can be remedied, but our extensive study of this very issue in chapter 4 on estimators and time domain equalizers demonstrates otherwise. In the case of time domain equalization, the estimator had to guess symbol $x[n]$, given a set of N observations that also contained traces of earlier and later symbols $x[n+1]$, $x[n+2]$, … etc. and $x[n-1]$, $x[n-2]$, … etc. In the spatial multiplex scenario described above, there are only two observations, $y_1[n]$ and $y_2[n]$, each containing traces of symbols $x_1[n]$ and $x_2[n]$. The obvious approach to estimating the transmit symbols $x_1[n]$ and $x_2[n]$ is to employ a zero-forcing, ZF, or minimum mean square error, MMSE, equalizer as shown in the figure above. Clearly, computing the coefficients of such an equalizer requires that we have estimated the coefficients of the channel matrix, H, at the receiver. All well designed communication systems provide reference or pilot signals in the transmitted waveform that allow the estimation of the channel with high reliability. Given this information, we want to show and cover the following topics.

→ We will show how a simple zero-forcing equalizer will recover or estimate the original transmit symbols for well and poorly behaved channel matrices. For this purpose, we define and quantify the quality of these matrices via their condition number. For well-behaved channels, the zero forcing equalizer can easily estimate the input symbols whereas for others, it won't.

→ We will introduce the use of the singular value decomposition, to better understand the behavior of the channel matrix.

→ With an understanding of the singular value decomposition, we will try to precode the input symbols to make it possible for the equalizer at the receiver to better estimate the transmit symbols. To do this, the receiver will have to send information about the estimated channel matrix back to the transmitter. Sending information back to the transmitter takes time and this scheme is only reasonable when the multipath channel remains stable for longer periods of time. Thus, it is preferable that neither the transmitter nor receiver are in motion. Of course, the overall frequency response can never remain completely static as frequency offset, phase noise and slow timing drift will change the channel even if the multipath environment is stable. This, however, we will be able to deal with, as we will find out soon.

Example 8.2. *Simple 2TX by 2RX Spatial Multiplexing*

In this first example, we will transmit a QPSK symbol stream, $x1[n]$, through transmit antenna 1 and a BPSK symbol stream, $x2[n]$, through transmit antenna 2.

```
% QPSK source, x1[n], and BPSK Source, x2[n]
NumSymbols = 200;
x1     = 0.7071*((randi([0 1],1,NumSymbols)*2 - 1) + 1j*(randi([0 1],1,NumSymbols)*2-1));
x2     = randi([0 1], 1, NumSymbols)*2 - 1;
X      = [x1; ...     % The x1[n] QPSK symbol stream
          x2];        % The x2[n] BPSK symbol stream
```

We will transmit the two symbol streams over three different channel configurations described by the channel matrices shown below. For the sake of simplicity, we keep the coefficients real valued.

$$H1 = \begin{bmatrix} -1 & 0.1 \\ 0.1 & -1 \end{bmatrix} \qquad H2 = \begin{bmatrix} -1 & -0.15 \\ 1 & -0.15 \end{bmatrix} \qquad H3 = \begin{bmatrix} -0.8 & 0.6 \\ 0.6 & -0.8 \end{bmatrix}$$

```
% Design the Channel Matrix.
ChannelSelection = 3;
switch(ChannelSelection)
    case 1; H  = [-1.0,  0.1 ; 0.1, -1.0 ];
    case 2; H  = [-1.0, -0.15; 1.0, -0.15];
    case 3; H  = [-0.8,  0.6 ; 0.6, -0.8 ];
    otherwise; error('Unsupported Channel Selection');
end
```

The observed signal vector, Y, consists of the received symbol streams summed with a certain amount of thermal antenna noise, V_N. Note that in previous examples we had used the variable V to represent the noise vector. We use the name V_N for now, as the nomenclature V is used by the singular value decomposition, which we will meet soon.

$$Y = \quad H \quad \cdot X + V_N$$
$$\begin{bmatrix} y_1 \\ y_2 \end{bmatrix} = \begin{bmatrix} h_{11} & h_{12} \\ h_{21} & h_{22} \end{bmatrix} \cdot \begin{bmatrix} x_1 \\ x_2 \end{bmatrix} + \begin{bmatrix} v_1 \\ v_2 \end{bmatrix}$$

```
% Pass the information through the Channel Matrix add noise to form the
NoisePower = 0.01;                                          %% observation Y
VN         = sqrt(NoisePower)*0.7071*(randn(2, NumSymbols) + 1j*randn(2, NumSymbols));
Y          = H*X + VN;    % The observation column vector
```

We will use the simple zero-forcing estimator = $inv(H)$.

$$\begin{bmatrix} \hat{x}_1 \\ \hat{x}_2 \end{bmatrix} = inv(H)(\begin{bmatrix} y_1 \\ y_2 \end{bmatrix} - \begin{bmatrix} v_1 \\ v_2 \end{bmatrix})$$

Remember the simple inverse of a 2x2 matrix from section 1.2.7.

$$inv(H) = inv\left(\begin{bmatrix} h_{11} & h_{12} \\ h_{21} & h_{22} \end{bmatrix} \right) = \frac{1}{h_{11}h_{22} - h_{12}h_{12}} \begin{bmatrix} h_{22} & -h_{12} \\ -h_{21} & h_{11} \end{bmatrix}$$

```
% Compute the Estimator Output
EstimatorMatrix = inv(H);  % The Estimator is simply the inverse of the Channel Matrix
X_Estimate      = EstimatorMatrix * Y;
```

Note the following:

→ We chose to use a QPSK signal at transmit antenna 1 and a BPSK signal at transmit antenna 2 to make it easier to see the performance of each estimator output.

→ Each matrix is crafted such that the signal power at each point of observation (at the receive antenna) will be close to unity. The matrix will not direct more power to either of the receive ports.

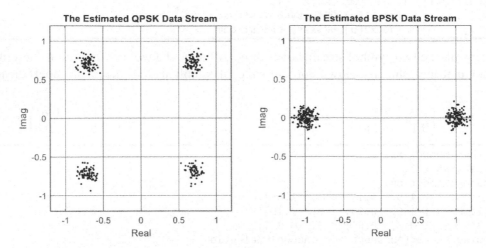

Figure 8-10: The Transmit Signal Estimations for the Channel Matrix H1

The MIMO scenario defined by the channel matrix $H1$ produces outstanding estimations of the transmit symbol streams $x1[n]$ and $x2[n]$. Just looking at the diagram, the signal to noise ratio of $10 \cdot log10(1.0/0.01) = 20$ dB is approximately correct for each estimation.

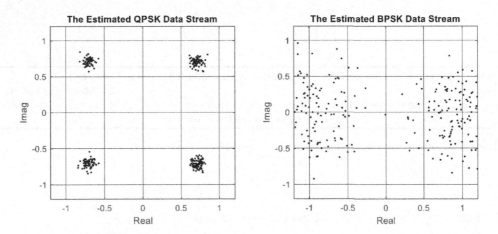

Figure 8-11: The Transmit Signal Estimations for the Channel Matrix H2

The MIMO scenario defined by the channel matrix $H2$ produces a very different picture. The reason for the inferior performance is due to the fact that the channel matrix is poorly behaved. Whereas the observations still feature equal amounts of power, power has in fact been taken away from the BPSK signal and reassigned to the QPSK signal. Notice that the constellation of our QPSK signal estimate is slightly tighter for this MIMO scenario as compared to the first. Note further that the estimator by itself is powerless to remedy the poorly behaved channel conditions.

However, as we will discover soon, there may not be a need to fix this problem as we can take advantage of the superior signal to noise ratio of the QPSK stream to transmit more information.

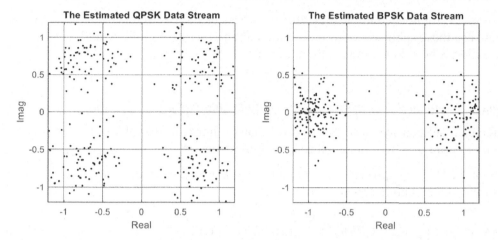

Figure 8-12: The Transmit Signal Estimations for the Channel Matrix H3

The MIMO scenario defined by the channel matrix $H3$ is the worst possible situation that we can encounter. Not just is the channel poorly behaved, but some aspect of the coefficients matrix cause noise to be distributed over both estimates. To understand exactly why the quality of the estimates are so different, we need to get a far better understanding of how the channel matrix acts to transform in the transmit symbols. The singular value decomposition discussed in chapter one section three is the ideal tool for this undertaking. It is highly recommended that you review this section before proceeding. Once we take the singular value decomposition of the channel matrix at the receiver, some of the results must be returned to the transmitter, which will then use that information to precode the transmit symbols and thereby improve the overall channel. The overall estimation / equalization process is thus distributed over both terminals.

8.3.1 Understanding Channel Matrices via the Singular Value Decomposition

Every transformation matrix may be broken into a cascade of rotational and scaling matrices. In the following expression, we show the decomposition for a 2x2 channel matrix H.

$$H = \begin{bmatrix} cos(\theta_1) & -sin(\theta_1) \\ sin(\theta_1) & cos(\theta_1) \end{bmatrix} \cdot \begin{bmatrix} \pm 1 & 0 \\ 0 & \pm 1 \end{bmatrix} \cdot \begin{bmatrix} \sigma_1 & 0 \\ 0 & \sigma_2 \end{bmatrix} \cdot \begin{bmatrix} cos(\theta_2) & -sin(\theta_2) \\ sin(\theta_2) & cos(\theta_2) \end{bmatrix} \cdot \begin{bmatrix} \pm 1 & 0 \\ 0 & \pm 1 \end{bmatrix}$$

We start off with a rotation by θ_1 followed by a possible reflection, which is in itself a simple scaling matrix. We then move on to a scaling matrix that features non-negative numbers we call singular values, θ_n. We finish as we had begun, with the product of a rotation by θ_2 and a possible reflection. In chapter one, we discovered that there are a couple of different ways matrices may be decomposed. For this application, the singular value decomposition is the tool of choice and decomposes the transformation (channel) matrix, H, as follows.

$$H = U \cdot \Sigma \cdot V^T$$

→ The first matrix U is the product of our first rotation and a possible reflection.

→ The second matrix Σ holds our singular values. Allowing reflection matrices to be integrated in the U and V^T matrices makes it possible for the singular values to be non-negative.
→ The third matrix V^T is the product of our second rotation and a possible reflection.

At the receiver, we now compute these three matrices using the singular value decomposition. We must now use these submatrices to fix the estimation process. As luck would have it, gaining this understanding is best done graphically rather than through dry math.

Example 8.2 *continued: Applying the Results of the Singular Value Decomposition*

The three channel matrices, *H*1, *H*2, and *H*3 that we introduced in the last example were in fact created by multiplying two rotational and one scaling matrix.

→ The matrix *H1* is the product of a rotation by ¾ π a scaling matrix with singular values of 1.1 and 0.9 followed by another rotation by ¼ π.

$$H1 = \begin{bmatrix} -1 & 0.1 \\ 0.1 & -1 \end{bmatrix} = \overset{U}{\begin{bmatrix} cos(0.75\pi) & -sin(0.75\pi) \\ sin(0.75\pi) & cos(0.75\pi) \end{bmatrix}} \cdot \overset{\Sigma}{\begin{bmatrix} 1.1 & 0 \\ 0 & 0.9 \end{bmatrix}} \cdot \overset{V^T}{\begin{bmatrix} cos(0.25\pi) & -sin(0.25\pi) \\ sin(0.25\pi) & cos(0.25\pi) \end{bmatrix}}$$

To visually appreciate how each one of these sub-matrices takes part in the overall transformation process, we create a special set of input vectors *X*. The individual transmit symbols in the vector are entirely real, for ease of plotting, and form a circle as follows.

$$X_i = \begin{bmatrix} Symbols\ on\ Tx\ Antenna\ 1 \\ Symbols\ on\ Tx\ Antenna\ 2 \end{bmatrix} = \begin{bmatrix} cos(\theta_i) \\ sin(\theta_i) \end{bmatrix} \ where\ \theta_i = 0, 0.1\pi...2\pi$$

We will highlight four of these input vectors each with their own unique marker, located at angles $\pi/4$, $3\pi/4$, $5\pi/4$ and $7\pi/4$.

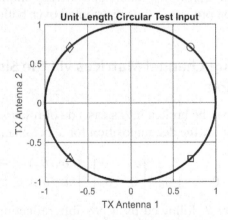

Figure 8-13: Test Input Vectors

The following figure shows the effect of V^T (leftmost plot), which is a rotation by 45 degrees in the positive, counter clockwise, direction of our test vector *X*. This is followed by the action of matrix Σ causing a slight compression and expansion (middle plot) along antenna port 2 and 1 respectively and a final rotation by +135 degrees (rightmost plot), due the multiplication by *U*. Keep track of the markers to see the different rotations.

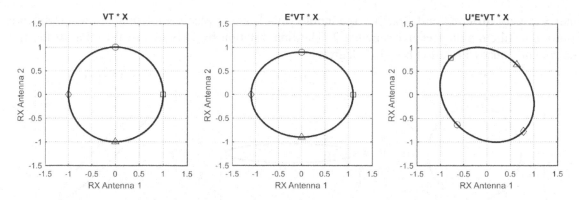

Figure 8-14: Progression of Input Vectors as they are multiplied by V^T, Σ **and** U

Applying the *inv(H)* at the receiver, simply reversed the process shown in the figure below. And as long as the signal to noise ratio is high, we need not worry. Let's take a look at the next plot that adds some noise to markers at $U\Sigma V^T \cdot X = HX$ (see rightmost figure above) to form the observation $Y = HX + V_N$. Now examine the final noisy estimate featuring markers that have been rotated and equalized back to their nominal positions. The size of the noise cloud has not changed appreciable and we can assume that this channel matrix will perform well in our spatial multiplexing scenario. And indeed, it is this channel matrix that leads to the high signal to noise ratio QPSK and BPSK sequences at the estimator output. See figure 8.10.

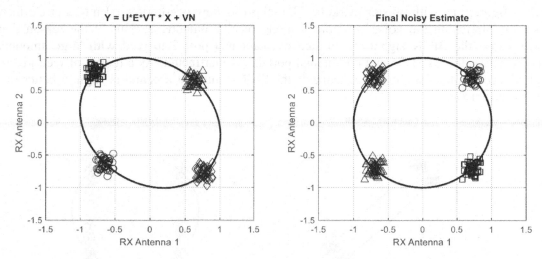

Figure 8-15: Observation Diagram and Final Noisy Estimate of the Original Circular Unit Length Input Vectors

→ Moving on to matrix *H2,* which is the product of a rotation by ¾ π, a scaling matrix with singular values of 1.4 and 0.2 followed by another rotation by 0 π, which reduces to the identity matrix.

$$H2 = \begin{bmatrix} -1 & -0.15 \\ 1 & -0.15 \end{bmatrix} = \overset{U}{\begin{bmatrix} cos(0.75\pi) & -sin(0.75\pi) \\ sin(0.75\pi) & cos(0.75\pi) \end{bmatrix}} \cdot \overset{\Sigma}{\begin{bmatrix} 1.4 & 0 \\ 0 & 0.2 \end{bmatrix}} \cdot \overset{V^T}{\begin{bmatrix} cos(0) & -sin(0) \\ sin(0) & cos(0) \end{bmatrix}}$$

The unit length circular input vectors X first encounter the identity matrix V^T, which leaves the input symbols entirely unaffected as seen in the leftmost plot below. This time around, the Σ

matrix is very different and stretches the input symbols along receive antenna 1 and significantly compresses them along receive antenna 2. The final rotation by U is identical to that of the first matrix and causes a rotation by +135 degrees.

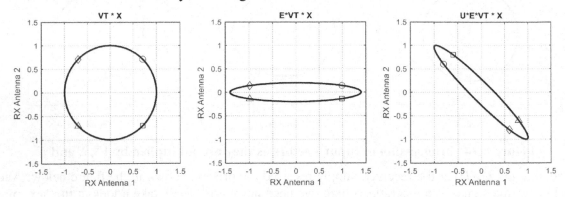

Figure 8-16: Progression of Input Vectors as they are multiplied by V^T, Σ and U

The next figure illustrates the addition of noise (to produce the observation vector Y), which is identical in power to the noise applied in the $H1$ scenario we looked at last. This time the estimator, $inv(H2)$, reverses only two processes: The rotation due to U and the compression due to Σ. Take a look at the observations Y in the left-most plot below. Reversing U rotates the observations by minus 135 degrees. The situation at this point looks a bit like the middle plot of the last figure, just with noisier markers. The ellipse is now decompressed into a circle thereby massively stretching the noise component along receive antenna 2. This is the reason that in figure 8.xxx, the BPSK signal transmitted over antenna port 2 arrived with large amounts of noise, whereas the QPSK signal on antenna port 1 was unaffected. As a matter of fact, given the original stretch of 1.4 along RX antenna 1, the SNR along that antenna is slightly better than it was for the $H1$ case.

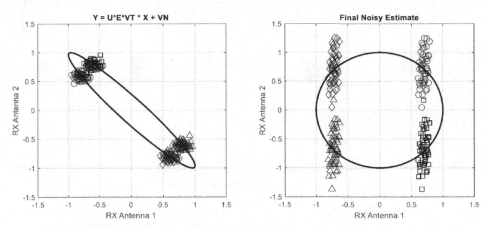

Figure 8-17: Observation Diagram and Final Noisy Estimate of the Original Circular Unit Length Input Vectors

8.3.2 Spatial Layers and the Matrix Condition Number

We have now found the first of two wrinkles in our spatial multiplexing exercise. The matrix Σ stretches and compresses the information transmitted by the two TX antennas. However, given that the channel forces a multiplication by the matrix V^T before we ever get to Σ, we technically can't speak of a stretching along antenna ports. The information that is stretched via Σ is the vector $V^T \cdot X$, which is a column vector with two entries and embodies the rotated input symbols. This information is stretched across what we refer to as spatial layers. As will become apparent when we look at the *H3* channel matrix scenario, it is imperative that we relay information about V^T back to the transmitter, which will premultiply future input symbol vectors, X, by $inv(V^T)$. Remember that $inv(V^T) = V$ as V^T is an orthonormal matrix. The observations reduce to the following.

$$Y = H \cdot VX + V_N = U\Sigma V^T \cdot VX + V_N = U\Sigma \cdot X + V_N$$

The input symbol streams $x_1[n]$ and $x_2[n]$ that reside in our input vector X are now directly exposed to the spatial layers and are scaled by the singular values σ_1 and σ_2 in Σ. The receiver knows both of these singular values and estimates the noise power. With this information at hand, the receiver figures out the signal to noise ratios across both spatial layers and suggest - to the transmitter - the right QAM constellation and amount of redundancy that both $x_1[n]$ and $x_2[n]$ should feature for successful demodulation.

8.3.3 The Matrix Condition Number

A channel matrix is said to be poorly conditioned, if relatively small inaccuracies in the matrix coefficient estimations, or small amounts of added noise, will cause large errors when the matrix is inverted to uncover the original transmit symbols. As we have seen in our example, if the Σ causes significant expansion along one spatial layer and compression along another, then large errors or noise boosting can occur during estimation. The larger the ratio of singular values in Σ, the worse the condition of the matrix. Fortunately for us, when the singular value decomposition produces the matrix Σ, the largest singular value is always positioned in the upper left corner, whereas the smallest singular value is located in the bottom right corner.

$$ChannelConditionNumber = k = \frac{\sigma_{max}}{\sigma_{min}}$$

Table 8-1: Condition Numbers for Channel Matrices H1, H2, and H3

Channel Matrix	Singular Values	Condition Number Linear	Condition Number dB
H1	[1.1 0.9]	1.222	0.871
H2	[1.4 0.2]	7	8.45
H3	[1.4 0.2]	7	8.45

8.3.4 Closed Loop Operation

In closed loop operation, the receiver has the opportunity to relay regarding the channel matrix that is has estimated back to the transmitter. Technically speaking, the transmitter could now boost the attenuated spatial layer and reduce power of the spatial layer stretched by the channel. This avoids the problem of excessive noise on either spatial layer when compensating for Σ. However, this may not be necessary, as the transmitter can take advantage of the boosted spatial layer to transmit more information. The improved SNR on that layer may support a higher QAM constellation or redundancies in the forward error correction scheme may be reduced to improve throughput. This assumes of course that we have the flexibility to encode information sent over spatial layer 1 differently that over spatial layer 2. We could even attempt to further boost the SNR of the high SNR transmission layer and forgo any transmission on the other. Clearly, there are quite a few choices here and the right approach will depend on your application.

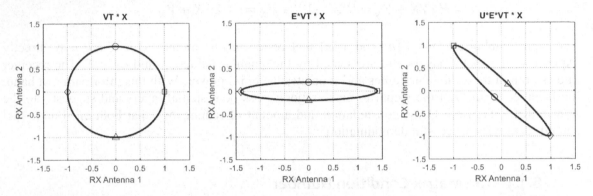

Figure 8-18: Progression of Input Vectors as they are multiplied by V^T, Σ and U

→ Moving on to matrix *H3*, which is the product of a rotation by ¾ π a scaling matrix with singular values of 1.4 and 0.2 followed by another rotation by 0.25 π, which reduces to the identity matrix.

$$H3 = \begin{bmatrix} -0.8 & 0.6 \\ 0.6 & -0.8 \end{bmatrix} = \begin{bmatrix} cos(0.75\pi) & -sin(0.75\pi) \\ sin(0.75\pi) & cos(0.75\pi) \end{bmatrix} \cdot \begin{bmatrix} 1.4 & 0 \\ 0 & 0.2 \end{bmatrix} \cdot \begin{bmatrix} cos(0.25\pi) & -sin(0.25\pi) \\ sin(0.25\pi) & cos(0.25\pi) \end{bmatrix}$$

The unique difference between channel matrices *H2* and *H3* is the fact that V^T is a rotation by +45 degrees rather than an identity matrix. The final rotation by 45 degrees puts the death nail into our estimation process. The estimator now overlays large amounts of noise on top of both the QPSK and BPSK transmit streams we saw in figure 8.12 earlier in our example. The observation, the left most plot of the next figure, looks suspiciously similar to that of the last example featuring channel matrix *H2*. If we now trace through the process of the estimator, $inv(H3)$, we undo the effects of U and Σ by multiplying by $inv(U)$ and $inv(\Sigma)$ respectively. For channel matrix H2, we did not need to proceed any further as V^T was the identity matrix. However, as can clearly be seen by the second plot of the figure below, our markers are 45 degrees away from their ideal positions. The final rotation $inv(V^T)$, rotates the noise onto both receive antenna ports, thus badly decreasing the signal to noise ratios along both spatial layers. It is for this reason that neither the QPSK nor the BPSK signals could be received with adequate SNR in the earlier part of our example.

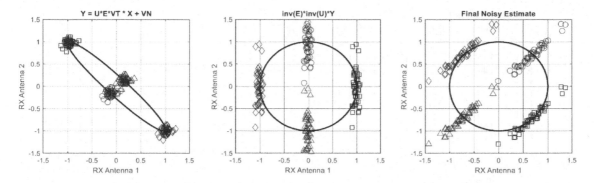

Figure 8-19: The Observation, *Y*, Midway Point through Final Estimation and the Final Noisy Estimate of the Original Circular Unit Length Input Vectors

With this knowledge we may summarize the key insights related to closed loop spatial multiplexing operation.

→ First and foremost, the receiver needs to relay information regarding V^T to the transmitter. The transmitter must premultiply the transmit symbols via $inv(V^T)$ or V in order to counteract the rotation due to V^T. Remember that V is an orthonormal matrix resulting and $inv(V^T) = V$.

→ Secondly, you may influence the matrix Σ to either balance the power on both spatial layers, move all power to the stronger spatial layer or do nothing at all. Some transmit specification are quite stringent on how much power each antenna can transmit and flexibility may be limited. In that case, we simply compute the signal to noise ratio of each spatial layer and then send independent symbols streams, $x_1[n]$ and $x_2[n]$, featuring different amounts of error redundancy over each spatial layer to optimize total throughput.

The next figure summarizes what we have learned in this section. The two input symbol streams $x_1[n]$ and $x_2[n]$ are rebalanced in power via the diagonal scaling matrix P. The potentially scaled input symbol streams now enter the two spatial layers ahead of matrix V, which cancels out the V^T rotation at the start of the channel matrix H. If the matrix P completely rebalanced Σ then the estimator will reduce to $inv(U) = U^T$, as the matrix U is also an orthonormal matrix. In real live, the estimator, or equalizer, will not simply become the matrix U^T. In real life, the channel matrix keeps changing due to unresolved frequency offset, phase noise and slow timing drift. However, these effects do not cause transmit information to be redistributed between antenna ports, and the application of the matrices P and V continue to be valid as long as the multipath characteristics remain static. Thus, as time progresses, the reference or pilot information in the signal is used to continually estimate what still remains of the channel, which is then equalizer before proceeding to demap the QAM symbols into bits.

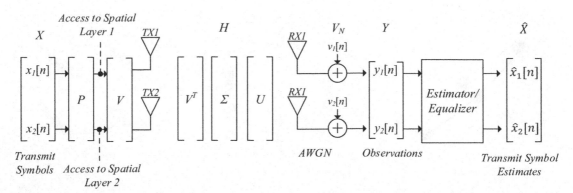

Figure 8-20: Precoding at the Transmitter for Closed Loop Spatial Multiplexing

In conclusion, premultiplying V at the transmitter ahead of the channel defined by matrix $H3$, will make the channel look like $H2$, where at least spatial layer one can be received with good signal to noise ratio. Rebalancing via P will be up to your application.

8.3.5 Channel Rank

Modern cellular communication systems will report a matrix rank back to the transmitter along with an indication regarding the matrix V^T. The rank of a matrix is defined as the maximum number of linearly independent row vectors in a matrix. In our situation, the rank of the channel is defined as the number of spatial layers that feature an SNR high enough to enable the most basic communication link. In LTE, the link has a large number of configurations that allow choices between QPSK, 16 QAM, 64 QAM as well as many options regarding the amount of FEC redundancy. The spatial layer is deemed usable as long as the most robust link configuration still works.

8.4 Open Loop Spatial Multiplexing

In the last section, we witnessed the challenges imposed by a 2TX by 2RX antenna, or 2x2, spatial multiplexing scenario on the estimator. Without the ability to relay precoding information back to the transmitter, the likelihood that one or both spatial streams are overwhelmed by noise cannot be ignored. This open loop type scenario is unavoidable in situations where the channel conditions change too quickly, making the time-consuming closed loop approach impractical. In this section we will take a look at how the minimum mean square error, or MMSE, optimal estimator performs in spatial multiplexing situations where additional receive antennas are available. Specifically, we will examine the MMSE estimator in 2x2, 2x3 and 2x4 spatial multiplexing situations.

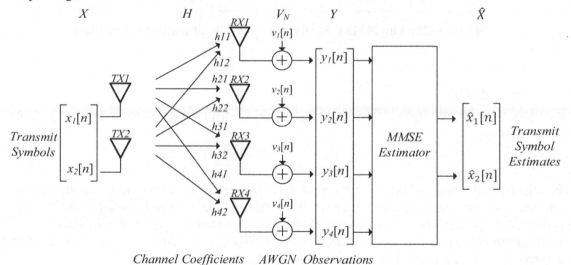

Figure 8-21: A 2TX by 4RX Antenna Spatial Multiplexing Scenario with MMSE Estimation

The configuration above can be expressed as a system of equation and its associated matrix expression.

$$y_1[n] = h_{11}x_1[n] + h_{12}x_2[n] + v_1[n]$$
$$y_2[n] = h_{21}x_1[n] + h_{22}x_2[n] + v_2[n]$$
$$y_3[n] = h_{31}x_1[n] + h_{32}x_2[n] + v_3[n]$$
$$y_4[n] = h_{41}x_1[n] + h_{42}x_2[n] + v_4[n]$$

$$Y = H \cdot X + V$$

$$\begin{bmatrix} y_1[n] \\ y_2[n] \\ y_3[n] \\ y_4[n] \end{bmatrix} = \begin{bmatrix} h_{11} & h_{12} \\ h_{21} & h_{22} \\ h_{31} & h_{32} \\ h_{41} & h_{42} \end{bmatrix} \cdot \begin{bmatrix} x_1[n] \\ x_2[n] \end{bmatrix} + \begin{bmatrix} v_1[n] \\ v_2[n] \\ v_3[n] \\ v_4[n] \end{bmatrix}$$

The MMSE optimal estimator is a pair of linear combiners that assume the following form.

$$\hat{X} = \begin{bmatrix} \hat{x}_1[n] \\ \hat{x}_2[n] \end{bmatrix} = C^T \cdot Y = \begin{bmatrix} c_{11} & c_{12} & c_{13} & c_{14} \\ c_{21} & c_{22} & c_{23} & c_{24} \end{bmatrix} \cdot \begin{bmatrix} y_1[n] \\ y_2[n] \\ y_3[n] \\ y_4[n] \end{bmatrix}$$

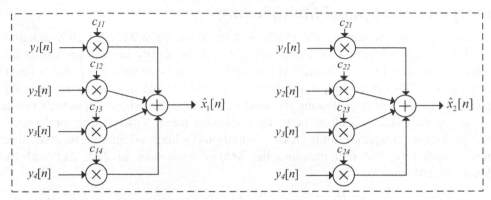

Figure 8-22: The MMSE Estimator as a Pair of Linear Combiners

Computing the MMSE Estimator Coefficients

We derived the MMSE optimal estimator coefficients in chapter 4 of this text. The equation for C^T is as follows.

$$C^T = R_x H^T \cdot inv(H R_x H^T + R_x)$$

The quantities $R_x = E(X[n] \cdot X[n]^T)$ and $R_v = E(V[n] \cdot V[n]^T)$ refer to the input and noise autocorrelation matrices. On the very reasonable assumption that the input symbols $x_1[n]$ and $x_2[n]$ are uncorrelated, the autocorrelation matrix of the input information will simplify into a diagonal matrix. Further, given that noise quantities $v_1[n]$, $v_2[n]$, and $v_3[n]$ are generated at different antennas, they are guaranteed to be uncorrelated, yielding the following expressions.

$$R_x = \begin{bmatrix} \sigma_{x1}^2 & 0 \\ 0 & \sigma_{x2}^2 \end{bmatrix} \quad R_v = \begin{bmatrix} \sigma_{v1}^2 & 0 & 0 \\ 0 & \sigma_{v2}^2 & 0 \\ 0 & 0 & \sigma_{v3}^2 \end{bmatrix}$$

As we have seen the format of the channel matrix H earlier, we can now turn our attention to a MatLab example, which allows us to transmit two separate QPSK symbol streams $x_1[n]$ and $x_2[n]$, over a 2x2, a 2x3 and a 2x4 spatial multiplexing channel. The channel coefficients will be complex with the real and imaginary components being assigned random values between -1.0 and 1.0. We compute the MMSE optimal estimator coefficient for each random channel matrix, transmit the symbol stream via our vector X, and finally determine the estimate of X. We then compute the error between X and $\hat{X}$ to arrive at a signal to noise ratio for each symbol stream.

What is our expectation? For the 2x2 spatial multiplexing scenario, we have already shown that the channel may either allow both, one or none of the transmit streams to arrive with good signal to noise ratio at the receiver output. It all depends on the condition of the channel matrix.

Our hope, is that as we add more receive antennas, the MMSE optimal estimator has more degrees of freedom to recover the transmit streams. Imagine that for a 2x3 spatial multiplexing configuration, the paths between the two TX antennas and the first two RX antennas make for a poorly conditioned matrix. But as we have paths to the third RX antenna, the MMSE optimal estimator can emphasize and deemphasize observations at the receive antenna any way it wants and thereby create a more favorable overall channel. We expect that as the number of RX antennas increases to four, this effect will become more and more pronounced.

In the MatLab simulation, we choose the antenna configuration and then run the communication link for many different channel matrices in order to get a statistical idea of how good the signal to noise radio of the estimate is.

```
%% 0. Let's define some of the quantities we need for the estimator.
%      Define the signal and noise powers
SigmaX  =   1;  % The rms value of each transmit stream x1[n] and x2[n]
SigmaV1 = 0.1;  % The rms noise value at receive antenna 1
SigmaV2 = 0.1;  % The rms noise value at receive antenna 2
SigmaV3 = 0.1;  % The rms noise value at receive antenna 3
SigmaV4 = 0.1;  % The rms noise value at receive antenna 4

Rx = [SigmaX^2,        0; ...
            0, SigmaX^2];
```

Next we setup a Monte Carlo simulation, in which we run two TX streams each with 1000 QPSK symbols, through 200 different random channel matrices that can be set up as the 2x2, 2x3 and 2x4 variety.

```
%% 1. Let's define a channel with random coefficients
NumRepetitions      = 200;
N                   = 1000; % The number of TX input symbols for x1[n] and x2[n]
Scenario            = 1;    % Scenario 1,2,3 - indicate 2x2, 2x3 and 2x4 channel cond
SNR_Array2          = zeros(2, NumRepetitions);

for Repetitions = 1:NumRepetitions
    % Generate QPSK streams for both x1[n] and x2[n]. X is a 2x1 column vector
    X           =      0.7071 * (2*randi([0,1], 2, N) - 1) ...
                    + 1j*0.7071 * (2*randi([0,1], 2, N) - 1);

    % Generate complex Gaussian noise with variances SigmaV1^2, SigmaV2^2, … SigmaV4^2
    V1   = SigmaV1 * sqrt(1/2) * (randn(1, N) + 1j*randn(1, N));
    V2   = SigmaV2 * sqrt(1/2) * (randn(1, N) + 1j*randn(1, N));
    V3   = SigmaV3 * sqrt(1/2) * (randn(1, N) + 1j*randn(1, N));
    V4   = SigmaV4 * sqrt(1/2) * (randn(1, N) + 1j*randn(1, N));

    % Generate the channel matrix for the scenario
    switch(Scenario)
```

```
        case 1      % 2x2 channel matrix
            H  = (2*rand(2,2) - 1) + 1j*(2*rand(2,2) - 1);
            Rv = [SigmaV1^2,          0; ...
                          0, SigmaV2^2];
            V  = [V1; V2];
        case 2      % 2x3 channel matrix
            H  = (2*rand(3,2) - 1) + 1j*(2*rand(3,2) - 1);
            Rv = [SigmaV1^2,          0,          0; ...
                          0, SigmaV2^2,          0; ...
                          0,          0, SigmaV3^2];
            V  = [V1; V2; V3];
        case 3      % 2x4 channel matrix
            H  = (2*rand(4,2) - 1) + 1j*(2*rand(4,2) - 1);


            Rv = [SigmaV1^2,          0,          0,          0; ...
                          0, SigmaV2^2,          0,          0; ...
                          0,          0, SigmaV3^2,          0; ...
                          0,          0,          0, SigmaV4^2];
            V  = [V1; V2; V3; V4];
        otherwise; error('Unsupported scenario.');
    end

    % Compute the Observation Vector Y
    Y = H*X + V;

    % Solve for the MMSE Coefficients (The Wiener Hopf equation)
    CT = Rx*H' * inv(H*Rx*H' + Rv);

    % Compute the Estimate of the transmit vector X
    X_Estimate = CT * Y;

    % Compute the error and the SNR
    Error                    = X_Estimate - X;
    SNR_X_Estimate1          = var(X_Estimate(1,:))/var(Error(1,:));
    SNR_X_Estimate2          = var(X_Estimate(2,:))/var(Error(2,:));
    SNR_Array2(:, Repetitions) = [SNR_X_Estimate1; SNR_X_Estimate2];
end

figure(1);
histogram(SNR_Array2(1,:), 10, 'facecolor', [1, 1, 1] );
xlabel('Linear SNR1'); axis([0 450 0 60]);
```

The histograms below illustrate the distribution of linear SNR values for one of the transmit symbol stream estimates. Note that the distribution is identical for both of the estimated symbol streams, $\hat{x}_1[n]$ and $\hat{x}_2[n]$. The progression of the histograms clearly illustrates the advantage of having additional receive antennas (and associated receive paths).

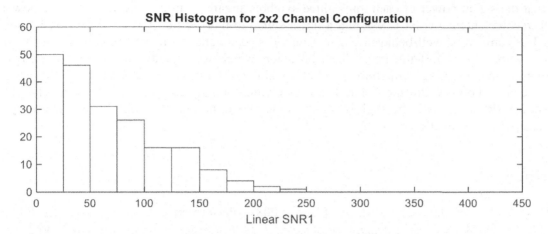

Figure 8-23: Poor SNR is Quite a Common Occurrence for the 2x2 Spatial Multiplexing Channel

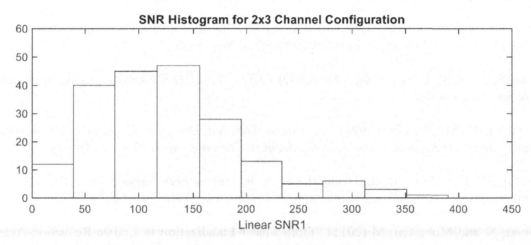

Figure 8-24: The SNR Histogram of the 2x3 Spatial Multiplexing Channel is Much Better

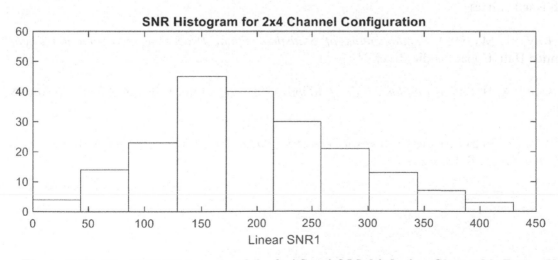

Figure 8-25: The SNR Histogram of the 2x4 Spatial Multiplexing Channel is Better Yet

Note that the signal power of each transmitted symbols stream is unity, whereas the noise power at each receive antenna is 0.01, yielding a linear signal to noise ratio of 100, which translates to 20dB. For some very well-behaved channel matrices, the signal to noise ratios will add as the MMSE optimal combines the paths. This behavior, which we already saw in our section on maximum ratio combining, will allow the total signal to noise ratio to accumulate to 300 and 400, which we indeed observe for the 2x3 and 2x4 spatial multiplexing configuration respectively. The combined SNR can actually be slightly larger as the maximum channel coefficient magnitude is equal to $sqrt(1+1) = sqrt(2) = 1.4142$.

References

[1] Alex, S. P. and Jalloul, L. M. A. (April 2008) "Performance Evaluation of MIMO in IEEE802.16e/WiMAX," *IEEE Journal of Selected Topics in Signal Processing*, vol.2, no.2, pp. 181–190

[2] Perahia, E. and Stacey, R. (2008) *Next Generation Wireless LANs – Throughput, Robustness, and Reliability in 802.11n*, Cambridge University Press, New York

[3] Sesia, S. , Toufik, I. and Baker, M. (2009) *LTE – The UMTS Long Term Evolution*, John Wiley & Sons, Chichester

[4] Alamouti, S. M. (October 1998) "A Simple Transmit Diversity Technique for Wireless Communications*", IEEE Journal on Selected Areas in Communications* 16(8): 1451–1458.

[5] Tarokh, V., Jafarkhani, H. and Calderbank, A. R. (July 1999). "Space–Time Block Codes from Orthogonal Designs", *IEEE Transactions on Information Theory*, 45(5): 744–765.

[6] Aubert, S. and Mohaisen, M (2011) "From Linear Equalization to Lattice-Reduction-Aided Sphere-Detector as an Answer to the MIMO Detection Problematic in Spatial Multiplexing Systems", Dr Miguel Almeida (Ed.), *Vehicular Technologies: Increasing Connectivity*, InTech (published online)

[7] Kay, S. M.(1993) *Fundamentals of Statistical signal Processing – Estimation Theory*, Prentice Hall, Upper Saddle River, NJ

[8] Sayed, A. H. (2003) *Fundamentals of Adaptive Filtering*, John Wiley & Sons, Hoboken, NJ, 1–48

[9] 3GPP, TS36.211 *Evolved Universal Terrestrial Radio Access (E-UTRA); Physical Channels and Modulation*, Release 8

Index

Made in the USA
Las Vegas, NV
04 November 2023